i	x-component of unit vector; coefficient of imaginary part
j	y-component of unit vector
j_{ij}	specific rate of momentum transfer of j-momentum in the i-th direction, kg $\cdot$ m/s^2
K	permeability of porous media $= \dfrac{\mu \mathbf{v}_o}{-\nabla \mathscr{P}}$, m^2
k	Kozeny constant of packed bed
k	z-component of unit vector
L	length of channel, or height of packed or fluidized bed, m
L_e	tortuous length in a packed bed, m
L_i	height of a packed bed at the point of incipient fluidization
ℓ	finite length of a plate, m
ℓ_w	length of a stationary wake, m
M	consistency factor (see Equation 2.10), kg/m $\cdot$ s$^{2-\alpha}$; Morton number $= \dfrac{\mu_c^4 g \Delta \varrho}{\varrho_c^2 \sigma^3}$
M'	modified consistency factor for pipe flow (see Equation 5.37), kg/m $\cdot$ s$^{2-\alpha'}$
m	exponent for wedge flow (see Equation 14.3) $= \dfrac{\beta}{2 - \beta}$
n	exponent in Equation 20.2
n	unit outward vector at surface
P_w	wetted perimeter, m
$\mathscr{P}$	dynamic pressure $= p + g\varrho h - p_0$, Pa
$\mathscr{P}_\infty$	dynamic pressure at infinity, Pa
$\mathscr{P}_0$	dynamic pressure at point of incidence, Pa
p	thermodynamic pressure, Pa; arbitrary exponent in correlating equation of Churchill and Usagi
p_o	reference (static) pressure at zero elevation, Pa
R	range of operability $= \dfrac{u_T}{u_i}$; radial spherical coordinate, m
Re	Reynolds number $\left(\text{for a channel} = \dfrac{4 A_x u_m \varrho}{\mu P_w}\right)$
Re$_D$	Reynolds number for a cylinder, disk or sphere $= \dfrac{D u_\infty \varrho}{\mu}$
Re$_\ell$	Reynolds number for a plate of length $\ell = \dfrac{\ell u_\infty \varrho}{\mu}$
Re$_T$	Reynolds number for terminal settling $= \dfrac{D u_T \varrho}{\mu}$
Re$_x$	Reynolds number at x in a boundary layer $= \dfrac{x u_\infty \varrho}{\mu}$
r	radial cylindrical coordinate, m
S_ℓ	strength of a line source, m^2/s
S_p	strength of a point source, m^3/s
Sr	Strouhal number $= \dfrac{\omega D}{u_\infty}$
t	time, s
U_x	velocity outside boundary layer, m/s
u	component of velocity in primary direction of
u_b	velocity for incipient bubbling, m/s
u_c	velocity at centerline or central plane, m/s
u_i	component of velocity in i-th direction, m/s
u_m	space-mean velocity, m/s
u_o	superficial velocity, m/s
u_T	terminal velocity of a particle, droplet or bubble, m/s
u_T^*	dimensionless terminal velocity $= u_T \left(\dfrac{\varrho_c^2}{\mu_c g \Delta \varrho}\right)^{1/3}$
u_∞	free-stream velocity, m/s

Viscous Flows

BUTTERWORTHS SERIES IN CHEMICAL ENGINEERING

SERIES EDITOR

HOWARD BRENNER
Massachusetts Institute of Technology

ADVISORY EDITORS

ANDREAS ACRIVOS
The City College of CUNY

JAMES E. BAILEY
California Institute of Technology

MANFRED MORARI
California Institute of Technology

E. BRUCE NAUMAN
Rensselaer Polytechnic Institute

ROBERT K. PRUD'HOMME
Princeton University

Viscous Flows
The Practical Use of Theory

Stuart Winston Churchill
The Carl V. S. Patterson Professor of Chemical Engineering
The University of Pennsylvania

Butterworths
Boston   London   Singapore   Sydney   Toronto   Wellington

Library of Congress Cataloging-in-Publication Data
Churchill, Stuart Winston, 1920–
  Viscous flows.
  (Butterworths series in chemical engineering)
  Bibliography: p.
  Includes index.
  1. Fluid dynamics.   I. Title.   II. Series.
QA911.C48   1988   532'.5   87-17854
ISBN 0-409-95185-4

British Library Cataloguing in Publication Data
Churchill, Stuart Winston
  Viscous flows: the practical use of theory.—
  (Butterworths series in chemical engineering).
  1. Fluid dynamics
  I. Title
  532'.051'02466   TP156.F6
  ISBN 0-409-95185-4

Butterworth Publishers
80 Montvale Avenue
Stoneham, MA 02180

10 9 8 7 6 5 4 3 2 1

Typeset in Hong Kong by Best-set Typesetter Ltd.

Printed in the United States of America

This book is dedicated
to Renate without whose
inspiration, encouragement,
and tolerance it would never
have been finished.

Contents

xii   *Contents*

Invocation

The following words of René Descartes from the last paragraph of *La Géométrie* (1637) [as translated by D. E. Smith and M. L. Latham, Dover, N.Y. (1954)] provided an inspiration for the contents of this book:

> But it is not my purpose to write a large book. I am trying rather to include much in a few words, as will perhaps be inferred from what I have done, if it is considered that, while reducing to a single construction all problems of one class, I have at the same time given a method transforming them into an infinity of others, and thus of solving each in an infinite number of ways....I hope that posterity will judge me kindly, not only as to the things which I have explained, but also to those which I have intentionally omitted so as to leave to others the pleasure of discovery.

The following words of the founder of my own university remain a valid assessment of the need to be selective and of the role of the pragmatic:

> As to their STUDIES, it would be well if they could be taught *every Thing* that is useful and *every Thing* that is ornamental: But Art is long, and their Time is short. It is therefore proposed that they learn those Things that are likely to be *most useful* and *most ornamental*, regard being had to the several Professions for which they are intended. (*Proposals Relating to the Education of Youth in Pennsylvania* by Benjamin Franklin, 1749.)

Preface

Fluid flow is of interest in nature, in physiology, and in a great variety of industrial applications ranging from the performance of aircraft to the processing of chemicals. Fluid motion and/or the mathematical challenge of the nonlinear differential equations required for its representation have attracted the interest of many of the greatest physicists and applied mathematicians of the past, including Leonardo da Vinci, Isaac Newton, the Bernoulli family, Leonhard Euler, Pierre-Simon de Laplace, Joseph-Louis Lagrange, Siméon-Denis Poisson, George Gabriel Stokes, Hermann von Helmholtz, Lord Rayleigh, and Osborne Reynolds.

PRIOR BOOKS

Many books continue to be written on fluid flow, usually under such titles as fluid mechanics, fluid dynamics, gas dynamics, aerodynamics, and hydrodynamics. They differ widely according to (1) their inclusion or exclusion of, or emphasis on ideal flows, inviscid flows, inertial flows, creeping flows, laminar flows, boundary-layer flows, turbulent flows, supersonic flows, external flows, internal flows, wave motions, and wakes; (2) their consideration or lack of consideration of incompressible or relatively incompressible fluids, compressible fluids, non-Newtonian fluids, and multiple phases; (3) their mathematical character—emphasizing or ignoring potential theory, perturbation theory, boundary-layer theory, matched asymptotic expansions, dimensional analysis, and numerical methods; and (4) their objective and approach, stretching from the purely theoretical with no concession to applications, to the philosophical, and to the purely empirical. As examples, Langlois [1] noted that "a book entitled *Slow Viscous Flow* could be anything from a treatise on weakly elliptic systems of partial differential equations to a plumber's manual," and Birkhoff [2] subtitled his monograph on hydrodynamics *A Study in Logic, Fact and Similitude*.

JUSTIFICATION FOR ANOTHER BOOK

The primary justification for another book on fluid flow is some degree of uniqueness. This book is significantly distinct in character and contents in its:

1. Extensive use of simple theories to predict, bound, and correlate fluid motions
2. Emphasis on the development of skill in deriving theoretical solutions for new problems by example and practice

3. Concern for the evaluation of the accuracy and range of validity of various theoretical solutions by comparison with experimental data
4. Use of asymptotic solutions to develop comprehensive correlating equations
5. Focus on fluid-mechanical behavior rather than on mathematical methods
6. Derivation of all solutions, including the required similarity transformations, from first principles
7. Objective of developing a physical sense of the general structure of fluid motions, and a quantitative sense for the magnitude of various dimensionless groupings

These distinctions are discussed in some detail below.

GENERAL CHARACTER

The general subtitle for this book implies a theoretical orientation but with the objective of solving practical problems. Theoretical models and concepts will be shown to be a more reliable guide for prediction, correlation, and generalization than any amount of empiricism. Emphasis is placed on identification or derivation of relevant models and conditions rather than on theory and mathematics for their own sake. Insofar as possible, the accuracy and range of applicability of the various theoretical models are evaluated by comparison with experimental data.

Although not stated explicitly in either the title or the subtitle, this book is intended to be used as a text, both in the classroom and for self-study. Accordingly, the material is ordered in terms of increasing difficulty and is organized to permit the omission of individual topics as required by constraints of time and previous preparation.

The objective is not merely to demonstrate the use of theoretical concepts but to develop facility and skill in using these concepts on future problems. Theoretical solutions are therefore generally derived in detail. Also, an extensive set of problems is provided for practice in such derivations. Some derivations and results that would otherwise have been included in the text have been deferred to these exercises. Solving a representative set of the problems is therefore essential for a complete and working knowledge of this material.

The organization of the book was largely predicated upon pedagogical considerations, but the particular flows to be examined were chosen for their practical importance, and the particular formulations and techniques for their usefulness rather than for their intrinsic interest. For the more complex flows a complete gamut of techniques is shown to be essential to cover all regimes. Some teachers and authors have asserted that students, at least undergraduates, should not be confronted by alternative expressions for the same range of behavior. That attitude does not recognize a responsibility to prepare students to read conflicting literature, to resolve discrepancies, and to make rational decisions.

This book thus differs significantly in objective and approach from those textbooks and handbooks that present results without derivation, explanation, or critique; and in content from those textbooks and monographs that focus on the existing theoretical solutions for some particular class of flow (such as

Schlichting [3] for boundary layers) or on some particular techniques of solution (such as Van Dyke [4] for perturbation methods).

SCOPE

Comprehensive textbooks on fluid flow necessarily cover much of the same material and differ primarily in arrangement, emphasis, depth, philosophy, applications, and mode of presentation.

The title of this book implies a concern with flow itself, as contrasted with the mechanics thereof or with some particular mathematical approach, and a restriction to flows that are primarily laminar (i.e., dominated by viscosity rather than by inertia or turbulence). Flows dominated by inertia have already been treated in a companion volume [5], and those dominated by turbulence will be described in a forthcoming one [6]. Multidimensional, laminar flow in channels, including developing flow, has arbitrarily been excluded from this volume out of considerations of length and specialized interest, and will also appear in a separate volume [7].

ORGANIZATION

As already stated, the order of subject matter was chosen primarily for pedagogical reasons. Students appear to learn more readily by first examining very simple phenomena and models, and then considering the effect of added complexities. Hence, in Part I, models are constructed and solved for individual one-dimensional flows using "shell" balances for fluid elements of finite depth. Thereafter, in Part II, the general three-dimensional equations describing the conservation of mass and the relationship between momentum and applied forces are derived, rearranged in various canonical forms, and reduced for special cases, such as those of Part I. Such reductions are useful in identifying all of the simplifications that may be implicit in restricted, direct derivations, such as those of Part I, but do not appear as effective in achieving physical understanding.

Part III is concerned with unconfined flows over plane surfaces and bluff bodies. The models for these two-dimensional flows are developed by direct derivation and by reduction. The use of dimensional analysis alone to determine the general and limiting behavior defined by these models is first demonstrated; then the derivation and use of similarity transformations and other techniques to obtain analytical solutions is covered. The resulting expressions for these flows are, however, shown to be valid only for limited regimes, and emphasis is accordingly placed on the development of comprehensive correlating equations for interpolation.

The flows through dispersed media that are treated in Part IV are not generally subject to exact analysis. The theoretical expressions of Parts I and III are, however, useful as a guide for the development of semitheoretical expressions.

The structure of the book thus has two overlapping elements: (1) an assembly of the pertinent theoretical solutions and comprehension correlations for the primarily laminar flows of greatest practical importance, and (2) the development of skillfulness in the use of modeling, dimensional analysis, similarity transformations, methods of solution, and techniques of correlation.

ACADEMIC LEVEL

The notes on which these books are based have evolved from a one-semester course on momentum transfer for seniors and first-year graduate students in chemical engineering (see Churchill [8]). Over a period of 35 years they have developed a life of their own through annual additions, deletions, and refinements in content and concept. The participants in these classes have almost always included a significant fraction of practicing engineers as well as students in contiguous fields such as aerospace engineering, bioengineering, chemistry, civil engineering, and mechanical engineering. The subject matter and presentation have been influenced by this distribution of special interests and backgrounds.

A primary consideration in this course has necessarily been the diversity of preparation of the students in thermodynamics, fluid mechanics, and applied mathematics, ranging from no acquaintance to superficial acquaintance to real understanding. My response, as indicated by the contents of this book, has been to start at a very elementary level and to proceed rapidly to the level of sophistication required by the fluid-mechanical topics themselves. This approach might be expected to involve considerable review and boredom for some students on some topics, but such reactions have rarely been evident. The degree of true comprehension achieved in undergraduate courses by even the best of students is not high. The increasing sophistication of the undergraduate curriculum has largely been at the expense of reinforcement by repetition and practice. Few students entering industry or graduate school have a *working* knowledge of even the elementary concepts of fluid flow. Hence, the degree of review in this book is not only acceptable but essential if understanding and facility in use, rather than mere exposure, are the goals.

In recent years these notes have been used, with surprising success, in a one-semester junior course providing a first exposure to fluid flow. A somewhat slower pace and some deletions of detail, particularly in Part II, were required, but the majority of the material was covered in addition to a selection of topics from the companion book on inertial flows [5] and the associated notes [6] on turbulent flows.

This book is therefore proposed for use, with some deletions, for a first course in fluid flow and, as is, for a first-year graduate or second-level undergraduate course.

CONCEPTS AND CONTENT

The Role of Theory. Most practitioners of engineering consider theoretical concepts to be difficult to understand, theoretical derivations to be the exclusive property of mathematicians and academicians, and theoretical results to be irrelevant and impractical. These misapprehensions, when they exist, are an indictment of modern undergraduate (or even graduate) education in engineering, which emphasizes theory (engineering science) but often does not develop skill or confidence in its use. In this book, theoretical concepts are shown to explain otherwise baffling behavior; limiting, illustrative, and even comprehensive results of great utility and generality are shown to be attainable from first principles by elementary derivations; and theoretical results are shown to be a better guide to correlation and generalization than any amount of empiricism. The reader is expected to develop a working knowledge, not merely an appreciation, of theory. By and large, contrary to the assertion of Gibbons,[1] this expectation has been realized with the many generations of students who have been exposed to the notes that preceded this book.

Approximate Models and Methods of Solution. The utility of a theoretical solution is not directly proportional to its complexity or generality. Solutions of first-order accuracy for limiting or highly idealized conditions are often sufficient for engineering purposes and may thereby even be preferable to more exact solutions of detailed or awkward form. Such approximate results are usually obtained by simplification of the model itself, but may arise directly from approximate formulations. Herein, solutions of varying degrees of exactness, generality, and complexity are compared with one another and with experimental data in terms of their practical utility.

Numerical Solutions. Modern computational facilities provide a possible means for solution of all differential models. However, the numerical methods utilized in such computations have not proven to be a panacea for fluid flow. First, the nonlinearity of many of the models, as contrasted with those for heat and mass transfer, poses difficulties that have not been entirely resolved. Second, the results attained generally consist of a tabulation of uncorrelated values that differ from experimental data only in their precision and uniformity. They may also be of unknown accuracy because of idealizations, known or unknown, in the model or arising from the method of solution.

Despite these limitations, numerical methods and results are contributing increasingly to fluid flow. In particular, they are supplanting the use of classical methods when the resulting solutions are not in closed form, and when some of the idealizations required to obtain an analytical solution can be avoided. Furthermore, these methods are an alternative to experimental values for testing analytical solutions, and act as a guide to interpolation between such solutions.

Such numerical results are utilized extensively in this book, and their contribution can be expected to increase rapidly in the next decade. Numerical methods, which are currently in a rapid state of development, and which constitute a separate and complete subject in themselves, are not described herein.

Analytical Solutions. Solutions in closed form generally have an unknown accuracy and a limited range of utility because of the necessary idealizations. Their range of validity, if any, is seldom defined by the solution itself. Limitations in range may occur owing to the increasing role of a neglected term or variable, or to a transition from one mode or state of flow (such as laminar) to another (such as turbulent). The theories of transitions and of multiple,

---

[1] "The power of instruction is seldom of much efficacy except in the happy dispositions where it is almost superfluous," according to Feynman et al. [9], p. 5.

stationary, or oscillatory states are much more complex and less successful than those for the singular states that bound such behavior.

Experimental Data.   In the past, experimental data were the first, and sometimes only, resource for the construction of correlations. The available theoretical solutions now provide a reliable structure for correlation, and the role of experimental data is thereby somewhat reduced. Such observations are still required to indicate the points or regions of transition from one state or mode to another, to confirm the physical existence of multiple states, and to guide interpolation between different theoretical solutions. In principle, experimental data should also be utilized to define the accuracy of theoretical results within their range of applicability, but their uncertainty is often so high that the converse may be a better premise.

Critical Comparisons.   An important but often neglected field of engineering is the critical comparison and analysis of experimental data, theoretical solutions, and numerically computed values. The book emphasizes such functions by illustration and through the problem sets.

Correlations.   A collateral function is the development or identification of comprehensive correlations that incorporate the insight gained by critical analyses. This book makes extensive use of the technique devised by the author and his associates for the development of correlating equations incorporating limiting solutions. This technique and the resulting correlations constitute a significant and unique feature of this book.

REFERENCES

1.   W. E. Langlois, *Slow Viscous Flow*, Macmillan, New York (1964).
2.   G. Birkhoff, *Hydrodynamics—A Study in Logic, Fact and Similitude*, Dover, New York (1955).
3.   H. Schlichting, *Boundary Layer Theory*, 7th ed., English transl. by J. Kestin, McGraw-Hill, New York (1979).
4.   M. Van Dyke, *Perturbation Methods in Fluid Mechanics*, Annotated ed., Parabolic Press, Stanford, CA (1975).
5.   S. W. Churchill, *The Practical Use of Theory in Fluid Flow. Book I. Inertial Flows*, Etaner Press, Thornton, PA (1980).
6.   S. W. Churchill, *The Practical Use of Theory in Fluid Flow. Book IV. Turbulent Flows*, Notes, The University of Pennsylvania (1981).
7.   S. W. Churchill, *The Practical Use of Theory in Fluid Flow. Book III. Laminar, Multidimensional Flows in Channels*, Notes, The University of Pennsylvania (1979).
8.   S. W. Churchill, "Theories, Correlations and Uncertainties for Waves, Gradients and Fluxes—a Course in Momentum and Energy Transfer," *Chem. Eng. Educ.*, 3 (Fall 1969) 178, 179, 181, 182.
9.   R. P. Feynman, R. B. Leighton, and M. Sands, *The Feynman Lectures*, Vol. II, Addison-Wesley, Reading, MA (1964).

Acknowledgments

I am indebted to many people for their assistance in one form or another in the preparation of this book; they include the following:

First in time, to Donald L. Katz who suggested that James G. Knudsen and I completely revise and modernize the graduate course in fluid mechanics and heat transfer in chemical engineering at the University of Michigan, thereby initiating the process that led to this book. Also to my colleagues James O. Wilkes, Joseph D. Godard, and J. Louis York who took part in the continuation of that effort.

To the graduate and undergraduate students in my classes on fluid flow at the University of Michigan and the University of Pennsylvania who tolerated incomplete segments of the manuscript as a text and made many constructive suggestions, and particularly to my doctoral students J. David Hellums, James O. Wilkes, and Dudley A. Saville at the University of Michigan, and Jai P. Gupta, Hiroyuki Ozoe, and Lance Collins at the University of Pennsylvania who each contributed in many different ways.

To my academic administrators Arthur E. Humphrey, Joseph Bordogna, Daniel D. Perlmutter, Alan L. Myers, John A. Quinn, and Douglas A. Lauffenburger who provided encouragement, assistance, and a receptive environment for my scholarly work.

To Patricia A. Lawrence and Charlotte Hill who typed and retyped the various versions of the manuscript.

To Yu Wang, who prepared most of the original sketches and plots, and to Renate Schultz who prepared some of those in Chapter 17.

Special thanks are due to Professor Alva D. Baer, University of Utah, for calling the work of Roscoe to my attention; to Professor Peter Rowe, University College, London, for his suggestions with respect to fluidized beds; and to Dr. Thomas Keene, E. I. Dupont, Wilmington, who called to my attention some of the recent work on packed beds.

To Dr. Charles J. Myers, Director, and his associates of the Towne Library of the University of Pennsylvania for their continued assistance.

Finally, to the following who responded promptly and generously to my request for reproducible drawings or tabulations of data:

Dr. Elmar Achenbach, Institut für Reaktorbauelemete, Jülich
Professor Koichi Asano, Tokyo Institute of Technology
Professor David V. Boger, The University of Melbourne
o. Professor Dr.- Ing. Heinz Brauer, Technische Universität
Professor Hsueh-Chia Chang, University of Notre Dame
Professor Ronald Darby, Texas A & M University
Professor M. E. Charles, University of Toronto

Professor Francis A. L. Dullien, University of Waterloo
Dr. George D. Fulford, Alcan International, Ltd., Kingston, Ontario
Professor L. B. Gibilaro, University College, London
Dr. Jacob E. Fromm, IBM Almaden Research Center
Professor Alvin E. Hamielec, McMaster University
Professor Yuji Kawamura, Hiroshima University
Dr. Dale L. Keairns, Westinghouse Electric Corp., Pittsburgh
Professor James G. Knudsen, Oregon State University
Professor Lambit U. Lilleleht, University of Virginia
Professor Dr.-Ing. A. Mersmann, Technische Universität München
Professor Arthur B. Metzner, University of Delaware
Professor Toshiro Miyahara, Okayama University
Professor J. D. Murray, University of Oxford
Professor Larry V. McIntire, Rice University
o. Professor Dr.-Ing. W. Stahl, Universität Karlsruhe (TH)
Professor Warren E. Stewart, University of Wisconsin
Professor Albin A. Szewczyk, University of Notre Dame
Professor Normand Thérien, Université de Sherbrooke
Professor Milton Van Dyke, Stanford University
Professor Dr. rer. nat. D. Vortmeyer, Technische Universität München
Dr. Marvin Warshay, NASA, Cleveland

PART I

One-Dimensional Laminar Flows

In steady, fully developed, laminar flow of an incompressible fluid through a straight channel, the fluid motion is confined to the direction of the channel; that is, it is *rectilinear*. This behavior can be modeled by a one-dimensional force-momentum balance in terms of one dependent variable (the velocity in the direction of the channel) and one or two independent variables (the coordinate or coordinates normal to the direction of flow). The *velocity field itself is thus one-dimensional* (i.e., only one independent variable is required) in geometries with axial or planar symmetry. One-dimensional velocity fields may also occur in rotating systems with axial symmetry.

A large fraction of the flows of most practical interest fall in this restricted class. Such flows may be treated by elementary techniques since the equations of conservation are ordinary differential equations and, except for some non-Newtonian fluids, are linear in the velocity and directly integrable.

This part is confined to these flows in which one-dimensional velocity fields occur. (Exceptions are made for a few pseudo-one-dimensional velocity fields.) More complex flows are considered in later sections of this book and in subsequent books in this series. Specifically, the general three-dimensional, unsteady equations of conservation are developed and examined in Part II, unconfined laminar flows are discussed in Part III, and flow through dispersed media is covered in Part IV. Multidimensional laminar flows and turbulent flows will be examined in separate volumes.

Chapter 1

Identification of Geometries and Dimensionless Variables

In this chapter, geometries in which the velocity field may be one-dimensional are first identified. Combining the variables and parameters in dimensionless groups simplifies the description of fluid motions. Important dimensionless groups that occur with one-dimensional velocity fields are therefore identified in what follows.

FLOWS THAT PRODUCE ONE-DIMENSIONAL VELOCITY FIELDS

One-dimensional or nearly one-dimensional velocity fields occur only in geometries with axial or planar symmetry. Some of these flows are listed here and sketched in Figure 1–1.

1. Forced flow through a round tube
2. Forced flow between parallel plates
3. Forced flow through the annulus between concentric round tubes of different diameters
4. Gravitational flow of a liquid film down an inclined or vertical plane
5. Gravitational flow of a liquid film down the inner or outer surface of a round vertical tube
6. Gravitational flow of a liquid through an inclined half-full round tube
7. Flow induced by the movement of one of a pair of parallel planes
8. Flow induced in a concentric annulus between round tubes by the axial movement of either the outer or the inner tube
9. Flow induced in a concentric annulus between round tubes by the axial rotation of either the outer or the inner tube
10. Flow induced in the cylindrical layer of fluid between a rotating circular disk and a parallel plane
11. Flow induced by the rotation of a central circular cylinder whose axis is perpendicular to parallel circular disks enclosing a thin cylindrical layer of fluid
12. Combined forced and induced flow between parallel plates

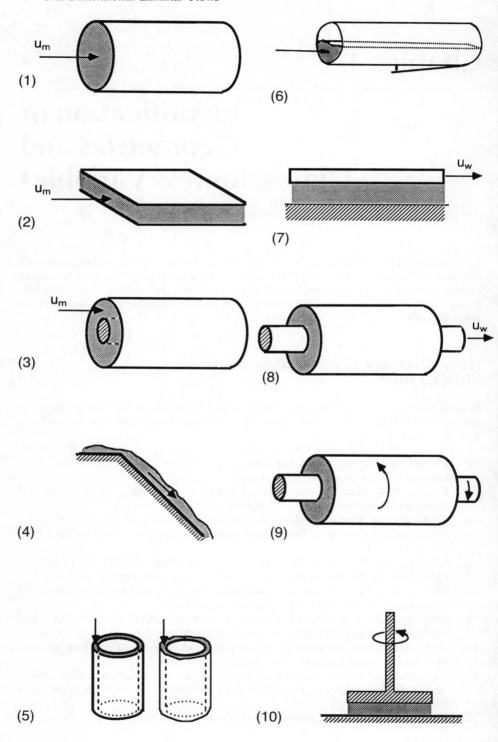

(1)

(2)

(3)

(4)

(5)

(6)

(7)

(8)

(9)

(10)

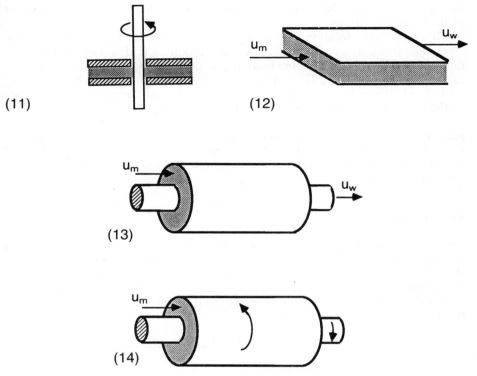

(11)   (12)

(13)

(14)

FIGURE 1-1   *Geometries and conditions that produce one-dimensional velocity fields.*

13. Combined forced and induced longitudinal flow in the annulus between concentric round tubes
14. Combined forced and rotationally induced flow in the annulus between concentric round tubes

    In addition, one-dimensional velocity fields are possible for the stratified flow of two or more fluids for cases 2, 4, and 10–12, and for annular flow with small density differences for cases 1, 3, 9, 13, and 14.

DIMENSIONAL CONSIDERATIONS

Each of the additive terms of an equation must have the same net dimensions. Then, if an equation is divided through by any of these additive terms, the additive terms of the modified equation must all be dimensionless (i.e., have no net dimension). It follows that any solution of the equation can be expressed in terms of dimensionless groups of variables. This consequence must prevail even if the original equation (or equations) relating the dimensional variables is unknown.

    The systematic determination of the minimum number of dimensionless groups required to describe a process from a simple listing of all of the presumed dimensional variables is called *dimensional analysis*. This technique is presumed to be known to the reader.

If an analytical solution of a model of a process can be attained, even in the form of an integral or an infinite series, the solution itself identifies the minimum number of dimensionless groups required to describe the behavior. Such is the case with all of the fluid flows described in Chapters 3–7. A technique is described in Chapter 11 that can be used to identify the minimum required number of dimensionless groups even when a model cannot be solved analytically.

The minimum number of dimensionless groups required to describe a process is obviously less, and usually far less, than the total number of dimensional variables. Hence, dimensional analysis may greatly reduce the number of experiments or numerical integrations required to produce the same amount of useful information.

Velocity Fields

The one-dimensional velocity fields described in Chapters 3–7 can all be modeled exactly. Furthermore, in almost all instances, an analytical solution is possible, thereby not only identifying the minimum number of dimensionless groups but the relationship between them. These several solutions, which involve many different variables, dimensionless groups, and functional relationships, can all be generalized as

$$U = \phi\{Y, \Pi_1, \Pi_2, \ldots\} \tag{1.1}$$

where $\phi\{\alpha\}$ = function of α
         U = dimensionless velocity
         Y = dimensionless coordinate normal to the direction of flow
         Π_n = dimensionless parameters

The appropriate scaling of the velocity and the identification of the dimensionless parameters for the several types of flow—forced, gravitational, and surface-induced—is summarized below in order to provide a framework and a commonality for the details which follow in Chapters 2–7.

After completing Chapter 7 it is suggested that Chapter 1 be reread to confirm the generality that was asserted in advance.

Throughout this book, except where stated otherwise, the coordinate will be scaled with respect to

$$D \equiv \frac{4A_x}{P_w} = \text{the hydraulic diameter in meters} \tag{1.2}$$

where A_x = cross-sectional area of fluid, m^2
         P_w = wetted perimeter of channel, m

Thus

$$Y = \frac{y}{D} = \frac{yP_w}{4A_x} \tag{1.3}$$

where y is the coordinate normal to the direction of flow in meters.

Forced Flows

The problem of forced flow can be stated in terms of the mean velocity produced by a specified pressure gradient, or in terms of the pressure gradient produced by a specified mean velocity. The latter choice is made here. Thus

$$u_m = \frac{w}{\varrho A_x} = \text{mean velocity in } x\text{-direction in meters per second} \quad (1.4)$$

where x = coordinate in direction of flow, m
          w = mass rate of flow, kg/s
          ϱ = density of fluid, kg/m^3

This mean velocity will be used to scale the local velocity. Thus

$$U = \frac{u}{u_m} \quad (1.5)$$

where u is the local velocity in the x-direction in meters per second.
     For flow through a round tube and between parallel plates, no other variables or parameters are required to describe the velocity distribution, and

$$\frac{u}{u_m} = \phi\left\{\frac{y}{D}\right\} \quad (1.6)$$

For a Newtonian fluid, as shown in Chapter 3,

$$\tau_w \equiv \mu\left(\frac{du}{dy}\right)_w \quad (1.7)$$

where the subscript w indicates the solid surface
       τ_w = shear stress *on solid surface in the direction of flow*, Pa or N/m^2
         or kg/m · s^2
       μ = (dynamic) viscosity, Pa · s or kg/m · s

Since for a specified D and u_m, u is a function only of y, Equation 1.7 implies that τ_w is uniform over the surface. It follows from Equations 1.6 and 1.7 that

$$\frac{\tau_w D}{\mu u_m} = B \quad (1.8)$$

where B is a constant, with one value for round tubes and another for parallel plates.
     An overall force balance on an element of fluid of length Δx, cross-sectional area A_x, and perimeter P_w, as shown in Figure 1–2, can be written as

$$\left[p - \left(p + \frac{dp}{dx}\Delta x\right)\right]A_x + g\varrho \sin\{\theta\}A_x\,\Delta x - \tau_w P_w\,\Delta x = 0 \quad (1.9)$$

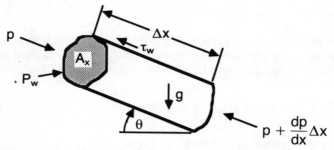

FIGURE 1–2   *Forces on a longitudinal segment of a fluid in a channel.*

which simplifies to

$$\tau_w = \frac{-dp}{dx}\frac{A_x}{P_w} + g\varrho \sin\{\theta\}\frac{A_x}{P_w} \tag{1.9A}$$

where p = thermodynamic pressure, Pa or kg/m · s^2
   g = acceleration or specific force due to gravity, m/s^2 or N/kg
   θ = *downward* angle of channel in direction of flow measured from
      horizontal, rad

Equation 1.9 implies that the pressure is uniform across the area A_x. It actually varies with elevation due to the hydrostatic pressure of the liquid, but the same variation occurs at x and $x + \Delta x$ and hence cancels out. This superficial complication can be avoided by defining a *dynamic or piezometric pressure* as

$$\mathscr{P} \equiv p + g\varrho h - p_0 \tag{1.10}$$

where h = elevation, m
   p_0 = reference pressure at $h = 0$, Pa

Then for constant ϱ and g,

$$\frac{dp}{dx} = \frac{d\mathscr{P}}{dx} - g\varrho\frac{dh}{dx} \tag{1.11}$$

That is, the total thermodynamic pressure gradient can be considered to be the sum of the gradient in the dynamic pressure and the gradient in the hydrostatic pressure $[-g\varrho(dh/dx)]$. In the absence of flow the total pressure varies only due to the hydrostatic force. Conversely, the effect of flow is represented completely by the gradient in the dynamic pressure. This latter result is demonstrated by substituting for dp/dx in Equation 1.9A from Equation 1.11 and noting that

$$\frac{dh}{dx} = -\sin\{\theta\} \tag{1.12}$$

The result is

$$\tau_w = -\frac{d\mathscr{P}}{dx}\frac{A_x}{P_w} = -\frac{d\mathscr{P}}{dx}\frac{D}{4} \tag{1.13}$$

$\mathscr{P}$ is uniform across A_x. All subsequent force and force-momentum balances will be written directly in terms of $\mathscr{P}$. If the total pressure p is of interest, it can be found from Equation 1.10.

The dimensionless group $\tau_w D/\mu u_m$ in Equation 1.8 does not yet have an accepted name or symbol. The group $D^2(-d\mathscr{P}/dx)/\mu u_m$ has been called the *Poiseuille number*[1] but does not have an accepted symbol. It seems appropriate to generalize this usage as

$$\text{Po} \equiv \text{Poiseuille number} = \frac{\tau_w D}{\mu u_m} \tag{1.14}$$

Equation 1.8 can then be rewritten as

$$\text{Po} = B \tag{1.15}$$

For flow through the annulus between two concentric round tubes, the aspect ratio in the y-direction,

$$\lambda = \frac{D_1}{D_2} \tag{1.16}$$

Where D_1 = inner diameter, m
D_2 = outer diameter, m

is a parameter. The results for round tubes and parallel plates are readily generalized for this case as

$$\frac{u}{u_m} = \phi\left\{\frac{y}{D}, \lambda\right\} \tag{1.17}$$

and

$$\text{Po} = B\{\lambda\} \tag{1.18}$$

However, since the shear stress differs on the inner and outer surfaces, Equation 1.18 must be applied separately for τ_{w1} and τ_{w2} or overall in terms of an area-weighted mean shear stress:

$$\tau_{wm} = \frac{\tau_{w1}D_1 + \tau_{w2}D_2}{D_1 + D_2} = \frac{D}{4}\left(-\frac{d\mathscr{P}}{dx}\right) \tag{1.19}$$

The functions $B_1\{\lambda\}$, $B_2\{\lambda\}$, and $B_m\{\lambda\}$ corresponding to the shear stresses τ_1, τ_2, and τ_m differ. The relationships between them are developed in Chapter 4.

---

[1] Named in honor of Jean-Louis-Marie Poiseuille (1799–1869), a French physician who studied flow in capillary tubes in order to simulate the circulation of the blood [1].

Open Gravitational Flows

The preceding results for forced flow between parallel plates and through a round tube are directly applicable to open flow of a liquid down an inclined plate or through an inclined half-full round tube, respectively, insofar as the drag of the air above the liquid is negligible and if the mean velocity is again considered as specified. This adaption can be made by noting that for open flows

$$\frac{dp}{dx} = 0 \tag{1.20}$$

Hence from Equations 1.11 and 1.12,

$$-\frac{d\mathscr{P}}{dx} = -g\varrho\frac{dh}{dx} = g\varrho\sin\{\theta\} \tag{1.21}$$

and from Equation 1.13,

$$\tau_w = \frac{g\varrho D}{4}\sin\{\theta\} \tag{1.22}$$

For these open flows $\theta > 0$, so $\sin\{\theta\} > 0$. The hydraulic diameter and thereby the depth and angle of inclination required to produce the mean velocity are defined by the combination of Equations 1.14, 1.15, and 1.22.

Longitudinally Induced Flows

For flows induced by the longitudinal motion of a surface, such as those sketched in Figures 1–1(7) and (8), the local velocity is scaled more conveniently by the velocity of the moving surface than by the mean velocity. Thus in this case

$$U \equiv \frac{u}{u_w} \tag{1.23}$$

where u_w is the velocity of a moving solid surface (m/s). It follows that

$$\frac{u}{u_w} = \phi\left\{\frac{y}{D}, \lambda\right\} \tag{1.24}$$

$$\frac{u_m}{u_w} = \phi\{\lambda\} \tag{1.25}$$

$$\text{Po} = \frac{\tau_w D}{\mu u_m} = B\{\lambda\} \tag{1.26}$$

where λ is again the ratio of diameters for annulus.

The overall force balances in Figure 1–3 indicate that the shear stress is equal on both surfaces in the planar case, but that in the annulus

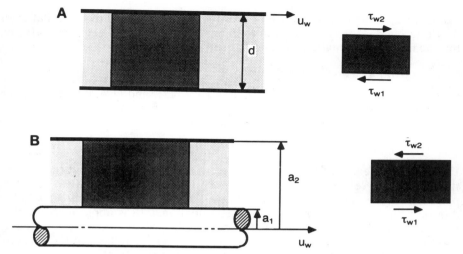

FIGURE 1–3   *Overall force balance on an element of fluid in longitudinal Couette flow: (A) in a planar channel; (B) in a circular annulus.*

$$\tau_{w1} D_1 = \tau_{w2} D_2 \tag{1.27}$$

Hence if τ_{w1} is determined, τ_{w2} follows from Equation 1.27.

Rotationally Induced Flows

For rotationally induced flows such as those sketched in Figures 1–1(9) to (11), the local velocity is again scaled most conveniently by the velocity of the moving surface, which in the rotational cases can be expressed as

$$u_w = \frac{D_\omega \omega}{2} \tag{1.28}$$

where D_ω = diameter of rotating surface, m
         ω = angular velocity of rotating surface, rad/s

It follows that

$$U = \frac{u}{U_w} = \frac{2u}{D_\omega \omega} = \phi \left\{ \frac{y}{D_\omega}, \lambda \right\} \tag{1.29}$$

$$\frac{u_m}{D_\omega \omega} = \phi \{ \lambda \} \tag{1.30}$$

$$\text{Po} = \frac{\tau_w D}{\mu u_m} = \phi \{ \lambda \} \tag{1.31}$$

For annular, rotationally induced flows, such as in Figure 1–1(9), Equation 1.31 must be applied separately to the two surfaces since the shear stresses are unequal according to Equation 1.17. For the motion induced in a thin flat layer by a rotating disk or cylinder, such as in Figures 1–1(10) and (11), respectively, the shear stress is equal on the two flat surfaces.

Combined Induced and Forced Flows

In flows such as those sketched in Figures 1–1, (12), and (13), either the mean velocity or the surface velocity can be used to scale the local velocity, but in any event the velocity ratio u_w/u_m is an additional parameter.

Two-Phase Flows

Either of the mean velocities may be used to scale the velocity, but the ratio of the two mean velocities and the ratio of the two viscosities are additional parameters. The shear stresses are not equal on the surfaces in contact with the different fluids, even in the planar case. Equation 1.18 is applicable to either surface with the appropriate mean velocity, viscosity, and the preceding additional parameters. The individual shear stresses are related to the pressure gradient and the mean shear stress through Equation 1.19 as before.

Generalities

The amount of information deduced in the previous paragraphs from simple dimensional considerations is remarkable. (Overall force balances were used to show the relationship between various dependent variables but were not involved directly in this analysis.) The major role of the derivations in Chapters 2–7 is merely to determine the unknown constants and functions. The structural similarity of all of these diverse cases is also remarkable. Further similarities are revealed when the functional behavior itself is determined.

It is apparent that the dimensionless velocity is independent of the viscosity in all of the single-phase flows, and depends only on the ratio of the viscosities for the two-phase flows. Likewise, the Poiseuille number, as defined for forced and gravitational flows, and its analog for induced flows depend only on the aspect, velocity, and viscosity ratios.

DIMENSIONLESS GROUPS DETERMINING ONE-DIMENSIONALITY

Dimensionless groups in addition to those already mentioned determine whether or not one-dimensional flow will occur. Some of these groups are examined next.

The Reynolds Number

The primary group characterizing the regime of flow and thereby determining whether laminar flow will occur is the *Reynolds number*,[1] here defined for an arbitrary geometry as

---

[1] Named after Osborne Reynolds (1842–1912), who discovered experimentally that the critical velocity for transition from laminar to turbulent flow in a pipe was inversely proportional to the diameter. He then deduced from dimensional considerations that proportionality to μ/ϱ was also necessary [2].

$$\text{Re} \equiv \frac{4A_x u_m \varrho}{P_w \mu} = \frac{D u_m \varrho}{\mu} = \frac{D u_m}{\nu} \tag{1.32}$$

where $\nu = \mu/\varrho$ is the kinematic viscosity (m^2/s). Sometimes in this book, as elsewhere, a characteristic dimension other than the hydraulic diameter and a characteristic velocity other than the mean velocity are used to define a Reynolds number. Hence caution is advised in interpreting such shorthand as *Re*. Herein subscripts will be used to designate any special definitions, and superscripts will be used to designate particular values of Re, such as the critical value for transition from laminar to turbulent flow.

The presence of μ in Equation 1.32 implies that this definition is limited to Newtonian fluids. The definition of a Reynolds number for non-Newtonian fluids is deferred to Chapter 5.

The Froude Number

The group that determines whether gravitational waves occur on the free surface of open-channel flows, and thereby disrupt one-dimensionality, is the *Froude number*,[1] herein defined for consistency[2] as

$$\text{Fr} \equiv \frac{u_m^2}{gD} \tag{1.33}$$

where $D = 4d$, m
        d = depth of fluid, m

The Weber Number

The group that characterizes the effect of surface tension and the consequent threshold for disturbance of one-dimensionality is the *Weber number*[3]

$$\text{We} \equiv \frac{\varrho u_m^2 D}{\sigma} \tag{1.34}$$

where σ is the surface tension (N/m or kg/s^2).

The Longitudinal Aspect Ratio

The attainment of fully developed flow depends on the dimensionless distance

$$X = \frac{x}{D} \tag{1.35}$$

The Aspect Ratio for Breadth

Independence from edge effects and hence from the related two-dimensionality depends on the aspect ratio in the direction normal to both x and y:

---

[1] Named in recognition of William Froude (1810–1879), an English engineer, who studied the influence of waves on the drag of ships [3].
[2] The thickness of the film d rather than the hydraulic diameter D has traditionally been used as the characteristic dimension in the Froude number for film flows.
[3] Named for Moritz G. Weber (1871–1951).

$$\beta = \frac{H}{D} \tag{1.36}$$

where H is the breadth of the surface or channel perpendicular to the direction of flow x and to the primary coordinate y (m).

ALTERNATIVE DIMENSIONLESS GROUPS

Additional independent dimensionless groups can have a significant effect or indicate a limitation on one-dimensional flow only insofar as they introduce a new variable, such as the surface tension in the Weber number.

New dimensionless groups can be constructed from the product of some power of two or more of the groups already mentioned. Such combined groups are not independent of the prior ones, and one of the combining groups must be dropped. Such alternative groups are useful if they result in a simplification of the functional relationships between the variables or provide a more explicit display of the variables of primary interest. One such variable is the *friction factor*. The shear stress on the wall is often expressed in terms of such a friction factor rather than in terms of the Poiseuille number. The *Stanton–Pannell friction factor*

$$f \equiv \frac{\tau_w}{\varrho u_m^2} \tag{1.37}$$

is used exclusively in this series of books. Other definitions, particularly the *Fanning friction factor*

$$f_F \equiv \frac{2\tau_w}{\varrho u_m^2} \tag{1.38}$$

and the *Darcy* or *Darcy–Weisbach friction factor*

$$f_{DW} = \frac{8\tau_w}{\varrho u_m^2} \tag{1.39}$$

are also widely used.[1] The subscript SP could be added to the friction factor defined by Equation 1.37. Then for comparison

$$f_{SP} = \frac{f_F}{2} = \frac{f_{DW}}{8} \tag{1.40}$$

---

[1] These names commemorate the early work of Stanton and Pannell [4], Fanning [5], Darcy [6], and Weisbach [7] on the shear stress in turbulent flow through pipes and channels. The exact precedence among these and other early workers with respect to the use of these particular factors is subject to uncertainty and controversy. For example, even earlier (in 1775), Chézy expressed the frictional resistance for open-channel flow in terms of a factor C, equivalent to $(gf)^{1/2}$, and now called the *Chézy coefficient* [8].

Unfortunately the subscripts in Equation 1.40 do not appear in the literature. Furthermore, velocities other than u_m are sometimes used. Hence great care must be exercised in interpreting numerical values of the friction factor from different sources. No decisive advantage can be cited for one factor relative to another. The advantage of f itself is only as a shorthand in equations. This advantage is largely overbalanced by the resultant confusion and uncertainty. In graphs and tables, and generally in equations, an unambiguous notation such as $\tau_w/\varrho u_m^2$ is preferable.

The Stanton–Pannell friction factor is related to the Poiseuille number and Reynolds number as follows:

$$f = \frac{\text{Po}}{\text{Re}} \tag{1.41}$$

A numerical coefficient would occur for the other two friction factors.

Replacement of Po by f in theoretical expressions or correlations for the shear stress corresponding to one-dimensional flows would not appear to have any advantage for laminar flows. However, this practice has generally been followed, apparently by extension of the somewhat more justifiable use of f for turbulent flows.

The replacement of Re by f as a criterion for the onset of the transition from laminar to turbulent flow may be advantageous in that a single value of f characterizes most systems, whereas different values of Re are observed.

SUMMARY

A review of the solutions in Chapters 3–7 after completing Part I will confirm the generalities asserted in Chapter 1 on the basis of dimensional analysis; for example, in all of these one-dimensional laminar flows the mean velocity is proportional to the driving force $(-d\mathscr{P}/dx$ and/or $u_w)$, and the local velocity is a function only of the fractional distance from one surface to another, as well as of aspect ratios and property ratios. A re-reading of Chapter 1 is suggested at that point, with specific reference to the various solutions, to consolidate mentally all of this detail.

PROBLEMS

1. The velocity u of a liquid draining through a hole in the bottom of a tank is presumed to be a function of the acceleration due to gravity, g (m/s); the liquid level, h (m); the diameter of the hole, D (m); the dynamic viscosity, μ (Pa · s); and the density, ϱ (kg/m^3).

   a. Determine the minimum set of dimensionless groups that can represent this behavior.
   b. Postulating that h, and in the laminar regime ϱ as well, occurs only as a multiplier of g, reduce the relationship as much as possible.
   c. Determine the functional relationship between the *orifice coefficient* $C_o \equiv u/(gh)^{1/2}$ and the Re $\equiv Du\varrho/\mu$ for the conditions of part (b).

2. The motion of a bubble rising through a fluid can be correlated in terms of $\bar{C}_t \equiv 2g(\varrho - \varrho_g)D/3\varrho u^2$ and $\mathrm{Re} \equiv Du\varrho/\mu$, where ϱ = density of liquid $(\mathrm{kg/m^3})$, ϱ_g = density of gas $(\mathrm{kg/m^3})$, μ = dynamic viscosity of liquid $(\mathrm{Pa \cdot s})$, and $\bar{C}_t$ = *mean overall drag coefficient*. Determine the functional form of the relationship between $\bar{C}_t$ and Re for

    a. $u \to 0$ (independence from ϱ but not from $\varrho - \varrho_g$)
    b. $u \to \infty$ (independence from μ)
    c. intermediate u (independence from D)

REFERENCES

1. J.-L.-M. Poiseuille, "Recherches expérimentales sur le mouvement des liquides dans les tubes de très petits diamètres," Mémoires présentés par divers savants à l'Académie Royale des Sciences de l'Institut de France, *Sci. Math. Phys.*, 9 (1846) 433; English transl. by W. H. Herschel, "Experimental Investigations upon the Flow of Liquids in Tubes of Very Small Diameter," *Rheol. Mem.*, E. C. Bingham, Ed., *1*, No. 1, Easton, PA (1940).
2. Osborne Reynolds, "An Experimental Investigation of the Circumstances Which Determine Whether the Motion of Water Shall Be Direct or Sinuous and the Law of Resistance in Parallel Channels," *Phil. Trans. Roy. Soc. (London)*, Ser. 3, *A174* (1883) 935, (*Scientific Papers*, Vol. II, Cambridge University Press (1901), p. 51); "On the Dynamical Theory of an Incompressible Viscous Fluid and the Determination of the Criterion," *Phil. Trans. Roy. Soc. (London)*, Ser. 3, *A186* (1896) 123 (*Scientific Papers*, Vol. II, Cambridge University Press (1901), p. 535).
3. William Froude, "Experiments with Models Capable of Application to Full-Sized Ships," *Trans. Inst. Naval Arch.*, *11* (1870) 88.
4. T. E. Stanton and J. R. Pannell, "Simularity of Motion Relative to the Surface Friction of Fluids," *Trans. Roy. Soc. (London)*, *A214* (1914) 199.
5. J. T. Fanning, *A Practical Treatise on Hydraulic and Water Supply Engineering*, Van Nostrand, New York (1893).
6. H. P. G. Darcy, "Recherches expérimentales rélatives au mouvement de l'eau dans les tuyaux," *Mém. Acad. Sci.*, *15* (1858) 141.
7. J. Weisbach, *Experimentale Hydraulik*, Leipzig (1855).
8. A. deChézy, "Manuscript Report on the Canal de l'Yvette," 1775, in *Mém. Classe des Sciences*, Paris (1913–1915), according to Herschell Clemens, "On the Origin of the Chézy Formula," *J. Assoc. Engng.*, *18* (1887) 363.

Chapter 2

Momentum Transfer, Viscosity, and Shear Stress

The behavior of fluids under a shearing stress is the determining factor in laminar flows. Such behavior for ordinary fluids is described and defined in quantitative terms in this chapter.

MOMENTUM TRANSFER BY MOLECULAR DIFFUSION

Consider steady, one-dimensional flow in the x-direction, with the velocity increasing with increasing y, as illustrated in Figure 2–1. Now consider the interchange of molecules by random motion (diffusion) across the plane $y = A$. Since the molecules below $y = A$ have a lesser component of velocity in the x-direction than those above $y = A$, this random process produces a net transfer of momentum in the direction of decreasing y. The rate of transfer in this simple flow has been shown by theoretical analyses (see, for example, Chapman and Cowling [1], Hirschfelder et al. [2], Bird et al. [3], and Guggenheim [4]) to be proportional to the negative of the velocity gradient. This relationship can be written as

Newton law

$$j_{yx} = -\mu \frac{du_x}{dy} \tag{2.1}$$

where j_{yx} = specific rate of transfer (per unit area) of x-momentum in the y-direction, or flux density of x-momentum in the y-direction, $kg/m \cdot s^2$
  u_x = component of velocity of fluid in the x-direction, m/s
  μ = proportionality constant = (dynamic) viscosity, $Pa \cdot s$ or $kg/m \cdot s$

The particular velocity gradient

$$\Gamma = \left| \frac{-du_x}{dy} \right| \tag{2.2}$$

17

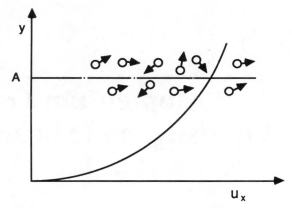

FIGURE 2–1   *Molecular transport of momentum in one-dimensional planar flow.*

is called the *rate of shear*. The viscosity, which can be seen to equal the specific rate of momentum transfer within a fluid due to a unit velocity gradient, is necessarily always greater than zero.

Equation 2.1 is known as *Newton's law* of *viscosity*.[1,2] Fluids for which Equation 2.1 holds, as illustrated in Figure 2–2 by the data of Boger (private communication) for an aqueous syrup, are called *Newtonian*. Most gases and simple liquids fall in this category.

By *Newton's second law of motion*,[3] the specific rate of transfer of momentum can be equated to and hence replaced by a hypothetical force per unit area or *shear stress*

$$\tau_{yx} = j_{yx} \tag{2.3}$$

where τ_{yx} is the shear stress (Pa or N/m^2 or $kg/m \cdot s^2$) in the *x*-direction applied on the fluid above *y* by the fluid below *y*.

This sign convention for the direction of application of τ_{yx} is arbitrary. Care must be taken in interpreting the direction of τ_{yx} (or merely τ) in various books, because the opposite sign is sometimes chosen. Consistency is essential if the correct sign is to be obtained in the final expressions obtained by substituting τ_{yx} for j_{yx}.

Substituting the expression for j_{yx} provided by Equation 2.1 in the force-momentum balance (Equation 2.3) gives

$$\tau_{yx} = -\mu \frac{du_x}{dy} \tag{2.4}$$

---

[1] "The resistance arising from the want of lubricity in the parts of a fluid, is, other things being equal, proportional to the velocity with which the parts of a fluid are separated from one another." (Newton [5], p. 385)

[2] Named after Isaac Newton (1642–1727), who invented the mathematical structure and discovered the "laws" upon which fluid mechanics is based.

[3] "The change of motion is proportional to the motive force impressed, and is in the direction of the right line in which the force is impressed." (Newton [5], p. 13)

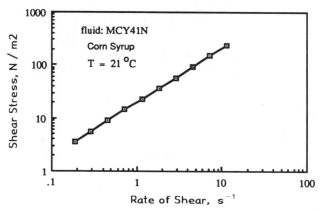

FIGURE 2-2   *Shear stress as a function of rate of shear for a Newtonian fluid. (Courtesy of D. V. Boger.)*

The distinction between Equations 2.1, 2.3, and 2.4 has frequently been overlooked in the literature (see Churchill [6]). *Equation 2.1 is a correlating equation or phenomenological expression for the one-dimensional rate of transfer of momentum due to molecular diffusion; Equation 2.3 is a one-dimensional balance of force and momentum; and Equation 2.4 is a specialized form of the one-dimensional balance of force and momentum for molecular transfer only.* Equation 2.4, rather than the more basic 2.1, is usually used for experimental determinations of the viscosity, since the shear stress can be measured or inferred more easily than the rate of momentum transfer. Equation 2.4 is also used to define an effective viscosity when mechanisms of momentum transfer other than isotropic molecular diffusion are involved.

Gases

The following expression for the viscosity of a gas can be derived (see, for example, Bird et al. [3]) from simple kinetic theory (elastic spheres without intermolecular forces):

$$\mu = \frac{2}{3\pi N d^2}\left(\frac{mRT}{\pi}\right)^{1/2} \tag{2.5}$$

where d = effective diameter of molecules, m
N = Avogadro's number = 6.023×10^{23} molecules/mol
m = molar mass, kg/mol
R = gas law constant = 8.3143 J/mol·K or $m^2 \cdot kg/mol \cdot s^2 \cdot K$

All gases are Newtonian; that is, their viscosity, as defined by Equation 2.1, is independent of the rate of shear. The effective diameter of a molecule must be determined experimentally. The viscosity of real gases is essentially independent of pressure, as predicted by Equation 2.5, at moderate pressures, and increases with temperature, although to a slightly higher power than the predicted half-power.

More complicated kinetic theories postulate a force field between the molecules (see, for example, Chapman and Cowling [1], Hirschfelder et al. [2], and Bird et al. [3], pp. 19–26). They provide a better prediction of viscosity but introduce additional empirical constants, and are therefore of questionable value for a priori predictions or for correlation (see Guggenheim [4] and Churchill [7]).

Empirical correlations for the effect of temperature and high pressure on the viscosity of gases in the form of the *law of corresponding states* are reproduced by Bird et al. [3], pp. 16–17. The effect of very low pressure, such that the mean free path is of the order of the least dimension of confinement, on the viscosity is described experimentally by Rasmussen [8]. The viscosities of a number of gases are listed in Table A.1 (see the Appendix).

Liquids

Liquids and solutions of low molecular weight are also Newtonian. The theory of liquids is yet inadequate to predict the viscosity with reliability (see, for example, Hirschfelder et al. [2] and Frenkel [9]). However a semitheoretical model has been developed by Ely and Hanley [10] for the prediction of the viscosity of nonpolar fluid mixtures over conditions ranging from a dilute gas to a dense liquid, and further improvements are to be expected. The viscosity *decreases* with temperature in contrast to the increase for gases. The viscosities of a number of liquids are listed in Table A.2 (in the Appendix).

NON-NEWTONIAN FLUIDS

Many suspensions, many solutions, and some pure liquids demonstrate a more complicated relationship between the observed shear stress and the rate of shear than indicated by Equation 2.4. In these fluids a mechanism other than diffusion, such as intertwining, contributes to the shear stress, and Equation 2.1 is not a satisfactory model. Such fluids are called *non-Newtonian*. The study of their behavior is called *rheology*.

Fluids for which the Shear Stress Varies Continuously with the Rate of Shear

The shear stress of some fluids changes continuously with the rate of shear and is essentially time independent. The application of shear to such fluids also produces a *normal stress*, which causes swelling or contraction when the fluid passes out of the end of a tube and therefore is important in some applications such as the extrusion of polymers. Further discussion of this aspect can be found in any book on *rheology*. Such fluids can be divided into two classes as follows.

Fluids for which the absolute value of the shear stress $\tau = |\tau_{xy}|$ increases less than linearly with the rate of shear, as illustrated in Figure 2–3 by the data of Boger (personal communication) for a solution of a drag-reducing polymer in a diluted syrup, are called *pseudoplastics*. (The data of Ashare [11] in problem 2–1 provide another example.) Values of the normal stress are included in Figure 2–3.

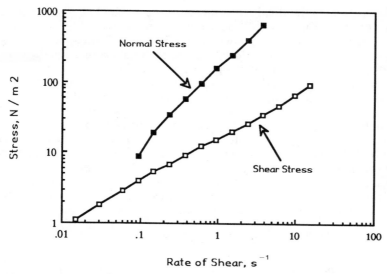

FIGURE 2–3   *Shear stress and normal stress as a function of rate of shear for a 0.5% weight solution of a polyacrylamide (Separan AP-30) in corn syrup (MCY 4 IN) with 13.85% weight added water at 23°C. (Courtesy of D. V. Boger.)*

Fluids such as napalm, cellulose acetate disssolved in acetone, paper-pulp suspensions, asphalts, polymer and rubber sols, polymer melts, adhesives, starch suspensions, and greases demonstrate this behavior. Pseudoplasticity is often attributed to the progressive alignment of asymmetrical molecules or suspended particles with increasing rates of shear, as compared to the random orientation that exists when the fluid is at rest. As sketched in Figure 2–4 in arithmetic coordinates, the behavior becomes more Newtonian with a limiting slope η_0 as the rate of shear approaches zero; it again approaches linearity, but with a lesser slope η_∞, as complete alignment of the molecules is attained at large rates of shear.

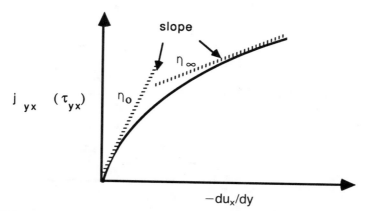

FIGURE 2–4   *Idealized pseudoplastic behavior in arithmetic coordinates.*

Fluids for which the shear stress increases more than linearly with the rate of shear, as illustrated in Figure 2–5 by the data of Boger (personal communication) for a suspension of starch in water, and in Figure 2–13 of problem 2–23 by the data of Metzner and Whitlock [12]. Such fluids were named *dilatants* by Reynolds [13], who discovered this behavior about 1885. He chose this name on the presumption that when dense suspensions are sheared at low rates the liquid lubricates the motion of one particle over another, but at high rates of stress the suspension "dilates" and the shear stress increases more than linearly. A modern interpretation is that a structure approaching that of a solid is formed by the particles as the rate of stress is increased. The limiting behavior of dilatants, per Figure 2–6, again in arithmetic coordinates, is analogous to that for pseudo-plastics except that the limiting slope η_∞ at large Γ is greater than the limiting slope η_0 at small Γ. Dilatant behavior is less common than pseudoplastic behavior but occurs in starch–water mixtures, wet beach sands, quicksands, mayonnaises, soaps, paints, and biological materials.

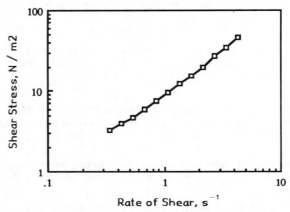

FIGURE 2–5   *Shear stress versus rate of shear for a suspension of starch in water at 23°C. (Courtesy of D. V. Boger.)*

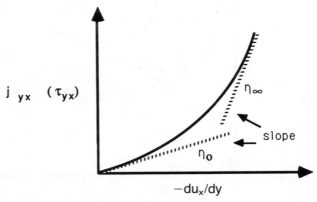

FIGURE 2–6   *Idealized dilatant behavior in arithmetic coordinates.*

The behavior of pseudoplastics and dilatants is usually represented by an effective or apparent viscosity, defined by analogy to Equation 2.1 as

$$\eta \equiv \frac{j_{yx}}{-du_x/dy} \tag{2.6}$$

The effective viscosity can be expressed in simpler but less explicit form as

$$\eta = \frac{\tau}{\Gamma} \tag{2.7}$$

where here Equation 2.2 implies that $\tau = |j_{yx}|$.

The value of η generally approaches a constant, η_0, at very small shear stresses (or rates of shear) and a constant, η_∞, at large shear stresses (or rates of shear), as suggested in Figures 2–4 and 2–6 and indicated more explicitly in Figures 2–7 and 2–8. Like μ, η is necessarily positive.

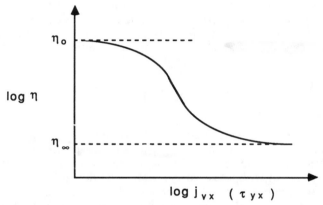

FIGURE 2–7   *Effective viscosity of a pseudoplastic fluid.*

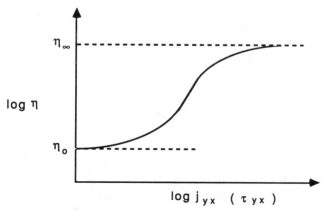

FIGURE 2–8   *Effective viscosity of a dilatant fluid.*

Models for Pseudoplastic and Dilatant Fluids

Churchill and Churchill [14] developed general empirical equations for the effective viscosity of pseudoplastic and dilatant fluids, based on the model of Churchill and Usagi [15, 16], which can be written either as

$$\left(\frac{\eta_0 - \eta_\infty}{\eta_0 - \eta}\right)^P = 1 + \left|\frac{\tau^*}{\tau}\right|^{pm} \tag{2.8}$$

or as

$$\left(\frac{\eta_0 - \eta_\infty}{\eta - \eta_\infty}\right)^P = 1 + \left|\frac{\tau}{\tau^*}\right|^{pm} \tag{2.9}$$

where τ^*, m, and p are arbitrary constants. The terms η_0 and η_∞ are also empirical constants but have the physical significance indicated in the previous paragraph. The values of τ^*, m, and p determined by Equation 2.9 will generally differ from those obtained by Equation 2.8 for the same set of data. (See problems 2 and 3.) Alternative correlating equations can be constructed by replacing τ and τ^* in Equations 2.8 and 2.9 by Γ and Γ^*, respectively.

The *power-law model* is a special case of Equations 2.8 and 2.9 (see problems 8 and 9). It can be expressed as

$$\tau_{yx} = -M \left|\frac{du_x}{dy}\right|^{\alpha-1} \frac{du_x}{dy} \tag{2.10}$$

where α is an arbitrary dimensionless constant called the *power-law index* and M is an arbitrary coefficient called the *consistency factor* with units of $Pa \cdot s^\alpha$ or $kg/m \cdot s^{2-\alpha}$. The higher the value of M, the more viscous is the fluid. For $\alpha = 1$ the fluid is Newtonian with a viscosity $\mu = M$. The deviation of α from unity is a measure of the degree of non-Newtonian behavior. Equation 2.10 has been widely used to represent the effective viscosity of pseudoplastic and dilatant fluids, in large part because of its mathematical tractability. Precedence in the use of this expression is subject to some uncertainty and controversy (see, for example, Nutting [17], Ostwald [18], and deWaele [19]). Equation 2.10 clearly cannot represent the limiting behavior at high and low shear stresses indicated in Figures 2–7 and 2–8, but it has been applied successfully in certain applications, such as laminar flow in pipes, that depend primarily on the maximum shear stress in the system.[1]

Equation 2.10 can also be written in less explicit form as

$$\tau = M\Gamma^\alpha \tag{2.11}$$

---

[1] Metzner [20] notes that, "Rheologists have long objected to the use of [Equation 2.10] on the basis that it is purely empirical, not being derived from any physical concepts. This criticism can hardly be considered valid in view of the fact that the 'physical concepts' upon which theoretical equations are frequently based consist of purely mechanical analogs such as springs, dashpots and blocks which at best have only vague equivalents in any real system....The objections to the use of [Equation 2.10] may therefore be considered to be of minor engineering interest at the present time."

and, by combination with Equation 2.7, as

$$\eta = M\Gamma^{\alpha-1} \tag{2.12}$$

and

$$\eta = M^{1/\alpha}\tau^{(\alpha-1)/\alpha} \tag{2.13}$$

The following derivation is adapted from Denn [21]. The shear stress τ is always observed to increase with Γ; that is,

$$\frac{d\tau}{d\Gamma} > 0 \tag{2.14}$$

Then from Equation 2.7

$$\frac{d(\eta\Gamma)}{d\Gamma} = \eta + \Gamma\frac{d\eta}{d\Gamma} > 0 \tag{2.15}$$

from which it follows that

$$\frac{d\ln\{\eta\}}{d\ln\{\Gamma\}} > -1 \tag{2.16}$$

From Equation 2.12

$$\frac{d\ln\{\eta\}}{d\ln\{\Gamma\}} > \alpha - 1 \tag{2.17}$$

Hence α is necessarily greater than zero. The smallest observed value of α (for molten polystyrene) is about 1/3. Equation 2.10 and many other models are special cases of the Churchill–Churchill model. Additional models are described in the problem set at the end of this chapter.

Fluids with a Yield Stress

One class of fluids, including toothpastes, oil-well drilling muds, sewage sludges, oil paints, margarines, shortenings, plastic melts, aqueous suspensions of clay, grain and paper pulps, chocolate syrups, and aqueous slurries of coal, peat, sand, cement, rock, chalk, and thorium oxide, require a finite shear stress to produce any motion. Such fluids are known as *Bingham plastics,* after E. C. Bingham [22] who first analyzed their flow in a circular pipe. Their behavior in terms of $\tau\{\Gamma\}$ is illustrated in Figure 2–9 by the data of Boger (private communication) for red mud II, a suspension of bauxite which is produced in vast quantities as a waste product of the Bayer process for aluminum. These muds are seen to have a finite yield stress at all concentrations, but also to demonstrate "shear thinning," at high concentrations. An additional example of both types of behavior is provided by the data of Thomas [23] in problem 25.

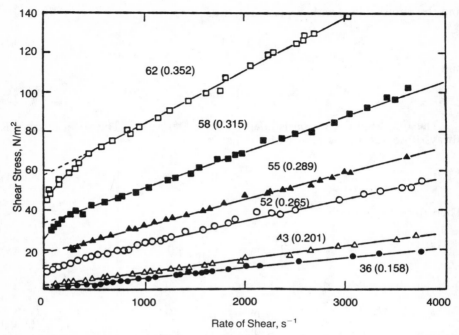

FIGURE 2–9   *Shear stress as a function of the rate of shear for suspensions of red mud (bauxite residue) in various concentrations. (Courtesy of D. V. Boger.)*

The variant behavior as represented by the dashed line, and the idealized behavior as represented by the solid line, are compared in Figure 2–10. The idealized behavior can be described by the expressions

$$\tau_{yx} = -\mu_0 \frac{du_x}{dy} \pm \tau_0 \qquad \text{for } |\tau_{yx}| > \tau_0 \qquad (2.18)$$

and

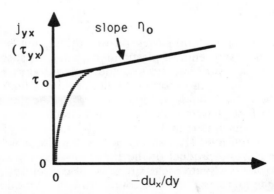

FIGURE 2–10   *Behavior of a Bingham plastic. ——— = idealized; · · · · · · = variant.*

$$\frac{du_x}{dy} = 0 \qquad \text{for } |\tau_{yx}| < \tau_0 \tag{2.19}$$

where τ_0 = *yield stress*, Pa
  μ_0 = effective viscosity for absolute values of the shear stress in excess of the *yield stress*, Pa·s or kg/m·s

The arbitrary sign preceding τ_0 in Equation 2.18 is chosen to be the same as the actual sign of τ_{yx}. Thus if $\tau_{yx} > 0$, the plus sign is chosen, and vice versa.

A Bingham plastic at rest is presumed to have a sufficiently rigid structure to resist any stress less than τ_0. When this stress is exceeded, the structure disintegrates and the material behaves like a Newtonian fluid. When the shear stress falls below τ_0, the rigid structure re-forms.

Fluids in which the Viscosity Is Time Dependent

The effective viscosity of some fluids depends on history as well as on the instantaneous rate of stress. Fluids that show a decrease in effective viscosity with time are called *thixotropic*. Fluids that show an increase in effective viscosity with time are said to be *rheopectic*, and those whose effective viscosity returns to the original value when the stress is released are called *viscoelastic*.

Milk, mayonnaise, greases, inks, and many suspensions are thixotropes. Paints should ideally be Bingham plastics, so they will not drain, and thixotropes, so they will flow more easily under brushing and then even out the brush marks. Suspensions of gypsum in water, and sols of vanadium pentoxide and bentonite are rheopectates. Rubber cement, flour dough, and bitumens are examples of viscoelastic materials.

The behavior of a thixotrope is illustrated in Figure 2–11. As the rate of stress is increased, the material may behave as a pseudoplastic or as a Bingham plastic. As the rate of stress is decreased, a different path may be followed, depending on the time rate of change of the rate of stress. If the rate of stress is maintained at a constant value, the shear stress may decrease. Rheopectates behave inversely, as illustrated in Figure 2–12.

A detailed treatment of non-Newtonian behavior, and particularly time-

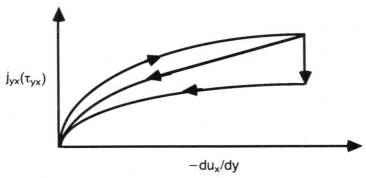

FIGURE 2–11   *Idealized behavior of a thixotropic fluid.*

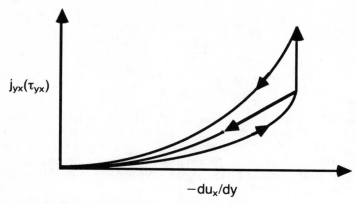

FIGURE 2–12   *Idealized behavior of a rheopectic fluid.*

dependent and two-dimensional behavior, is beyond the scope of this monograph. The books by Skelland [24], Fredrickson [25], Lodge [26], Wilkinson [27], Coleman et al. [28], Astarita and Marrucci [29], Van Wazer et al. [30], and Bird et al. [31] are recommended for further reading.

APPLICATIONS

One-dimensional applications of the above expressions relating the momentum flux density or the shear stress to the rate of stress are described in Chapters 3–7.

SUMMARY

The behavior of fluids in flow is characterized primarily by their viscosity. In this chapter the viscosities of gases and various types of liquids were described quantitatively. Many of the liquids encountered in chemical processing demonstrate non-Newtonian behavior. Models for several important classes of such liquids were examined.

PROBLEMS

1. The following data were selected from those reported by Ashare [11] for the effective viscosity of a 5% solution of polystyrene with a molar mass of 1.3×10^3 in Arochlor 1248:

τ (Pa)	η (Pa·s)
1.96	739.20
2.83	742.72
20.43	767.28
30.75	732.92
45.28	681.05
64.85	615.39
95.59	572.39

124.10	468.87
194.85	298.49
243.16	230.77
265.53	200.17
313.04	148.90
371.73	111.56
469.56	70.63
553.41	52.52
721.11	34.30
883.22	26.51
1185.08	22.44
1565.20	18.70

Develop a representation for these values in terms of Equation 2.13.

2. Develop a representation for the values of problem 1 in terms of Equation 2.8.

3. Develop a representation for the values of problem 1 in terms of Equation 2.9.

4. Develop a representation for the values of problem 1 in terms of the Ellis model (see Bird et al. [3], p. 14):

$$\Gamma = (A + B\tau^{\alpha-1})\tau \qquad (2.20)$$

Here A, B, and α are arbitrary constants.

5. Develop a representation for the values of problem 1 in terms of the Powell–Eyring model (see Skelland [24], p. 8):

$$\tau = +M\Gamma + A \ \text{arcsinh}\left\{\frac{\Gamma}{B}\right\} \qquad (2.21)$$

(Note: $\text{arcsinh}\{x\} = \ln\{x + \sqrt{x^2 + 1.}\}$)

6. Develop a representation for the values of problem 1 in terms of the Reiner–Philippoff model (see Bird et al. [3], p. 14):

$$\Gamma = \tau\left(\eta_\infty + \frac{\eta_0 - \eta_\infty}{1 + (\tau/\tau^*)^2}\right)^{-1} \qquad (2.22)$$

Here τ^*, η_∞, and η_0 are arbitrary constants.

7. Develop a representation for the values of problem 1 in terms of the Meter equation (see Churchill and Churchill [14]):

$$\eta = \eta_\infty + \frac{\eta_0 - \eta_\infty}{1 + (\tau/\tau^*)^m} \qquad (2.23)$$

Here η_0, η_∞, τ^*, and m are arbitrary constants.

8. Replace τ and τ^* in Equation 2.8 by Γ and Γ^*, respectively, and use it to correlate the data of problem 1.

9. Repeat problem 8 for Equation 2.9.

10. Compare the success of the representations of problems 1–9 for the data of problem 1. Interpret.
11. **a.** Develop a model for the effective viscosity of pseudoplastic fluids in terms of the Churchill–Usagi model [15, 16] (1) by taking the qth root of the sum of the qth powers of η_∞ and the power-law model to obtain an expression for intermediate and large τ_{yx}, and (2) by taking the pth root of the sum of the pth powers of η_0 and the expression derived in (1).
    b. Compare the expression of a(2) with Equations 2.8 and 2.9.
    c. Develop a representation for the data of problem 1 in terms of this new model.
    d. Reduce to the special case of $q = 1$ and compare with Equations 2.8 and 2.9.
    e. Reduce to the special case of $p = 1$ and compare with Equations 2.8 and 2.9.
12. Repeat problem 11 by first combining η_0 with the power-law model and then combining the resulting expression with η_∞.
13. The effective viscosity of a pseudoplastic fluid at small shear stresses can be represented approximately by the equation

$$\eta = 0.3 - 0.23\tau^{1/10} \tag{2.24}$$

where η is the effective viscosity in pascal seconds and τ is the shear stress in pascals. (Note that the two numerical coefficients of Equation 2.24 are not dimensionless.) The effective viscosity approaches zero at very large shear stresses, and is 0.033 Pa·s at a shear stress of 14 Pa. Use the Churchill–Churchill model to develop an empirical correlating equation conforming to these three conditions.
14. Determine the values of η_0, η_∞, τ^*, m, and n for which Equations 2.8 and 2.9 reduce to Equations 2.10, 2.20, 2.22, and 2.23.
15. Relate m in Equations 2.8 and 2.9 to α in Equation 2.13 by utilizing limiting conditions.
16. Repeat problems 8–10 using the values plotted in Figure 2–8 of Denn [21].
17. Repeat problems 1–10 using the values tabulated in problem 1 of Denn [21].
18. Replot the data of Figure 2–3 as tabulated below as η versus τ and η versus Γ.

Γ (s^{-1})	τ (N/m^2)
0.0150	1.10
0.0299	1.79
0.0597	2.85
0.0946	3.95
0.150	5.31
0.238	6.80
0.377	9.10
0.597	12.2
0.946	15.1

1.50	19.8
2.38	25.4
3.77	34.1
5.97	47.0
9.46	66.1
15.00	93.5

19. Correlate the data of problem 18 in terms of $\tau\{\Gamma\}$, $\eta\{\Gamma\}$, and $\eta\{\tau\}$.
20. Replot the data of Figure 2–5, which are tabulated below as η versus τ and η versus Γ.

Γ (s^{-1})	τ (N/m^2)
0.341	3.21
0.430	4.00
0.541	4.78
0.682	5.9
0.858	7.4
1.08	9.32
1.36	12.3
1.71	15.5
2.15	20.2
2.71	27.0
3.41	34.6
4.30	45.5

21. Correlate the data of problem 20 in terms of $\tau\{\Gamma\}$, $\eta\{\Gamma\}$, and $\eta\{\tau\}$.
22. For the data of Figure 2–9:

    a. Determine μ_0 and τ_0 according to Equations 2.18 and 2.19.
    b. Correlate in terms of Equation 2.13.
    c. Correlate in terms of Equation 2.20.

23. Develop correlating equations for each set of data presented in Figure 2–13.
24. Develop correlating equations for each set of data in Figure 2–14, which are tabulated below per Thomas (private communication).

ϱ (gm/cm^3)	D (in)	τ_w (lb$_{12}$/ft^2)	$(du/d\Gamma) \times 10^3$ (s^{-1})
1.38	0.944		
		0.952	0.672
		0.760	0.232
		1.067	1.100
		0.817	0.381
		0.958	0.812
		1.197	1.680
		1.115	1.400
1.38	0.304	2.065	6.48
		1.70	4.30

		1.885	4.92
		1.51	3.057
		1.994	5.30
		1.240	2.09
		1.32	2.34
		1.59	3.49
		2.02	5.40
		1.482	2.80
		1.050	1.38
		1.885	5.30
		1.960	6.34
		1.902	6.08
1.40	0.124	1.478	3.407
		1.113	1.616
		0.931	0.973
		0.748	0.521
		1.293	2.420
		3.31	13.22
		2.75	11.11
		2.295	8.52
		1.81	5.42
		3.75	15.70

25. Develop a correlating equation for the data of Figure 2–15.
26. Determine the viscosity of the fluid of Figure 2–2.

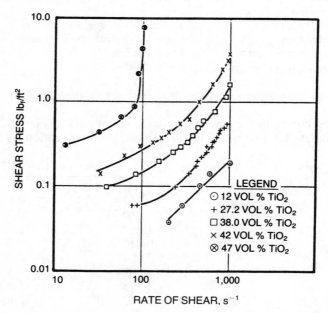

FIGURE 2–13 *Shear stress as a function of rate of shear for suspensions of TiO_2 in water. (From Metzner and Whitlock [12].)*

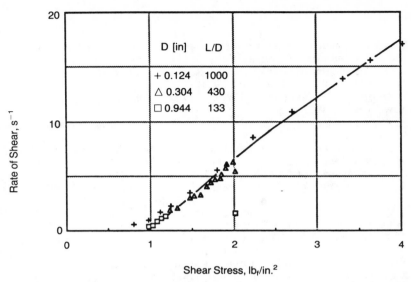

FIGURE 2–14   *Rate of shear as a function of shear stress for suspensions of kaolin. (Data from Thomas [23].)*

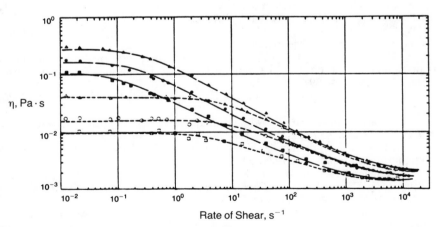

FIGURE 2–15   *Apparent viscosity of fresh and shear-degraded aqueous solutions of Separan AP-30 as a function of rate of shear. (From Chang and Darby [32].)*

Concentration, parts per million by weight	100	250	500
Fresh	■	●	▲
Degraded	□	○	△

REFERENCES

1. Sidney Chapman and T. G. Cowling, *Mathematical Theory of Non-Uniform Gases*, 2nd ed., Cambridge University Press (1951).
2. J. O. Hirschfelder, C. F. Curtiss, and R. B. Bird, *Molecular Theory of Gases and Liquids*, John Wiley, New York (1954).
3. R. B. Bird, E. N. Lightfoot, and W. E. Stewart, *Transport Phenomena*, John Wiley, New York (1960).
4. E. A. Guggenheim, *Elements of the Kinetic Theory of Gases*, Pergamon, New York (1960), p. 46.
5. Isaac Newton, *Principia* S. Pepys, London (1686); English transl. of 2nd ed. (1713) by Andrew Motte (1729), Vol. I, *The Motion of Bodies*; revised transl. by Florian Cajori, University of California Press, Berkeley (1966).
6. S. W. Churchill, *The Interpretation and Use of Rate Data—The Rate Process Concept*, rev. printing, Hemisphere, Washington, D.C. (1979).
7. S. W. Churchill, "Choosing between Theory and Experiment," *Chem. Eng. Progr.*, *66* (July 1970) 86; *67* (January 1971) 8.
8. R. E. H. Rasmussen, "Über die Strömung von Gasen in engen Kanälen," *Ann. Phys.*, *29* (1937) 665; English transl., The Flow of Gases in Narrow Channels, NACA TM 1301, Washington, D.C. (August 1951).
9. J. Frenkel, *Kinetic Theory of Liquids*, Dover, New York (1955).
10. J. F. Ely and H. J. M. Hanley, "Prediction of Transport Properties, 1. Viscosity of Fluids and Mixtures," *Ind. Eng. Chem. Fundam.*, *20* (1981) 323.
11. E. Ashare, *Rheological Properties of Monodisperse Polystyrene Solutions*, Ph.D. Thesis, University of Wisconsin, Madison (1968).
12. A. B. Metzner and Malcomb Whitlock, "The Flow Behavior of Concentrated (Dilatant) Suspensions," *Trans. Soc. Rheology*, *2* (1958) 239.
13. Osborne Reynolds, "On the Dilatancy of Media Composed of Rigid Particles in Contact. With Experimental Illustrations," *Phil. Mag.*, *8* (1885) 20 [*Scientific Papers*, Vol. II, Cambridge University Press (1901), p. 203].
14. S. W. Churchill and R. U. Churchill, "A General Model for the Effective Viscosity of Pseudoplastic and Dilatant Fluids," *Rheologica Acta*, *14* (1975) 404.
15. S. W. Churchill and R. Usagi, "A General Expression for the Correlation of Rates of Transfer and Other Phenomena," *AIChE J.*, *18* (1972) 1121.
16. S. W. Churchill and R. Usagi, "A Standardized Procedure for the Production of Correlations in the Form of a Common Empirical Equation," *Ind. Eng. Chem. Fundam.*, *13* (1974) 39.
17. P. G. Nutting, "A New General Law of Deformation," *J. Franklin Inst.*, *19* (1926) 679.
18. W. Ostwald, "Über die Geschwindigkeitsfunktion der Viskosität disperser Systeme," *Kolloid Z.*, *36* (1925) 99, 157, 248.
19. A. deWaele, "Die Änderung der Viskosität mit der Schergeschwindigkeit disperser Systeme," *Kolloid Z.*, *36* (1925) 332.
20. A. B. Metzner, "Non-Newtonian Technology: Fluid Mechanics, Mixing and Heat Transfer," p. 77 in *Advances in Chemical Engineering*, Vol. 5, T. B. Drew and J. W. Hooper, Jr., Eds., Academic Press, New York (1956).
21. M. M. Denn, *Process Fluid Mechanics*, Prentice-Hall, Englewood Cliffs, NJ (1980).
22. E. C. Bingham, *Fluidity and Plasticity*, McGraw-Hill, New York (1922).
23. D. G. Thomas, "Non-Newtonian Suspensions, Part II, Turbulent Transport Characteristics," *Ind. Eng. Chem.*, *55* (1963) 27.
24. A. H. P. Skelland, *Non-Newtonian Flow and Heat Transfer*, John Wiley, New York (1967).
25. A. G. Fredrickson, *Principles and Applications of Rheology*, Prentice-Hall, Englewood Cliffs, NJ (1964).

26. A. S. Lodge, *Elastic Liquids*, Academic Press, New York (1964).
27. W. L. Wilkinson, *Non-Newtonian Fluids*, Pergamon, New York (1960).
28. B. D. Coleman, H. Markowitz, and W. Noll, *Viscometric Flows of Non-Newtonian Fluids*, Springer-Verlag, Berlin (1966).
29. G. Astarita and G. Marrucci, *Principles of Non-Newtonian Fluid Mechanics*, McGraw-Hill, London (1974).
30. J. R. Van Wazer, J. W. Lyons, K. Y. Kim, and R. E. Colwell, *Viscosity and Flow Measurement—A Laboratory Handbook of Rheology*, Interscience, New York (1963).
31. R. B. Bird, R. C. Armstrong, and O. Hassager, *Dynamics of Polymeric Liquids*, Vols. I, II, John Wiley, New York (1977).
32. H.-F. D. Chang and R. Darby, "Effect of Shear Degradation on the Rheological Properties of Dilute Drag-Reducing Polymer Solutions," *J. Rheology*, 27 (1983) 77.

Chapter 3

Newtonian Flow between Parallel Plates

Flow between parallel plates of infinite breadth represents a hypothetical situation. The principal value of solutions for this geometry is as asymptotes for flow in open and closed rectangular channels as the breadth-to-height ratio β increases, and for flow through the annulus between concentric round tubes as the inner-diameter-to-outer-diameter ratio λ approaches unity. Within this book, flow between infinite parallel plates serves as a stepping stone to more complex situations.

SINGLE-PHASE FLOW

Development of the Model

The variables used to describe the flow between parallel plates are shown in Figure 3–1A. Although the plates are postulated to be of infinite breadth, this breadth will initially be represented for dimensional convenience by the finite quantity H. A force-momentum balance on a finite element (the control volume) of height y above the lower plate, breadth H, and length Δx is indicated for the isolated element in Figure 3–1B. The net dynamic pressure force in the x-direction minus the shear force imposed on the fluid by the lower plate is seen to equal the flux of x-momentum out of the upper surface of the fluid element. This balance can be written as

$$\left[\mathscr{P} - \left(\mathscr{P} + \frac{d\mathscr{P}}{dx}\,\Delta x\right)\right]Hy - \tau_w H\,\Delta x = j_{yx}H\,\Delta x \tag{3.1}$$

Here, as in Chapters 1 and 2,

$\mathscr{P} = p + g\varrho h - p_0 =$ dynamic pressure, Pa or $kg/m \cdot s^2$
$p =$ total pressure, Pa or $kg/m \cdot s^2$
$p_0 =$ pressure at $h = 0$, Pa or $kg/ms \cdot s^2$
$g =$ acceleration or specific force due to gravity $= 9.806\ 65$ m/s^2 or N/kg at standard conditions (sea level)
$h =$ elevation above some reference plane, m
$x =$ distance in direction of flow, m

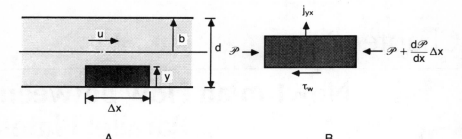

A   B

FIGURE 3–1   *Force-momentum balance for flow between parallel plates: (A) control volume; (B) force-momentum balance on control volume.*

  y = distance from bottom plate, m
  H = (unlimited) breadth of plates, m
  $\tau_w \equiv -\tau_{0x}$ = shear stress in x-direction imposed by fluid on the bottom
    plate, Pa or kg/m·s^2
  j_{yx} = flux density of x-momentum in y-direction, kg/m·s^2

The symbol τ_w is used hereafter for $-\tau_{0x}$ for simplicity and to give a positive value.
  Equation 3.1 can be simplified and rearranged as

$$j_{yx} = -y\frac{d\mathscr{P}}{dx} - \tau_w \tag{3.1A}$$

Substituting for j_{yx} from Equation 2.1 gives

$$\mu\frac{du}{dy} = y\frac{d\mathscr{P}}{dx} + \tau_w \tag{3.2}$$

From symmetry

$$\frac{du}{dy} = 0 \qquad \text{at } y = b \tag{3.3}$$

where b is the half-distance in meters between the plates. From Equations 3.2 and 3.3,

$$\tau_w = b\left(-\frac{d\mathscr{P}}{dx}\right) \tag{3.4}$$

Equation 3.4 can be recognized as an overall force balance and could have been derived directly by expanding the shaded element in Figure 3–1B to the full height $d = 2b$. Using Equation 3.4 to eliminate $d\mathscr{P}/dx$ from Equations 3.1A and 3.2 gives

$$j_{yx} = \left(\frac{y}{b} - 1\right)\tau_w \tag{3.5}$$

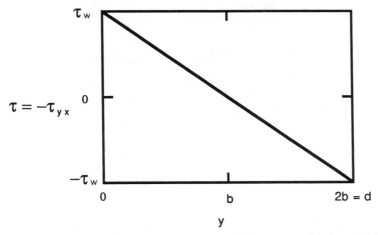

FIGURE 3–2 *Shear stress on plane of fluid above control volume in Figure 3.1A.*

and

$$\mu\frac{du}{dy} = \left(1 - \frac{y}{b}\right)\tau_w \tag{3.6}$$

Equation 3.6 is the starting point for subsequent derivations, but Equation 3.5 is useful for interpretation of the variation of the momentum flux density or equivalent shear stress within the fluid. In recognition of Equation 2.3, Equation 3.5 can be rewritten as

$$\frac{\tau}{\tau_w} \equiv \frac{\tau_{yx}}{(\tau_{yx})_{y=0}} = \frac{j_{yx}}{(j_{yx})_{y=0}} = 1 - \frac{y}{b} \tag{3.7}$$

where $\tau \equiv -\tau_{yx}$ is the shear stress in the negative x-direction imposed on the fluid above y by the fluid below y (Pa or $kg/m \cdot s^2$). As with τ_w, the symbol τ is used as a shorthand for $-\tau_{yx}$, and to give a positive value.

The shear stress τ within the fluid is observed from Equation 3.7 and in Figure 3–2 to vary linearly from τ_w at $y = 0$ to zero to $y = b$ to $-\tau_w$ at $y = 2b$. Such a linear variation in the shear stress or flux density of momentum in the fluid is observed in all one-dimensional pure forced and pure induced flows.

Derivation of Solution

Integrating Equation 3.6 from $u = 0$ at $y = 0$ to any y produces the following expression for the velocity distribution in terms of the shear stress:

$$u = \frac{y\tau_w}{\mu}\left(1 - \frac{y}{2b}\right) \tag{3.8}$$

Integrating once more gives an expression for the mean velocity:

$$u_m = \frac{1}{b} \int_0^b u \, dy = \frac{b\tau_w}{3\mu} \tag{3.9}$$

Combining Equations 3.8 and 3.9 gives an expression for the velocity distribution in terms of the mean velocity:

$$\frac{u}{u_m} = \frac{3y}{b}\left(1 - \frac{y}{2b}\right) \tag{3.10}$$

Setting $y = b$ indicates that the ratio of the central velocity to the mean value is

$$\frac{u_c}{u_m} = \frac{3}{2} \tag{3.11}$$

where u_c is the velocity in meters per second at the central plane ($y = h/2$).

The anticipated symmetry of the velocity distribution between parallel plates is more apparent if Equation 3.10 is recast in the form

$$\frac{u}{u_m} = \frac{3}{2}\left[1 - \left(\frac{z}{b}\right)^2\right] \tag{3.12}$$

where $z = \pm(b - y)$ is the distance in meters from the central plane.

Combining Equation 3.12 with 3.11 gives

$$\frac{u}{u_c} = 1 - \left(\frac{z}{b}\right)^2 \tag{3.13}$$

which will subsequently be found useful for comparison with other geometries. It also follows that the variation in the shear stress has the following simpler representation in terms of z:

$$\frac{\tau}{\tau_w} = \frac{\tau_{zx}}{(\tau_{zx})_{z=\pm b}} = \frac{z}{b} \tag{3.14}$$

For flow in a rectangular duct the hydraulic diameter

$$D = \frac{8Hb}{2H + 4b} = \frac{4Hb}{H + 2b} \tag{3.15}$$

For parallel plates as $H \to \infty$,

$$D \to 4b \tag{3.16}$$

Hence from Equation 3.9

$$\mathrm{Po} = f\mathrm{Re} = \frac{D\tau_w}{\mu u_m} = 12 \tag{3.17}$$

Experimental Confirmation of Solution

The experimental data of Whan and Rothfus [1] for $H/b = 40$ (or $H/D = 10$) are plotted in Figures 3–3 and 3–4. Good agreement of the measured velocities with Equation 3.13 may be noted in Figure 3–3 despite the finite experimental value of H/D. Good agreement of the measured pressure drops with Equation 3.17

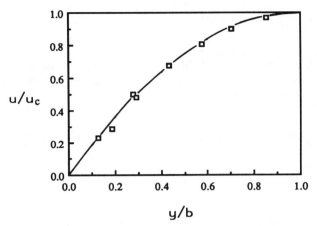

FIGURE 3–3   *Experimental velocity distribution for laminar flow at Re = 2053 in a wide rectangular channel with H/b = 40. Curve represents Equation 3.13. (Data read from plot of Whan and Rothfus [1].)*

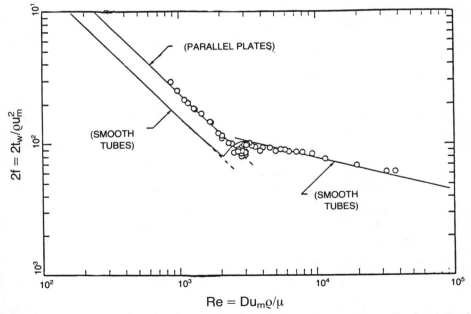

FIGURE 3–4   *Experimental pressure drops for flow through a wide rectangular channel with H/b = 40. (From Whan and Rothfus [1].)*

may similarly be observed in Figure 3–4 (also see Figure 4–12). In this plot the hydraulic diameter is computed from Equation 3.15. The use of the value given by Equation 3.16 would shift the data points negligibly (5%) to the right. The critical value of Re for the onset of the transition from laminar to turbulent motion appears to be from 2400 to 2800. Others have observed critical values over this same range (see Eckert and Irvine [2]).

Alternative Methods

Bird et al. [3], pp. 37, 62, employ two alternative methods for derivation of solutions such as the one above. Both start with a force-momentum balance on an element of finite length Δx but differential width dy as shown in Figure 3–5. This balance can be written as

$$\left[\mathscr{P} - \left(\mathscr{P} + \frac{d\mathscr{P}}{dx}\Delta x\right)\right] H\, dy + \left[j_{yx} - \left(j_{yx} + \frac{dj_{yx}}{dy}dy\right)\right] H\, \Delta x = 0 \quad (3.18)$$

which reduces to

$$\frac{dj_{yx}}{dy} = -\frac{d\mathscr{P}}{dx} \quad (3.18A)$$

In one method, Equation 3.18A is integrated to obtain

$$j_{yx} = -y\frac{d\mathscr{P}}{dx} + C \quad (3.19)$$

Combination with Equation 2.1 then gives

$$-\mu\frac{du}{dy} = -y\frac{d\mathscr{P}}{dx} + C \quad (3.20)$$

which differs from Equation 3.2 only in that the arbitrary constant C has not yet been identified as $-\tau_w$. Their alternative method involves substitution from Equation 2.1 in Equation 3.18, which gives

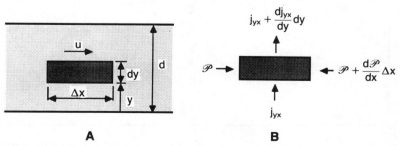

FIGURE 3–5 *Differential force-momentum balance for flow between parallel plates: (A) differential control volume; (B) force-momentum balance on differential element.*

$$-\mu\frac{d^2u}{dy^2} = -\frac{d\mathscr{P}}{dx} \tag{3.21}$$

which on integration produces Equation 3.20. These two procedures utilizing a differential element require one more step and do not appear to have any advantage for one-dimensional velocity fields over the one utilized here involving a finite element in y as well as x.

CONCURRENT, STRATIFIED, HORIZONTAL FLOW OF TWO IMMISCIBLE FLUIDS

Development of Model

Consider two fluids in concurrent flow between horizontal parallel plates as illustrated in Figure 3–6. A force-momentum balance on the shaded element can be written as

$$\left[\mathscr{P} - \left(\mathscr{P} + \frac{d\mathscr{P}}{dx}\Delta x\right)\right]Hy - \tau_{w1}H\,\Delta x = j_{yx}H\,\Delta x \tag{3.22}$$

which simplifies to

$$j_{yx} = y\left(-\frac{d\mathscr{P}}{dx}\right) - \tau_{w1} \tag{3.22A}$$

where τ_{w1} is the shear stress of lower fluid 1 on lower plate 1. A subscript identifying the plate is necessary since the shear stress differs on the two plates. Equation 3.22A holds for $0 \le y \le d$ (i.e., for either fluid) if the appropriate expression is used for j_{yx}. Substituting τ_{w2} for j_{yx} at $y = d$ or writing an overall force balance for the full element $bh\,\Delta x$ produces

$$\tau_{w1} + \tau_{w2} = d\left(\frac{-d\mathscr{P}}{dx}\right) \tag{3.23}$$

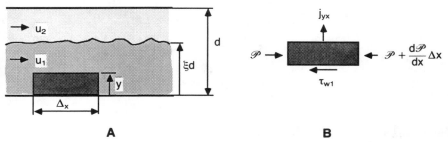

A   **B**

FIGURE 3–6 *Force-momentum balance for stratified flow of two immiscible fluids between parallel plates: (A) control volume; (B) force-momentum balance on control volume.*

Substituting for $-d\mathscr{P}/dx$ in Equation 3.22A from 3.23 gives

$$j_{yx} = \frac{y}{d}\tau_{w2} - \left(1 - \frac{y}{d}\right)\tau_{w1} \tag{3.24}$$

Then substituting for j_{yx} from Equation 2.1 gives

$$-\mu\frac{du}{dy} = \frac{y}{d}\tau_{w2} - \left(1 - \frac{y}{d}\right)\tau_{w1} \tag{3.25}$$

Derivation of Solution

Integrating Equation 3.25 for the lower fluid, with $\mu = \mu_1$, from $u = 0$ at $y = 0$ to some y still in fluid 1 gives

$$\mu_1 u_1 = -\frac{y^2}{2d}\tau_{w2} + y\left(1 - \frac{y}{2d}\right)\tau_{w1} \tag{3.26}$$

Similarly, integrating Equation 3.25 for the upper fluid, with $\mu = \mu_2$, from $u = 0$ at $y = d$ to some y in fluid 2 gives

$$\mu_2 u_2 = \left(\frac{d^2 - y^2}{2}\right)\tau_{w2} - \frac{(d - y)^2}{2d}\tau_{w1} \tag{3.27}$$

Equating the velocities from Equations 3.26 and 3.27 at the interface, here defined as

$$y_i = d\zeta \tag{3.28}$$

where y_i = height of interface, m
          ζ = fractional height of fluid 1

gives

$$\frac{-\zeta^2\tau_{w2}}{\mu_1} + \frac{\zeta\tau_{w1}(2 - \zeta)}{\mu_1} = \frac{(1 - \zeta^2)\tau_{w2}}{\mu_2} - \frac{(1 - \zeta)^2\tau_{w1}}{\mu_2} \tag{3.29}$$

or

$$\frac{\tau_{w1}}{\tau_{w2}} = \frac{\zeta^2 + (1 - \zeta^2)\xi}{\zeta(2 - \zeta) + (1 - \zeta)^2\xi} \tag{3.29A}$$

where the ratio of viscosities is

$$\xi = \frac{\mu_1}{\mu_2} \tag{3.30}$$

The mean velocity of fluid 1 can be obtained by integrating Equation 3.26 from $y = 0$ to $y = \zeta d$ as follows:

$$u_{m1} = \frac{1}{\zeta d} \int_0^{\zeta d} u_1 \, dy = \frac{1}{\zeta} \int_0^{\zeta} u_1 d\left(\frac{y}{d}\right) \tag{3.31}$$

The result, after rearrangement, is

$$\begin{aligned}
\frac{\mu_1 u_{m1}}{d\tau_{w1}} &= \frac{\zeta(3 - \zeta)}{6} - \frac{\zeta^2 \tau_{w2}}{6\tau_{w1}} \\
&= \frac{\zeta(3 - \zeta)}{6} - \frac{\zeta^2}{6}\left(\frac{\zeta(2 - \zeta) + (1 - \zeta)^2\xi}{\zeta^2 + (1 - \zeta^2)\xi}\right) \\
&= \frac{\zeta}{6}\left(\frac{\zeta^2 + (3 + \zeta)(1 - \zeta)\xi}{\zeta^2 + (1 - \zeta^2)\xi}\right)
\end{aligned} \tag{3.32}$$

The mean velocity of fluid 2 can similarly be obtained by integrating Equation 3.27 from $y = \zeta d$ to $y = d$. This result can be arranged as

$$\frac{\mu_2 u_{m2}}{d\tau_{w2}} = \frac{1 - \zeta}{6}\left(\frac{\zeta(4 - \zeta) + (1 - \zeta)^2\xi}{\zeta(2 - \zeta) + (1 - \zeta)^2\xi}\right) \tag{3.33}$$

The shear stresses in Equations 3.32 and 3.33 can be replaced by the gradient of the dynamic pressure through Equations 3.23 and 3.29A, leading to

$$\frac{d^2}{\mu_1 u_{m1}}\left(\frac{-d\mathscr{P}}{dx}\right) = \frac{12}{\zeta}\left(\frac{\zeta + (1 - \zeta)\xi}{\zeta^2 + (1 - \zeta)(3 + \zeta)\xi}\right) \tag{3.34}$$

and

$$\frac{d^2}{\mu_2 u_{m2}}\left(-\frac{d\mathscr{P}}{dx}\right) = \frac{12}{1 - \zeta}\left(\frac{\zeta + (1 - \zeta)\xi}{\zeta(4 - \zeta) + (1 - \zeta)^2\xi}\right) \tag{3.35}$$

The equivalent of this solution was apparently first derived by Russell and Charles [4].

Interpretation of Solution

Equations 3.32 and 3.34 reduce to those for a single fluid for $\xi = 1$. Equations 3.33 and 3.35 reduce similarly for $\xi = 0$. They are applicable for a gas and a liquid as well as for two liquids for any value of ξ for which $0 < \xi < 1$.

The left-hand sides of Equations 3.34 and 3.35 are equivalent to $f\,\text{Re} = \text{Po}$ for fluid 1 and fluid 2, respectively. The left-hand sides of Equation 3.32 and 3.33 are equivalent to $2/f\,\text{Re} = 2/\text{Po}$ for fluid 1 and surface 1, and for fluid 2 and surface 2, respectively.

Application

The foregoing expressions suggest that the rate of flow of a fluid between parallel plates can be increased by the introduction of a second, less viscous fluid. As an illustration, the lower (denser) fluid 1 will be assumed to be less viscous; that is, ξ will be assumed to be less than unity. (This situation corresponds, for example, to water beneath a very viscous crude oil.) Then the volumetric rate of flow of upper fluid 2, as obtained from Equation 3.35 is

$$V_2 = u_{m2}(1 - \zeta)Hd$$
$$= \frac{d^3H(1 - \zeta)^2}{12\mu_2}\left(\frac{\zeta(4 - \zeta) + (1 - \zeta)^2\xi}{\zeta + (1 - \zeta)\xi}\right)\left(-\frac{d\mathscr{P}}{dx}\right) \tag{3.36}$$

whereas the volumetric rate of flow of fluid 2 alone, obtained by setting $\zeta = 0$, is

$$V_{20} = \frac{d^3H}{12\mu_2}\left(-\frac{d\mathscr{P}}{dx}\right) \tag{3.37}$$

The ratio of these rates is

$$\frac{V_2}{V_{20}} = \frac{(1 - \zeta)^2[\zeta(4 - \zeta) + (1 - \zeta)^2\xi]}{\zeta + (1 - \zeta^2)\xi} \tag{3.38}$$

In the limit as $\xi \to 0$,

$$\frac{V_2}{V_{20}} \Rightarrow (1 - \zeta)^2(4 - \zeta) \tag{3.39}$$

A fourfold increase in V_2 is seen to be possible with a thin but finite layer of a much less viscous fluid 1. The corresponding ratio of flow of fluid 1 to fluid 2 is

$$\frac{V_1}{V_2} = \frac{\zeta^2[\zeta^2 + (1 - \zeta)(3 + \zeta)\xi]}{(1 - \zeta)^2[\zeta(4 - \zeta) + (1 - \zeta)^2\xi]\xi} \tag{3.40}$$

In the limit as $\xi \to 0$,

$$\frac{V_1}{V_2} \Rightarrow \frac{\zeta^3}{(1 - \zeta)^2(4 - \zeta)\xi} \tag{3.41}$$

The ratio of the required pressure gradients for equal rates of flow is the inverse of the ratio of flows for a given pressure gradient. The optimum fractional thickness of fluid 1 for any viscosity ratio ξ can be obtained by equating to zero the derivative of V_2/V_{20} with respect to ζ. The corresponding ratio of V_1/V_2 can then be found from Equation 3.40 (see problem 6).

Experimental Confirmation

Charles and Lilleheht [5] experimentally confirmed their own theoretical predictions for two-phase concurrent flow in a rectangular channel with $H/b = 15.91$ and $\xi = 0.1878$, as shown in Figures 3–7 to 3–9. It may be inferred from their

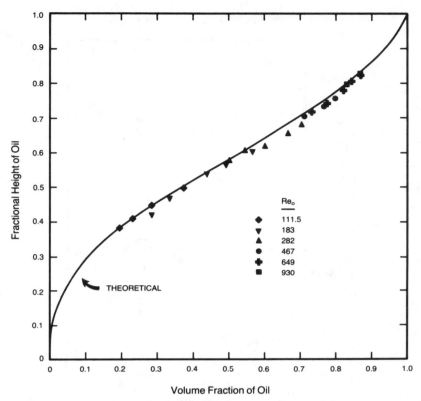

FIGURE 3-7 *Comparison of theoretical and experimental depths of oil for stratified, concurrent, laminar flow with water in a wide rectangular channel. $H/b = 15.91$; $\xi = 0.1878$, and $Re_0 =$ superficial Reynolds number for oil. (From Charles and Lilleleht [5].)*

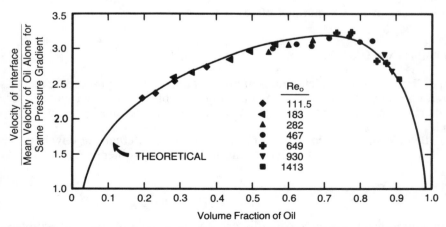

FIGURE 3-8 *Comparison of theoretical and experimental velocities at interface for stratified, concurrent, laminar flow of oil and water in a wide rectangular channel. Same conditions as for Figure 3-7. (From Charles and Lilleleht [5].)*

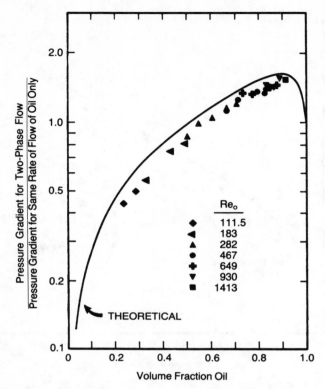

FIGURE 3–9   *Comparison of theoretical and experimental pressure gradient (or flow) ratios for stratified, concurrent, laminar flow of oil and water in a wide rectangular channel. Same conditions as for Figure 3–7. (From Charles and Lilleleht [5].)*

results that similar confidence can be placed in the predictions for the somewhat simpler case of infinite parallel plates.

Limitations and Extensions

Upper limits on the predicted rates of flow obtained from the preceding analysis are imposed by the onset of turbulence in either phase. An even more severe limit arises from the onset of instability of the planar interface between the two fluids.

    The preceding solutions are not applicable for countercurrent flow between parallel plates, which is, however, considered in the problem set.

SUMMARY

Expressions were developed in this chapter for single-phase and two-phase concurrent flow between parallel plates. For two fluids the fractional height of the lower fluid and the ratio of the viscosities are added parameters. Although infinite parallel plates constitute an idealized geometry, the results provide a useful stepping stone as well as an asymptote for flow in rectangular ducts and circular, concentric annuli.

An important application of the results of this chapter is the possible increase in the rate of flow of a very viscous fluid by the concurrent flow of a second less viscous fluid.

PROBLEMS

1. Derive Equations 3.10 and 3.17 starting with a force-momentum balance on an element located on the central plane.
2. Is the solution for concurrent flow of two immiscible fluids applicable for the limiting case of no flow of one or the other of the fluids? Explain.
3. Reexpress the solution for the velocity distribution in two-phase concurrent flow in terms of u_1/u_{m1} and u_2/u_{m2} as a function of y/d, ξ, and ζ only. Prepare plots of the velocity distribution for $\xi = 0.001, 0.01, 0.1, 1, 10, 100,$ and 1000 with $\zeta = 0.1, 0.5,$ and 0.9.
4. Prepare a plot of $(f\mathrm{Re})_1$ versus ζ for the values of ξ in problem 3.
5. Repeat problem 4 for $(f\mathrm{Re})_2$.
6. Derive an expression for the optimum value of ζ for a maximum V_2/V_{20}. Plot this value of ζ and the corresponding values of V_2/V_{20} and V_1/V_2 versus ξ.
7. Plot V_2/V_{20} versus ζ for the values of ξ in problem 3.
8. Plot V_2/V_{20} versus V_1/V_2 for the values of ξ in problem 3.
9. Derive expressions for the velocity and location of the interface in concurrent flow in terms of

   a. $-d\mathscr{P}/dx$
   b. u_m, ζ, and ξ only

10. Derive an expression for the maximum velocity and its location in concurrent flow. Will the maximum always occur in the less viscous phase? Explain.
11. Equation 3.39 indicates that the maximum rate of flow of fluid 2 occurs in the limit as $\zeta \to 0$. However, this limit corresponds to pure phase 2, for which the rate of flow should be unchanged. Explain.
12. Could a stream of air be used to increase the forced flow of a liquid between parallel plates? Explain.
13. Derive expressions for Po for the individual phases in concurrent flow in terms of hydraulic diameters for the individual phases.
14. Show that the solution for concurrent flow reduces to that for single-phase flow for $\xi = 1$.
15. Derive a solution for horizontal countercurrent flow of two immiscible fluids.
16. Are the expressions in this chapter for concurrent flow applicable for inclined plates? Explain.
17. Oil, with a viscosity of 60 mPa·s and specific gravity of 0.9, and water are pumped concurrently between parallel plates.

   a.   Calculate the required ratio of flows for $\zeta = 0.01$ and the corresponding ratio of the Reynolds numbers for the two phases.
   b.   Calculate ζ for equal rates of flow and the corresponding ratio of Reynolds numbers for the two phases.
   c.   Calculate the ratio of the shear stresses on the two walls for parts a and b.
   d.   Calculate V_2/V_{20} for parts a and b.

18. Plot the variation in shear stress in concurrent flow.
19. Derive an expression for the location of the plane of zero shear stress in concurrent flow.
20. Reexpress the solution for concurrent flow in terms of distance from the interface. Interpret the results.
21. Reexpress the solution for concurrent flow in terms of distance from the plane of zero shear stress. Interpret the results.
22. Derive an expression for the shear stress at the central plane in concurrent flow. Interpret the result.
23. Water and *n*-pentane are pumped simultaneously in fully developed laminar flow between parallel plates.

   a.   Determine the ratio of the rate of flow if the interface is halfway between the plates.
   b.   Determine the location of the interface if the fluids are pumped at equal rates.

	μ (mPa·s)	ϱ (Mg/m^3)
water	1.0	1.0
n-pentane	0.25	0.62

REFERENCES

1. G. A. Whan and R. R. Rothfus, "Characteristics of Transition Flow between Parallel Plates," *AIChE J.*, 5 (1959) 204.
2. E. R. G. Eckert and T. F. Irvine, Jr., "Incompressible Friction Factor, Transition and Hydrodynamic Entrance-Length Studies of Ducts with Triangular and Rectangular Cross Sections," *Proc. Fifth Midwestern Conf. on Fluid Mech.*, University of Michigan Press, Ann Arbor (1957), p. 122.
3. R. B. Bird, E. N. Lightfoot, and W. E. Stewart, *Transport Phenomena*, John Wiley, New York (1960).
4. T. W. F. Russell and M. E. Charles, "The Effect of the Less Viscous Fluid in the Laminar Flow of Two Immiscible Fluids," *Can. J. Chem. Eng.*, 37 (1959) 18.
5. M. E. Charles and L. U. Lilleleht, "Co-current Stratified Laminar Flow of Two Immiscible Liquids in a Rectangular Conduit," *Can. J. Chem. Eng. 43* (1965) 110.

Chapter 4

Newtonian Flow in Round Tubes and Circular Annuli

Most applications of fluid flow in closed channels occur in round tubes. The derivations and results closely follow those for flow between parallel plates with the slight added complication of curvature.

Flow through the annulus between concentric circular tubes is important in many applications, particularly on the outer passage of double-pipe heat exchangers.

SINGLE-PHASE FLOW IN ROUND TUBES

Development of the Model

The geometry and variables used here for flow in a round tube are shown in Figure 4–1. A force and momentum balance on the central cylindrical element of Figure 4–1A takes the form

$$\left[\mathscr{P} - \left(\mathscr{P} + \frac{d\mathscr{P}}{dx} \Delta x \right) \right] \pi r^2 = j_{rx} \cdot 2\pi r \, \Delta x \tag{4.1}$$

which reduces to

$$j_{rx} = \frac{r}{2} \left(-\frac{d\mathscr{P}}{dx} \right) \tag{4.2}$$

where r = radial distance from axis of tube

j_{rx} = flux density of x-momentum in r-direction, kg/m·s² or Pa

The other symbols are defined in Chapter 3. Substituting τ_w for j_{rx} at $r = a$ or directly formulating an overall force balance gives

$$\tau_w = \frac{a}{2} \left(-\frac{d\mathscr{P}}{dx} \right) \tag{4.3}$$

where a is the radius of the tube in meters. Then from Equations 4.2 and 4.3,

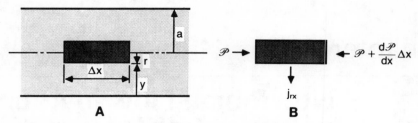

FIGURE 4-1   *Control volume for force-momentum balance for flow in a round tube: (A) control volume; (B) force-momentum balance on control volume.*

$$\frac{\tau}{\tau_w} = \frac{\tau_{rx}}{\tau_{ax}} = \frac{j_{rx}}{j_{ax}} = \frac{r}{a} \tag{4.4}$$

The shear stress or momentum flux density within the fluid thus varies linearly from the center line to the wall of the tube. This variation may be recognized as equivalent to that of Equation 3.14 for parallel plates. Thus curvature does not affect the distribution of the shear stress.

Finally, combining Equations 4.2 and 4.3 and substituting for j_{rx} per Equation 2.1 gives

$$-\mu \frac{du}{dr} = \frac{r \tau_w}{a} \tag{4.5}$$

Derivation of the Solution

Integration of Equation 4.5 from $u = 0$ at $r = a$ to any r gives

$$u = \frac{a \tau_w}{2\mu} \left[1 - \left(\frac{r}{a}\right)^2 \right] \tag{4.6}$$

Integrating with respect to the differential area $2\pi r\, dr$, as indicated in Figure 4-2, gives the mean velocity in terms of the shear stress at the wall:

$$u_m \equiv \int_0^a u \cdot 2\pi r\, dr \bigg/ \int_0^a 2\pi r\, dr = \int_0^1 u\, d\left(\frac{r}{a}\right)^2$$

$$= \frac{a \tau_w}{2\mu} \int_0^1 \left[1 - \left(\frac{r}{a}\right)^2 \right] d\left(\frac{r}{a}\right)^2 = \frac{a \tau_w}{4\mu} \tag{4.7}$$

Substituting from Equation 4.3 gives the mean velocity in terms of the gradient of the dynamic pressure:

$$u_m = \frac{a^2}{8\mu}\left(-\frac{d\mathscr{P}}{dx}\right) \tag{4.8}$$

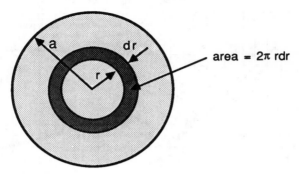

FIGURE 4–2 *Differential area of uniform velocity for calculation of mean velocity in a round tube.*

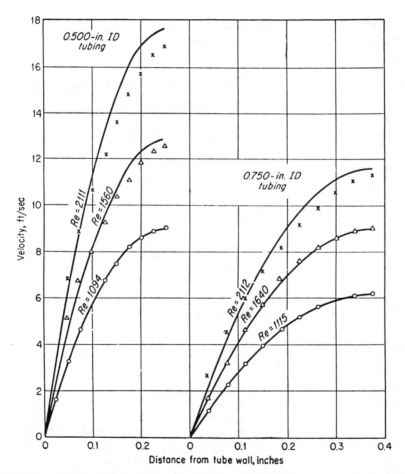

FIGURE 4–3 *Comparison of theoretical and experimental velocity distributions for laminar flow through round tubes. Curves represent Equation 4.6. (Data from Senecal and Rothfus [6] as replotted by Knudsen and Katz [5], p. 87.)*

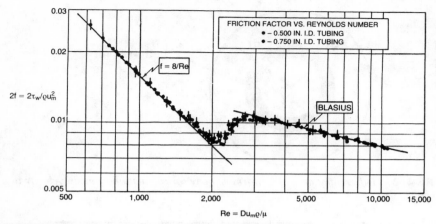

FIGURE 4–4   *Comparison of theoretical and experimental pressure drops for flow through round tubes. (From Senecal and Rothfus [6].)*

Combining Equations 4.6 and 4.7 gives the velocity distribution in terms of the mean velocity:

$$\frac{u}{u_m} = 2\left[1 - \left(\frac{r}{a}\right)^2\right] \tag{4.9}$$

It follows that

$$\frac{u_c}{u_m} = 2 \tag{4.10}$$

$$\frac{u}{u_c} = 1 - \left(\frac{r}{a}\right)^2 \tag{4.11}$$

and

$$\mathrm{Po} = f\mathrm{Re} = 8 \tag{4.12}$$

The ratio of the central (maximum) velocity to the mean velocity differs from that for parallel planes, but the velocity distribution in terms of the central velocity, as given by Equation 4.11, is equivalent to Equation 3.13. Po has two-thirds the value found for parallel planes.

Equation 4.8 is called *Poiseuille's law*, and Equation 4.9 is called the *Poiseuille velocity distribution* in honor of J.-L.-M. Poiseuille (see footnote, p. 9). However, other early investigators, including Stokes [1], Hagen [2], Hagenbach [3], and Jacobson [4], also contributed to the development of the present formulation for fully developed laminar flow in a round tube.

Equations 4.9 and 4.12 have been confirmed experimentally in many investigations. Illustrations of such results are shown in Figure 4–3 from

Knudsen and Katz [5] and Figure 4–4 from Senecal and Rothfus [6]. Laminar flow appears in Figure 4–4 to be stable up to Re = 2000. Other investigators have observed values in the range 1800–2200, and the nominal accepted value is 2100.

CONCURRENT ANNULAR TWO-PHASE FLOW IN A ROUND TUBE

Stratified flow of two fluids in a circular tube produces a two-dimensional velocity field except for equal depths (see problem 32). Consideration of such flow for other depths is deferred to a companion volume [7]. If the densities of two fluids are nearly equal or if the ratio of the velocities is very high, it may be possible to establish annular flow with nearly radial symmetry. The use of a thin outer film of a less viscous fluid, such as water, to decrease the pumping requirements for a more viscous fluid, such as crude oil, was proposed as early as 1904 [8].

Development of the Model

The notation for the derivation here is shown in Figure 4–5. The outer fluid is designated by subscript 1 and the inner fluid by subscript 2. The radial location of the interface is designated as

$$r_i = \zeta a \tag{4.13}$$

The fractional areas occupied by inner fluid 2 and outer fluid 1 are thus ζ^2 and $1 - \zeta^2$, respectively.

Equation 4.5 is still applicable for the central elemental volume insofar as the appropriate viscosity is used. The shear stress distribution given by Equation 4.4 is also still valid.

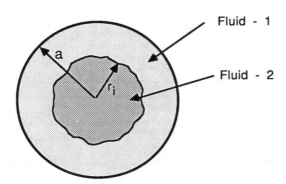

FIGURE 4–5   *Notation for derivations for two-phase annular flow through a round tube.*

Derivation of the Solution

Integrating from $u = 0$ at $r = a$ to any r still in fluid 1 gives

$$u_1 = \frac{\tau_w}{2a\mu_1}(a^2 - r^2) \tag{4.14}$$

Similarly, integrating from $u = u_c$ at $r = 0$ to any r still in fluid 2 gives

$$u_2 = u_c - \frac{r^2\tau_w}{2a\mu_2} \tag{4.15}$$

Equating these velocities at $r = a\zeta$ gives the following expression for the velocity at the center line:

$$\frac{a\tau_w}{2\mu_1}(1 - \zeta^2) = u_c - \frac{a\zeta^2\tau_w}{2\mu_2} \tag{4.16}$$

or

$$u_c = \frac{a\tau_w}{2\mu_1}(1 - \zeta^2 + \xi\zeta^2) \tag{4.16A}$$

where ξ is defined by Equation 3.30. Substitution of Equation 4.16A in 4.15 produces

$$u_2 = \frac{a\tau_w}{2\mu_2}\left[\frac{1 - \zeta^2}{\xi} + \zeta^2 - \left(\frac{r}{a}\right)^2\right] \tag{4.17}$$

The mean velocities of the two phases are obtained by integration with respect to $(r/a)^2$ over the appropriate limits. Thus

$$u_{m1} = \frac{1}{1 - \zeta^2}\int_{\zeta^2}^{1} u_1 \, d\left(\frac{r}{a}\right)^2 = \frac{a\tau_w}{4\mu_1}(1 - \zeta^2) \tag{4.18}$$

and

$$u_{m2} = \frac{1}{\zeta^2}\int_{0}^{\zeta^2} u_2 \, d\left(\frac{r}{a}\right)^2 = \frac{a\tau_w}{4\mu_1}[\xi\zeta^2 + 2(1 - \zeta^2)] \tag{4.19}$$

The ratio of the rate of flow of fluid 2 to that in the absence of fluid 1 (with $\zeta = 1$) is

$$\frac{V_2}{V_{20}} = \frac{u_{m2}\pi a^2\zeta^2}{(u_{m2})_{\zeta=1}\pi a^2} = \zeta^2\left(\zeta^2 + 2\frac{1 - \zeta^2}{\xi}\right) \tag{4.20}$$

and the ratio of the volumetric rates of flow of the two fluids is

$$\frac{V_1}{V_2} = \frac{u_{m1}\pi a^2(1 - \zeta^2)}{u_{m2}\pi a^2 \zeta^2} = \frac{(1 - \zeta^2)^2}{\zeta^2[\xi\zeta^2 + 2(1 - \zeta^2)]} \tag{4.21}$$

The maximum value of V_2/V_{20}, as found by setting the derivative with respect to ζ to zero, occurs at

$$\zeta^2 = \frac{1}{2 - \xi} \tag{4.22}$$

and is

$$\left(\frac{V_2}{V_{20}}\right)_{max} = \frac{1}{\xi(2 - \xi)} \tag{4.23}$$

The corresponding ratio of flows is

$$\left(\frac{V_1}{V_2}\right)_{max(V_2/V_{20})} = \frac{(1 - \xi)^2}{2 - \xi} \tag{4.24}$$

As $\xi \to 0$,

$$\zeta^2 \to \frac{1}{2} \tag{4.25}$$

$$\left(\frac{V_2}{V_{20}}\right)_{max} \to \frac{1}{2\xi} \tag{4.26}$$

and

$$\left(\frac{V_1}{V_2}\right)_{max(V_2/V_{20})} \to \frac{1}{2} \tag{4.27}$$

Since the volumetric rate of flow is proportional to the shear stress at the wall and hence to the pressure gradient, Equation 4.24 can also be considered to give the ratio of the required pressure drop for single-phase flow of fluid 2 to that for the same flow of fluid 2 in concurrent flow.

Russell and Charles [9] apparently first developed a solution for annular flow. They presented results graphically for the rate of flow of fluid 2 per unit pressure gradient as a function of ζ for $\xi = 0.001, 0.01, 0.1$, and 1, and for the relative pressure gradient and pumping power as a function of ζ for $\xi \to 0$. They concluded that the pumping power is a minimum for $\xi = 0.786$ ($\zeta^2 = 0.618$) and that the pressure gradient is a minimum for $\zeta = 0.717$ ($\zeta^2 = 0.514$). As indicated by Equation 4.25, the theoretical value of ζ for the minimum in the pressure gradient is $(1/2)^{1/2} = 0.707$.

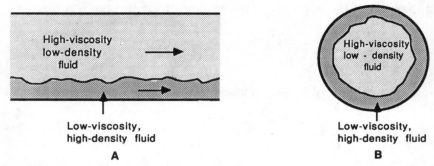

FIGURE 4–6   *Comparison of stratified and annular two-phase flow through channels: (A) stratified flow between parallel plates; (B) annular flow through a round tube.*

Applications

If $\xi = 0.01$, an approximately 50-fold increase in the rate of flow of fluid 2 or an approximately 50-fold decrease in the pressure drop for the same flow can be accomplished at the expense of pumping one extra volume of fluid 1 for every two volumes of fluid 2. Annular flow in a round tube is thus much more effective than stratified flow between parallel plates because, as illustrated in Figure 4–6, the fluid with the higher viscosity is not in contact with the wall.

The general solution is presumably also applicable for a fluid of low viscosity, such as air, in the core with an outer film of liquid, such as water (see problem 5).

Experimental Confirmation of Solution

Charles [10] carried out extensive tests with crude oil and water. He observed pressure gradient ratios as high as 15, which is higher than possible with stratified flow (see problem 32) but considerably less than the values predicted for idealized annular flow. These results may indicate nonconcentric annular flow or nonwetting, as sketched in Figure 4–7 by Charles et al. [11] who analyzed several sets of experimental data and reached a similar conclusion.

FLOW THROUGH A CIRCULAR ANNULUS

The derivation and results for flow through the annulus between concentric circular tubes are related to those for flow between parallel plates and to those for single- and two-phase flow through a round tube. However, as shown below, this system has some interesting characteristics that differ from the previous ones.

Development of the Model

The velocity in an annulus goes to zero on both the inner and outer walls and through a maximum at an intermediate radius, but, contrary to the two previous

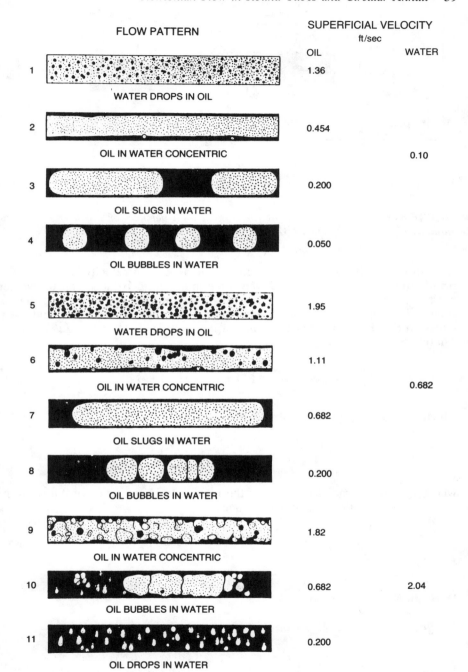

FIGURE 4–7   *Different types of two-phase flow in a round tube observed by Charles et al. [11] for oil with a viscosity of 16.8 centipoise and water.*

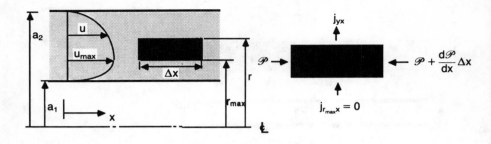

A   **B**

FIGURE 4–8   *Control volume for force-momentum balance for a concentric circular annulus: (A) control volume; (B) force-momentum balance on control volume.*

cases of single-phase flow, the location of the maximum velocity is not known a priori from symmetry. Even so, the force-momentum balance can be written most simply for an elemental volume based on this unknown location, as sketched in Figure 4–8. The momentum flux through or the equivalent shear stress on the fluid is zero at the point of maximum velocity since du/dr is zero. Therefore,

$$\left[\mathscr{P} - \left(\mathscr{P} + \frac{d\mathscr{P}}{dx}\Delta x\right)\right]\pi(r^2 - r^2_{\max}) = j_{rx}2\pi r\,\Delta x \qquad (4.28)$$

which can be simplified to

$$j_{rx} = \frac{r^2 - r^2_{\max}}{2r}\left(-\frac{d\mathscr{P}}{dx}\right) \qquad (4.28A)$$

where $r_{\max}$ is the radial location of the maximum in the velocity. Substituting for j_{rx} from Equation 2.1 in 4.28A gives

$$-\mu\frac{du}{dr} = \frac{r^2 - r^2_{\max}}{2r}\left(-\frac{d\mathscr{P}}{dx}\right) \qquad (4.29)$$

Derivation of the Solution

Integrating Equation 4.29 from $u = 0$ at $r = a_2$ gives

$$u = \frac{1}{\mu}\left(\frac{a_2^2 - r^2}{4} - \frac{r^2_{\max}}{2}\ln\left\{\frac{a_2}{r}\right\}\right)\left(-\frac{d\mathscr{P}}{dx}\right) \qquad (4.30)$$

The condition that $u = 0$ at $r = a_1$ then yields the following expression for the location of the maximum velocity:

$$r_{max}^2 = \frac{a_2^2 - a_1^2}{2 \ln\{a_2/a_1\}} \tag{4.31}$$

The dimensionless location r_{max}/a_2 is a function only of the aspect ratio of the annulus and, hence, is independent of the rate of flow.

The mean velocity is obtained by integrating Equation 4.30 with respect to r^2 as follows:

$$u_m = \frac{1}{a_2^2 - a_1^2} \int_{a_1^2}^{a_2^2} u \, dr^2$$

$$= \frac{1}{8\mu}\left(-\frac{d\mathscr{P}}{dx}\right)(a_2^2 + a_1^2 - 2r_{max}^2) \tag{4.32}$$

Then from Equations 4.30 and 4.32,

$$\frac{u}{u_m} = \frac{2(a_2^2 - r^2 - 2r_{max}^2 \ln\{a_2/r\})}{a_2^2 + a_1^2 - 2r_{max}^2} \tag{4.33}$$

Finally, substituting for r_{max} from Equation 4.31 leads to

$$\frac{u}{u_m} = \frac{2[1 - (r/a_2)^2 - (1 - \lambda^2)\ln\{a_2/r\}/\ln\{1/\lambda\}]}{1 + \lambda^2 - (1 - \lambda^2)/\ln\{1/\lambda\}} \tag{4.34}$$

where, as in Chapter 1, $\lambda = a_1/a_2 = D_1/D_2$ is the aspect ratio of the annulus. Equation 4.34 obviously reduces to that for a round tube of radius a_2 as $\lambda \to 0$. It also reduces to that for parallel plates as $\lambda \to 1$, but expansion of the logarithmic terms is necessary to obtain that limit (see problem 14).

The hydraulic diameter for this geometry is

$$D = \frac{4\pi(a_2^2 - a_1^2)}{2\pi(a_1 + a_2)} = 2(a_2 - a_1) \tag{4.35}$$

Therefore from Equations 4.32 and 4.31

$$\text{Po} = \text{Re}\,f = \frac{8(a_2 - a_1)^2}{a_2^2 + a_1^2 - 2r_{max}^2} = \frac{8(1 - \lambda)^2}{1 + \lambda^2 - (1 - \lambda^2)/\ln\{1/\lambda\}} \tag{4.36}$$

Equation 4.36 also obviously reduces to the prior result for a round tube as $\lambda \to 0$. This result is somewhat surprising, since the velocity goes to zero at the inner wall for any finite value of λ. Again, expansion of the logarithmic term is necessary to obtain the limiting value of 12 for Po as $\lambda \to 1$.

The shear stress on the two surfaces can be related to each other and to the pressure gradient by making a force-momentum balance from r_{max} to a_2 and from a_1 to r_{max}, yielding

$$2a_2\tau_{w2} = (a_2^2 - r_{max}^2)\left(\frac{-d\mathscr{P}}{dx}\right) \tag{4.37}$$

and

$$2a_1\tau_{w1} = (r_{\max}^2 - a_1^2)\left(\frac{-d\mathscr{P}}{dx}\right) \tag{4.38}$$

It follows from Equations 4.31 and 4.36–4.38 that

$$\frac{\tau_{w2}}{\tau_{w1}} = \frac{a_1(a_2^2 - r_{\max}^2)}{a_2(r_{\max}^2 - a_1^2)} = \frac{\lambda(2\ln\{1/\lambda\} - 1 + \lambda^2)}{1 - \lambda^2 - 2\lambda^2\ln\{1/\lambda\}} \tag{4.39}$$

and

$$a_2\tau_{w2} + a_1\tau_{w1} = \left(\frac{a_2^2 - a_1^2}{2}\right)\left(-\frac{d\mathscr{P}}{dx}\right) \tag{4.40}$$

The equivalent of this solution for annuli was apparently first derived by Lamb [12], p. 555.

Meter and Bird [13] suggested defining a special Reynolds number for annuli as

$$\mathrm{Re}_B = \frac{D_2}{1 - \lambda}\left[1 + \lambda^2 - \frac{1 - \lambda^2}{\ln\{1/\lambda\}}\right]\frac{u_m}{\nu} \tag{4.41}$$

Then

$$\mathrm{Po}_B = f\mathrm{Re}_B = 8 \tag{4.42}$$

for all λ.

Alternative Representation

Rothfus et al. [14] suggested representation of experimental data for the pressure drop in both laminar and turbulent flow in terms of a friction factor and Reynolds number for the region outside $r_{\max}$, i.e., the use of

$$f_2 \equiv \frac{\tau_{w2}}{\varrho u_m^2} \tag{4.43}$$

and

$$\mathrm{Re}_2 \equiv \frac{D_2 u_m \varrho}{\mu} \tag{4.44}$$

where

$$D_2 = \frac{4\pi(a_2^2 - r_{\max}^2)}{2\pi a_2} = \frac{2(a_2^2 - r_{\max}^2)}{a_2} \tag{4.45}$$

It follows from Equation 4.36 that

$$f_2 \text{Re}_2 = \frac{8(a_2^2 - r_{max}^2)^2}{a_2^2(a_2^2 + a_1^2 - 2r_{max}^2)} = \frac{16(2\ln\{1/\lambda\} - 1 + \lambda^2)}{(1 + \lambda^2)\ln\{1/\lambda\} - 1 + \lambda^2} \quad (4.46)$$

Note that u_m, the mean velocity for the whole cross section of the annulus, is utilized in Equations 4.43 and 4.44 rather than u_{m2}, the mean velocity for the region outside r_{max}. It follows from Equation 4.39 that

$$\frac{f_1}{f_2} = \frac{\tau_{w1}}{\tau_{w2}} = \frac{a_2(r_{max}^2 - a_1^2)}{a_1(a_2^2 - r_{max}^2)} = \frac{1 - \lambda^2 - 2\lambda^2\ln\{1/\lambda\}}{\lambda(2\ln\{1/\lambda\} - 1 + \lambda^2)} \quad (4.47)$$

where

$$f_1 \equiv \frac{\tau_{w1}}{\varrho u_m^2} \quad (4.48)$$

Experimental Confirmation of Solution

Local velocities measured by Rothfus et al. [14] are compared in Figures 4–9 and 4–10 with the prediction of Equation 4.34. Reasonable agreement may be noted in Figure 4–9 for Re = 1250 and λ = 0.650. For Re = 1850 and λ = 0.162, as shown in Figure 4–10, the data deviate somewhat from the predicted values based on the mean velocity obtained by integration of the local values. These data are, however, represented satisfactorily by Equation 4.33 using the **observed** value of r_{max}. This latter result is attributed by the experimenters to the onset of the transition from laminar to turbulent flow below Re = 1850. However, the general deviation of their measured pressure drops from the predictions for the whole range of both laminar and turbulent flow, as shown in Figure 4–11, suggests that experimental error in the determination of u_m may

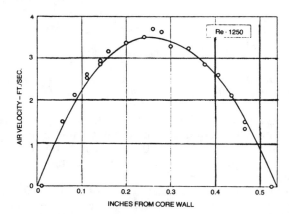

FIGURE 4–9 *Comparison of theoretical and experimental velocity distribution in laminar flow through a circular annulus with $a_1 = 1.0$ in., $a_1/a_2 = 0.650$, and Re = 1280. (From Rothfus et al. [14].)*

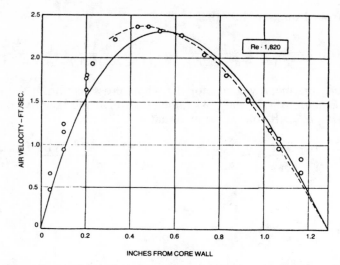

FIGURE 4–10 *Comparison of theoretical and experimental velocity distribution in laminar flow, near the point of transition, through a circular annulus with $a_1 = 0.25$ in., $a_1/a_2 = 0.162$, and Re = 1820. The solid curve represents the complete theoretical solution and the dashed curve represents Equation 4.33 with the observed value of r_{max}. (From Rothfus et al. [14].)*

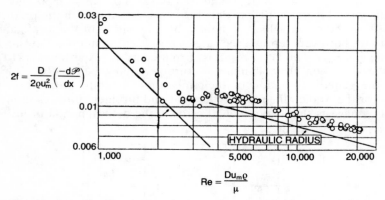

FIGURE 4–11 *Comparison of theoretical and experimental pressure drops in flow through a circular annulus with $a_1 = 0.25$ in. and $a_1/a_2 = 0.162$. (From Rothfus et al. [14].)*

instead be responsible. Prengle and Rothfus [15] determined velocity profiles using dye as a tracer. They found agreement between their observations and the theoretical predictions only for Re < 700 and concluded that the discrepancies at higher Re were due to the development of local instability prior to complete transition at Re = 2000.

Pressure drops measured by Walker et al. [16] are plotted in Figure 4–12 for a series of aspect ratios in the form of f_2 versus Re_2. The lines labeled *LAMINAR EQN* represent Equation 4.46, and those labeled *SMOOTH TUBES* represent the Blasius equation (see, for example, Churchill [17]) modified for the outer region as

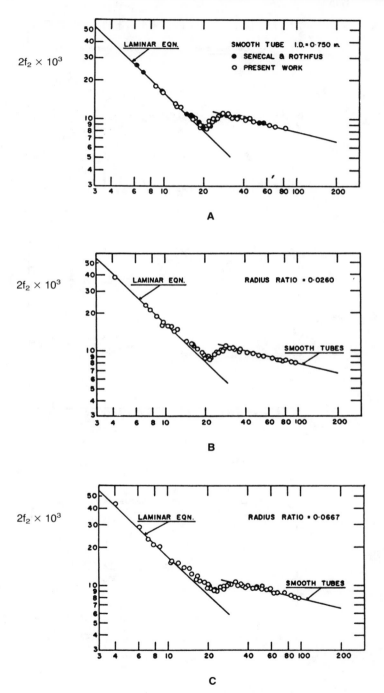

FIGURE 4–12   *Comparison of theoretical and experimental friction factors for round tubes, parallel plates, and the outer surface of circular annuli. (From Walker et al. [16].)*

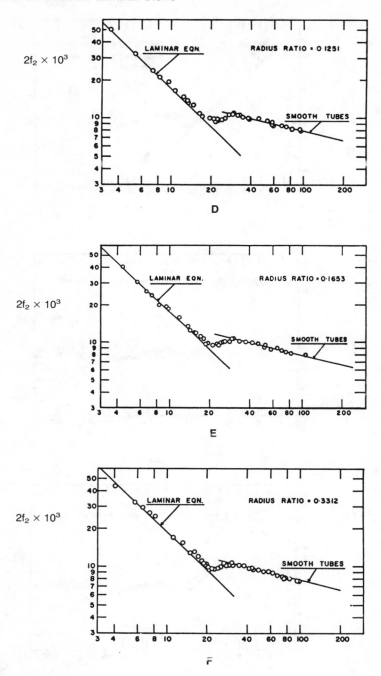

FIGURE 4–12 *(continued)*

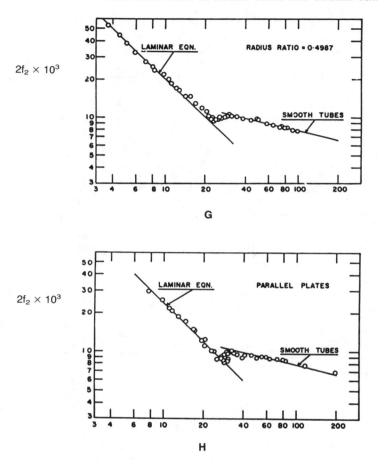

FIGURE 4–12   *(concluded)*

$$f_2 = \frac{0.0395}{\mathrm{Re}_2^{1/4}} \tag{4.49}$$

There is excellent agreement for each of the eight aspect ratios. Transition appears to occur at $\mathrm{Re}_2 \cong 2100$ in each case. However, a gradual increase might be expected for values of λ larger than those included in this investigation.

A graphical correlation by Knudsen and Katz [5], p. 98, of the experimental values of a number of investigators for various aspect ratios in the form of

$$f\left(1 + \lambda^2 - \frac{1 - \lambda^2}{\ln\{1/\lambda\}}\right)\bigg/(1 - \lambda)^2$$

versus Re, as suggested by Equation 4.36, is reproduced in Figure 4–13. There is good agreement, but the experimental values for small Re fall predominantly below the prediction.

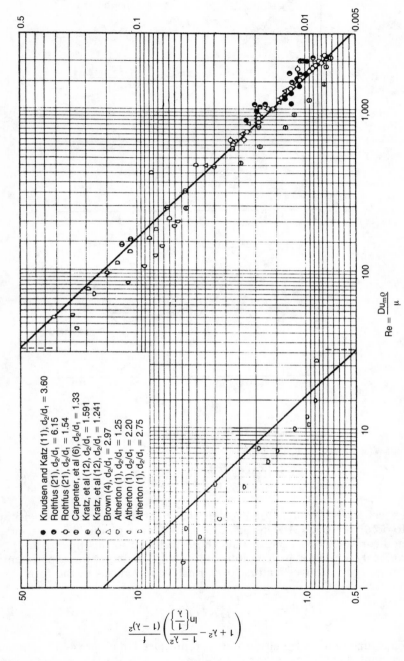

FIGURE 4–13 *Comparison of theoretical and experimental pressure drops for laminar flow through circular annuli. (From Knudsen and Katz [5], p. 98.)*

Nootbar and Kintner [18] observed friction factors 10% less than those predicted for $0.64 \leq \lambda \leq 0.91$ and attribute the discrepancy to some unaccounted effect of curvature. Some support for this contention is provided by Figure 4–13.

SUMMARY

Solutions for single-phase flow through round tubes and annuli and for two-phase annular flow were derived in this chapter. The solutions for single-phase flow have the same basic structure as for parallel plates but are somewhat more complicated because of the variation of the shear stress and the differential area for flow with radius. For an annulus the shear stress is different on the two surfaces. Only the outer fluid contacts the wall in two-phase annular flow, offering the possibility for reducing the cost of pumping a fluid of high viscosity by the addition of a second immiscible fluid of lower viscosity. This practical problem was explored in some detail.

Experimental confirmation was demonstrated for most of the theoretical results. However, the cited advantage of introducing a second immiscible fluid of lower viscosity is apparently not always realized, owing to the failure to achieve truly annular flow.

PROBLEMS

1. From the following data of Poiseuille [19] for flow through capillary tubes, calculate the viscosity of water and compare with the value given in handbooks.

Length of tube, cm	10.05
Average diameter, cm	0.0141
Temperature, °C	10.0
Volume of water, cm^3	13.341
Time, s	3505.75
Pressure drop, mm Hg	385.87

2. Reexpress the solution for single-phase flow in a round tube in terms of y, the distance from the wall, and a, the radius. Compare the results with the solution for parallel plates in terms of y and b.
3. Derive a solution for single-phase flow in a round tube in terms of y, the distance from the wall, and a, the radius.
4. Derive a complete solution for countercurrent annular flow.
5. The solution for annular flow is presumably applicable for air flowing inside a thin layer of water. Specialize the solution for that limiting case of $\xi \rightarrow \infty$.
6. Derive an approximate solution for annular flow with $\xi \rightarrow 0$, assuming plug flow of the inner fluid and

$$\tau_w = \frac{\mu_1 u_c}{a(1 - \zeta)} \tag{4.50}$$

(see Chilton [20]). Compare this approximate result with the general solution and with the asymptotic solution for $\xi \to 0$.

7. Prepare a plot of u_1/u_{1m} versus r/a for $\zeta = 0, 0.1, 0.5, 0.9,$ and 1.

8. Prepare a plot of u_2/u_{2m} versus r/a for $\zeta = 0.5$ and $\xi = 0.001, 0.01, 0.1, 1, 10,$ and 100.

9. Prepare a plot of $\mu_1 u/\tau_w a$ versus r/a for $\xi = 0.01$ and $\zeta = 0, 0.1, 0.5, 0.9,$ and 1.

10. Prepare a plot of $\mu_1 u/\tau_w a$ versus r/a for $\zeta = 0.5$ and $\xi = 0.001, 0.01, 0.1, 1, 10,$ and 100.

11. Prepare a plot of V_2/V_{20} versus ζ for $\xi = 0.001, 0.01, 0.1, 1, 10,$ and 100.

12. Prepare a plot of V_1/V_2 versus ζ for $\xi = 0.001, 0.01, 0.1, 1, 10,$ and 100.

13. Prepare a plot of V_2/V_{20} versus V_1/V_2 for $\zeta = 0.1, 0.5,$ and 0.9.

14. **a.** Derive an approximate expression for u/u_m for $\lambda \to 1$ from Equation 4.34 and show that it reduces to Equation 3.10 for $\lambda = 1$.
    b. Derive an approximate expression for u/u_m for $\lambda \to 0$ from Equation 4.34 and show that it reduces to Equation 4.9 for $\lambda = 0$.

15. Plot u/u_m versus r/a_2 from Equation 4.35 for $\lambda = 1, 10^{-3}, 10^{-6}, 10^{-9},$ and 0. Explain physically how Equation 4.35 can reduce to Equation 4.9 as $\lambda \to 0$.

16. Derive an expression for u_{max}/u_m from Equation 4.34 and show that it reduces to 2 as $\lambda \to 0$ and to 3/2 as $\lambda \to 1$. Plot u_{max}/u_m versus λ.

17. Plot r_{max}/a_2 versus λ.

18. Plot Po versus λ.

19. Derive an approximate expression for Po for $\lambda \to 1$ from Equation 4.36 and show that it reduces to 12 for $\lambda = 1$.

20. Derive an approximate expression for Po for $\lambda \to 0$ from Equation 4.36 and show that it reduces to 8 for $\lambda = 0$.

21. Derive expressions for Po_1 in terms of r_{max} and λ.

22. Derive an expression for Re_2/Re in terms of λ and plot. Interpret.

23. Prepare a common plot of Po, Po_1, and Po_2 versus λ. Interpret.

24. Rederive the solution for an annulus starting from a force-momentum balance on an element adjacent to the outer wall.

25. Rederive the solution for an annulus starting from a force-momentum balance on an element adjacent to the inner wall.

26. Rederive the solution for an annulus starting from a force-momentum balance on an element adjacent to and inside the plane of no shear.

27. **a.** Calculate the local shear stress on the wall of a tube through which a liquid is flowing at a rate of 315 cm^3/s at a point where the temperature profile resulting from the heat transfer is as follows.

r/a	T (°C)	μ (mPa·s)
0	15.6	56
0.2	19.4	48
0.4	32.8	38
0.6	54.4	19
0.8	85.0	8.8
0.9	103.3	5.7
1.0	123.3	3.7

The viscosity corresponding to the temperature is also given. The inside diameter of the tube is 50.4 mm. The effect of the change in density can be assumed to be of second order.

b. What percentage error would be made by using Poiseuille's law for constant viscosity with the viscosity evaluated at the temperature of the wall?

28. Under what circumstances can the rate of flow of oil through a pipeline be increased for a given pressure drop by the introduction of water?

29. Derive an asymptotic form for Equation 4.34 for small but finite λ.

30. Derive an asymptotic form for Equation 4.34 for small but finite δ ≡ 1 − λ.

31. a. Derive from first principles an asymptotic solution for the central flow in a round tube of a very, very viscous liquid inside a very, very thin outer film of a less viscous fluid of the same density. The following simplifications are suggested: (1) negligible curvature in the outer film; (2) a negligible velocity gradient in the inner fluid. Express the solution in terms of the ratio of the volumetric rates of flow of the two fluids and the ratio of the volumetric rates of flow of the inner fluid with and without the thin film of the outer fluid.

b. Compare the solution of part a with the exact solution and define its range of applicability.

32. Derive a solution for the stratified concurrent flow of two immiscible fluids such that their cross sections are equal.

33. Determine the critical Reynolds number for transition from laminar to turbulent flow in an annulus as a function of λ, based on the premise that f is the same as for a round pipe, i.e., equal to 8/2100. Compare this result with the experimental data of Figure 4–12.

REFERENCES

1. G. Stokes, "On the Theories of the Internal Friction of Fluids in Motion, and of the Equilibrium and Motion of Elastic Solids," *Trans. Camb. Phil. Soc.*, *8* (1845) 287, (*Mathematical and Physical Papers*, Vol. I, Cambridge University Press (1880) p. 75).
2. G. Hagen, "Über die Bewegung des Wassers in engen zylinderschen Rohren," *Poggendorff's Ann. Phys. Chem.*, *46* (1839) 423.
3. E. Hagenbach "Über die Bestimmung der Zähigkeit einer Flüssigkeit durch den Ausfluss aus Rohren," *Poggendorff's Ann. Phys. Chem.*, *Ser. 2, 109* (1860) 385.
4. H. Jacobson, *Hemodynamics, Archiv. Anat. Physiol.*, (1860) 80; (1862) 683; (1867) 224.
5. J. G. Knudsen and D. L. Katz, *Fluid Dynamics and Heat Transfer*, McGraw-Hill, New York (1958).
6. V. E. Senecal and R. R. Rothfus, "Transition Flow of Fluids in Smooth Tubes," *Chem. Eng. Progr.*, *49* (1953) 533.
7. S. W. Churchill, *The Practical Use of Theory in Fluid Flow. Book III. Laminar Multidimensional Flows in Channels*, Notes, The University of Pennsylvania (1979).
8. J. D. Isaacs and J. B. Speed, U.S. Patent 759,374 (1904).
9. T. W. F. Russell and M. E. Charles, "The Effect of the Less Viscous Fluid in the Laminar Flow of Two Immiscible Fluids," *Can. J. Chem. Eng.*, *37* (1959) 18.
10. M. E. Charles, "The Reduction of Pressure Gradients in Oil Pipelines: Experi-

mental Results for the Stratified Flow of a Heavy Crude Oil and Water," *Trans. Soc. Pet. Engrs.*, *63* (1960) 306.

11. M. E. Charles, G. W. Govier, and G. W. Hodgson, "The Horizontal Pipeline Flow of Equal Density Oil—Water Mixtures," *Can. J. Chem. Eng.*, *39* (1961) 27.
12. Horace Lamb, *Hydrodynamics*, Dover, New York (1945).
13. D. M. Meter and R. B. Bird, "Turbulent Newtonian Flow in Annuli," *AIChE J.*, *7* (1961) 41.
14. R. R. Rothfus, C. C. Monrad, and V. E. Senecal, "Velocity Distribution and Fluid Friction in Smooth Concentric Annuli," *Ind. Eng. Chem.*, *42* (1950) 2511.
15. R. S. Prengle and R. R. Rothfus, "Transition Phenomena in Pipes and Annular Cross Sections," *Ind. Eng. Chem.*, *47* (1955) 379.
16. J. E. Walker, G. A. Whan, and R. R. Rothfus, "Fluid Friction in Noncircular Ducts," *AIChE J.*, *3* (1957) 484.
17. S. W. Churchill, *The Practical Use of Theory in Fluid Flow. Book IV. Turbulent Flows*, Notes, The University of Pennsylvania (1981).
18. R. F. Nootbar and R. C. Kintner, "Fluid Friction in Annuli of Small Clearance," *Proc. Second Midwestern Conf. on Fluid Mech.*, Ohio State University Studies-Engineering Series, *21*, No. 3, Columbus (September 1952), p. 185.
19. J.-L.-M. Poiseuille, "Recherches expérimentales sur le mouvement des liquides dans les tubes de très petits diamètres," *Mémoires présentés par divers savants a l'Académie Royale des Sciences de l'Institut de France, Sci. Math. Phys.*, *9* (1846) 433; English transl. by W. H. Herschel, "Experimental Investigations upon the Flow of Liquids in Tubes of Very Small Diameter," *Rheol. Mem.*, E. C. Bingham, Ed., *1*, No. 1, Easton, PA (1940).
20. E. G. Chilton, "Letter to Editor," *Can. J. Chem. Eng.*, *37* (1959) 127.

Chapter 5

Non-Newtonian Flow through Channels

The behavior of non-Newtonian fluids in flow through channels is of both practical and intrinsic interest. In addition, flow through a round tube is often used to measure the effective viscosity of a non-Newtonian fluid. Attention here, as in Chapter 2, is confined to fluids with purely viscous behavior. The derivations are limited to flow through round tubes. Flows in annuli and between parallel plates are deferred to the problem set. Equations 4.2–4.4 are applicable for non-Newtonian as well as Newtonian fluids and provide the basis for the developments below.

PSEUDOPLASTICS AND DILATANTS

The power-law model, Equation 2.10, is used here as an approximation for pseudoplastic and dilatant behavior. The validity of this simplification is examined in problems 44 and 45. Combining Equations 4.4 and 2.10 gives

$$\frac{r\tau_w}{a} = M \left| \frac{du}{dr} \right|^{\alpha - 1} \left(-\frac{du}{dr} \right) \tag{5.1}$$

Taking the $1/\alpha$ root of both sides to obtain $-du/dr$, and integrating from $u = 0$ at $r = a$, gives

$$u = \left(\frac{\tau_w}{Ma} \right)^{1/\alpha} \int_r^a r^{1/\alpha} \, dr$$

$$= \left(\frac{\tau_w}{Ma} \right)^{1/\alpha} \left(\frac{\alpha}{1 + \alpha} \right) (a^{(1+\alpha)/\alpha} - r^{(1+\alpha)/\alpha}) \tag{5.2}$$

A second integration gives the mean velocity:

$$u_m = \int_0^1 ua \left(\frac{r}{a} \right)^2 = \left(\frac{\tau_w}{M} \right)^{1/\alpha} \left(\frac{\alpha}{1 + 3\alpha} \right) a \tag{5.3}$$

From Equations 5.2 and 5.3,

73

$$\frac{u}{u_m} = \left(\frac{1 + 3\alpha}{1 + \alpha}\right)\left(1 - \left(\frac{r}{a}\right)^{(1+\alpha)/\alpha}\right) \tag{5.4}$$

$$\frac{u_c}{u_m} = \frac{1 + 3\alpha}{1 + \alpha} \tag{5.5}$$

$$\tau_w = M\left(\frac{(1 + 3\alpha)u_m}{a\alpha}\right)^\alpha \tag{5.6}$$

and

$$f \equiv \frac{\tau_w}{\varrho u_m^2} = \frac{M}{\varrho u_m^{2-\alpha}}\left(\frac{1 + 3\alpha}{\alpha a}\right)^\alpha \tag{5.7}$$

Equation 5.7 can be forced into the form of Poiseuille's law (Equation 4.8) by defining an effective Reynolds number as

$$\mathrm{Re}_\alpha = 2^{3-\alpha}\left(\frac{\alpha}{1 + 3\alpha}\right)^\alpha \frac{D^\alpha u_m^{2-\alpha}\varrho}{M} \tag{5.8}$$

For $\alpha = 1$ and $M = \mu$, Equations 5.1–5.8 reduce to their counterparts in Chapter 4.

Velocity profiles corresponding to Equation 5.4 are shown in Figure 5–1 (after Metzner [1], p. 108). The profile progresses from flat (plug flow) at $\alpha = 0$ to parabolic (Poiseuille flow) at $\alpha = 1$ to pointed at $\alpha \to \infty$.

Experimental values for a solution of polyacrylamite in water are seen in

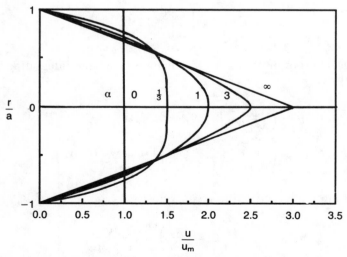

FIGURE 5–1   *Velocity profiles for the laminar flow of power-law fluids through a round tube.*

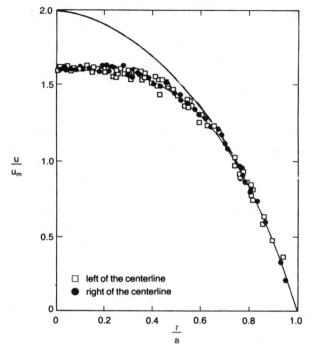

FIGURE 5–2 *Comparison of experimental velocity for the laminar flow of an aqueous solution of polyacrylamide through a round tube with the theoretical solution for a power-law fluid. The lower curve represents Equation 5.4 with $\alpha = 0.48$; the upper curve represents Equation 4.9. (After Denn [2], p. 193.)*

Figure 5–2 after Denn [2], p. 193, to be in excellent agreement with Equation 5.4 for $\alpha = 0.48$.

BINGHAM PLASTICS

For fluids whose behavior can be represented by Equations 2.18 and 2.19, no motion will occur unless the shear stress exceeds τ_o. As indicated by Equation 4.4, the maximum shear stress in a round tube occurs at the wall. Hence no motion will occur in such a fluid unless τ_w exceeds τ_o. If $\tau_w > \tau_o$, then τ_{rx}, which varies linearly, according to Equation 4.4, from zero at the center to τ_w at $r = a$, will equal τ_o at some critical radius

$$r_o \equiv a \frac{\tau_o}{\tau_w} \tag{5.9}$$

Then for $r \geq r_o$, from Equation 2.18,

$$\tau_{rx} = \frac{r\tau_w}{a} = -\mu_o \frac{du}{dr} + \tau_o \tag{5.10}$$

which gives after integration

$$u = \frac{\tau_w}{a\mu_o} \int_r^a r\, dr - \frac{\tau_o}{\mu_o} \int_r^a dr$$

$$= \frac{\tau_w a}{2\mu_o} \left[1 - \left(\frac{r}{a}\right)^2 - 2\frac{\tau_o}{\tau_w}\left(1 - \frac{r}{a}\right) \right] \tag{5.11}$$

For $r \leq r_o$, $du/dr = 0$. The resulting uniform velocity is obtained by setting $r = r_o$ in Equation 5.11, and then substituting for r_o/a from Equation 5.9, yielding

$$u = u_c = \frac{\tau_w a}{2\mu_o}\left(1 - \frac{\tau_o}{\tau_w}\right)^2 \tag{5.12}$$

The corresponding mean velocity across the entire tube is

$$u_m = \int_0^1 u\, d\left(\frac{r}{a}\right)^2$$

$$= u_c\left(\frac{\tau_o}{\tau_w}\right)^2 + \frac{\tau_w a}{2\mu_o}\int_{(\tau_o/\tau_w)^2}^1 \left(1 - 2\frac{\tau_o}{\tau_w} + \frac{ar\tau_o}{a\tau_w} - \left(\frac{r}{a}\right)^2\right) d\left(\frac{r}{a}\right)^2$$

$$= \frac{\tau_w a}{4\mu_o}\left(1 - \frac{4\tau_o}{3\tau_w} + \frac{1}{3}\left(\frac{\tau_o}{\tau_w}\right)^4\right) \tag{5.13}$$

Then from Equations 5.11 and 5.13,

$$\frac{u}{u_m} = \frac{2[1 - 2\tau_o/\tau_w + 2r\tau_o/a\tau_w - (r/a)^2]}{1 - 4\tau_o/3\tau_w + \frac{1}{3}(\tau_o/\tau_w)^4} \qquad \text{for } 1 \geq \frac{r}{a} \geq \frac{\tau_o}{\tau_w} \tag{5.14}$$

and from Equations 5.12 and 5.13,

$$\frac{u}{u_m} = \frac{2(1 - \tau_o/\tau_w)}{1 - 4\tau_o/3\tau_w + \frac{1}{3}(\tau_o/\tau_w)^4} \qquad \text{for } \frac{r}{a} \leq \frac{\tau_o}{\tau_w} \tag{5.15}$$

Equation 5.13 is known as the *Buckingham–Reiner equation*. Following Perkins and Glick (according to Metzner [1]) the Buckingham–Reiner equation [3, 4] can be reexpressed as

$$\frac{8}{f\text{Re}_o} = 1 - \frac{4\text{He}}{3f\text{Re}_o^2} + \frac{1}{3}\left(\frac{\text{He}}{f\text{Re}_o^2}\right)^4 \tag{5.16}$$

where

$$\text{Re}_o \equiv \frac{Du_m\varrho}{\mu_o} \tag{5.17}$$

and the Hedstrom number [5] is

$$\text{He} \equiv \frac{\tau_o D^2 \varrho}{\mu_o^2} \qquad (5.18)$$

Thus $f\text{Re}_o = \tau_w D/\mu_o u_m$ is an explicit function of $\text{He}/f\text{Re}_o^2 = \tau_o/\tau_w$ only. This functionality is inconvenient if u_m is specified, since τ_w is in both groups. Equation 5.16 can, however, be regrouped as

$$\frac{8}{f\text{Re}_o} = 1 - \frac{4(\text{He}/\text{Re}_o)}{3(f\text{Re}_o)} + \frac{1}{3}\frac{(\text{He}/\text{Re}_o)^4}{(f\text{Re}_o)^4} \qquad (5.19)$$

in which $f\text{Re}_o$ is an implicit function of $\text{He}/\text{Re}_o = \tau_o D/\mu_o u_m$ only. (These two relationships are explored in problems 5 and 6.) The conventional plot of Equation 5.16 as f versus Re_o with He as a parameter is shown in Figure 5–3. The friction factor is seen to increase above that for Newtonian flow as He increases. Experimental data of Thomas [6] for several suspensions of thoria are seen in Figure 5–4 to have the behavior predicted in Figure 5–3, indicating that they follow the Bingham-plastic model.

The velocity field, as given by Equations 5.14 and 5.15, is similarly noted to be a function of $\text{He}/f\text{Re}_o^2 = \tau_o/\tau_w$ as well as of r/a. Figure 5–5 is a plot of this relationship. The profile is seen to vary from parabolic (Poiseuille flow) to flat (plug flow) as $\text{He}/f\text{Re}_o^2$ varies from zero to unity. (No flow occurs for $\text{He} > f\text{Re}_o^2$.)

GENERAL VISCOUS FLUIDS

A general viscous fluid is here defined as homogeneous, with its shear stress an unknown but unique function of the rate of stress or, conversely,

$$-\frac{du}{dr} = \phi\{\tau\} \qquad (5.20)$$

where, again for simplicity,

$$\tau \equiv \tau_{rx} \qquad (5.21)$$

The volumetric rate of flow in a round tube is

$$V = \int_0^a u \cdot 2\pi r \, dr = \pi \int_0^{a^2} u \, dr^2 \qquad (5.22)$$

Equation 5.22 can be integrated by parts as follows:

$$V = \pi \left(ur^2 - \int r^2 \, du \right)\Big|_{\substack{r=0 \\ u=u_c}}^{\substack{r=a \\ u=0}} = \pi \int_0^a r^2 \left(-\frac{du}{dr} \right) dr \qquad (5.23)$$

Substituting for $-du/dr$ from Equation 5.20 gives

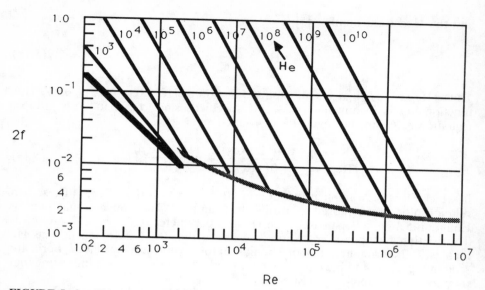

FIGURE 5-3   *Friction factor for the flow of a Bingham plastic through a round tube. (After Hedstrom [5].)*

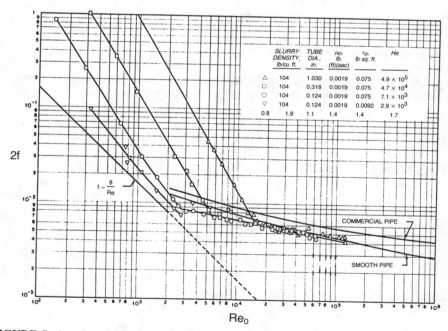

FIGURE 5-4   *Comparison of theoretical and experimental friction factors for the flow of suspensions of thoria through a round tube. (From Thomas [6].)*

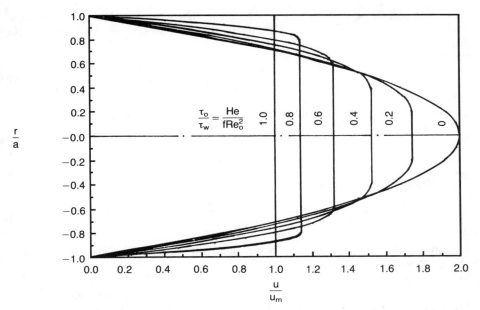

FIGURE 5-5 *Velocity profiles for the laminar flow of a Bingham plastic through a round tube.*

$$V = \pi \int_0^a r^2 \phi\{\tau\}\, dr \tag{5.24}$$

Equation 4.4 is still applicable; hence

$$dr = \frac{a\,d\tau}{\tau_w} \tag{5.25}$$

Substituting for r and dr in 5.24 from 4.4 and 5.25 gives

$$V = \frac{\pi a^2}{\tau_w^3} \int_0^{\tau_w} \tau^2 \phi\{\tau\}\, d\tau \tag{5.26}$$

Differentiating with respect to τ_w and using the Leibniz rule gives

$$\frac{d(V\tau_w^3)}{d\tau_w} = \pi a^3 \tau_w^2\, \phi\{\tau_w\} \tag{5.27}$$

Equation 5.27 in one form or another is known by various names, such as the *Rabinowitsch–Mooney equation* [7, 8]. It can be rearranged as

$$\phi\{\tau_w\} = \frac{1}{\pi a^3 \tau_w^2} \frac{d(V\tau_w^3)}{d\tau_w} = \frac{\tau_w}{\pi a^2} \frac{dV}{d\tau_w} + \frac{3V}{\pi a^3} \tag{5.28}$$

Thus $\phi\{\tau_w\}$ can be determined as a function of τ_w from measured values of V and τ_w (or through Equation 4.3 from measured values of $-d\mathscr{P}/dx$). From Equations 2.6, 5.20, and 5.21,

$$\eta \equiv \frac{\tau_{rx}}{-du/dr} = \frac{\tau}{\phi\{\tau\}} \tag{5.29}$$

Equation 5.29 obviously holds for the particular case of τ_w; that is,

$$\eta = \frac{\tau_w}{\phi\{\tau_w\}} \tag{5.30}$$

Combining Equations 5.27 and 5.30 then gives

$$\eta = \frac{\pi a^3 \tau_w^3}{d(V\tau_w^3)/d\tau_w} = \frac{\pi a^3}{4} \frac{d(\tau_w^4)}{d(V\tau_w^3)} \tag{5.31}$$

Equation 5.31 provides a relationship between the effective viscosity and the shear stress from measured values of V and τ_w. η can be determined from the varying slope of a plot of $V\tau_w^3$ versus τ_w^4. However, a better procedure (see, for example, Churchill [9]) is to construct an equal-area curve through a plot of $(4/\pi a^2)\Delta(V\tau_w^3)/\Delta(\tau_w^4)$ versus τ_w^2, giving $1/\eta$ directly as a function of τ_w^4.

Substituting from Equation 5.29 for $\phi\{\tau\}$ in Equation 5.26 gives

$$V = \frac{\pi a^3}{\tau_w^3} \int_0^{\tau_w} \frac{\tau_w^3}{\eta} d\tau = \frac{\pi a^3}{4\tau_w^3} \int_0^{\tau_w^4} \frac{d(\tau^4)}{\eta} \tag{5.32}$$

Equation 5.32 can be used to calculate the volumetric rate of flow of a general viscous fluid for either a specified τ_w or $-d\mathscr{P}/dx$, using a correlating equation for $\eta\{\tau\}$. Equation 5.32 is analogous to Equation 5.13 for a Bingham plastic and to Equation 4.7 for a Newtonian fluid.

The application of Equations 5.31 and 5.32 is illustrated in problems 9–13.

THE METZNER–REED MODEL

Metzner and Reed [10] (also see Metzner [1]) utilized the Rabinowitsch–Mooney equation to develop a practical representation for pipe flow as follows. Equation 5.28 is first rewritten in terms of $(du/dr)_{r=a}$, u_m, and τ_w for fixed D as

$$-\left(\frac{du}{dr}\right)_{r=a} = \frac{2\tau_w}{D} \frac{du_m}{d\tau_w} + \frac{6u_m}{D} \tag{5.33}$$

and then as

$$\left(-\frac{du}{dr}\right)_{r=a} = \frac{8u_m}{D}\left[\frac{3}{4} + \frac{1}{4}\frac{d\ln\{8u_m/D\}}{d\ln\{\tau_w\}}\right] \tag{5.34}$$

$$= \frac{8u_m}{D}\left(\frac{1 + 3\alpha'}{4\alpha'}\right) \tag{5.35}$$

where

$$\alpha' = \frac{d \ln \{\tau_w\}}{d \ln \{8u_m/D\}} \tag{5.36}$$

For most fluids a plot of $\ln \{\tau_w\}$ versus $\ln \{u_m\}$ is linear over a wide range. In that event Equation 5.36 can be integrated to obtain

$$\tau_w = M' \left(\frac{8u_m}{D}\right)^{\alpha'} \tag{5.37}$$

where M' is a constant of integration. The values of α' and M' can both be obtained from the previously mentioned log-log plot. If the plot is not linear, α' and M' can be determined for any particular $8u_m/D$ from the tangent at that point. A better method, as mentioned earlier for the determination of η from Equation 5.31, is to construct an equal-area curve through a plot of $\Delta \ln \{\tau_w\}/\Delta \ln\{8u_m/D\}$ versus $\ln \{8u_m/D\}$. The curve gives α' as a function of $8u_m/D$. The corresponding values of M' can then be obtained from Equation 5.37. (The inclusion of $8/D$ in the derivative of Equation 5.36 is unessential, but results in the conventional definition of M' as the constant of integration.)

Eliminating $8u_m/D$ between Equations 5.37 and 5.35 gives

$$\tau_w = M' \left(\frac{4\alpha'}{1 + 3\alpha'}\right)^{\alpha'} \left(-\frac{du}{dr}\right)^{\alpha'}_{r=a} \tag{5.38}$$

If α' is constant with a value of unity, Equation 5.37 reduces to the Newtonian relationship (Equation 4.5) for $r = a$. If α' is constant with a value other than unity, the power-law model (Equation 2.10) is obtained with

$$\alpha = \alpha' \tag{5.39}$$

and

$$M = M' \left(\frac{4\alpha'}{1 + 3\alpha'}\right)^{\alpha'} \tag{5.40}$$

According to Metzner and Reed, α' characterizes the degree of non-Newtonian behavior by its deviation above or below unity, and M', rather than M, characterizes the viscosity of the fluid.

The friction factor corresponding to Equation 5.37 is

$$f = \frac{M'}{\varrho u_m^2} \left(\frac{8u_m}{D}\right)^{\alpha'} = \frac{M' 8^{\alpha'}}{\varrho D^{\alpha'} u_m^{2-\alpha'}} \tag{5.41}$$

If Equation 5.41 is forced to conform to Poiseuille's law, the resulting effective Reynolds number is

$$\mathrm{Re}' \equiv \frac{D^{\alpha'} u_m^{2-\alpha'} \varrho}{M' 8^{\alpha'-1}} \tag{5.42}$$

Equation 5.42 suggests that an effective viscosity can be defined as

$$\mu' = M'8^{\alpha'-1} \tag{5.43}$$

It may be inferred that any fluid that does not follow Poiseuille's law in terms of Re′ in the laminar region is thixotropic, rheopectic, or nonwetting. The predicted conformity to Equation 4.12 in terms of Re′ is demonstrated in Figures 5–6 to 5–8. Figure 5–8 is a corrected version from Dodge and Metzner [11]. The fluids and conditions for the data in Figures 5–6 and 5–7 are given in Table 5.1. The critical value of Re for stable laminar flow can be noted in Figure 5–8 to be approximately 2100, just as for Newtonian flow.

The fluids in Table 5.1 include several that might be expected to behave as Bingham plastics. The Bingham-plastic model and the power-law model might appear to be incompatible. However, the velocity field in Figure 5–3 for $\alpha = 1/3$ closely resembles that of Figure 5–5 for $\tau_o/\tau_w = 0$. Probably either model could be used for data correlation and for design calculations.

When written in terms of the pressure drop and the volumetric rate of flow, Equation 5.37 becomes

$$-\frac{d\mathscr{P}}{dx} = \frac{32\,\mu'}{D^{3\alpha'+1}}\left(\frac{4V}{\pi}\right)^{\alpha'} \tag{5.44}$$

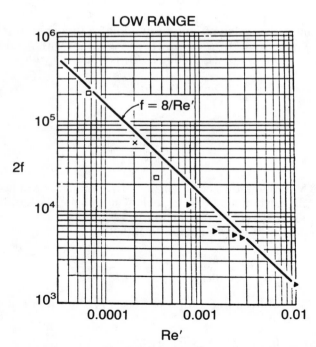

FIGURE 5–6   *Comparison of experimental data for non-Newtonian fluids with correlation of Metzner and Reed [10]. Low range of Re′.*

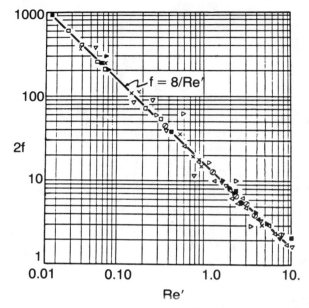

FIGURE 5-7 *Comparison of experimental data for non-Newtonian fluids with correlation of Metzner and Reed [10]. Intermediate range of Re'.*

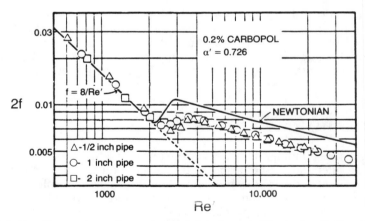

FIGURE 5-8 *Comparison of experimental data for non-Newtonian fluids with correlation of Metzner and Reed [10]. High range of Re'. (From Dodge and Metzner [11].)*

For a Newtonian fluid the pressure gradient is proportional to the rate of flow and inversely proportional to the fourth power of the diameter. However, for a non-Newtonian fluid with small α' the pressure gradient is almost independent of the rate of flow and decreases only linearly with diameter. Design and operational considerations are therefore radically different.

In cases where a distinction can be made between α' and α, and M' and M,

Table 5.1
Rheological constants for fluids of Figures 5–5 and 5–6. *(From Metzner and Reed [10].)*

Symbol used in	Nominal pipe size, in.	Composition of fluid	Rheological constants	
			α'	μ'
+	1	23.3% Illinois yellow clay in water	0.229	0.183
⊕	⅞ and 1½	0.67% Carboxy-methyl-cellulose (CMC) in water	0.716	0.121
⊖	⅞ and 1½	1.5% CMC in water	0.554	0.920
⊘	⅞ and 1½	3.0% CMC in water	0.566	2.80
⊗	⅞, 1½ and 2	33% Lime water	0.171	0.983
◁	⅞ and 1½	10% Napalm in kerosine	0.520	1.18
▼	8, 10 and 12	4% Paper pulp in water	0.575	6.13
△	¾ and 1½	54.3% Cement rock in water	0.153	0.331
▲	4	18.6% Solids, Mississippi clay in water	0.022	0.105
●	¾ and 1¼	14.3% Clay in water	0.350	0.0344
▷	¾ and 1¼	21.2% Clay in water	0.335	0.0855
×	¾ and 1¼	25.0% Clay in water	0.185	0.204
▽	¾ and 1¼	31.9% Clay in water	0.251	0.414
□	¾ and 1¼	36.8% Clay in water	0.176	1.07
■	¾ and 1¼	40.4% Clay in water	0.132	2.30
▶	⅛, ¼, ½ and 2	23% Lime in water	0.178	1.04

the use of the primed pair is recommended for flow in round pipes since they are defined operationally for that case. For other geometries and particularly for translation from one geometry to another, the use of α and M may be preferable since their definition is independent of apparatus. Fortunately α' and M' are effectively constant for most materials, and this distinction does not often need to be made.

OTHER FORCED FLOWS

Solutions corresponding to the foregoing can readily be derived for flow between parallel plates (see problems 20–24) and for an annulus between two concentric circular tubes (see problems 25–29). Analytical solutions for two-phase flows of one or two non-Newtonian fluids are also possible, although they become quite detailed (see problems 30–35).

SUMMARY

In this chapter expressions have been derived for the velocity distribution and pressure drop in round tubes for the fully developed flow of power-law, Bingham-plastic, and general viscous fluids. The development leading to the solution for the latter case also provides a rigorous method for the determination of the consistency factor and the power-law index from simple measurements of the pressure drop as a function of the volumetric rate of flow.

Similar results are readily attainable for flow between parallel plates and in circular annuli, but are left as exercises.

PROBLEMS

1. What is the Poiseuille number corresponding to Equation 5.7?
2. What is the effective viscosity corresponding to Equation 5.8?
3. Rationalize the effective viscosity of problem 2 with that of Equation 5.1.
4. Compare the solution for a Bingham plastic with that for concurrent annular flow with $\xi \to 0$.
5. Prepare a plot of $f\mathrm{Re}_o$ versus $\mathrm{He}/f\mathrm{Re}_o^2$ according to Equation 5.16.
6. Prepare a plot of $f\mathrm{Re}_o$ versus $\mathrm{He}/\mathrm{Re}_o$ according to Equation 5.19.
7. Replot the data in Figure 5-4 in the form of problem 5.
8. Replot the data in Figure 5-4 in the form of problem 6.
9. G. E. Alves [12] reported the following data for laminar flow of a slurry of density 72.5 $\mathrm{lb/ft}^3$ through a 10-ft-long horizontal section of 0.255-in.-I.D. stainless steel pipe:

Volumetric Rate of Flow *(ft³/s × 10⁵)*	$-\Delta\mathscr{P}$ *(lb$_f$/in.²)*
0	3.9
18.7	5.2
37.4	6.11
93.5	7.15
224	9.75
374	12.75
430	14.3
486	18.2
655	33.5
935	59.2

a. Assuming that the fluid behaves as a Bingham plastic, determine μ_o and τ_o.
b. Determine the effective viscosity as a function of τ using Equation 5.31. Plot $\ln\{\eta\}$ versus $\ln\{\tau\}$ according to Equation 2.13 and interpret the results.
c. Compare the data with Figure 5-3.
d. Compare the data with the plot of problem 5.
e. Compare the data with the plot of problem 6.
f. Compute the rate of flow of the same slurry through 20 ft of a horizontal 0.5-in.-I.D. pipe for a pressure drop of 30 psi, using the results of parts a and b. Compare and explain.
g. Compute the pressure drop for the flow of 2 gal/min of the same slurry through 20 ft of horizontal 0.5-in.-I.D. pipe using the results of parts a and b, respectively. Compare and explain.
10. Plot $\ln\{8u_m/D\}$ versus $\ln\{\tau_w\}$ for the data of problem 9. Determine α', M', and μ' as functions of $8u_m/D$. Compare the data with Equation 5.44 and Figures 5-6 to 5-8.
11. Calculate the rate of flow at 523 K of the pseudoplastic fluid whose properties are given below if 20 MPa is imposed over a 100-mm-long segment of 3.175-mm tube.

$$\eta = \frac{Me^{E/RT}}{(1 + (\tau/\tau^*)^2)}$$

$$M = 6.3 \times 10^{-14} \text{ Pa} \cdot \text{s} \tag{5.45}$$

$$\frac{E}{R} = 19,500 \text{ K}$$

$$\tau^* = 160 \text{ MPa}$$

$$\varrho = 1.09 \text{ Mg/m}^3 \tag{5.46}$$

12. The data below are given by Skelland [13], p. 37, for the pressure drop as a function of the rate of flow for a fluid with a density of 70 lb/ft^3.

D(in.)	L(in.)	w(lb/hr)	$-\Delta\mathscr{P}(lb_f/in.^2)$
0.05	5	0.894	445
0.05	5	0.129	167
0.10	10	3.435	333
0.10	10	0.642	125
0.15	10	8.21	185
0.15	10	1.16	55.5

   a. Using Equation 5.31 determine η versus τ.
   b. Correlate $\eta = \phi\{\tau\}$ using the Churchill–Churchill model (Equation 2.8 or 2.9).
   c. Determine α', M', and μ' as functions of $8u_m/D$.
   d. Compare the data with Equation 5.44 and Figures 5.6–5.8.

13. Determine α', M', and μ' as functions of $8u_m/D$ for one of the sets of data in Figure 5–4. Test the results against Equation 5.44.
14. Explain how a fluid that follows the Bingham-plastic model can also follow Equation 5.37.
15. Show that Equation 5.31 gives the viscosity for a Newtonian fluid.
16. Prepare a dimensionless plot of the pressure gradient versus the volumetric rate of flow in a smooth pipe for the fluid of problem 1 of Chapter 2. Do not include the pressure gradient and the rate of flow in the same dimensionless group. [*Hint*: Use a reference viscosity such as η_o.]
17. Calculate the mean velocity that will produce a shear stress of 172.4 Pa on the wall of a 76-mm tube with the fluid of problem 1 of Chapter 2.
18. Calculate the pressure gradient necessary to pump the fluid of problem 1 of Chapter 2 through a 50-mm pipe at a mean velocity of 30 mm/s.
19. Develop expressions analogous to Equations 5.31 and 5.32 suitable for experimental measurements of pressure drop for different tube diameters at a fixed volumetric rate of flow.
20. Derive a solution for flow between parallel plates analogous to Equations 5.4 and 5.7.
21. Determine the effective Reynolds number and friction factor for the solutions of problem 20.
22. Derive a solution for flow between parallel plates analogous to Equations 5.14–5.19.

23. Derive expressions for flow between parallel plates equivalent to Equations 5.31 and 5.32.
24. Derive expressions for flow between parallel plates analogous to Equations 5.33–5.44.
25. Derive a solution for flow in an annulus analogous to Equations 5.4 and 5.7.
26. Determine the effective Reynolds number and friction factor for the solution of problem 25.
27. Derive a solution for flow in an annulus analogous to Equations 5.14–5.18.
28. Derive an expression for an annulus analogous to Equation 5.31.
29. Derive expressions for an annulus analogous to Equations 5.33–5.44. (See Lescarboura et al. [14] and Hanks and Larsen [15]).
30. Derive a solution for flow of a pseudoplastic fluid concurrent with a Newtonian fluid between parallel plates.
31. Derive a solution for flow of a Bingham plastic concurrent with a Newtonian fluid between parallel plates.
32. Derive a solution for the annular flow of a pseudoplastic fluid inside a Newtonian fluid.
33. Derive a solution for the annular flow of a pseudoplastic fluid outside a Newtonian fluid.
34. Repeat problem 32 for a Bingham plastic.
35. Repeat problem 33 for a Bingham plastic.
36. What is the minimum diameter for drainage of a Bingham plastic from a vertical tube?
37. Derive an expression for the friction factor and velocity distribution for an Ellis fluid flowing between parallel plates (see problem 4 of Chapter 2).
38. Repeat problem 37 for flow in a round tube.
39. The power-law model provides a reasonable prediction for the pressure drop in a pipe for real fluids even through it predicts $\eta\{0\} \to 0$ and $\eta\{\infty\} \to \infty$. Explain.
40. Eliminating $8u_m/d$ between Equations 5.35 and 5.37 gives

$$\tau_w = M' \left(\frac{4\alpha'}{1 + 3\alpha'}\right)^{\alpha'} \left(\frac{-du}{dr}\right)^{\alpha'}_{r=a} \tag{5.47}$$

which differs from Equation 2.10. Explain the significance of this difference.
41. Repeat the derivation of Equations 5.33–5.37 in terms of V rather than u_m.
42. Determine $\phi\{\tau\}$ of Equation 5.20 for

    a. a Newtonian fluid
    b. a power-law fluid
    c. a Bingham plastic

43. Determine $V\{\tau_w\}$ according to Equation 5.32 for the fluids of problem 42. Check your results with the solutions for $u_m\{\tau_w\}$ given earlier.
44. Determine f for the fluid of problem 1 of Chapter 2 as a function of Re and compare with the solution for the power-law regime only.
45. Determine f as a function of Re for a fluid whose behavior can be described approximately by three discrete regimes:

a. $\eta = \eta_0$ for $\tau \leq \tau_1$
b. $\eta = \eta_0(\tau_1/\tau)^\alpha$ for $\tau_1 \leq \tau \leq \tau_2$
c. $\eta = \eta_\infty = \eta_0(\tau_1/\tau_2)^\alpha$ for $\tau > \tau_2$

Compare the results with the solution for regime (b) only.

46. Determine $\eta\{\tau\}$ and correlate the data of Alves et al. [16] for

   a. 4.46% napalm (their Figures 9 and 13)
   b. 24.8% cellulose acetate (their Figures 10, 11, and 14)
   c. 23% lime slurry (their Figures 12 and 15)

47. Determine the total kinetic energy of a power-law fluid in a round tube.
48. Derive the values tabulated in problem 24, Chapter 2, for $D = 0.125$ in. from the actual data below from Thomas (private communication).

$\dfrac{8V}{g_c D}$	$-\dfrac{D\,\Delta\mathcal{P}}{4L}$
$\dfrac{lb_f \cdot s}{lb \cdot ft}$	$\dfrac{lb_f}{ft^2}$
74.5	1.478
29.9	1.113
15.9	0.931
6.98	0.748
49.1	1.293
372	3.31
290.5	2.75
207.0	2.295
123.7	1.81
685	8.94
540	4.50
453	3.75
606	6.38
690	8.65
644	7.74
568	5.49
590	6.39
377	3.12
500	4.03

REFERENCES

1. A. B. Metzner, "Non-Newtonian Technology: Fluid Mechanics, Mixing and Heat Transfer," p. 77 in *Advances in Chemical Engineering*, Vol. 5, T. B. Drew and J. W. Hooper, Jr., Eds., Academic Press, New York (1956).
2. M. M. Denn, *Process Fluid Mechanics*, Prentice-Hall, Englewood Cliffs, NJ (1980).
3. E. Buckingham, "On Plastic Flow through Capillary Tubes," *Proc. ASTM*, *21* (1921) 1154, 1157.
4. Markus Reiner, *Deformation and Flow–An Elementary Introduction to Theoretical Rheology*, Lewis, London (1949).

5. B. O. A. Hedstrom, "Flow of Plastics Materials in Pipes," *Ind. Eng. Chem., 44* (1952) 651.
6. D. G. Thomas, "Heat and Momentum Transport Characteristics of Non-Newtonian Aqueous Thorium Oxide Suspensions," *AIChE J., 6* (1960) 631.
7. B. Rabinowitsch, "Über die Viskosität and Elastizität von Solen," *Z. Phys. Chem., A145* (1929) 1.
8. M. Mooney, "Explicit Formulas for Slip and Fluidity," *J. Rheol., 2* (1931) 210.
9. S. W. Churchill, *The Interpretation and Use of Rate Data–The Rate Process Concept*, rev. printing, Hemisphere, Washington, D.C. (1979).
10. A. B. Metzner and J. C. Reed, "Flow of Non-Newtonian Fluids—Correlation of Laminar, Transition and Turbulent-Flow Regions," *AIChE J., 1* (1955) 434.
11. D. W. Dodge and A. B. Metzner, "Turbulent Flow of Non-Newtonian Systems," *AIChE J., 5* (1959) 189.
12. G. E. Alves, "Non-Newtonian Flow," Chap. 7 in *Fluid and Particle Mechanics*, C. E. Lapple, Ed., University of Delaware, Newark (1951), p. 115.
13. A. H. P. Skelland, *Non-Newtonian Flow and Heat Transfer*, John Wiley, New York (1967).
14. J. A. Lescarboura, F. J. Eichstadt, and G. W. Swift, "A Generalized Differentiation Method for Interpreting Rheological Data of Time-Independent Fluids," *AIChE J., 13* (1967) 169.
15. R. W. Hanks and K. M. Larsen, "The Flow of Power-Law Non-Newtonian Fluids in Concentric Annuli," *Ind. Eng. Chem. Fundam., 18* (1979) 33.
16. G. E. Alves, D. F. Boucher, and R. L. Pigford, "Pipeline Design for Non-Newtonian Flow," *Chem. Eng. Progr., 48* (1952) 385.

Chapter 6

Thin Films and Other Open, Gravitational Flows

The fully developed flow of a liquid down an inclined or vertical plane due to gravity produces a thin film with a one-dimensional velocity field insofar as rippling or turbulence do not occur. Such flows represent the limiting case of a river or spillway of uniform depth as the ratio of the width to the depth increases and the effect of the sidewalls thereby becomes negligible. Film flows are also widely used to accomplish continuous heat or mass transfer. The velocity distribution in a film on the inside or outside of a vertical, cylindrical surface may also closely approach one-dimensionality.

When the drag of the gas phase on the liquid surface and rippling are negligible, the flow is equivalent to a confined forced flow, but with the surface of zero shear due to symmetry replaced by the gas–liquid interface of negligible shear. Flow in an inclined, half-full, round tube similarly approaches one-dimensionality.

In view of this analogy between forced and free flows, solutions for gravitational flows can be adapted from the corresponding ones for forced flow, as indicated in Chapter 1, in which it was noted that for the flow of a liquid with a free surface

$$\frac{dp}{dy} \cong 0 \qquad\qquad (1.20)$$

and

$$\frac{d\mathscr{P}}{dx} = g\varrho\frac{dh}{dx} = -g\varrho\sin\{\theta\} \qquad\qquad (1.21)$$

Hence it is only necessary to replace $-d\mathscr{P}/dx$ with $g\varrho\sin\{\theta\}$. Rederivation of some of the solutions in order to examine the effect of the idealizations is, however, instructive.

This chapter is limited to the regime of laminar, gravitational flows. Developing gravitational flows are examined in Book III [1] and turbulent gravitational flows, which are of greater practical importance than laminar ones, are examined in a companion volume [2].

91

FIGURE 6–1   *Control volume for force-momentum balance for flow down an inclined plane: (A) control volume; (B) force-momentum balance on control volume.*

NEWTONIAN FLOW DOWN AN INCLINED PLANE

Force-Momentum Balance

The flow of a liquid film of depth d down an inclined plane is illustrated in Figure 6–1. A force-momentum balance on the shaded element of depth $d - y$, breadth H, and length Δx can be written as

$$g\varrho \sin\{\theta\}\, H(d - y)\, \Delta x + j_{yx} H \Delta x - \tau_a H \Delta x = 0 \qquad (6.1)$$

which simplifies to

$$j_{yx} = -g\varrho\,(d - y)\sin\{\theta\} + \tau_a \qquad (6.1A)$$

where τ_a is the shear stress of air on the free surface of the liquid (Pa or kg/m · s²). If τ_a is assumed to be negligible, Equation 6.1A reduces to

$$j_{yx} = -g\varrho\,(d - y)\sin\{\theta\} \qquad (6.2)$$

Derivation

Substituting for j_{yx} from Equation 2.1 in 6.2 gives

$$\mu\frac{du}{dy} = g\varrho(d - y)\sin\{\theta\} \qquad (6.3)$$

Integrating from $u = 0$ at $y = 0$ gives

$$u = \frac{g\varrho \sin\{\theta\}}{\mu}\left(dy - \frac{y^2}{2}\right) \qquad (6.4)$$

The velocity of the free surface is

$$u_d = (u)_{y=d} = \frac{g\varrho\, d^2 \sin\{\theta\}}{2\mu} \qquad (6.5)$$

and the mean velocity is

$$u_m = \frac{1}{d} \int_0^d u \, dy = \frac{g\varrho \, d^2 \sin\{\theta\}}{3\mu} \tag{6.6}$$

Hence

$$\frac{u}{u_m} = 3\frac{y}{d}\left(1 - \frac{y}{2d}\right) \tag{6.7}$$

$$\frac{u_d}{u_m} = \frac{3}{2} \tag{6.8}$$

and

$$\frac{u}{u_d} = 2\frac{y}{d}\left(1 - \frac{y}{2d}\right) \tag{6.9}$$

The foregoing solution is usually attributed to Nusselt [3], who in 1916 carried out the derivation in connection with film-type condensation. However, these results are a special case of the series solution developed in 1910 by Hopf [4] for two-dimensional flow in an open, inclined, rectangular channel (see [1]).

Adaption from Solution for Forced Flow

Substitution of d for b and, when it occurs, $dg\varrho \sin\{\theta\}$ for τ_w in Equations 3.8–3.11 generates Equations 6.4–6.8 as anticipated.

Reexpansion of Solution in Terms of the Volumetric Rate of Flow

In practical applications the volumetric rate of flow V or the volumetric rate of flow per unit breadth $\tilde{v}$ is usually specified rather than the mean velocity or depth, which become dependent variables. The foregoing results then take the following form. From continuity

$$V = \tilde{v} H = d H u_m \tag{6.10}$$

Then from Equation 6.6,

$$\tilde{v} = \frac{g\varrho \, d^3 \sin\{\theta\}}{3\mu} \tag{6.11}$$

or

$$d = \left(\frac{3\mu\tilde{v}}{g\varrho \sin\{\theta\}}\right)^{1/3} \tag{6.11A}$$

Substituting for d from Equation 6.6 in 6.11 and rearranging give

$$u_m = \left(\frac{g\varrho \bar{v}^2 \sin\{\theta\}}{3\mu}\right)^{1/3} \tag{6.12}$$

This dependence of d on $\bar{v}^{1/3}$ and of u_m on $\bar{v}^{2/3}$ might not have been anticipated from purely qualitative considerations.

Comparison of Different Forms of Solution

Equations 6.6 and 6.11A indicate that the depth d is proportional to $u_m^{1/2}$ and to $\bar{v}^{1/3}$, respectively. These different powers, depending on the arbitrary choice of the independent variable, also occur in terms of the standard dimensionless groups. For flow down an inclined plane

$$D = \lim_{H \to \infty} \frac{4dH}{2d + H} = 4d \tag{6.13}$$

Hence

$$\text{Re} \equiv \frac{Du_m\varrho}{\mu} = \frac{4d\,u_m}{v} \tag{6.14}$$

Using prior expressions, one can express the Reynolds number in terms of d, u_m, or $\bar{v}$ as follows:

$$\text{Re} = \frac{4g\,d^3 \sin\{\theta\}}{3v^2} \tag{6.15}$$

$$= 4\left(\frac{3u_m^3}{vg\sin\{\theta\}}\right)^{1/2} \tag{6.16}$$

and

$$= \frac{4\bar{v}}{v} \tag{6.17}$$

Thus, for example, Re is proportional to $g\sin\{\theta\}$ for a specified d, is inversely proportional to $(g\sin\{\theta\})^{1/2}$ for a specified u_m, and is independent of $g\sin\{\theta\}$ for a specified $\bar{v}$.

From Equation 6.5,

$$\tau_w = \mu\left(\frac{du}{dy}\right)_{y=0} = g\varrho d\sin\{\theta\} \tag{6.18}$$

and

$$f \equiv \frac{\tau_w}{\varrho u_m^2} = \frac{g\,d\sin\{\theta\}}{u_m^2} \tag{6.19}$$

Alternatively, in terms of d, u_m, or $\bar{v}$,

$$f = \frac{9v^2}{gd^3 \sin\{\theta\}} \tag{6.20}$$

$$= \left(\frac{3g\,v\sin\{\theta\}}{u_m^3}\right)^{1/2} \tag{6.21}$$

and

$$= \frac{3v}{\bar{v}} \tag{6.22}$$

Again, different power dependencies on μ and $g\sin\{\theta\}$ are to be observed, and, again the simplest result is in terms of $\bar{v}$.

The Froude number defined by Equation 1.33 can correspondingly be expressed in terms of d, u_m, and $\bar{v}$ as

$$Fr = \frac{g\,d^3\sin^2\{\theta\}}{36v^2} \tag{6.23}$$

$$= \left(\frac{u_m^3\sin\{\theta\}}{48g\,v}\right)^{1/2} \tag{6.24}$$

and

$$= \frac{\bar{v}\sin\{\theta\}}{12v} \tag{6.25}$$

Eliminating $\bar{v}$ between Equations 6.17 and 6.22 gives

$$f = \frac{12}{Re} \tag{6.26}$$

which is equivalent to

$$Po = 12 \tag{6.27}$$

Comparison of the several expressions for Re and Fr reveals that for the flow of a laminar film

$$Fr = \frac{\sin\{\theta\}\,Re}{48} \tag{6.28}$$

Thus Fr and Re are not independent for laminar flow. Equation 6.26 can therefore alternatively be expressed as

$$f = \frac{\sin\{\theta\}}{4\,Fr} \tag{6.29}$$

Caution is also suggested in the interpretation of the preceding dimensionless interrelationships, since Re and Fr, as well as f, may be defined differently elsewhere.[1]

The dimensionless group

$$\mathscr{D} = d\left(\frac{g\sin\{\theta\}}{\nu^2}\right)^{1/3} \tag{6.30}$$

is known as the *Nusselt film thickness*. Then, from Equations 6.30 and 6.15,

$$\mathscr{D} = \left(\frac{3\mathrm{Re}}{4}\right)^{1/3} \tag{6.31}$$

from Equations 6.31 and 6.28,

$$\mathscr{D} = \left(\frac{36\,\mathrm{Fr}}{\sin\{\theta\}}\right)^{1/3} \tag{6.32}$$

and from Equations 6.31 and 6.26 or 6.32 and 6.29,

$$\mathscr{D} = \left(\frac{9}{f}\right)^{1/3} \tag{6.33}$$

Expressions such as Equations 6.6–6.12 which are written in terms of the primitive variables, are thereby more reliable and useful working relationships than Equations 6.27–6.33. However, values of Re, f, and Fr provide useful criteria for the range of applicability of the solutions in this chapter. Such criteria are examined in subsequent sections.

The moral of this extensive comparison of different forms of such a simple solution is that one should be very meticulous in specifying which variable or group of variables is maintained constant when stating the power-dependence on μ, ϱ, g, or $\sin\{\theta\}$.

Experimental Confirmation

Fulford [5] compiled some 1013 experimental values of the film thickness. Those values in the regime of flow near the point of transition to turbulent motion are plotted as $\mathscr{D}$ versus Re/4 in Figure 6–2. Reasonable agreement with Equation 6.31 (labelled line-1 in Fulford's plot) may be noted for Re up to about 1400, with a wide scatter of values above that expression for Re $>$ 1400. The other numbered lines and curves in Figure 6–2 represent theoretical and empirical expressions, which are considered subsequently or not at all.

---

[1] For example, Fulford [5] defines the Reynolds number as $du_m\varrho/\mu$, the Froude number as $u_m/(gd)^{1/2}$ and the Weber number as $u_m(\varrho d/\sigma)^{1/2}$. Also, Dukler [6] called the slightly more generalized group $d[g\sin\{\theta\}\,(1-\varrho_2/\varrho)/\nu^2]^{1/3}$, where ϱ_2 is the density of the adjacent fluid, the Nusselt film thickness.

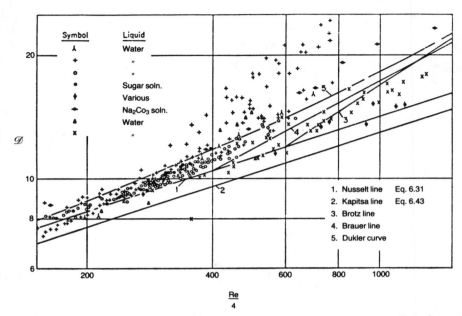

FIGURE 6–2 *Comparison of theoretical and experimental depths of flowing films on inclined plates. (From Fulford [5].)*

Limits of Applicability of Solutions

Film flow may be smooth and laminar, wavy but still laminar, or turbulent. The waviness may be due to gravity or surface tension or both. The onset of gravity waves is characterized by the Froude number; the onset of capillary waves by the Weber number. The onset of turbulence is characterized by the Reynolds number. This section is indebted to the thorough review of the subject by Fulford [5].

Kapitsa [7] used stability theory to derive the following expression for the *critical Reynolds number* Re^* at the onset of waviness in a flowing film:

$$Re^* = 2.44 \left(\frac{\sigma^3 \varrho}{g \mu^4 \sin\{\theta\}} \right)^{1/11} \qquad (6.34)$$

The dimensionless group in brackets on the right side of Equation 6.34 can be recognized as equal to $Fr\, Re^4/We^3 \sin\{\theta\}$ or, utilizing Equation 6.28, as $Re^5/48\, We^3$, where for a film the Weber number defined by Equation 1.34 is $4\, \varrho u_m^2 d/\sigma$. Hence, insofar as the Hopf-Nusselt solution is valid, Equation 6.34 can be rewritten as

$$Re^* = \frac{2.69}{We^{1/2}} \qquad (6.35)$$

Brauer [8] correlated experimental data for the critical Reynolds number in water and aqueous solutions of ethylene glycol with the expression

$$\text{Re}^* = 1.224\left(\frac{\sigma\varrho}{g\mu^4\sin\{\theta\}}\right)^{1/10} \tag{6.36}$$

which, by the foregoing procedure, can be rewritten as

$$\text{Re}^* = \frac{0.691}{\text{We}^{3/5}} \tag{6.37}$$

Grimley [9] correlated experimental data for vertical columns with the expression

$$\text{Re}^* = 0.86\left(\frac{\sigma^3\varrho}{g\mu^4}\right)^{1/8} \tag{6.38}$$

which can be rewritten for the Hopf-Nusselt solution as

$$\text{Re}^* = \frac{0.184}{\text{We}} \tag{6.39}$$

For water $\sigma^3\varrho/g\mu^4 \simeq 4.1 \times 10^{10}$. For this value Equations 6.34, 6.36, and 6.38 give 22.5, 14.1, and 18.2, respectively, for Re* in a vertical column. On the other hand, Jackson [10] asserts that a Froude number (as defined by Equation 6.23) of 0.25 is the effective criterion for wave motion; from Equation 6.28 it then follows that

$$\text{Re}^* = \frac{12}{\sin\{\theta\}} \tag{6.40}$$

This result implies that surface tension is not a factor. None of the above expressions 6.34, 6.36, 6.38, or 6.40 agree closely with the experimental data for water, which are plotted vs. the angle of inclination from the vertical in Figure 6–3, but Equation 6.40 appears to be the best overall.

The first-order stability analysis of Kapitsa [7] indicates that wave motion does not change the mean thickness of the film. Levich [11] extended this analysis and determined a time-mean film thickness

$$\bar{d} = \left(\frac{2.4\nu\bar{v}}{g\sin\{\theta\}}\right)^{1/3} \tag{6.41}$$

which is 0.93 times that given by Equation 6.11A for a smooth film. The corresponding time-mean friction factor is

$$\bar{f} = \frac{48}{5\text{Re}} \tag{6.42}$$

and the corresponding time-mean Nusselt film thickness is

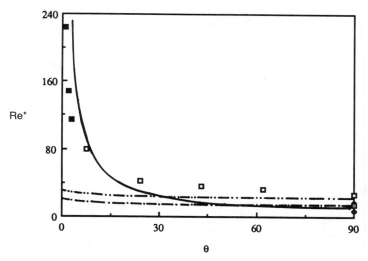

FIGURE 6–3 *Comparison of experimental Reynolds numbers for the onset of wave motion in a film flowing down an inclined plane with theoretical predictions.* □ = *Fulford (1962);* ■ = *Binnie (1959);* ▲ = *Binnie (1952);* □ = *Brauer (1956);* ◇ = *Portalski (1960);* ——— = *Jackson, Equation 6.40;* —··— = *Kapitza, Equation 6.34;* —·— = *Brauer, Equation 6.36. (After Fulford [5].)*

$$\bar{\mathcal{D}} = \left(\frac{3\mathrm{Re}}{5}\right)^{1/3} \tag{6.43}$$

Equation 6.43 is included in Figure 6–2. The data of Feind [12], as represented by the + symbol, are seen to shift downward from Equation 6.31 toward Equation 6.43 as Re decreases below 800. Jackson [10] observed a reduction in film thickness and an increase in u_d/u_m from 1.5 to 2.2 as Re increased from 12 to 108, then a decrease to about 1.8 as Re increased to 2000. Similar behavior has been observed by others (see Fulford [5]). It may be concluded that wave motion causes small deviations from the theoretical solution for a smooth laminar film in the range $20 < \mathrm{Re} < 800$.

Fulford [5] tabulated the critical Reynolds number determined by 26 independent observers for the onset of turbulent motion. These values range from 576 to 2360, with a predominance of values at about 1600. The point of transition was also seen in Figure 6–2 to cover a wide range. Dukler [6] suggested that transition in a thin film might be expected to take place more gradually than in a channel since the laminar sublayer may occupy a large fraction of the entire film thickness even in fully developed turbulent flow.

Fulford [5] has noted that the miniscus on the sidewalls of a channel may provide a significant perturbation in the total flow. If the fluid wets the walls, the velocity near the wall is higher than midway between the walls. This higher velocity coupled with the greater thickness increases the total flow over that predicted for a smooth film.

The series solution of Hopf [4] for flow in a rectangular channel can be reduced to the following approximate expression for large H/d:

$$u_m = \frac{gd^2 \sin\{\theta\}}{3\nu}\left(1 + \frac{2d}{H}\right)\left(1 - 0.63\frac{d}{H}\right) \tag{6.44}$$

As $H/d \to \infty$, Equation 6.44 reduces to Equation 6.6. The terms in parentheses represent a first-order correction for edge effects. For most films these corrections are negligible (see problem 45).

Effect of the Adjoining Phase

Fallah et al. [13] have shown that the effect of the adjoining phase in the limiting case of negligible interfacial drag can be represented by replacing $g\varrho$ by $g(\varrho - \varrho_2)$, where ϱ_2 is the density of the adjoining phase. Detailed reviews of work on the effect of an adjoining phase in the case of significant interfacial drag are provided by Fulford [5], Blass [14], and Taitel et al. [15]. Concurrent flow, countercurrent flow, and a stationary fluid are all considered. Derivation of some of these solutions is suggested in problems 15 and 31–36.

OTHER FREE FLOWS

The foregoing results for a vertical plate are applicable as a first approximation for flow down the inside or outside wall of a vertical round tube insofar as the thickness of the film is small relative to the radius of the tube (see problems 20 and 23–29).

The flow of a liquid film down the inner wall of a column with counter-current or concurrent flow of a gas is widely used for simultaneous heat and mass transfers. The behavior of such *wetted-wall columns*, as sketched in Figure 6–4, is examined in problems 15–30 and 34–35. One special feature of wetted-wall columns is *flooding*, which occurs when the mean velocity of the liquid becomes zero (see problem 36).

The slightest inclination of such a column results in a progressively uneven layer of liquid around the tube and hence deviations from the solutions proposed in problems 15–30 and 34–36 (see, for example, Wilson [16]).

The velocity field in inclined channels is indeed inherently two-dimensional except for a round tube that is just half-full of liquid. The concept of an equivalent diameter can be used to estimate the volumetric rate of flow in other open channels, but the accuracy of this method is questionable since it gives a 50% overestimate for the limiting case of an inclined flat plate (see problem 37).

Non-Newtonian flows on inclined plates and the surfaces of vertical round tubes represent extensions of the foregoing results and those of Chapter 5 (see problems 38–44). The experimental data of Thérien et al. [17] for 12 polymeric solutions show good agreement with the following solution for a power-law fluid (see problem 39):

$$d = \left[\left(\frac{M}{\varrho g \sin\{\theta\}}\right)^{1/\alpha} \frac{\tilde{v}(2\alpha + 1)}{\alpha}\right]^{\alpha/(2\alpha+1)} \tag{6.45}$$

This agreement is illustrated in Figure 6–5 for aqueous solutions of Methocel. The rheological properties of these solutions are given in Table 6.1.

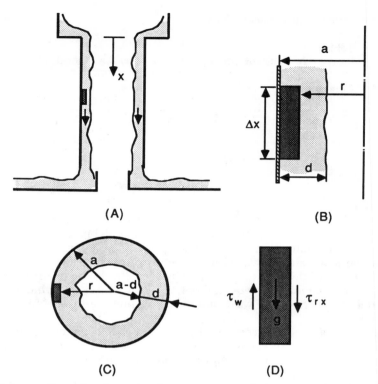

FIGURE 6–4 *Control volume for force-momentum balance in a liquid stream in a wetted-wall column: (A) wetted-wall column; (B) horizontal view of control volume; (C) vertical view of control volume; (D) force-momentum balance on control volume.*

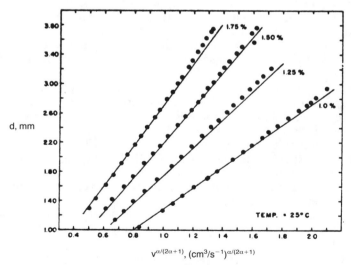

$$v^{\alpha/(2\alpha+1)}, (cm^3/s^{-1})^{\alpha/(2\alpha+1)}$$

FIGURE 6–5 *Comparison of measured and predicted thickness of aqueous solutions of Methocel flowing down an inclined plane at 25°C. (from Thérien et al. [17]). The lines represent Equation 6.45 with the constants in Table 6.1.*

Table 6.1
Power-Law Constants for Aqueous Solutions of Methocel [17]

Concentration %	α	$M(Pa \cdot s^{\alpha} \times 10^{-3})$
1.00	0.777–0.854	458–589
1.25	0.763–0.828	1051–1252
1.50	0.262–0.881	1627–2169
1.75	0.801–0.856	2976–3248

SUMMARY

The previous solution for laminar flow between parallel plates was shown to be applicable for a laminar film on an inclined plane. This adapted solution may be expressed in many alternative forms, some of which were illustrated here. The range of validity of this solution is limited by edge effects and by the formation of gravitational or capillary waves, as well as by the onset of turbulent motion. Fortunately, the effects of wave motion and finite width are minor in most applications.

Thin films are also encountered in wetted-wall columns. The effects of curvature are ordinarily negligible in applications, but the concurrent or countercurrent flow of a gas stream may have a major effect.

PROBLEMS

1. The volumetric rate of flow down an inclined plane is to be doubled. Assuming the flow remains laminar, what will be the corresponding change in depth, surface velocity, shear stress at the wall, Reynolds number, Froude number, Weber number, friction factor, and $\mathscr{D}$?
2. Derive expressions for the momentum and kinetic energy in terms of the mean velocity for fully developed, laminar Newtonian flow down an inclined plate.
3. A turbid oil with a viscosity of 1 Pa $\cdot$ s and a density of 0.86 Mg/m^2 is flowing down a spillway with an inclination of $2°$. The surface velocity is observed to be 0.3 m/s. Estimate Re, the depth, and the mass rate of flow per unit width.
4. Calculate and plot d and u_m for water versus θ for Re = 10, 100, and 1000.
5. Is a plot of experimental data in the form of $\mathscr{D}$ versus Re equivalent to a plot of f versus Re? Explain.
6. Is a plot of experimental data in the form of $\mathscr{D}$ versus Fr equivalent to a plot of f versus Re? Explain.
7. Prepare a plot of d and $\bar{v}$ versus the angle of inclination at which water will flow down a plate without rippling. Indicate your chosen criterion for the onset of rippling.
8. Repeat problem 7 for the onset of turbulent motion.
9. Determine the relationships between the Reynolds, Froude, and Weber numbers used here and those used by Fulford [5]. What relative advantages, if any, do these two sets of definitions have?

10. Explain the different power dependencies of the critical Reynolds number for wave motion on the Weber number as indicated by Equations 6.35, 6.37, and 6.39.

11. Explain physically how the surface tension causes wave motion on an inclined plate.

12. Explain physically how the gravitational force causes wave motion on an inclined plate.

13. Derive a relationship for the depth of liquid at which rippling begins as a function of physical properties and $g \sin\{\theta\}$ only, using Equations (a) 6.34, (b) 6.35, (c) 6.38, and (d) 6.40. Compare with the plot of problem 7.

14. Repeat problem 13 for the onset of turbulent motion.

15. Determine a minimal set of dimensionless groups that defines the thickness of a film inside a wetted-wall column with

   a. forced countercurrent flow of gas
   b. forced concurrent flow of gas
   c. induced flow of gas

16. Reduce the solution of problem 15 for the limiting case of negligible viscosity for the gas.

17. Sketch the velocity profile in the liquid and gas phases for the conditions of problems 15 and 16.

18. Derive a *complete* solution for a liquid film inside a wetted-wall column, assuming the viscosity of the gas is negligible, but taking curvature into account.

19. Reduce the solution of problem 18 for the limiting case of $d \to a$.

20. Expand the solution of problem 18 in a power series in d/a and show that this series converges to the solution for a flat plate as $d/a \to 0$.

21. Repeat problem 18 for a film outside the column.

22. Repeat problem 19 for a film outside the column.

23. Repeat problem 20 for a film outside the column.

24. Using the result of problem 18 or 20, determine the values of d/a that lead to a 10% error in the surface velocity and a 10% error in the film thickness for a specified value of $\bar{v}$ when the solution for an inclined plane is used as an approximation.

25. Repeat problem 24 for an outer film.

26. Plot $\mu\bar{v}/g\rho d^3$ versus d/a using the result of problem 18.

27. Plot u_d/u_m versus d/a using the result of problem 18.

28. Repeat problem 24 for a film outside the column.

29. Repeat problem 25 for a film outside the column.

30. Determine d, $\bar{v}$, v, and u_m for a film of water inside a 100-mm column at Re = 10, 100, and 1000.

31. Derive a solution for the ratio of the volumetric rates of simultaneous flow of water and a less dense oil at equal depths down an inclined plane.

32. Derive a solution for the ratio of the depths of oil and water at simultaneous equal volumetric rates of flow down an inclined plane.

33. Derive a general solution for the simultaneous flow of two liquids of different density and viscosity down an inclined plane.

34. Derive a solution for the ratio of the volumetric rates of flow of liquid and induced gas in a wetted-wall column, neglecting the viscosity of the gas.

35. Develop a solution for the gravitational flow of a film of liquid and the forced downward flow of gas inside a round tube, using the results of Chapter 4.
36. Derive a solution for the onset of flooding in a wetted-wall column due to the upward flow of gas.
37. Develop an expression for an equivalent diameter for a rectangular channel such that Equation 4.12 as applied to a rectangle reduces to Equation 6.19 for $H/d \to \infty$.
38. Derive a solution for the flow of a Bingham plastic down an inclined plate.
39. Derive a solution for the flow of a power-law fluid down an inclined plate.
40. Derive an expression equivalent to Equation 5.32 for flow down an inclined plate.
41. Derive the analog of Equations 5.33–5.44 for flow down an inclined plate.
42. **a.** A paint with a density of 0.80 mg/m^3 is known to behave as a Bingham plastic with a yield strength of 5 Pa. Determine the maximum depth that can be applied without running.
    b. If twice the allowable depth is applied, calculate the surface and mean velocities, assuming $\mu_o = 0.4$ Pa $\cdot$ s.

43. Calculate the depth at which a polymer solution whose effective viscosity can be represented by

$$\frac{723.5}{740 - \eta} = 1 + \left(\frac{172.4}{\tau}\right)^{2.6} \tag{6.46}$$

where η is in pascal seconds and τ is in pascals will flow at a rate of 0.7 kg/m $\cdot$ s down a plane inclined at 10°. Check the Reynolds number to see if laminar flow will prevail. ($\varrho = 0.90$ Mg/m^3).
44. The viscosity of a solution of 10% wt napalm in kerosene has been correlated in terms of the power-law model with $\alpha = 0.52$, $M = 8.93$ mPa $\cdot$ s$^{0.52}$. Calculate the rate of flow (m^3/m $\cdot$ s) down a 45° plane at a depth of 6 mm.
45. Calculate the breadth-to-depth ratio for which Equation 6.6 is 5% in error.
46. Experimental data are obtained for the depth of liquids, with various surface tensions, densities, and viscosities, running down inclined plates of various inclinations and breadths at various volumetric rates of flow.

    a. What dimensionless groups would you suggest for correlation of this data?
    b. What dimensionless groups would you suggest as independent variables, dependent variables, and parameters?
    c. What asymptotic relationships would you expect? Where would you expect the data to fall relative to these asymptotes?

REFERENCES

1. S. W. Churchill, *The Practical Use of Theory in Fluid Flow. Book III. Laminar, Multidimensional Flows in Channels*, Notes, The University of Pennsylvania (1979).

2. S. W. Churchill, *The Practical Use of Theory in Fluid Flow. Book IV. Turbulent Flows*, Notes, The University of Pennsylvania (1981).
3. Wilhelm Nusselt, "Die Oberflächenkondensation des Wasserdampfes," *Z. Ver. Deut. Ing.*, *60* (1916) 541, 569.
4. L. Hopf, "Turbulenz bei einem Flusse," *Ann. Phys.*, Ser. 4, *32* (1910) 777.
5. G. D. Fulford, "The Flow of Liquids in Thin Films," p. 151 in *Advances in Chemical Engineering*, Vol. 5, Academic Press, New York (1964).
6. A. E. Dukler, "Dynamics of Vertical Falling Film Systems," *Chem. Eng. Progr.*, 55, (October 1959) 62.
7. P. L. Kapitsa, "Volnovoe Techenie Tonkikh Sloev Vyaskoi Zhidkosti. I, II, *Zh. Eksperim. i Teor. Fiz. 18* (1948) 3, 19 (Wavy Flow of Thin Layers of Viscous Liquids. I, II).
8. Heinz Bräuer, "Strömung und Wärmeubergang bei Rieselfilmen," *Ver. Deut. Ing. Forschungsheft 457* (1956).
9. S. S. Grimley, "Liquid Flow Conditions in Packed Columns," *Trans. Inst. Chem. Engr. (London)*, *23* (1945) 228.
10. M. L. Jackson, "Liquid Films in Viscous Flow," *AIChE J.*, *1* (1955) 231.
11. V. G. Levich, *Fiziko-Khimicheskaya Gidrodinamika*, 2nd ed., Fizmatgiz, Moscow (1959); English transl., *Physicochemical Hydrodynamics*, Prentice-Hall, Englewood Cliffs, NJ (1962), p. 688.
12. K. Feind, "Strömungsuntersuchungen bei Gegenstrom von Rieselfilmen und Gas in lotrechten Rohren," *Ver. Deut. Ing. Forschungsheft 481* (1960).
13. R. Fallah, T. G. Hunter, and A. W. Nash, "The Application of Physico-Chemical Principles to the Design of Liquid-Liquid Contact Equipment. Part III. Isothermal Flow in Liquid Wetted-Wall Systems," *J. Soc. Chem. Ind. (London)*, *53* (1934) 369T.
14. Eckhart Blass, "Gas/Film-Strömung in Rohren," *Chem.-Ing.-Tech.*, *49* (1977) 95; English transl., "Gas/Film Flow in Tubes," *Int. Chem. Eng.*, *19* (1979) 183.
15. Y. Taitel, D. Bornea, and A. E. Dukler, "Modelling Flow Pattern Transitions for Steady Upward Gas-Liquid Flow in Vertical Tubes," *AIChE J.*, *26* (1980) 346.
16. S. D. R. Wilson, "Flow of a Liquid Film down a Slightly Inclined Tube," *AIChE J.*, *20* (1974) 408.
17. N. Thérien, B. Coupal, and J. N. Corneille, "Vérification expérimental de l'épaisser du film pour des liquides non-Newtoniens s'écoulant par gravité sur un plan incliné," *Can. J. Chem. Eng.*, *48* (1970) 17.

Chapter 7

Couette Flows

Fluid motion generated by the movement of one surface of a channel or confined region is examined in this chapter. Such flows are generally classified as *Couette flows*[1]. These flows are utilized primarily in the laboratory for the measurement of the viscosity. However, the results are applicable to lubrication theory and to pumps for very viscous fluids.

The fluid motions generated jointly by a moving surface and a pressure gradient are also examined in this chapter. Such processes are generally called *Couette–Poiseuille flows*. Applications occur in viscous screw pumps and fluid-driven motors.

PLANAR COUETTE FLOW OF A NEWTONIAN FLUID

Development of the Model

The motion between a fixed and a moving plate separated by distance d, as illustrated in Figure 7–1, is called *planar Couette* flow or simply *Couette flow*. A force-momentum balance on an element of height y, breadth H, and length Δx can be written as

$$-\tau_w H \Delta x = j_{yx} H \Delta x = 0 \tag{7.1}$$

or

$$-j_{yx} = \tau_w \tag{7.1A}$$

Replacing j_{yx} through Equation 2.1 gives

$$\mu \frac{du}{dy} = \tau_w \tag{7.2}$$

---

[1] Couette [1] in 1890 studied experimentally the motion in a thin layer of liquid between two concentric cylinders when one was rotated. Reynolds [2] in 1886 had systematically studied the motion induced in a fluid by one plane moving parallel to another. Newton [3], p. 385, had earlier noted that "If a solid cylinder infinitely long in an uniform and infinite fluid, revolves with an uniform motion about an axis given in position, and the fluid is forced around by only this impulse of the cylinder, and every part of the fluid continues uniformly in its motion: I say that the periodic times of the parts of the fluid are as their distances from the axis of the cylinder."

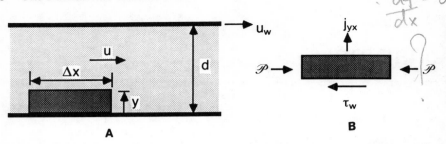

FIGURE 7–1   *Control volume for force-momentum balance for planar Couette flow: (A) control volume; (B) force-momentum balance on control volume.*

Derivation of the Solution

Equation 7.2 can be integrated from $u = 0$ at $y = 0$ to give

$$u = \frac{y\,\tau_w}{\mu} \tag{7.3}$$

Since $u = u_w$ at $y = d$,

$$u_w = \frac{d\,\tau_w}{\mu} \tag{7.4}$$

Also

$$u_m = \frac{1}{d}\int_0^d u\,dy = \frac{1}{d}\int_0^d \frac{y\,\tau_w}{\mu}\,dy = \frac{d\,\tau_w}{2\mu} \tag{7.5}$$

Therefore

$$\frac{u}{u_m} = \frac{2y}{d} \tag{7.6}$$

$$\frac{u_w}{u_m} = 2 \tag{7.7}$$

and

$$\frac{u}{u_w} = \frac{y}{d} \tag{7.8}$$

From Equations 7.2 and 7.8 it is apparent that the effective shear stress within the fluid is constant and that the velocity varies linearly between the plates. This simplicity suggests that such motion is ideal for measuring the viscosity of fluids.

Canonical Form

The friction factor is

$$f = \frac{\tau_w}{\varrho u_m^2} = \frac{2\mu}{d\varrho u_m} = \frac{4\mu}{d\varrho u_w} \tag{7.9}$$

The hydraulic diameter is

$$D = \lim_{H \to \infty} = \frac{4Hd}{2d + 2H} = 2d \tag{7.10}$$

just as for Poiseuille flow in this geometry. Hence

$$\text{Re} = \frac{2du_m\varrho}{\mu} = \frac{du_w\varrho}{\mu} \tag{7.11}$$

and

$$\text{Po} = f\text{Re} = 4 \tag{7.12}$$

where Po has the generalized definition given by Equation 1.14.

Experimental Confirmation

Reichardt [4] confirmed the linearity of the velocity profile as shown in Figure 7–2. He found laminar flow to persist up to Re = 1500. In these experiments the plates were moving at equal velocities, u_w, in opposite directions.

PLANAR COUETTE FLOW OF NON-NEWTONIAN FLUIDS

The shear stress generated by a fluid between a moving plate and a fixed plate is constant for a non-Newtonian fluid as well as for a Newtonian fluid. The Bingham-plastic and power-law models will be used to illustrate the modification of the rest of the solution.

Bingham Plastic

For $\tau_w > \tau_o$, $\tau_{yx} = -\tau_w$, and Equation 2.18 becomes

$$\tau_w = \mu_o \frac{du}{dy} + \tau_o \tag{7.13}$$

Integration from $u = 0$ at $y = 0$ gives the velocity distribution

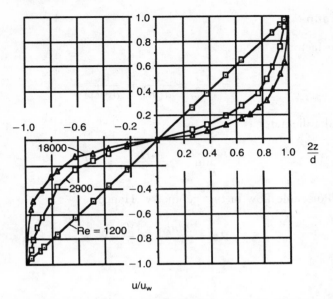

FIGURE 7-2   *Experimental confirmation of linear velocity distribution in laminar regime of planar Couette flow. Two plates moving in opposite directions were used; z is measured from the central plane. (After Reichardt [4].)*

$$u = \frac{(\tau_w - \tau_0)\, y}{\mu_0} \tag{7.14}$$

Hence

$$u_w = \frac{(\tau_w - \tau_0)\, d}{\mu_0} \tag{7.15}$$

or

$$\tau_w = \frac{\mu_0 u_w}{d} + \tau_0 \tag{7.15A}$$

If a plot of experimental data for τ_w versus u_p can be represented by a straight line, μ_o is defined by the slope and τ_o by the intercept.

Integrating the velocity distribution given by Equation 7.14 yields

$$u_m = \frac{1}{d} \int_0^d u \, dy = \frac{d(\tau_w - \tau_0)}{2\mu_0} \tag{7.16}$$

It follows that

$$\frac{u_w}{u_m} = 2 \tag{7.17}$$

and

$$\frac{u}{u_w} = \frac{y}{d}$$

(7.18)

just as for a Newtonian fluid. However,

$$f = \frac{\tau_w}{\varrho u_m^2} = \left(\frac{2\mu_o u_m}{d} + \tau_o\right)\frac{1}{\varrho u_m^2} = \frac{2\mu_o}{\varrho d u_m} + \frac{\tau_o}{\varrho u_m^2}$$
$$= \frac{4\mu_o}{\varrho d u_w} + \frac{4\tau_o}{\varrho u_w^2}$$

(7.19)

The special Reynolds number defined by Equation 5.17 here takes the form

$$\text{Re}_o = \frac{2 d u_m \varrho}{\mu_o} = \frac{d u_w \varrho}{\mu_o}$$

(7.20)

It follows that

$$f\,\text{Re}_o = 4\left(1 + \frac{d\tau_o}{\mu_o u_w}\right)$$

(7.21)

Power-Law Fluids

For fluids whose behavior can be approximated be Equation 2.10, Equation 7.2 is replaced by

$$\tau_w = M\left(\frac{du}{dy}\right)^\alpha$$

(7.22)

which can be rearranged as

$$\frac{du}{dy} = \left(\frac{\tau_w}{M}\right)^{1/\alpha}$$

(7.22A)

Integration from $u = 0$ at $y = 0$ gives

$$u = \left(\frac{\tau_w}{M}\right)^{1/\alpha} y$$

(7.23)

Since $u = u_w$ at $y = d$,

$$u_w = \left(\frac{\tau_w}{M}\right)^{1/\alpha} d$$

(7.24)

Also

$$u_m = \frac{1}{d} \int_0^d u \, dy = \frac{d}{2} \left(\frac{\tau_w}{M}\right)^{1/\alpha} \tag{7.25}$$

It follows that again

$$\frac{u_w}{u_m} = 2 \tag{7.26}$$

and that

$$\frac{u}{u_w} = \frac{y}{d} \tag{7.27}$$

If a logarithmic plot of experimental values of τ_w versus u_w can be represented by a straight line, α may be determined from the slope, and thereafter M.

From Equations 7.25,

$$f \equiv \frac{\tau_w}{\varrho u_m^2} = \left(\frac{2}{d}\right)^\alpha \frac{M}{\varrho u_m^{2-\alpha}} = \frac{2^\alpha M}{\varrho d^\alpha u_m^{2-\alpha}} \tag{7.28}$$

A special Reynolds number

$$\mathrm{Re}_{\alpha 0} = \frac{\varrho d^\alpha u_w^{2-\alpha}}{M} = \frac{2^{2-\alpha} \varrho d^\alpha u_m^{2-\alpha}}{M} \tag{7.29}$$

forces conformity of Equation 7.28 with Equation 7.12 for a Newtonian fluid.

General Formulation

Procedures analogous to that represented by Equations 5.20–5.32 and 5.33–5.44 for forced flow in a round tube can also be developed for plane Couette flow (see problems 11 and 12).

LONGITUDINAL ANNULAR COUETTE FLOW

Development of the Model

A Couette-type flow occurs when a cylindrical rod is pulled through a concentric housing, as shown in Figure 7–3. If end effects are neglected, a force-momentum balance on the shaded element of fluid can be written as

$$\tau_{w1} 2\pi a_1 \, \Delta x = j_{rx} 2\pi r \, \Delta x \tag{7.30}$$

or

$$j_{rx} = \frac{a_1 \tau_{w1}}{r} \tag{7.30A}$$

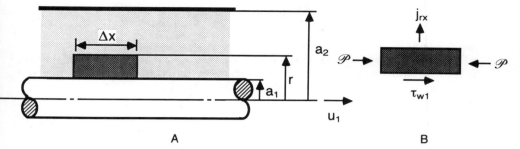

FIGURE 7–3 *Control volume for force-momentum balance for longitudinal annular Couette flow: (A) control volume; (B) force-momentum balance on control volume.*

where a_1 is the radius of the inner surface in meters.

Substituting from Equation 2.1 with y replaced by $-r$ gives

$$\mu\frac{du}{dr} = \frac{a_1\tau_{w1}}{r} \tag{7.31}$$

Derivation of the Solution

Integrating from $u = 0$ at $r = a_2$, the radius of the outer surface, gives

$$u = \frac{a_1\tau_{w1}}{\mu}\ln\left\{\frac{a_2}{r}\right\} \tag{7.32}$$

Since the fluid has the velocity u_w of the inner rod at $r = a_1$:

$$u_w = \frac{a_1\tau_{w1}}{\mu}\ln\left\{\frac{a_2}{a_1}\right\} \tag{7.33}$$

and

$$\frac{u}{u_w} = \frac{\ln\{a_2/r\}}{\ln\{a_2/a_1\}} \tag{7.34}$$

The mean velocity is

$$u_m = \frac{1}{\pi(a_2^2 - a_1^2)}\int_{a_1^2}^{a_2^2} u \cdot 2\pi r\,dr$$

$$= u_w\left(\frac{1}{2\ln\{a_2/a_1\}} - \frac{1}{(a_2/a_1)^2 - 1}\right) \tag{7.35}$$

From a force balance over the entire cross section, the shear stress on the outer surface is

$$\tau_{w2} = \frac{a_1 \tau_{w1}}{a_2} = \frac{\mu u_w}{a_2 \ln\{a_2/a_1\}} \qquad (7.36)$$

However, in applications the volumetric rate of flow induced by the motion of the rod and the total drag force on the rod are of more interest than the quantities given by Equations 7.32 and 7.36. They are

$$V = \pi(a_2^2 - a_1^2)u_m = \pi u_w \left(\frac{a_2^2 - a_1^2}{2 \ln\{a_2/a_1\}} - a_1^2 \right) \qquad (7.37)$$

and

$$F = 2\pi a_1 l \tau_{w1} = \frac{2\pi \mu u_w l}{\ln\{a_2/a_1\}} \qquad (7.38)$$

where F = total force, N
$\quad\quad\quad l$ = length of rod within annulus, m

PLANAR ROTATIONAL COUETTE FLOW

Derivation of the Solution

Consider a disk rotating adjacent to a parallel surface, as shown in Figure 7–4. Insofar as edge effects can be neglected, the motion at any radius corresponds to that of planar Couette flow. That is, the tangential velocity at any radial distance r is

$$u = \frac{y u_w}{d} = \frac{y r \omega}{d} = \frac{2\pi r y \Omega}{d} \qquad (7.39)$$

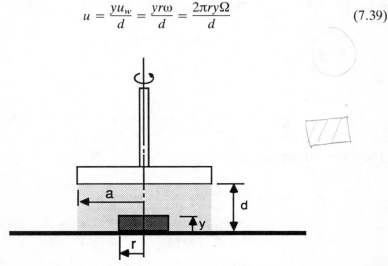

FIGURE 7–4   *Control volume for force-momentum balance for planar rotational Couette flow.*

where y = distance from fixed surface, m
$\quad\quad d$ = gap between disk and fixed surface, m
$\quad\quad u_w$ = local velocity of rotating disk, m/s
$\quad\quad \omega$ = angular velocity of disk, rad/s
$\quad\quad \Omega$ = rate of rotation, s^{-1}

The shear stress on the disk in the fluid and on the fixed surface at any r is correspondingly

$$\tau_w = \frac{\mu u_w}{d} = \frac{\mu r\omega}{d} = \frac{2\pi\mu r\Omega}{d} \qquad (7.40)$$

The torque on the shaft due to the shear stress on the disk is

$$\mathscr{T} = \int_0^a r\tau_w\{r\}\, 2\pi r\, dr = \frac{\pi\mu a^4\omega}{2d} = \frac{\pi^2\mu a^4\Omega}{d} \qquad (7.41)$$

where $\mathscr{T}$ = torque, N·m
$\quad\quad a$ = radius of disk, m

Canonical Form

Equation 7.41 can be arranged in a dimensionless form analogous to prior expressions by defining a Reynolds number based on the linear velocity of the edge of the plate,

$$\mathrm{Re}_{\mathscr{T}} = \frac{2d\omega a\varrho}{\mu} \qquad (7.42)$$

and a torque friction factor

$$f_{\mathscr{T}} = \frac{\mathscr{T}}{\varrho a^2\omega^2\pi a^2 a} = \frac{\mathscr{T}}{\pi a^5\varrho\omega^2} \qquad (7.43)$$

Equation 7.41 then can be expressed as

$$\mathrm{Po}_{\mathscr{T}} = f_{\mathscr{T}}\,\mathrm{Re}_{\mathscr{T}} = 1 \qquad (7.44)$$

Experimental Confirmation

Zumbush, according to Schultz-Grunow [5], confirmed Equation 7.41 experimentally for $a^2\omega\varrho/\mu < 10^4$ with $d/a = 0.0199$. This corresponds to $\mathrm{Re}_{\mathscr{T}} < 4 \times 10^3$. On the other hand, Ellenberger and Fortuin [6], using values of d/a from 0.004 to 0.04, found Equation 7.41 to be applicable up to $\Omega d^2/\nu \cong 2$.

Application to Viscometry

A device equivalent to that sketched in Figure 7–4 is used to measure the viscosity of liquids. The viscosity is given in terms of measured values of the torque and speed of rotation, i.e., by Equation 7.41 rearranged as

$$\mu = \frac{d\mathcal{T}}{\pi^2 a^4 \Omega} \tag{7.45}$$

The constants in the Bingham-plastic and power-law models can in turn be determined from this device (see problems 15–17).

ANNULAR ROTATIONAL COUETTE FLOW

Development of the Model

Consider the motion generated by a cylinder rotating inside a fixed, concentric cylindrical housing, as sketched in Figure 7–5. To balance the torque on the shaded element, one must have

$$\tau_{w1} \cdot 2\pi a_1 \Delta x \cdot a_1 = \tau_{r\theta} \cdot 2\pi r \Delta x \cdot r \tag{7.46}$$

or

$$\tau_{r\theta} = \frac{a_1^2 \tau_{w1}}{r^2} \tag{7.46A}$$

The analog of Equation 2.4 for tangential motion in cylindrical coordinates (see, for example, Chapter 8) is

$$\tau_{r\theta} = -\mu r \frac{d}{dr}\left(\frac{u_\theta}{r}\right) \tag{7.47}$$

Combining Equations 7.46A and 7.47 gives

$$-\mu r^3 \frac{d}{dr}\left(\frac{u_\theta}{r}\right) = a_1^2 \tau_{w1} \tag{7.48}$$

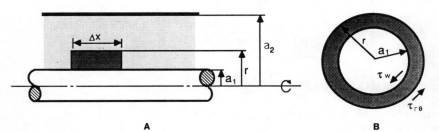

FIGURE 7–5   *Control volume for force-momentum balance for annular rotational Couette flow: (A) side view; (B) end view.*

For simplicity, the double subscript $r\theta$ will be omitted from now on.

Derivation of the Solution

Integrating Equation 7.48 with the boundary condition $u = 0$ at $r = a_2$ gives

$$u = \frac{r\tau_{w1}}{2\mu}\left[\left(\frac{a_1}{r}\right)^2 - \left(\frac{a_1}{a_2}\right)^2\right] \tag{7.49}$$

At $r = a_1$, $u = u_1 = a_1\omega = 2\pi a_1\Omega$ and

$$\Omega = \frac{\tau_{w1}}{4\pi\mu}\left(1 - \left(\frac{a_1}{a_2}\right)^2\right) \tag{7.50}$$

Therefore

$$\frac{u}{2\pi a_1\,\Omega} = \frac{a_2/r - r/a_2}{a_2/a_1 - a_1/a_2} \tag{7.51}$$

and

$$\mathscr{T} = 2\pi a_1^2 l\tau_{w1} = \frac{8\pi^2\mu\,la_1^2\Omega}{1 - (a_1/a_2)^2} \tag{7.52}$$

Rotation of Outer Cylinder

For a rotating outer cylinder and a fixed inner cylinder the velocity distribution is

$$\frac{u}{2\pi a_2\Omega} = \frac{r/a_1 - a_1/r}{a_2/a_1 - a_1/a_2} \tag{7.53}$$

and the torque is still given by Equation 7.52.

Rotation of Cylinder in Infinite Media

A solution for a cylinder of radius a, rotating in a fluid of infinite extent can be obtained by letting $a_2 \to \infty$ in Equations 7.51 and 7.52. The resulting velocity distribution is

$$u = \frac{2\pi a_1^2\Omega}{r} \tag{7.54}$$

and the torque is

$$\mathscr{T} = 8\pi^2\mu\,la_1^2\Omega \tag{7.55}$$

Limiting Behavior

Equations 7.50, 7.51, and 7.53 degenerate to those for plane Couette flow if a_2 and r are replaced by $a_1 + d$ and $y + a_1$, respectively, and y/a_1 and d/a_1 are allowed to approach zero (see problem 18).

Stability

When the outer cylinder is rotated, the centrifugal force stabilizes the flow. Couette [1] determined the following criterion for stability from experiments with a small gap:

$$\Omega < \frac{151\,\mu}{\varrho(a_2 - a_1)a_2} \tag{7.56}$$

Schlichting [7] derived the following theoretical criterion for stability

$$\Omega < \frac{(1.05 \times 10^4)\,\mu}{\varrho\,a_2^2} \tag{7.57}$$

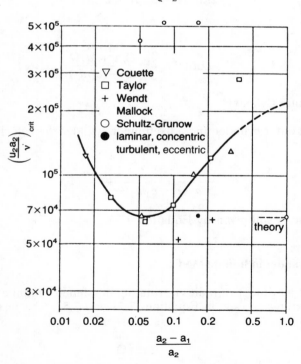

FIGURE 7–6   *Stability limits for rotation of the outer cylinder of an annulus. a_2 = radius of outer cylinder; $a_2 - a_1$ = width of annular gap; $\bigcirc$ = Schultz-Grunow, laminar and concentric; $\bullet$ = Schultz-Grunow, turbulent and slightly eccentric. Older measurements by Couette, Taylor, Wendt, and Mallock; more recent ones by Schultz-Grunow [9]. (From Schlichting [8], p. 430.)*

for acceleration of the outer cylinder in the limiting case of $(a_2 - a_1)/a_2 \rightarrow 1$ (also see [8], p. 429). Various measurements are plotted in Figure 7–6. The open circles represent the measurements of Schultz-Grunow [9] in which particular care was taken to minimize vibrations and eccentricity and to obtain a steady state. He concluded from these measurements that this flow is stable for all steady rates of circulation under ideal conditions.

For rotation of the inner cylinder, the centrifugal force is destabilizing. Prandtl [10] developed the following approximate expression for the critical rate of circulation:

$$\Omega = \frac{6.57\,\mu}{\varrho(a_2 - a_1)^{3/2}a_1^{1/2}} \tag{7.58}$$

This limit is for the formation of laminar *Taylor vortices* [11].

Application to Viscometry

Rotational annular Couette flow is frequently used to determine the viscosity of liquids. With a Couette-type viscometer, the torque on the outer rotating cylinder is measured as a function of the speed of rotation. With a *Stormer* or *Searle* viscometer the torque on the inner, rotating cylinder is measured. Because of the greater stable rate of rotation, a fixed inner cylinder provides a wider potential range of operating conditions and is therefore preferable.

Equation 7.46A can be integrated for the Bingham-plastic model (see problem 22). The result for a fixed inner cylinder is [12]

$$\mathcal{T} = \frac{8\pi^2\,la_1^2}{1 - (a_1/a_2)^2}\left(\mu_0\,\Omega + \tau_0\ln\left\{\frac{a_2}{a_1}\right\}\right) \tag{7.59}$$

This expression, known as the *Reiner–Riwlin equation*, permits μ_0 and τ_0 to be determined from experimental measurements of $\mathcal{T}\{\Omega\}$.

Integration of Equation 7.46A is not feasible for fluids that have a more complex rheological behavior. A rotating viscometer can be used for such fluids only if the gap is sufficiently small so that curvature can be neglected.

COUETTE FLOW IN A GAP OF VARYING DEPTH

In a lubricated bearing the spacing between the shaft and the journal must necessarily vary if the shaft is to be supported. Such behavior can be idealized in terms of a flat surface moving with respect to a fixed bevelled surface as illustrated in Figure 7–7, with the spacing given by

$$d = d_1 - x\sin\{\theta\} \tag{7.60}$$

As a result of the variable spacing the momentum of the fluid changes with x. Neglecting this change as a first approximation permits a force-momentum balance on the shaded element to be written as

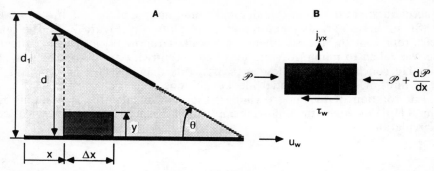

FIGURE 7–7   *Control volume for force-momentum balance for longitudinal Couette flow in a gap of varying depth: (A) control volume; (B) force-momentum balance on control volume.*

$$-\tau_w \Delta x + \mathscr{P}_x \Delta x = \left(\mathscr{P} + \frac{d\mathscr{P}}{dx}\Delta x\right)y + j_{yx}\Delta x \qquad (7.61)$$

Substituting for j_{yx} from Equation 2.1 and simplifying give

$$\mu\frac{du}{dy} = \tau_w + \frac{d\mathscr{P}}{dx}y \qquad (7.62)$$

which can be integrated from $u = u_w$ at $y = 0$ to give

$$\mu(u_w - u) = -y\tau_w - \frac{y^2}{2}\frac{d\mathscr{P}}{dx} \qquad (7.63)$$

From the condition $u = 0$ at $y = d$,

$$\tau_w = -\frac{\mu u_w}{d} - \frac{d\mathscr{P}}{dx}\frac{d}{2} \qquad (7.64)$$

From Equations 7.63 and 7.64,

$$u = u_w\left(1 - \frac{y}{d}\right) + \frac{y(d - y)}{2\mu}\left(-\frac{d\mathscr{P}}{dx}\right) \qquad (7.65)$$

The volumetric flow per unit of breadth is

$$\bar{v} = \int_0^d u\,dy = \frac{du_w}{2} + \frac{d^3}{12\mu}\left(-\frac{d\mathscr{P}}{dx}\right) \qquad (7.66)$$

Substituting for d from Equation 7.60 and rearranging give

$$-\frac{d\mathscr{P}}{dx} = \frac{6\mu u_w}{(d_1 - x\sin\{\theta\})^2} - \frac{12\mu\bar{v}}{(d_1 - x\sin\{\theta\})^3} \qquad (7.67)$$

Integrating from $\mathscr{P} = 0$ at $x = 0$ gives

$$\mathscr{P} = \frac{6\mu u_w}{\sin\{\theta\}}\left(\frac{1}{d_1 - x\sin\{\theta\}} - \frac{1}{d_1}\right)$$
$$- \frac{6\mu\tilde{v}}{\sin\{\theta\}}\left(\frac{1}{(d_1 - x\sin\{\theta\})^2} - \frac{1}{d_1^2}\right) \tag{7.68}$$

At $x = l$, $\mathscr{P} = 0$ as well. Designating

$$d_2 = d_1 - l\sin\{\theta\} \tag{7.69}$$

then gives

$$\tilde{v} = u_w\left(\frac{1}{d_2} - \frac{1}{d_1}\right)\Big/\left(\frac{1}{d_2^2} - \frac{1}{d_1^2}\right) = \frac{d_1 d_2 u_w}{d_1 + d_2} \tag{7.70}$$

Also

$$\mathscr{P} = \frac{6\mu u_w}{\sin\{\theta\}}\left(\frac{1}{d_1 - x\sin\{\theta\}} - \frac{1}{d_1}\right.$$
$$\left. - \frac{d_1 d_2}{d_1 + d_2}\left(\frac{1}{(d_1 - x\sin\{\theta\})^2} - \frac{1}{d_1^2}\right)\right) \tag{7.71}$$

Eliminating $\sin\{\theta\} = (d_1 - d_2)/l$ gives

$$\mathscr{P} = \frac{6\mu u_w x(l - x)(d_1 - d_2)}{l(d_1 + d_2)[d_1 - (x/l)(d_1 - d_2)]^2} \tag{7.72}$$

This pressure distribution is illustrated in Figure 7–8. By differentiating Equation 7.72 one can see that the maximum pressure occurs at

$$x = \frac{ld_1}{d_1 + d_2} \tag{7.73}$$

The total normal force is

$$F = \int_0^l \mathscr{P}H \, dx = \frac{6\mu u_w l^2 H}{(d_1 - d_2)^2}\left[\ln\left\{\frac{d_1}{d_2}\right\} - 2\left(\frac{d_1 - d_2}{d_1 + d_2}\right)\right] \tag{7.74}$$

where H is the breadth (or circumference). The maximum value of F occurs for $d_1/d_2 \cong 2.2$. Therefore

$$F_{\max} \cong \frac{0.16\mu u_w l^2 H}{d_2^2} \tag{7.75}$$

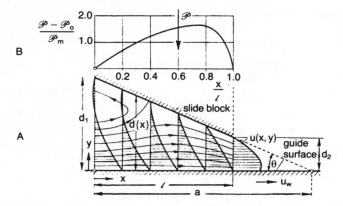

FIGURE 7-8   *Computed characteristics for Couette flow in a gap of varying depth with a/l = 1.57: (A) streamlines; (B) pressure distribution. (From Schlichting [8], p. 98.)*

The preceding derivation was first carried out by Reynolds [2]. Sommerfeld [13] carried out the corresponding derivation for a circular journal and an eccentric circular shaft. (See, for example, Kaufmann [14], p. 249f, or Lamb [15], p. 883f.)

PLANAR COUETTE–POISEUILLE FLOW

Combined forced and induced flow is called *Couette–Poiseuille flow*. It is utilized in pumping very viscous fluids and in lubrication. Attention here is limited to the simple case of parallel plates and a Newtonian fluid. Other cases are examined in the problem set.

Derivation of the Solution

Equation 3.2 with τ_{w1}, the shear stress on the fixed wall, substituted for τ_w is applicable for the shaded element in Figure 7–9; that is,

$$\mu\frac{du}{dy} = \tau_{w1} + y\frac{d\mathscr{P}}{dx} \tag{7.76}$$

Integrating from $u = 0$ at $y = 0$ gives

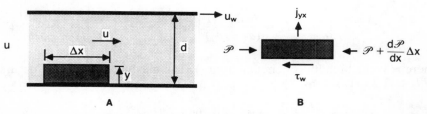

FIGURE 7-9   *Control volume for force-momentum balance for planar Couette-Poiseuille flow: (A) control volume; (B) force-momentum balance over control volume.*

$$\mu u = y\tau_{w1} + \frac{y^2}{2}\frac{d\mathscr{P}}{dx} \tag{7.77}$$

The boundary condition $u = u_w$ at $y = d$ then requires

$$\mu u_w = d\tau_{w1} + \frac{d^2}{2}\frac{d\mathscr{P}}{dx} \tag{7.78}$$

Eliminating τ_{w1} between Equations 7.77 and 7.78 produces

$$\frac{u}{u_w} = \frac{y}{d}\left[1 + \frac{d^2}{2\mu u_w}\left(1 - \frac{y}{d}\right)\left(-\frac{d\mathscr{P}}{dx}\right)\right] \tag{7.79}$$

Then

$$\frac{u_m}{u_w} = \frac{1}{d}\int_0^d \frac{u}{u_w}\,dy = \frac{1}{2} + \frac{d^2}{12\mu u_w}\left(-\frac{d\mathscr{P}}{dx}\right) \tag{7.80}$$

and

$$\tilde{v} = du_m = du_w\left[\frac{1}{2} + \frac{d^2}{12\mu u_w}\left(-\frac{d\mathscr{P}}{dx}\right)\right] \tag{7.81}$$

It follows that a moving wall of length l can produce a maximum pressure *increase* (with $\tilde{v} = 0$) of

$$(\Delta\mathscr{P})_{\max} = \frac{6\mu u_w l}{d^2} \tag{7.82}$$

Interpretation of the Solution

The velocity distribution, as given by Equation 7.79, consists of a linear term plus a parabolic term with the dimensionless weighting factor:

$$\psi = \frac{d^2}{2\mu u_w}\left(-\frac{\partial\mathscr{P}}{\partial x}\right) \tag{7.83}$$

The distribution is plotted in Figure 7–10 with this weighting term as a parameter. It is apparent from Figure 7–10 or by setting du/dy to zero that back flow occurs over part of the cross section if an adverse (positive) pressure gradient exists of magnitude greater than $2\mu u_w/d^2$.

The variation of the shear stress within the fluid is

$$\tau = \tau_{yx} = \mu\frac{du}{dy} = \frac{\mu u_w}{d}\left[1 + \frac{d^2}{2\mu u_w}\left(1 - \frac{2y}{d}\right)\left(-\frac{d\mathscr{P}}{dx}\right)\right] \tag{7.84}$$

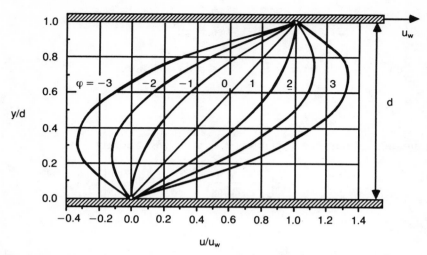

FIGURE 7–10   *Computed velocity distribution for planar Couette-Poiseuille flow.*

This behavior is somewhat surprising in that the shear stress and the velocity gradient are zero for

$$-\frac{d\mathscr{P}}{dx} = \frac{2\mu u_w/d^2}{(2y/d) - 1} \tag{7.85}$$

and hence at the moving surface ($y = d$) for

$$-\frac{d\mathscr{P}}{dx} = \frac{2\mu u_w}{d^2} \tag{7.86}$$

With the adverse pressure gradient

$$\frac{d\mathscr{P}}{dx} = \frac{2\mu u_w}{d^2} \tag{7.87}$$

the shear stress and velocity gradient are zero at the fixed wall ($y = 0$). For

$$\left|\frac{d\mathscr{P}}{dx}\right| > \frac{2\mu u_w}{d^2} \tag{7.88}$$

the shear stress is zero at some intermediate location. The shear stress at the central plane, $y = d/2$, is equal to $\mu u_w/d$ for all pressure gradients.

Couette–Poiseuille Pumping

Considered as a pump, the work accomplished in planar Couette–Poiseuille flow is $(d\mathscr{P}/dx)\,l\bar{v}H$, whereas the work expended in moving the upper surface is $\tau_{w2}Hlu_w$. Therefore, the fractional efficiency is

$$E = \frac{(d\mathscr{P}/dx)\tilde{v}H}{\tau_{w2}Hlu_w} = \frac{\tilde{v}(d\mathscr{P}/dx)}{u_w\tau_{w2}} \tag{7.89}$$

From an overall force balance

$$\left(-\frac{d\mathscr{P}}{dx}\right)d + \tau_{w2} = \tau_{w1} \tag{7.90}$$

Substituting for τ_{w2} from Equation 7.84, for τ_{w1} from 7.78, and for $\tilde{v}$ from 7.81 gives, after simplifying,

$$E = \frac{\psi(3 - \psi)}{3(1 + \psi)} \tag{7.91}$$

This efficiency is zero at $\psi = 0$, increases to a maximum of 1/3 at $\psi = 1$, and decreases to zero at $\psi = 3$.

Couette–Poiseuille Motor

For $\psi < -1$ the flowing fluid acts as a motor moving the plate. The fractional efficiency as a motor is

$$E' = \frac{1}{E} = \frac{3(1 + \psi)}{\psi(3 - \psi)} \tag{7.92}$$

This efficiency is zero at $\psi = -1$, increases to a maximum of 1/3 at $\psi = -3$, and slowly decreases as $\psi \to -\infty$.

Applications

Couette–Poiseuille pumps and extruders usually have the form of an annular channel or rotating screw, such as those illustrated in problems 34 and 36, rather than a moving plate such as in Figure 7–9.

SUMMARY

Couette flows were shown to occur in a number of geometries. Planar motion is simple and directly applicable for determination of the viscosity of Newtonian fluids or the effective viscosity of non-Newtonian fluids. However, this motion is difficult to establish experimentally. Rotational Couette flow is easier to establish experimentally, but the solutions are much more complicated. These complications restrict the applicability of rotational Couette flow for the characterization of non-Newtonian behavior except in the limit of gaps thin enough to allow the neglect of curvature.

Couette flows are useful for pumping liquids without a pressure drop; Couette–Poiseuille flows are useful when common devices are not feasible owing to very high viscosity.

PROBLEMS

1. Translate the quotation from Newton (footnote on the first page of the chapter) into an equation or equations. Compare with the complete solution (Equations 7.54 and 7.55).
2. Reexpress the solution for plane Couette flow of a Bingham plastic in terms of the Hedstrom number.
3. Define a special Reynolds number for a Bingham plastic such that Equation 7.12 holds for plane Couette flow.
4. Compare the Reynolds number defined by Equation 7.29 with that of Equation 5.42.
5. Compare the shear stress on the wall in plane Couette flow with that for forced flow between parallel plates with

   a. the same u_m
   b. the same maximum velocity

6. Two immiscible liquids are confined between parallel plates, one of which is moving at velocity u_w. Derive a solution for the velocity distribution, the shear stress on the plates, and the volumetric rate of flow as a function of the fractional depth and the physical properties.
7. A pump for polymeric solutions consists of an endless belt between two pulleys. The distance between the two pulleys is l, and the width of the belt is b. Derive an expression for the power required to pump without change of head a

   a. Newtonian fluid
   b. power-law fluid

8. Calculate the linear velocity of the belt and the horsepower required to pump the fluids of the following problems with the belt of problem 7 at a rate of $0.80 \, \text{m}^3/\text{s}$ if $l = 150$ mm, $b = 25$ mm, and $h = 1.6$ mm:

   a. problem 9 of Chapter 5
   b. problem 11 of Chapter 5
   c. problem 12 of Chapter 5
   d. problem 43 of Chapter 6
   e. problem 44 of Chapter 6
   f. water

9. Calculate the shear stress on a plate moving at 15 mm/s at a distance of 25 mm from a stationary plate for the fluids of problem 8.
10. Derive expressions analogous to Equations 5.20–5.32 for plane Couette flow.
11. Derive expressions analogous to Equations 5.33–5.44 for plane Couette flow.
12. Show that Equations 7.32–7.37 degenerate to those for plane Couette flow as $a_2/a_1 \rightarrow 1$.
13. Define D, Re, and f for longitudinal annular Couette flow and derive a relationship for f and fRe.

14. Calculate F and u_m for longitudinal annular Couette flow of the fluids of problem 8 at a rate of 0.80 m³/s if $l = 150$ mm, $a_1 = 4$ mm, and $a_2 = 5.6$ mm. Compare the results with those of problem 8.

15. Explain how the constants in the Bingham-plastic model can be obtained from measurements of $\mathcal{T}\{\Omega\}$ with a rotating disk viscometer.

16. Explain how the constants in the power-law model can be obtained from measurements of $\mathcal{T}\{\Omega\}$ with a rotating disk viscometer.

17. Can the rotating disk viscometer be used to determine the effective viscosity of general non-Newtonian fluids? Explain.

18. Show that Equations 7.50, 7.51, and 7.53 degenerate to those for plane Couette flow as $a_2/a_1 \to 1$.

19. Derive the analog of Equations 7.49–7.53 for rotation of both cylinders.

20. Derive expressions for u_m, f, and $f\mathrm{Re}$ for annular rotational Couette flow for rotation of

    a. the outer cylinder
    b. the inner cylinder
    c. both cylinders

21. Derive Equation 7.53 from a force-momentum balance.
22. Derive Equation 7.59.
23. Derive the analog of Equation 7.59 for rotation of the inner cylinder.
24. Why is a rotational, annular viscometer unsatisfactory for a general non-Newtonian fluid?
25. Calculate the torque on a cylinder 75 mm in diameter turning at 30 rps in a housing with a 3-mm gap for the fluids of problem 8.
26. Calculate the torque per meter on a 48-mm cylinder rotating inside a 50-mm housing at 40 rps for an oil in the gap with $\mu = 80$ mPa · s. Check to see if laminar flow will exist.
27. Define Re for rotational annular Couette flow and reexpress the criteria of Couette, Schlichting, and Prandtl in terms of this group.
28. Using Equations 7.65–7.67 plot u/u_m versus y/d with x/l and d_2/d_1 as parameters.
29. Derive expressions for the maximum and mean pressures corresponding to Equation 7.72.
30. Derive an expression for the local shear stress distribution in planar Couette–Poiseuille flow in terms of the shear stress at the walls.
31. Derive expressions for the friction factors in planar Couette–Poiseuille flow for the

    a. moving surface
    b. fixed surface
    c. mean value for the two surfaces

32. Prove that the local shear stress in planar Couette–Poiseuille flow is always finite for $|d\mathcal{P}/dx| < 2\mu u_w/d^2$.
33. Plot the efficiency of a planar Couette–Poiseuille pump and of a planar Couette–Poiseuille motor versus ψ.
34. A pump consists of a rotating cylinder 100 mm in diameter and 150 mm long inside a stationary cylinder 150 mm in diameter as shown in Figure

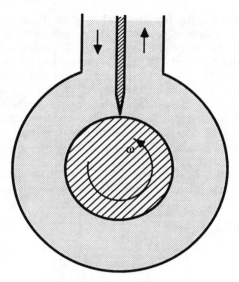

FIGURE 7–11   *Annular Couette pump.*

7–11. Derive expressions for the rate of flow, horsepower, and efficiency as a function of the speed of rotation and the pressure gradient. End effects may be neglected. As a first approximation, curvature may also be neglected.

35.   Calculate the pumping rate and horsepower for the following fluids, using the pump in problem 34 with $\Omega = 30$ s^{-1}:

    a.   water
    b.   the napalm solution of problem 44 of Chapter 6
    c.   the polymer solution of problem 43 of Chapter 6
    d.   a fluid for which $\eta = 815/(1 + 160 \ \tau^2)$
    e.   a 0.09% wt aqueous solution of carboxymethylcellulose whose viscosity Shah et al. [16] found could be approximated by the power law with $M = 0.92 \times 10^{-3}$ lb$_f \cdot$ s$^{0.72}$/ft^2 and $\alpha = 0.72$

36.   A screw pump or extruder consists of a fixed barrel and a rotating screw as shown in Figure 7–12. The screwshaft does not advance axially.

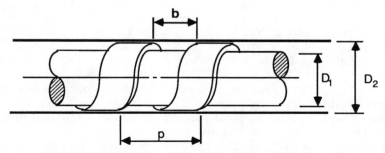

FIGURE 7–12   *Screw pump with a rectangular thread.*

a. Assuming the fluid behavior can be approximated by planar Couette–Poiseuille flow and that $D_2 - D_1 \ll b$ and p, show that the pumping rate of a Newtonian fluid can be approximated by

$$V = \left(\frac{d^3(\mathscr{P}_1 - \mathscr{P}_2)}{12\mu Sl} + \frac{\pi d D_1 \Omega}{2}\right) b$$

where $d = (D_2 - D_1)/2$, m
Ω = rate of rotation, s^{-1}
$S = [(\pi D_1/p)^2 + 1]^{1/2}\{1 + (\pi D_1/p)^2(d/D_1)[(\pi D_1/p)^2 + 1)]^{-1}\}$
l = length of pump, m
$\mathscr{P}_1$ = inlet pressure, Pa
$\mathscr{P}_2$ = outlet pressure, Pa

Note and evaluate any assumptions that have been made.

b. What speed of rotation and horsepower are required to pump a liquid of $1000 \text{ Pa} \cdot \text{s}$ at 3×10^{-7} m³/s from 0.1 to 1.0 MPa if $l = 600$ mm, $D_2 = 25$ mm, $D_1 = 18.75$ mm, $b = 25$ mm, and $p = 40$ mm/turn.

37. Redefine Re and f for planar rotational Couette flow, based on the mean velocity, and determine Po. Compare with Equation 7.44.
38. Redefine Re and f and determine Po for annular rotational Couette flow.
39. Derive a solution for flow down an inclined belt moving opposite to the gravitational flow.
40. Compare the criteria of Zumbush with that of Ellenberger and Fortuin for the upper limit of applicability of Equation 7.41.

REFERENCES

1. M. Couette, "Sur un nouvel appareil pour l'étude du frottement des fluids," *Compt. Rend. Acad. Sci. Paris*, *107* (1888) 388; "Etudes sur le frottement des liquides," *Ann. Chem. Phys.*, Ser. 6, *21* (1890) 433.
2. Osborne Reynolds, "On the Theory of Lubrication and its Application to Mr. Beauchamp Tower's Experiments Including an Experimental Determination of the Viscosity of Olive Oil," *Phil. Trans. Roy. Soc. (London)*, *A177* (1886) 157, (*Scientific Papers*, Vol. II, Cambridge University Press (1901), p. 228).
3. Isaac Newton, *Principia*, Vol. I, *The Motion of Bodies*, S. Pepys, London (1686); English transl. of 2nd ed. (1713) by Andrew Motte (1729), revised transl. by Florian Cajori, University of California Press, Berkeley (1966).
4. H. Reichardt, "Über die Geschwindigkeitsverteilung in einer geradlinigen turbulenten Couette-Strömung," *Z. Angew. Math. Mech.*, *36*, Sonderheft, (1956) S26.
5. F. Schultz-Grunow, "Der Reibungswiderstand rotierender Scheiben in Gehäusen," *Z. Angew. Math. Mech.*, *15* (1935) 191.
6. J. Ellenberger and J. M. N. Fortuin, "A Criterion for Purely Tangential Laminar Flow in the Cone-and-Plate Rheometer and the Parallel-Plate Rheometer," *Chem. Eng. Sci.*, *40* (1985) 111.
7. H. Schlichting, "Über die Stabilität der Couette-Strömung," *Ann. Phys.*, 5 (1932) 905.
8. H. Schlichting, *Boundary Layer Theory*, English transl. by J. Kestin, 4th ed., McGraw-Hill, New York (1960).

9. F. Schultz-Grunow, "Für Stabilität der Couette-Strömung," *Z. Angew. Math. Mech.*, *39* (1959) 101.
10. L. Prandtl, "Einfluss stabilisierender Kräfte auf die Turbulenz," *Vorträge Gebiet Aerodynamik verwandter Gebiete*, Aachen (1929), p. 1; *Essentials of Fluid Dynamics*, Hafner, New York (1952), p. 132.
11. G. I. Taylor, "Stability of a Viscous Liquid Contained between Two Rotating Cylinders," *Phil. Trans.*, *A223* (1923) 289.
12. M. Reiner and R. Riwlin, "Die Theorie der Strömung einer elastischen Flüssigkeiten Couette-Apparat," *Kolloid Z.*, *43* (1927) 1.
13. A. Sommerfeld, "Zur hydrodynamischen Theorie der Schmiermittelreibung," *Z. Math. Phys.*, *50* (1904) 97.
14. W. Kaufmann, *Fluid Mechanics*, English transl. by E. G. Chilton, McGraw-Hill, New York (1963).
15. Horace Lamb, *Hydrodynamics*, Dover, New York (1945).
16. M. J. Shah, E. E. Petersen, and A. Acrivos, "Heat Transfer from a Cylinder to a Power-Law Non-Newtonian Fluid," *AIChE J.*, *8* (1962) 542.

PART II

The General
Equations of Motion

In Chapters 2 through 7 of Part I, expressions were developed for one-dimensional and quasi-one-dimensional, steady laminar flow by constructing one-dimensional mass balances and one-dimensional force-momentum balances for each individual case. Bird et al. [1],[1] p. 35, call these *shell* balances.

An alternative is to develop very general non-steady-state, three-dimensional balances. These partial differential equations can then be simplified, transformed, reduced, and integrated for particular cases. This procedure is usually more convenient than deriving balances from scratch for multidimensional flow, and has the advantage of revealing explicitly the assumptions and simplifications that may be implicit and unrecognized in a derivation for a specific case. Also, the fundamental relationship between various specific flows is emphasized. On the other hand, the general approach has the disadvantage of introducing unnecessary complexity into simple problems and of obscuring the physical principles with mathematical symbols and manipulations.

The development of solutions for fluid flow by the derivation and reduction of the general equations of motion is emphasized by Bird et al. [1] and, even more so, by Slattery [2]. Also, with this approach, attention is naturally focused on these flows for which solutions can be attained in closed form. By contrast, the emphasis in this series of books is on the use of theory to derive solutions for those flows of greatest practical importance. The fluid motions most encountered in practice are steady on-the-mean and confined to one dimension. Therefore, the direct derivation of balances for each individual case is feasible and perhaps preferable. It is desirable to be able to use both methods. Hence, the general equations are considered in this section, their derivations in Chapter 8, special cases and formulations in Chapter 9, and exact solutions in Chapter 10. The use of the general equations of motion to derive approximate solutions is illustrated in Parts III and IV and, to a lesser extent, in the companion volumes of this series [3, 4].

---

[1] The references in this Introduction are included with those at the end of Chapter 8.

Chapter 8

Derivation of the General Mass and Force-Momentum Balances

In this chapter an equation is first derived in *Cartesian*[1] (x, y, z) coordinates for the conservation of mass in unsteady, three-dimensional fluid motion. Several alternative notations and forms are illustrated, and various specialized cases are examined. This differential mass balance is also expressed in cylindrical and spherical coordinates.

The same procedure is followed for the conservation of momentum, particular attention being given to the special case of a *Newtonian* fluid.

THE MASS BALANCE

Derivation

Consider a small rectangular parallelepiped with sides Δx, Δy, and Δz, as shown in Figure 8–1. The velocity vector **v** of the fluid motion can be represented by its components in the directions, x, y, and z, that is, as

$$\mathbf{v} = iu_x + ju_y + ku_z \tag{8.1}$$

where u_x = component of velocity in x-direction, m/s
   u_y = component of velocity in y-direction, m/s
   u_z = component of velocity in z-direction, m/s
   i = x-component of unit vector
   j = y-component of unit vector
   k = z-component of unit vector

---

[1] Named after René Descartes (1596–1650), a French physicist, mathematician, and philosopher whose many accomplishments include the creation of analytical geometry.

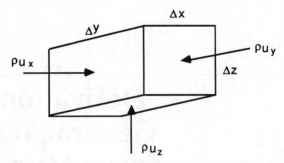

FIGURE 8–1   *Volumetric element for mass balance in Cartesian coordinates.*

The mass rate of flow entering the face at x is seen to be $(\varrho u)_x \, \Delta y \, \Delta z$, and the mass rate of flow leaving the face at $x + \Delta x$ is $(\varrho u_x)_{x+\Delta x} \, \Delta y \, \Delta z$.[1] The net influx in the x-direction per unit volume and unit time is therefore

$$\lim_{\Delta x \to 0} \left(\frac{(\varrho u_x)_x \, \Delta y \, \Delta z - (\varrho u_x)_{x+\Delta x} \, \Delta y \, \Delta z}{\Delta x \, \Delta y \, \Delta z} \right) \equiv -\frac{\partial(\varrho u_x)}{\partial x} \qquad (8.2)$$

Similar expressions follow for the y- and z-directions. The rate of accumulation of mass within the volume is $\Delta x \, \Delta y \, \Delta z \, (\partial \varrho / \partial t)$. Hence the conservation of mass requires that

$$-\left[\frac{\partial(\varrho u_x)}{\partial x} + \frac{\partial(\varrho u_y)}{\partial y} + \frac{\partial(\varrho u_z)}{\partial z} \right] = \frac{\partial \varrho}{\partial t} \qquad (8.3)$$

Equation 8.3 and the subsequent forms of the general mass balance for a fluid derived in this chapter are sometimes called the *continuity equation* since they imply the continuity of matter, as contrasted with a discrete molecular representation.

Alternative Notations

Equation 8.3 can be written more compactly in *Cartesian tensor notation* as

$$-\frac{\partial(\varrho u_i)}{\partial x_i} = \frac{\partial \varrho}{\partial t} \qquad (8.4)$$

where the *repeated subscript i* here implies summation of the x, y, and z terms, and in *vector notation* as

$$-\nabla \cdot (\varrho \mathbf{v}) = \frac{\partial \varrho}{\partial t} \qquad (8.5)$$

For the coordinate system of Figure 8–1 the *vectorial operator* ∇, called *del*, is

---

[1] An alternative form for this term is $(\varrho u_x)_x + (\partial/\partial x)(\varrho u_x) \, \Delta x$.

$$\nabla \equiv i\frac{\partial}{\partial x} + j\frac{\partial}{\partial y} + k\frac{\partial}{\partial z} \tag{8.6}$$

where i, j, and k again are the components of a unit vector. Although Equation 8.6 is expressed in x, y, z coordinates, the operator ∇ can be considered to represent all coordinate systems.

The *scalar (dot) product* of the operator ∇ with another vector (here $\varrho\mathbf{v}$) is called the *divergence of that vector* and produces a scalar equal to the net efflux per unit volume and unit time due to that vector. The divergence of the mass velocity vector $\varrho\mathbf{v}$ is thus the net efflux of mass per unit volume and time due to the flow. *This concept is very useful and efficient in constructing balances involving a number of inputs, as illustrated subsequently in the derivation of the force-momentum balance, and should become second nature to all engineering scientists.* The sum of the negatives of the divergences of the various vector inputs can simply be equated to the rate of accumulation. As previously implied, the vector representation has the additional advantages of generality for all coordinate systems.

Lagrangian Form

Expanding the derivatives in Equation 8.3 and rearranging give

$$-\varrho\left(\frac{\partial u_x}{\partial x} + \frac{\partial u_y}{\partial y} + \frac{\partial u_z}{\partial z}\right) = \frac{\partial \varrho}{\partial t} + u_x\frac{\partial \varrho}{\partial x} + u_y\frac{\partial \varrho}{\partial y} + u_z\frac{\partial \varrho}{\partial z} \tag{8.7}$$

which in Cartesian tensor notation is

$$-\varrho\frac{\partial x_i}{\partial x_i} = \frac{\partial \varrho}{\partial t} + u_i\frac{\partial \varrho}{\partial x_i} \tag{8.8}$$

In vector notation Equation 8.7 is written

$$-\varrho\nabla \cdot \mathbf{v} = \frac{D\varrho}{Dt} \tag{8.9}$$

where the operator

$$\frac{D}{Dt} = \frac{\partial}{\partial t} + (\mathbf{v} \cdot \nabla) \tag{8.10}$$

In the x, y, z coordinate system

$$\frac{D}{Dt} = \frac{\partial}{\partial t} + u_x\frac{\partial}{\partial x} + u_y\frac{\partial}{\partial y} + u_z\frac{\partial}{\partial z} \tag{8.10A}$$

The operator D/Dt is called the *substantial derivative* and has the physical interpretation of the rate of change of some property of a fixed mass of material as it moves through time and space. In Equation 8.9 the changing property is the density. Because of its efficiency and general applicability in fluid mechanics, *the use of this operator should also become second nature*.

Equations 8.3–8.5 are in the *Eulerian*[1] form, and Equations 8.7–8.9 in the *Lagrangian*[2] form.[3] The Eulerian form was obtained by making a balance of some property (in this case mass) within a fixed volume in space. The Lagrangian form is obtained if a balance of some property (in this case volume or density) is made for a fixed mass as it moves through time and space. Although Equation 8.7 was obtained here from Equation 8.3 by mathematical operations, it could have been derived directly using the foregoing concept (see problem 1). Engineers are said to "take the Eulerian point of view" and to "live in an Eulerian world," whereas scientists ordinarily "take the Lagrangian point of view" and "live in a Lagrangian world." This difference in perspective occurs because the treatment of steady-state processes (which are the most common concern of engineers) is generally simpler in fixed coordinates, whereas the treatment of batch processes (which are the most common concern of scientists) is generally simpler for a fixed mass, as illustrated in the following special cases.

Special Cases

Steady State

For this important case Equation 8.5 reduces to

$$\nabla \cdot (\varrho \mathbf{v}) = 0 \tag{8.11}$$

This result is less obvious from Equation 8.9, since $D\varrho/Dt \neq 0$ and must be expanded and combined with the terms on the left side.

Constant Density

On the other hand, for this equally important case, Equation 8.9 gives directly

$$\nabla \cdot \mathbf{v} = 0 \tag{8.12}$$

whereas Equation 8.5 requires slightly more reduction to yield the same result. Although the density of gases varies greatly and that of liquids varies somewhat with temperature and pressure, the assumption of constant density is a reasonable approximation in most applications.

Other Coordinate Systems

Many engineering operations occur in flows through a pipe or a cylindrical vessel. In such cases cylindrical coordinates are more convenient than rec-

---

[1] Named after Leonhard Euler (1707–1783), a Swiss mathematician, and the father of *rational mechanics*.
[2] Named after Joseph-Louis Lagrange (1736–1813), a French mathematician, who can be considered the father of *mathematical analysis*.
[3] In true Lagrangian form the velocities are usually expressed as derivatives of distance with time, i.e., $\partial u_x/\partial t = \partial^2 x/\partial^2 t$, etc., as described by, for example, Cambel and Jennings [5].

tangular ones. For example, in laminar flow through a pipe, the velocity in the axial direction is a function of radius only (Equation 4.11), but in rectangular coordinates it is a function of both x and y.

Cylindrical Coordinates

The continuity equation can be rederived directly for a cylindrical volume in terms of the coordinates r, θ, and z and the velocity components u_r, u_θ, and u_z, as illustrated in Figure 8–2 (see problem 2a). The result is

$$-\left[\frac{1}{r}\frac{\partial(\varrho r u_r)}{\partial r} + \frac{1}{r}\frac{\partial(\varrho u_\theta)}{\partial \theta} + \frac{\partial(\varrho u_z)}{\partial z}\right] = \frac{\partial \varrho}{\partial t} \tag{8.13}$$

Alternatively Equation 8.13 can be obtained from Equation 8.3 using the relationships

$$x = r\cos\{\theta\} \tag{8.14}$$

$$y = r\sin\{\theta\} \tag{8.15}$$

$$z = z \tag{8.16}$$

$$u_r = u_x\cos\{\theta\} + u_y\sin\{\theta\} \tag{8.17}$$

$$u_\theta = -u_x\sin\{\theta\} + u_y\cos\{\theta\} \tag{8.18}$$

$$u_z = u_z \tag{8.19}$$

(see problem 2b). Of course, Equation 8.13 can also be inferred from Equation 8.5, which is applicable for all coordinate systems if the appropriate components of the divergence in cylindrical coordinates are used.

The expanded Lagrangian form of Equation 8.13 is

$$-\varrho\left[\frac{1}{r}\frac{\partial(ru_r)}{\partial r} + \frac{1}{r}\frac{\partial u_\theta}{\partial \theta} + \frac{\partial u_z}{\partial z}\right] = \frac{\partial \varrho}{\partial t} + u_r\frac{\partial \varrho}{\partial r} + \frac{u_\theta}{r}\frac{\partial \varrho}{\partial \theta} + u_z\frac{\partial \varrho}{\partial z} \tag{8.20}$$

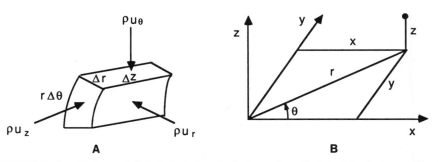

FIGURE 8–2 *Volumetric element for mass balance in cylindrical (polar) coordinates: (A) volumetric element; (B) relationship between cylindrical (polar) and Cartesian coordinates.*

Equation 8.20 can also be obtained from Equation 8.7 using Equations 8.14–8.19 or from 8.9 using the cylindrical form of the vector operators (see problem 3).

Spherical Coordinates

The relationship between rectangular and spherical coordinates, as defined in Figure 8–3, can be written as

$$x = R \sin \theta \cos \phi \tag{8.21}$$

$$y = R \sin \theta \sin \phi \tag{8.22}$$

$$z = R \cos \theta \tag{8.23}$$

$$u_R = u_x \sin \theta \cos \phi + u_y \sin \theta \sin \phi + v_z \cos \theta \tag{8.24}$$

$$u_\theta = u_x \cos \theta \cos \phi + u_y \cos \theta \sin \phi - v_z \sin \theta \tag{8.25}$$

$$u_\phi = -u_x \sin \phi + u_y \cos \phi q \tag{8.26}$$

Here the braces { } symbolizing *functions* have been deleted from the trigonometric functions in the interest of compactness. This choice of the spherical angular coordinates, θ, the *polar* or *cone angle*, and ϕ, the *azimuthal* or *bearing angle*, which is inconsistent with the choice of θ in cylindrical coordinates, was made to conform with the majority of the literature of fluid mechanics. In these coordinates the continuity equation can be shown (see problem 4) to have the Eulerian form

$$-\left[\frac{1}{R^2} \frac{\partial(R^2 \varrho u_r)}{\partial R} + \frac{1}{R \sin \theta} \frac{\partial(\varrho u_\theta \sin \theta)}{\partial \theta} + \frac{1}{R \sin \theta} \frac{\partial(\varrho u_\phi)}{\partial \phi} \right] = \frac{\partial \varrho}{\partial t} \tag{8.27}$$

and the Lagrangian form

$$-\varrho \left[\frac{1}{R^2} \frac{\partial(R^2 u_R)}{\partial R} + \frac{1}{R \sin \theta} \frac{\partial(u_\theta \sin \theta)}{\partial \theta} + \frac{1}{R \sin \{\theta\}} \frac{\partial u_\phi}{\partial \phi} \right]$$

$$= \frac{\partial \varrho}{\partial t} + u_R \frac{\partial \varrho}{\partial R} + \frac{u_\theta}{R} \frac{\partial \varrho}{\partial \theta} + \frac{u_\phi}{R \sin \{\theta\}} \frac{\partial \varrho}{\partial \phi} \tag{8.28}$$

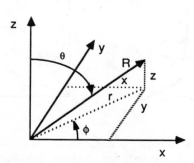

FIGURE 8–3 *Relationship between spherical and Cartesian coordinates.*

Additional Coordinate Systems

Other coordinates (see, for example, Hughes and Gaylord [6]) are advantageous in special geometries, such as flow in noncircular tubing or in a helical coil, but the previous three systems cover the most important applications.

THE FORCE-MOMENTUM BALANCE

Derivation

According to Newton's second law of motion [7][1] the time rate of change of momentum is equal to the net sum of the applied forces. For a flowing fluid, the time rate of change of momentum in an incremental volume can be equated to the net influx of momentum by gross flow, the net influx by molecular motion (diffusion), and the net sum of the pressure, gravitational, and other operative forces. For simplicity the components of momentum and force in the three orthogonal coordinate directions can be balanced separately. Consider first the forces and fluxes of momentum in the x-direction relative to the incremental volume $\Delta x\, \Delta y\, \Delta z$ of Figure 8–1. The rate of accumulation of x-momentum per unit volume is simply $\partial(\varrho u_x)/\partial t$. The net influx of x-momentum per unit time and volume due to flow in the x-direction is

$$\lim_{\Delta x \to 0} \left(\frac{(\varrho u_x u_x)_x\, \Delta y\, \Delta z - (\varrho u_x u_x)_{x+\Delta x}\, \Delta y\, \Delta z}{\Delta x\, \Delta y\, \Delta z} \right) \equiv -\frac{\partial(\varrho u_x^2)}{\partial x} \qquad (8.29)$$

Here again, the subscript outside the parentheses indicates location. The net influx of x-momentum per unit time and volume due to flow in the y-direction is similarly

$$\lim_{\Delta y \to 0} \left(\frac{(\varrho u_x u_y)_y\, \Delta x\, \Delta z - (\varrho u_x u_y)_{y+\Delta y}\, \Delta x\, \Delta z}{\Delta x\, \Delta y\, \Delta z} \right) \equiv -\frac{\partial(\varrho u_x u_y)}{\partial y} \qquad (8.30)$$

It follows that the net influx of x-momentum per unit time and volume due to flow in all three directions is

$$-\left(\frac{\partial(\varrho u_x^2)}{\partial x} + \frac{\partial(\varrho u_x u_y)}{\partial y} + \frac{\partial(\varrho u_x u_z)}{\partial z} \right)$$

This expression can be recognized as $-\nabla \cdot (\varrho u_x \mathbf{v})$ and could simply have been written down on the basis of the previously noted physical interpretation of the divergence.

---

[1] See footnote 3, p. 18.

The flux of x-momentum entering a unit area of the ith face of the incremental volume in Figure 8–4 *by molecular motion* can be symbolized by the tensor j_{ix}.[1] Or, more generally,

j_{ij} = specific rate of transfer or flux density of j-momentum in the
$\quad\quad$ i-direction due to melocular motion, kg/m·s^2

The net influx of x-momentum per unit time and volume due to molecular motion is then $-\nabla \cdot j_{ix}$.

If the pressure is assumed to be *isotropic* (uniform in all directions)[2] the resulting net pressure force per unit volume in the x-direction is

$$\lim_{\Delta x \to 0} \left(\frac{p_x \, \Delta y \, \Delta z - p_{x+\Delta x} \, \Delta y \, \Delta z}{\Delta x \, \Delta y \, \Delta z} \right) \equiv -\frac{\partial p}{\partial x} \tag{8.32}$$

The force per unit volume in the x-direction due to gravity is simply the x-component of the gravitational vector **g**, in meters per second per second or newtons per kilogram, times the density (i.e., ϱg_x).

These are the principal forces in most processing applications. Hence a momentum and force balance in the x-direction can be written as

$$\frac{\partial(\varrho u_x)}{\partial t} = -\nabla \cdot (\varrho u_x \mathbf{v}) - \nabla \cdot j_{ix} - \frac{\partial p}{\partial x} + \varrho g_x + \cdots \tag{8.33}$$

The undefined terms on the right side of Equation 8.33 indicate other, unspecified forces such as those due to electrical and magnetic fields. Equation 8.33 can be simplified somewhat as follows. Expanding the derivatives involving u_x gives

$$\varrho \frac{\partial u_x}{\partial t} + u_x \frac{\partial \varrho}{\partial t} = -u_x \nabla \cdot (\varrho \mathbf{v}) - (\mathbf{v}\varrho) \cdot \nabla u_x - \nabla \cdot j_{ix}$$

$$-\frac{\partial p}{\partial x} + \varrho g_x + \cdots \tag{8.34}$$

Then eliminating the terms adjacent to the equality sign by Equation 8.5 and rearranging gives

$$\varrho \left(\frac{\partial u_x}{\partial t} + \mathbf{v} \cdot \nabla u_x \right) = -\nabla \cdot j_{ix} - \frac{\partial p}{\partial x} + \varrho g_x + \cdots \tag{8.35}$$

---

[1] The effect of molecular motion on momentum transfer has generally been expressed indirectly as a shear stress (force per unit area). It follows that for this interpretation

$$j_{ij} = \tau_{ij} \tag{8.31}$$

where τ_{ij} is the shear stress in the i-direction on the j-plane due to molecular motion (Pa or kg/m·s^2). The transfer interpretation will, however, be utilized here just as in Part I.

[2] This is a reasonable approximation except possibly for very extreme rates of change, such as in detonation waves, or for very dilute gases, such as in plasmas or under high vacuum.

The momentum and force balances for the y- and z-directions follow simply by substituting y for z and x in Equation 8.35.

Alternative Notations

All three of these balances can be represented in *Cartesian tensor form* as

$$\varrho\left(\frac{\partial u_j}{\partial t} + u_i\frac{\partial u_j}{\partial x_i}\right) = -\frac{\partial j_{ij}}{\partial x_i} - \frac{\partial p}{\partial x_j} + \varrho g_j + \cdots \tag{8.36}$$

and in *vector form* as

$$\varrho\left(\frac{\partial \mathbf{v}}{\partial t} + (\mathbf{v} \cdot \nabla)\mathbf{v}\right) = -\nabla \cdot j_{ij} - \nabla p + \varrho\mathbf{g} + \cdots \tag{8.37}$$

Here the term $\nabla \cdot j_{ij}$ has a special meaning, which may be inferred from Equation 8.35 (see problem 5).

Alternative Forms

Equation 8.33 was derived from an Eulerian point of view. However, it is apparent that the left side of Equation 8.35 is the substantial derivative of u_x. Hence the Lagrangian forms of Equations 8.35–8.37 are

$$\varrho\frac{Du_x}{Dt} = -\nabla \cdot j_{ix} - \frac{\partial p}{\partial x} + \varrho g_x + \cdots \tag{8.38}$$

$$\varrho\frac{Du_j}{Dt} = -\frac{\partial j_{ij}}{\partial x_i} - \frac{\partial p}{\partial x_j} + \varrho g_j + \cdots \tag{8.39}$$

$$\varrho\frac{D\mathbf{v}}{Dt} = -\nabla \cdot j_{ij} + \nabla p + \varrho\mathbf{g} + \cdots \tag{8.40}$$

The form of these equations in other coordinate systems will be considered after substitution of an expression for j_{ij}.

MOLECULAR TRANSFER MOMENTUM FOR NEWTONIAN FLUIDS

For *Newtonian fluids* the rate of momentum transfer by molecular motions can be represented approximately by the empirical expressions[1]

---

[1] These expressions can be derived by statistical mechanics, as illustrated by Chapman and Cowling [8]. Higher-order terms are discussed by Tsien [9], Hirschfelder et al. [10], and Karim and Rosenhead [11].

$$j_{ij} = -\mu\left(\frac{\partial u_i}{\partial x_j} + \frac{\partial u_j}{\partial x_i}\right) \qquad i \neq j \tag{8.41}$$

$$= -2\mu\left(\frac{\partial u_i}{\partial x_j} - \frac{1}{3}\nabla \cdot \mathbf{v}\right) \qquad i = j \tag{8.42}$$

For constant density ($\nabla \cdot \mathbf{v} = 0$), Equation 8.41 holds for $i = j$ as well as $i \neq j$. Equation 2.1 is a special case of Equation 8.41 for one-dimensional motion.

Expressions for the shear stress and normal stress due to molecular motion were originally derived by Stokes [12], based on a presumed analogy to the deformation of elastic solids. These expressions are equivalent to Equations 8.41 and 8.42. A modern derivation may be found in Schlichting [13], pp. 49–61.

Neither Equations 8.41 and 8.42 nor the resulting force-momentum balances, such as Equation 8.46, have ever been confirmed experimentally in the general sense. However, as Schlichting [13], p. 66, notes, the solutions that have been obtained for laminar flow under highly simplified conditions agree with experiments, and hence the general validity of these equations "can hardly be doubted."

THE FORCE-MOMENTUM BALANCE FOR NEWTONIAN FLUIDS

Derivation

From Equations 8.41 and 8.42 for the x-direction,

$$-\nabla \cdot j_{ix} = \frac{\partial}{\partial x}\left(2\mu\left(\frac{\partial u_x}{\partial x} - \frac{1}{3}\nabla \cdot \mathbf{v}\right)\right) + \frac{\partial}{\partial y}\left(\mu\left(\frac{\partial u_x}{\partial y} + \frac{\partial u_y}{\partial x}\right)\right)$$
$$+ \frac{\partial}{\partial z}\left(\mu\left(\frac{\partial u_x}{\partial z} + \frac{\partial u_z}{\partial x}\right)\right) \tag{8.43}$$

By regrouping, the terms in this expression can be written more compactly as

$$-\nabla \cdot j_{ix} = \nabla \cdot \mu\left(\frac{\partial \mathbf{v}}{\partial x} + \nabla u_x\right) - \frac{2}{3}\frac{\partial}{\partial x}(\mu\nabla \cdot \mathbf{v}) \tag{8.44}$$

where

$$\frac{\partial \mathbf{v}}{\partial x} = i\frac{\partial u_x}{\partial x} + j\frac{\partial u_y}{\partial y} + k\frac{\partial u_z}{\partial z} \tag{8.45}$$

Introducing Equation 8.44 in Equation 8.38 gives

$$\varrho\frac{Du_x}{Dt} = \varrho g_x - \frac{\partial p}{\partial x} + \nabla \cdot \mu\left(\frac{\partial \mathbf{v}}{\partial x} + \nabla u_x\right) - \frac{2}{3}\frac{\partial}{\partial x}(\mu\nabla \cdot \mathbf{v}) + \cdots \tag{8.46}$$

Again the y- and z-equations can be obtained simply by replacing x with y and z, respectively.

Special Cases

The following special cases may be obtained by reduction, again, for simplicity, for the x-direction only.

Constant μ

$$\varrho \frac{Du_x}{Dt} = \varrho g_x - \frac{\partial p}{\partial x} + \mu \left[\nabla \cdot \left(\frac{\partial \mathbf{v}}{\partial x} + \nabla u_x \right) - \frac{2}{3} \frac{\partial}{\partial x} (\nabla \cdot \mathbf{v}) \right] + \cdots \quad (8.47)$$

Since

$$\nabla \cdot \frac{\partial \mathbf{v}}{\partial x} = \frac{\partial}{\partial x} (\nabla \cdot \mathbf{v}) \quad (8.48)$$

then

$$\varrho \frac{Du_x}{Dt} = \varrho g_x - \frac{\partial p}{\partial x} + \mu \left(\nabla^2 u_x + \frac{1}{3} \frac{\partial}{\partial x} (\nabla \cdot \mathbf{v}) \right) + \cdots \quad (8.49)$$

where, in Cartesian coordinates, the *Laplacian operator*[1]

$$\nabla^2 = \frac{\partial^2}{\partial x^2} + \frac{\partial^2}{\partial y^2} + \frac{\partial^2}{\partial z^2} \quad (8.50)$$

As noted in Chapter 2, the viscosity of most fluids varies strongly with temperature but only moderately with pressure. Thus this idealization is strictly valid only for essentially isothermal flows.

Constant ϱ

$$\varrho \frac{Du_x}{Dt} = \varrho g_x - \frac{\partial p}{\partial x} + \nabla \cdot \mu \left(\frac{\partial \mathbf{v}}{\partial x} + \nabla u_x \right) + \cdots \quad (8.51)$$

This idealization is generally valid for liquids. Also, because of the great simplification that results, the assumption of a constant, mean density is often justifiable as an approximation for gases.

Constant ϱ *and* $\mathbf{g}$

Except in space flights and some geophysical problems, the assumption of a constant gravitational force per unit mass is a good approximation. Two special

---

[1] Named after Pierre-Simon de Laplace (1749–1827), a French mathematician whose greatest contributions were in celestial mechanics (in which the *Laplacian operator* and *Laplace equation* first arose) and probability.

variables are often introduced to take advantage of the postulate of constant ϱ and $\mathbf{g}$.

First, a dynamic pressure may be defined according to Equation 1.10. Then since

$$\varrho g_x = -\varrho g \frac{\partial h}{\partial x} \tag{8.52}$$

where $g = |\mathbf{g}|$, m/s^2 or N/kg and h = elevation (i.e., distance in the direction opposite to $\mathbf{g}$) in meters,

$$\varrho g_x - \frac{\partial p}{\partial x} = -\varrho g \frac{\partial h}{\partial x} - \frac{\partial \mathscr{P}}{\partial x} + g \varrho \frac{\partial h}{\partial x} = -\frac{\partial \mathscr{P}}{\partial x} \tag{8.53}$$

and

$$\varrho \frac{Du_x}{Dt} = -\frac{\partial \mathscr{P}}{\partial x} + \nabla \cdot \mu \left(\frac{\partial \mathbf{v}}{\partial x} + \nabla u_x \right) + \cdots \tag{8.54}$$

The dynamic pressure $\mathscr{P}$ in Equation 8.54 depends only on the velocity, but the thermodynamic pressure p in Equation 8.51 depends on the elevation as well.

Alternatively, a *piezometric head* $\hbar$ may be defined for constant ϱ as

$$\hbar = h + \frac{p}{\varrho g} \tag{8.55}$$

leading to

$$\varrho \frac{Du_x}{Dt} = -g \varrho \hbar + \nabla \cdot \mu \left(\frac{\partial \mathbf{v}}{\partial x} + \nabla u_x \right) + \cdots \tag{8.56}$$

In formulations such as Equation 8.56, in which the pressure gradient has been eliminated, and in which ϱ is constant, it is convenient to combine μ and ϱ as

$$\nu = \frac{\mu}{\varrho} = \text{kinematic viscosity, m}^2\text{/s}$$

Thus Equation 8.56 becomes

$$\frac{Du_x}{Dt} = -g \hbar + \nabla \cdot \nu \left(\frac{\partial \mathbf{v}}{\partial x} + \nabla u_x \right) \tag{8.56A}$$

Constant μ *and* ϱ

$$\varrho \frac{Du_x}{Dt} = \varrho g_x - \frac{\partial p}{\partial x} + \mu \nabla^2 u_x + \cdots \tag{8.57}$$

Negligible μ

$$\varrho \frac{Du_x}{Dt} = \varrho g_x - \frac{\partial p}{\partial x} \tag{8.58}$$

The equations for the y- and z-directions corresponding to Equations 8.47–8.58 may again be obtained simply by substituting y and z, respectively, for x. These balances can also be written in Cartesian tensor and vector forms (see problem 6 or Hughes and Gaylord [6]).

Equation 8.58 is named after Euler, who derived it in 1755 [14]. Equation 8.57 and the corresponding pair for the y- and z- directions are known as the *Navier–Stokes equations* after Navier,[1] who first derived this equation in 1822 [15] on the basis of intermolecular arguments, and Stokes,[2] who first derived the more general form of Equation 8.49 in 1845 for compressible flow and without such molecular hypotheses [12]. According to Dryden et al. [16], Poisson [17] in 1831 independently carried out essentially the same derivation as Navier, and St. Venant [18] in 1843 carried out the same derivation as Stokes.

Other Coordinate Systems

Equation 8.46 and the corresponding expressions for the y- and z-directions can be transformed to cylindrical and spherical coordinates using 8.14–8.19 and 8.21–8.26, respectively (see problem 7). The results follow.

Cylindrical Coordinates

$$\varrho \left[\frac{Du_r}{Dt} - \frac{u_\theta^2}{r} \right] = \varrho g_r - \frac{\partial p}{\partial r} + \frac{\partial}{\partial r}\left[2\mu\left(\frac{\partial u_r}{\partial r} - \frac{1}{3}\nabla \cdot \mathbf{v} \right) \right]$$

$$+ \frac{1}{r}\frac{\partial}{\partial \theta}\left[\mu\left(\frac{1}{r}\frac{\partial u_r}{\partial \theta} + \frac{\partial v_\theta}{\partial r} - \frac{v_\theta}{r} \right) \right] + \frac{\partial}{\partial z}\left[\mu\left(\frac{\partial u_r}{\partial z} + \frac{\partial u_z}{\partial r} \right) \right]$$

$$+ \frac{2\mu}{r}\left[\frac{\partial u_r}{\partial r} - \frac{1}{r}\frac{\partial u_\theta}{\partial \theta} - \frac{u_r}{r} \right] + \cdots \tag{8.59}$$

$$\varrho \left[\frac{Du_\theta}{Dt} + \frac{u_r u_\theta}{r} \right] = \varrho g_\theta - \frac{1}{r}\frac{\partial \varrho}{\partial \theta} + \frac{\partial}{\partial r}\left[\mu\left(\frac{1}{r}\frac{\partial u_r}{\partial \theta} + \frac{\partial u_\theta}{\partial r} - \frac{u_\theta}{r} \right) \right]$$

$$+ \frac{1}{r}\frac{\partial}{\partial \theta}\left[2\mu\left(\frac{1}{r}\frac{\partial u_r}{\partial \theta} - \frac{1}{3}\nabla \cdot \mathbf{v} \right) \right] + \frac{\partial}{\partial z}\left[\mu\left(\frac{1}{r}\frac{\partial u_z}{\partial z} + \frac{\partial u_\theta}{\partial z} \right) \right]$$

$$+ \frac{2\mu}{r}\left[\frac{1}{r}\frac{\partial u_r}{\partial \theta} + \frac{\partial u_\theta}{\partial r} - \frac{u_\theta}{r} \right] + \cdots \tag{8.60}$$

and

---

[1] Claude-Louis Marie Henri Navier (1785–1836), a French civil engineer, was one of the first to develop a theory of elasticity.
[2] George Gabriel Stokes (1819–1903) was an Irish mathematician and physicist who made extensive contributions to fluid mechanics.

$$\varrho\frac{Du_z}{Dt} = \varrho g_z - \frac{\partial p}{\partial z} + \frac{1}{r}\frac{\partial}{\partial r}\left[\mu r\left(\frac{\partial u_r}{\partial z} + \frac{\partial u_z}{\partial r}\right)\right]$$

$$+ \frac{1}{r}\frac{\partial}{\partial\theta}\left[\mu\left(\frac{1}{r}\frac{\partial u_z}{\partial\theta} + \frac{\partial u_\theta}{\partial z}\right)\right]$$

$$+ \frac{\partial}{\partial z}\left[2\mu\left(\frac{\partial u_z}{\partial z} - \frac{1}{3}\nabla\cdot\mathbf{v}\right)\right] + \cdots \tag{8.61}$$

The form of the operators D/Dt and ∇ in cylindrical coordinates can be inferred from Equation 8.20 (also see problem 3).

The term $\varrho u_\theta^2/r$ in Equation 8.59 is the flux density of r-momentum due to flow in the θ-direction and arises naturally when the y-analog of Equation 8.38 or 8.46 is transformed into cylindrical coordinates. It is sometimes called the *centrifugal force*. The term $\varrho u_r u_\theta/r$ in Equation 8.60 is the flux density of θ-momentum due to flow in the r-direction and similarly arises naturally upon transformation of coordinates. It is sometimes called the *Coriolis force*.[1] The Coriolis force causes the opposing directions of circulation of the trade winds in the northern and southern hemispheres of the earth, and may also be important in flow over oblique rotating surfaces. These two "forces" are implicit in the former equations, which is a good reason for not interpreting fluxes of momentum as forces, or flux densities of momentum as shear stresses. Also, if Equations 8.59 and 8.60 or the corresponding "shell" balances were derived directly in cylindrical coordinates the centrifugal and Coriolis terms would need to be introduced heuristically.

The reduced expressions for constant viscosity, constant density, constant viscosity *and* density, and negligible viscosity can readily be obtained from the previous general expressions (see problem 9). For that purpose, in cylindrical coordinates,

$$\nabla^2 = \frac{\partial^2}{\partial r^2} + \frac{1}{r}\frac{\partial}{\partial r} + \frac{1}{r^2}\frac{\partial^2}{\partial\theta^2} + \frac{\partial^2}{\partial z^2} \tag{8.62}$$

Spherical Coordinates

$$\varrho\left[\frac{Du_R}{Dt} - \frac{u_\theta^2 + u_\phi^2}{R}\right] = \varrho g_R - \frac{\partial p}{\partial R} + \frac{\partial}{\partial R}\left[2\mu\left(\frac{\partial u_R}{\partial R} - \frac{1}{3}\nabla\cdot\mathbf{v}\right)\right]$$

$$+ \frac{1}{R}\frac{\partial}{\partial\theta}\left[\mu\left(\frac{R\partial(u_\theta/R)}{\partial R} + \frac{1}{R}\frac{\partial u_R}{\partial\theta}\right)\right]$$

$$+ \frac{1}{R\sin\theta}\frac{\partial}{\partial\phi}\left[\mu\left(\frac{1}{R\sin\theta}\frac{\partial u_R}{\partial\phi} + R\frac{\partial(u_\phi/R)}{\partial R}\right)\right]$$

$$+ \frac{\mu}{R}\left[4\frac{\partial u_R}{\partial R} - \frac{2}{R}\frac{\partial u_\theta}{\partial\theta} - \frac{4u_R}{R} - \frac{2}{R\sin\theta}\frac{\partial u_\phi}{\partial\phi}\right]$$

---

[1] Named after Gaspard Gustave de Coriolis (1792–1843), a French physicist who developed the first modern definitions of kinetic energy and work and who recognized the significance of the forces that effect a rotating fluid.

$$\left. - \frac{2u_\theta \cot\theta}{R} + R\cot\theta \frac{\partial(u_\theta/R)}{\partial R} + \frac{\cot\theta}{R}\frac{\partial u_R}{\partial\theta} \right]$$

$$+ \cdots \tag{8.63}$$

$$\varrho\left[\frac{Du_\theta}{Dt} + \frac{u_R u_\theta}{R} - \frac{u_\phi^2 \cot\theta}{R}\right] = \varrho g_\theta - \frac{1}{R}\frac{\partial p}{\partial\theta} + \frac{\partial}{\partial R}\left[\mu\left(R\frac{\partial(u_\theta/R)}{\partial R} + \frac{1}{R}\frac{\partial u_R}{\partial\theta}\right)\right]$$

$$+ \frac{1}{R}\frac{\partial}{\partial\theta}\left[2\mu\left(\frac{1}{R}\frac{\partial u_\theta}{\partial\theta} + \frac{u_R}{R} - \frac{1}{3}\nabla\cdot\mathbf{v}\right)\right]$$

$$+ \frac{1}{R\sin\theta}\frac{\partial}{\partial\phi}\left[\mu\left(\frac{\sin\theta}{R}\frac{\partial(u_\phi/\sin\theta)}{\partial\theta} + \frac{1}{R\sin\theta}\frac{\partial u_\theta}{\partial\phi}\right)\right]$$

$$+ \frac{\mu}{R}\left[2\left(\frac{1}{R}\frac{\partial u_\theta}{\partial\theta} - \frac{1}{R\sin\theta}\frac{\partial u_\phi}{\partial\phi} - \frac{u_\theta\cot\theta}{R}\right)\cot\theta\right.$$

$$\left. + 3\left(R\frac{\partial(u_\theta/R)}{\partial R} + \frac{1}{R}\frac{\partial u_R}{\partial\theta}\right)\right] + \cdots \tag{8.64}$$

and

$$\varrho\left[\frac{Du_\phi}{Dt} + \frac{u_\phi u_R}{R} + \frac{u_\theta u_\phi \cot\theta}{R}\right] = \varrho g_\phi - \frac{1}{R\sin\theta}\frac{\partial p}{\partial\phi}$$

$$+ \frac{\partial}{\partial R}\left[\mu\left(\frac{1}{R\sin\theta}\frac{\partial u_R}{\partial\phi} + R\frac{\partial(u_\phi/R)}{\partial R}\right)\right]$$

$$+ \frac{1}{R}\frac{\partial}{\partial\theta}\left[\frac{\mu}{R}\left(\sin\theta\frac{\partial(u_\phi/\sin\theta)}{\partial\theta} + \frac{1}{\sin\theta}\frac{\partial u_\theta}{\partial\phi}\right)\right]$$

$$+ \frac{1}{R\sin\theta}\frac{\partial}{\partial\phi}\left(2\mu\left(\frac{1}{R\sin\theta}\frac{\partial u_\phi}{\partial\phi} + \frac{u_R}{R}\right.\right.$$

$$\left.\left. + \frac{u_\theta\cot\theta}{R} - \frac{1}{3}\nabla\cdot\mathbf{v}\right)\right]$$

$$+ \frac{\mu}{R}\left[\frac{2}{R}\left(\sin\theta\frac{\partial(u_\phi/\sin\theta)}{\partial\theta} + \frac{1}{\sin\theta}\frac{\partial u_\theta}{\partial\phi}\right)\cot\theta\right.$$

$$\left. + 3\left(\frac{1}{R\sin\theta}\frac{\partial u_R}{\partial\phi} + R\frac{\partial(u_\phi/R)}{\partial R}\right)\right] + \cdots \tag{8.65}$$

Again the braces indicating function have been deleted from the trigonometric functions in the interest of compactness. The form of the substantial derivative and of the divergence in spherical coordinates can be inferred from Equation 8.28 (see problem 4c).

Expressions for the reduced cases can readily be obtained from the foregoing general expressions (see problem 8). For that purpose, in spherical coordinates,

$$\nabla^2 = \frac{1}{R^2}\frac{\partial}{\partial R}\left(R^2\frac{\partial}{\partial R}\right) + \frac{1}{R^2\sin\theta}\frac{\partial}{\partial\theta}\left(\sin\theta\frac{\partial}{\partial\theta}\right) + \frac{1}{R^2\sin^2\theta}\frac{\partial^2}{\partial\phi^2} \tag{8.66}$$

Additional Coordinate Systems

Expressions for the force-momentum balances in general orthogonal curvilinear coordinate systems are provided by Hughes and Gaylord [6].

Eulerian Form

The foregoing expressions for the force-momentum balances for a Newtonian fluid are all in Lagrangian form. They can readily be converted to Eulerian form by reversing the procedure illustrated for the equation of continuity (see problem 11).

INTEGRAL FORMULATIONS

Integral expressions for the conservation of mass and momentum can also be formulated. These expressions are more convenient than the differential ones in describing the overall consequences of fluid flow (e.g., the pressure drop), as compared to the local behavior (e.g., the velocity field). The differential formulations can be derived, as shown here, from the integral formulations, and vice versa.

The differential equations of conservation are utilized exclusively throughout this book because their derivation, manipulation, and solution appear to have a greater physical and lesser mathematical basis than the equivalent integral forms. The following derivations are therefore included only for completeness and to provide a stepping-stone to integral treatments of fluid flow such as in Shames [20] and Bird et al. [1], Chapter 7. This section may be omitted with no loss of material prerequisite to the rest of the book.

Conservation of Mass

The basis for the integral formulations of the laws of conservation is the *transport theorem of Reynolds* [21][1]

$$\frac{\partial}{\partial t} \int \int_{V_C} \int \eta \varrho \, dV = \frac{\mathscr{D} N}{\mathscr{D} t} - \int \int_{S_C} \eta \varrho (\mathbf{v} \cdot \mathbf{n}) \, dS \qquad (8.66)$$

where N = the amount of an extensive property such as the mass in V
   η = the local mass concentration of N
   V = volume, m^3
   V_C = arbitrary *control volume*, m^3
   S = surface area, m^2
   S_C = *control surface* corresponding to V_C, m^2
   $\mathbf{n}$ = unit (outward) vector normal to S

---

[1] Named after Osborne Reynolds (see Chapter 1).

This expression equates the time rate of change of N, as obtained by integrating $\varrho\eta$ over V_C, with the total rate of generation of N in V_C minus the net efflux of N, as obtained by integrating $\varrho\eta(\eta \cdot v)$ over S_C.

For mass mass, $\mathscr{D}\mathscr{N}/\mathscr{D}t = 0$ and $\eta = 1$. Hence, the integral equation for the conservation of mass is

$$\frac{\partial}{\partial t} \int\int\int_{V_C} \varrho \, dV = - \int\int_{S_C} \varrho(\mathbf{v} \cdot \mathbf{n}) \, dS \tag{8.67}$$

The *divergence theorem of Gauss*[1] is

$$\int\int_{S_C} (A \cdot \eta) \, dS = \int\int\int_{V_C} (\nabla \cdot A) \, dV \tag{8.68}$$

where A may be a scalar (with the symbol for a *dot product* then eliminated), vector, or tensor. Equation 8.68 with $A = \varrho v$ can be used to transform Equation 8.67 to

$$\frac{\partial}{\partial t} \int\int\int_{V_C} \varrho \, dV = - \int\int\int_{V_C} [\nabla \cdot (\varrho v)] \, dV \tag{8.69}$$

which can be rewritten as

$$\int\int\int_{V_C} \left[\frac{\partial \varrho}{\partial t} + \nabla \cdot (\varrho v)\right] dV = 0 \tag{8.69A}$$

Since the volume of integration is arbitrary, the integrand must be zero, yielding Equation 8.5.

Conservation of Momentum

The momentum per unit mass is $\mathbf{v}$, hence Equation 8.66 becomes

$$\frac{\mathscr{D}M}{\mathscr{D}t} = \frac{\partial}{\partial t} \int\int\int_{V_C} \varrho v \, dV - \int\int_{S_C} \varrho \mathbf{n}(\mathbf{v} \cdot \mathbf{v}) \, dS \tag{8.70}$$

where M is the total momentum of fluid within V_C ($\text{kg} \cdot \text{m/s}^2$). From *Newton's second law of motion*, the total rate of generation of momentum is equal to the sum of the applied forces acting on the control volume; that is,

$$\frac{\mathscr{D}M}{\mathscr{D}t} = \int\int_{S_C} \varrho\eta \, dS + \int\int\int_{V_C} \varrho g \, dV + \int\int_{S_C} j_{ij} \, dS + \cdots \tag{8.71}$$

---

[1] Named after Karl Friedrich Gauss (1777–1855) of Germany, generally considered to be the greatest mathematician of all time.

Equating the right sides of Equations 8.70 and 8.71 and replacing the surface integrals with volume integrals through Equation 8.68 give, after rearranging,

$$\int \int \int_{V_C} \left[\frac{\partial}{\partial t}(\varrho v) + \nabla(\varrho v \cdot v) + \nabla p - \varrho g + \nabla \cdot j_{ij} + \cdots \right] dV = 0 \quad (8.72)$$

Again, since the volume of integration is arbitrary, the integrand must vanish everywhere; that is,

$$\frac{\partial(\varrho v)}{\partial t} + \nabla(\varrho v \cdot v) = -\nabla p + \varrho g - \nabla \cdot j_{ij} + \cdots = 0 \quad (8.73)$$

Expanding the left side of Equation 8.73, subtracting Equation 8.5, and recombining the remaining terms on the left side finally give Equation 8.40.

THE ENERGY BALANCE

The motion of a fluid is completely described by the foregoing equations for the conservation of mass and momentum insofar as the variation of the physical properties of the fluid, particularly the density and viscosity, with temperature is negligible. If the effect of temperature on the physical properties is significant, the temperature distribution within the flowing fluid must be defined by an energy balance. Significant temperature variations within a fluid may arise from chemical reactions, heat transfer from the surroundings, compression or expansion of the fluid, and viscous dissipation. Extreme conditions, such as those near a space vehicle during its reentry into the atmosphere, are required if the latter two effects are to be significant. Only essentially isothermal flows are considered in this book. Hence, the energy balance for a fluid will be omitted. For a detailed discussion of the energy balance for a flowing fluid, see, for example, Bird et al. [1], Chapter 10, or Churchill [22], Chapter 1.

SUMMARY

The equations developed in this chapter serve as a formal starting point for the derivation of analytical and numerical solutions for problems in fluid mechanics. Memorization of all of these detailed expressions is not necessary, although at least the basic forms generally become part of the mental database of those who work in the field. A physical understanding of the role of the various terms and also of the idealizations inherent in the various reduced forms is essential to anyone dealing even casually with fluid mechanics.

Manipulation and reduction of these general equations will be found to be clearly adventageous as compared to derivation of specialized mass and force-momentum balances for the two-and three-dimensional laminar flows examined in Part III. The turbulent flows described in a companion volume [4] are inherently unsteady and three-dimensional. Hence, it might be inferred that the general approach would be essential. However, no solutions have yet been attained for this three-dimensional, unsteady behavior without introducing some empiricism. Most practical applications involve laminar and turbulent flows that are stationary and one-dimensional on-the-mean. One-dimensional time-averaged representations for these reduced cases can be constructed by shell balances as readily as by reduction of the general equations. Flows through porous and dispersed media are inherently multidimensional even in the laminar regime, but can be treated as quasi-one-dimensional in most practical cases.

Special forms of the equations of motion that are useful in particular applications are explored in Chapter 9, and exact and asymptotic solutions are covered in Chapter 10.

PROBLEMS

1. Derive Equation 8.7 by considering the change in volume of a mass of fluid as it moves through time and space.
2. Derive Equation 8.13

   a. by making a mass balance according to Figure 8–2
   b. from Equation 8.3 using Equations 8.14–8.19

3. a. Derive Equation 8.20 from Equation 8.7 using Equations 8.14–8.19.
   b. Determine the form of the divergence and the substantial derivative in cylindrical coordinates from Equation 8.20.

4. a. Derive Equations 8.27 or 8.28 from Equations 8.3 or 8.7 using Equations 8.21–8.26.
   b. Derive Equation 8.27 by making a mass balance according to Figure 8–3.
   c. Determine the form of the divergence and the substantial derivative in spherical coordinates from Equations 8.27 or 8.28.

5. Expand $\nabla \cdot j_{ij}$ in Equation 8.37 by referring to Equation 8.35.
6. Express Equations 8.49, 8.51, 8.54, 8.56–8.58 in

   a. Cartesian tensor form
   b. vector form

7. Derive Equations 8.59–8.61 starting from Equation 8.46

   a. for constant μ and ϱ
   b. for constant μ
   c. for constant ϱ
   d. in general

8. Derive Equations 8.63–8.65 starting from Equation 8.46

   a. for constant μ and ϱ
   b. for constant μ
   c. for constant ϱ
   d. in general

9. Reduce Equations 8.59–8.61 and Equations 8.63–8.65 for

    a.   constant μ
    b.   constant ϱ
    c.   constant μ and ϱ
    d.   negligible viscosity

    c can be checked against the expressions given by Bird et al. [1], pp. 85, 87, and a–d can be checked against the expressions given by Hughes and Gaylord [6] and others.

10. Interpret the individual terms on the left side of Equations 8.63–8.65. Why are these fluxes not included in the substantial derivative?

11. Convert the terms on the left side of the following Equations to Eulerian form:

    a.   8.39
    b.   8.40
    c.   8.46, 8.49, and 8.51
    d.   8.59–8.61
    e.   8.63–8.65

12. Reduce the three-dimensional, unsteady equations of motion to those governing fully developed, steady, laminar flow of a Newtonian fluid in the following geometries. Identify and justify each simplification:

    a.   between two flat plates
    b.   in a tube
    c.   down an inclined plate
    d.   radially between two concentric cylindrical surfaces; also, derive a relationship between the pressure drop and the mass rate of flow
    e.   radially between two concentric spherical surfaces
    f.   adjacent to a plate moving parallel to a fixed plate

13. Reduce the general equations as far as possible for one-dimensional, steady (laminar) flow of a compressible gas with constant viscosity.

14. Two infinite, horizontal parallel plates are separated by a distance d. This space is occupied by a Newtonian fluid of constant density and viscosity. Both plates and the fluid are initially at rest. The top plate is then rapidly accelerated to a constant velocity u_p. Show by reduction of the equations of motion that the transient fluid motion can be represented by $\partial u/\partial t = v(\partial^2 u/\partial y^2)$ and the appropriate boundary conditions.

15. Determine the components of the shear stress corresponding to Equations 8.41 and 8.42 in

    a.   cylindrical coordinates
    b.   spherical coordinates

16. Under what conditions is the following equality valid?

$$\nabla \cdot \left(\mu \frac{\partial \mathbf{v}}{\partial x} \right) = \frac{\partial}{\partial x} \left(\mu \nabla \cdot \mathbf{v} \right)$$

REFERENCES

1. R. B. Bird, W. E. Stewart, and E. N. Lightfoot, *Transport Phenomena*, John Wiley, New York (1960).
2. J. C. Slattery, *Momentum, Energy and Mass Transfer in Continuum*, McGraw-Hill, New York (1972).
3. S. W. Churchill, *The Practical Use of Theory in Fluid Flow. Book III. Laminar, Multidimensional Flows in Channels*, Notes, The University of Pennsylvania (1979).
4. S. W. Churchill, *The Practical Use of Theory in Fluid Flow. Book IV. Turbulent Flows*, Notes, The University of Pennsylvania (1981).
5. A. B. Cambel and B. G. Jennings, *Gas Dynamics*, McGraw-Hill, New York (1958), p. 210.
6. W. F. Hughes and E. W. Gaylord, *Basic Equations of Engineering Science*, McGraw-Hill, New York (1964).
7. Isaac Newton, *Principia*, Vol. I, *The Motion of Bodies*, S. Pepys, London (1686); English transl. of 2nd ed. (1713) by Andrew Motte (1729); revised transl. by Florian Cajori, University of California Press, Berkeley, (1966), p. 13.
8. S. Chapman and T. G. Cowling, *The Mathematical Theory of Non-Uniform Gases*, Cambridge University Press (1939), Chap. 3.
9. H. S. Tsien, "The Equations of Gas Dynamics," p. 12 in *Fundamentals of Gas Dynamics*, Section A, Princeton University Press, Princeton, NJ (1958).
10. J. O. Hirschfelder, C. F. Curtiss, and R. B. Bird, *The Molecular Theory of Gases and Liquids*, John Wiley, New York (1954), pp. 503, 647.
11. S. M. Karim and L. Rosenhead, "The Second Coefficient of Viscosity of Liquids and Gases," *Rev. Mod. Phys.*, *24* (1952) 108.
12. G. G. Stokes, "On the Theories of Internal Friction of Fluids in Motion, and of the Equilibrium and Motion of Elastic Solids," *Trans. Camb. Phil. Soc.*, *8* (1845) 287, (*Mathematical and Physical Papers*, Vol. I, Cambridge University Press (1880) p. 75).
13. H. Schlichting, *Boundary Layer Theory*, 7th ed., English transl. by J. Kestin, McGraw-Hill, New York (1979).
14. L. Euler, "Principes généraux du mouvement des fluides," *Hist. Acad. Berlin* (1755).
15. C.-L. M. N. Navier, "Mémoire sur les lois du mouvement des fluides," *Mém. Acad. Roy. Sci.*, *6* (1822) 389.
16. H. L. Dryden, F. P. Murnaghan, and H. Bateman, *Hydrodynamics*, Dover, New York (1956), p. 91.
17. S. D. Poisson, "Mémoire sur les équations générales de l'equilibre et du mouvement des corps solides élastique et des fluides," *J. Ecole Polytechnique*, *13* (1831) 139.
18. A.-J.-C. B. de Saint-Venant, "Note à joindre au mémoire sur la dynamique des fluids, présenté le 14 avril 1834," *Compt. Rend. Acad. Sci., Paris*, *17* (1843) 1240.
19. G. Coriolis, "Sur l'établissement de la formule qui donne la figure des remous et de la correction à y introduire," *Ann. Ponts et Chausses*, *1* (1836) 314.
20. I. H. Shames, *Mechanics of Fluids*, 2nd ed., McGraw-Hill, New York (1982).
21. Osborne Reynolds, "On the Dynamical Theory of an Incompressible Viscous Fluid and the Determination of the Criterion," *Phil. Trans. Roy. Soc. (London)*, Ser. 3, A186 (1896) 123, (*Scientific Papers*, Vol. II, Cambridge University Press (1901) p. 535).
22. S. W. Churchill, *The Practical Use of Theory in Heat Transfer—Laminar Forced Convection*, Notes, The University of Pennsylvania (1977).

Chapter 9

Modified Forms of the General Mass and Force-Momentum Balances

Several modified forms of the basic equations of conservation of mass and momentum given, in general form, in Chapter 8 have been found useful in interpreting fluid flows and in obtaining solutions. A few such modified representations are examined in this chapter. Some of these modified forms require minor constraints such as constant density, and/or constant viscosity, and one- or two-dimensionality. Others are applicable only for the *limiting conditions* of *creeping flow* (negligible inertia) or of *potential flow* (for a fluid with zero viscosity under conditions of zero *rotation*).

Several of the modified forms of the equations of conservation establish a distinct class of behavior in the mathematical sense and have their own associated techniques of solution. Indeed, fluid mechanics can be classified mathematically in terms of these modified forms more readily than in terms of the idealizations resulting in Equations 8.47–8.58. In addition, and perhaps of equal importance, several of the modified forms provide physical insight concerning fluid-mechanical behavior.

Attention is first given to those modified formulations that are applicable to general conditions of flow, except in most instances for the mild restrictions of two-dimensionality and constant density, viscosity, and acceleration due to gravity. The specialization of these forms for *creeping* flow is next considered, and finally the specialization of these forms and the introduction of further useful modifications for *inviscid* and *potential* flows.

The restrictions of two-dimensionality, a steady state, negligible viscosity, and negligible inertia each preclude turbulent motion.

SPECIAL FORMULATIONS FOR GENERAL CONDITIONS

The Stream Function

The equations of motion for *constant density, viscosity, and acceleration due to gravity, and two-dimensional, rectilinear motion* can be reduced from three coupled equations (continuity, x-momentum, and y-momentum) in three dependent variables (u_x, u_y, and ϕ) to a single higher-order equation in only one dependent variable by the introduction of the *Lagrange stream function* $\psi\{t, x, y\}$, defined by

$$u_x \equiv -\frac{\partial \psi}{\partial y} \tag{9.1}$$

and

$$u_y \equiv \frac{\partial \psi}{\partial x} \tag{9.2}$$

The dimensions of $\psi\{t, x, y\}$ are square meters per second. This stream function,[1] which was introduced by Lagrange[2] [1] in 1781, but whose fluid-mechanical interpretation awaited Rankine [2] 83 years later, satisfies the equation of continuity exactly.

For a *steady-state flow*, a constant value of the stream function corresponds to a *streakline* (the path of a particle of fluid) and is called a *streamline*. The average velocity between any two streamlines is equal to the difference of the values of the corresponding stream functions divided by the distance between them (see problem 1). Thus the stream function provides a quantitative measure of the rate of flow as well as a picture of the motion.

The solid surfaces that confine the flow constitute the limiting stream functions. The stream function is defined in terms of derivatives, and hence its absolute value in steady-state flow depends on an arbitrary choice of an additive constant. This constant is usually made zero at one of the boundaries.

The mathematical reduction mentioned earlier can be outlined as follows. Substituting ψ for u and v in the x-momentum equation (Equation 8.57) gives

$$\varrho\left(\frac{\partial^2 \psi}{\partial t \partial y} - \frac{\partial \psi}{\partial y}\frac{\partial^2 \psi}{\partial x \partial y} + \frac{\partial \psi}{\partial x}\frac{\partial^2 \psi}{\partial y^2}\right) = -g_x\varrho + \frac{\partial p}{\partial x} + \mu\left(\frac{\partial^3 \psi}{\partial x^2 \partial y} + \frac{\partial^3 \psi}{\partial y^3}\right) \tag{9.3}$$

Differentiating this expression with respect to y and subtracting the analogous expression for momentum transfer in the y-direction, after differentiation with respect to x, eliminates the gravitational and pressure terms and yields

$$\frac{\partial^3 \psi}{\partial y \partial t \partial y} + \frac{\partial^3 \psi}{\partial x \partial t \partial x} - \frac{\partial \psi}{\partial y}\left(\frac{\partial^3 \psi}{\partial y \partial x \partial y} + \frac{\partial^3 \psi}{\partial x^3}\right) + \frac{\partial \psi}{\partial x}\left(\frac{\partial^3 \psi}{\partial x \partial y \partial x} + \frac{\partial^3 \psi}{\partial x^3}\right)$$
$$= v\left(\frac{\partial^4 \psi}{\partial x^4} + \frac{\partial^4 \psi}{\partial y \partial x \partial x \partial y} + \frac{\partial^4 \psi}{\partial x \partial y \partial y \partial x} + \frac{\partial^4 \psi}{\partial y^4}\right) \tag{9.4}$$

which, through the use of the ∇ operator, can be written more compactly as

---

[1] Some authors, particularly in German, use the opposite sign in defining ψ. See, for example, Kaufmann [3], p. 157, and Schlichting [4], p. 74. Hence, care must be exercised in adapting results from this literature.

[2] Named after Joseph-Louis Lagrange (see Chapter 8).

$$\frac{\partial(\nabla^2\psi)}{\partial t} - \frac{\partial\psi}{\partial y}\frac{\partial(\nabla^2\psi)}{\partial x} + \frac{\partial\psi}{\partial x}\frac{\partial(\nabla^2\psi)}{\partial y} = \nu\nabla^4\psi \tag{9.5}$$

Equation 9.5 is a precursor of the subsequently derived *vorticity transport equation*. The problem of fluid motion is thus reduced from that of solving three coupled equations in u_x, u_y, and p to that of solving a single, fourth-order, partial differential equation in ψ. This consolidation is achieved at the expense of a higher-order equation. The four boundary conditions of ψ for both x and y must be derived from those for u_x and u_y.

In *polar coordinates* with flow in the r- and θ-directions only, the Lagrange stream function is defined by

$$u_r \equiv -\frac{1}{r}\frac{\partial\psi}{\partial\theta} \tag{9.6}$$

and

$$u_\theta \equiv \frac{\partial\psi}{\partial r} \tag{9.7}$$

In *axially symmetric flow* about the z-axis a special stream function $\tilde{\psi}$ with dimensions of cubic meters per second can be defined in polar coordinates by the equations

$$u_r \equiv \frac{1}{r}\frac{\partial\tilde{\psi}}{\partial z} \tag{9.8}$$

and

$$u_z \equiv -\frac{1}{r}\frac{\partial\tilde{\psi}}{\partial r} \tag{9.9}$$

$\tilde{\psi}$ was first derived in 1842 by Stokes[1] [5] and is known as the *Stokes stream function*. In some applications the Stokes stream function is more conveniently written in *spherical coordinates* as follows:

$$u_R = \frac{-1}{R^2\sin\{\theta\}}\frac{\partial\tilde{\psi}}{\partial\theta} \tag{9.10}$$

and

$$u_\theta = \frac{1}{R\sin\{\theta\}}\frac{\partial\tilde{\psi}}{\partial R} \tag{9.11}$$

Stream functions for compressible flow can be defined by multiplying the right side of all previous definitions by ϱ_0/ϱ, where ϱ_0 is a reference density. For example, Equation 9.1 becomes

---

[1] See Chapter 8.

$$u_x \equiv -\frac{\varrho_0}{\varrho}\frac{\partial \psi}{\partial y} \tag{9.12}$$

Stream-function representations such as Equations 9.3–9.5 serve as the starting point for many of the derivations in subsequent chapters. Even when the solution is carried out in terms of the primitive variables (u_x, u_y, and p), the stream function is often calculated and plotted to display the motion. The only significant restriction on the use of a stream function is to two-dimensional flow.

The Vorticity

The equations of motion for constant density, viscosity, and gravity can be expressed even more simply in terms of the *vorticity*,[1] a vector defined as

$$\begin{array}{c} \text{or} \\ \mathbf{\Omega} \equiv \nabla \times \mathbf{v} = \text{curl } \mathbf{v} \end{array} \tag{9.13}$$

Also, the use of the vorticity is not restricted to two-dimensional flows.

As indicated by Equation 9.13, this new vector is the *vector* (or *cross*) *product* of the ∇ operator and the velocity, which is generally symbolized by *curl* **v**. In rectilinear coordinates

$$\mathbf{\Omega} = \mathbf{i}\left(\frac{\partial u_z}{\partial y} - \frac{\partial u_y}{\partial z}\right) + \mathbf{j}\left(\frac{\partial u_x}{\partial z} - \frac{\partial u_z}{\partial x}\right) + \mathbf{k}\left(\frac{\partial u_y}{\partial x} - \frac{\partial u_x}{\partial y}\right) \tag{9.14}$$

These three components of the vorticity vector represent twice the angular velocity about the x-, y-, and z-axes, respectively.[2]

The x-, y-, and z-components corresponding to Equation 8.57 can be rewritten in terms of the components of $\mathbf{\Omega}$ by taking cross-derivatives and subtracting to eliminate the pressure and gravitational terms, thereby obtaining (see problem 8)

$$\frac{D\Omega_i}{Dt} = \Omega_i \cdot \nabla u_i + \nu \nabla^2 \Omega_i + \cdots \tag{9.15}$$

The second term on the right side of Equation 9.15 represents the rate of viscous dissipation of vorticity.

In cylindrical coordinates (see Figure 8–2) the components of the vorticity are

---

[1] In the German literature the vorticity is usually defined as the negative of half of the quantity defined by Equation 9.13. Also, in the European literature the symbol *rot* (for rotation) is used instead of the symbol *curl*. (See, for example, Kaufmann [3], p. 146, and Schlichting [4], p. 73.)

[2] Defining the vorticity as $(1/2)\nabla \times \mathbf{v}$, as mentioned in footnote 1, makes it numerically equal to the angular velocity.

$$\Omega_r = \frac{1}{r}\frac{\partial u_z}{\partial \theta} - \frac{\partial u_\theta}{\partial z} \tag{9.16}$$

$$\Omega_\theta = \frac{\partial u_r}{\partial z} - \frac{\partial u_z}{\partial r} \tag{9.17}$$

and

$$\Omega_z = \frac{1}{r}\frac{\partial}{\partial r}(ru_\theta) - \frac{1}{r}\frac{\partial u_r}{\partial \theta} \tag{9.18}$$

In spherical coordinates (see Figure 8–3) they are

$$\Omega_R = \frac{1}{R\sin\{\theta\}}\frac{\partial}{\partial \theta}\left(u_\phi \sin\{\theta\} - \frac{\partial u_\theta}{\partial \phi}\right) \tag{9.19}$$

$$\Omega_\theta = \frac{1}{R\sin\{\theta\}}\frac{\partial u_R}{\partial \phi} - \frac{1}{R}\frac{\partial}{\partial R}(Ru_\phi) \tag{9.20}$$

and

$$\Omega_\phi = \frac{1}{R}\frac{\partial}{\partial R}(Ru_\theta) - \frac{1}{R}\frac{\partial u_R}{\partial \theta} \tag{9.21}$$

Equation 9.15, of course, represents these two coordinate systems as well.

Equation 9.15 appears superficially to be far simpler than the combination of Equations 8.9 and 8.57. However, the substantial derivatives of Ω_i contain u_j and u_k implicitly as well as u_i. The resolution of this difficulty by means of the stream function for two-dimensional flows is considered immediately next. Later, three-dimensional flows in terms of the vector potential are considered.

In two-dimensional flow in the x- and y-directions, only the z-component of the vorticity vector

$$\Omega_z = \frac{\partial u_y}{\partial x} - \frac{\partial u_x}{\partial y} \tag{9.22}$$

is finite. From Equations 9.1 and 9.2 it follows that

$$\Omega_z = \frac{\partial^2 \psi}{\partial y^2} + \frac{\partial^2 \psi}{\partial x^2} = \nabla^2 \psi \tag{9.23}$$

Therefore, Equation 9.5 can be written in terms of Ω_z, u_x, and u_y as

$$\frac{\partial \Omega_z}{\partial t} + u\frac{\partial \Omega_z}{\partial x} + v\frac{\partial \Omega_z}{\partial y} = \nu\nabla^2\Omega_z \tag{9.24}$$

or as

$$\frac{D\Omega_z}{Dt} = \nu\nabla^2\Omega_z \qquad (9.25)$$

Equation 9.25, which is equivalent to Equation 9.5, is called the *vorticity transfer equation*. It indicates that in two-dimensional flow the substantial change in the vorticity with time and flow (the left side) is simply equal to the rate of dissipation of vorticity due to molecular motion (the right side). The implicit occurrence of u_x and u_y in Equation 9.25 requires that Equations 9.1, 9.2, and 9.17 be solved simultaneously with 9.25. This representation, although still involved, constitutes a simpler problem mathematically than the original one, and has often been utilized to obtain solutions in applicable situations, such as flow across a cylinder. The satisfaction of the boundary conditions in Ω_i, on the basis of the original ones in u_x and u_y, sometimes poses a difficulty with numerical methods of integration.

The Vector Potential

Lamb [6], p. 208, proposed the use of a quantity, $\mathbf{A}$, later called the *vector potential*, defined by

$$\mathbf{v} \equiv \nabla \times \mathbf{A} \qquad (9.26)$$

Substituting $\mathbf{v}$ from Equation 9.26 in 9.13 gives

$$\Omega = \nabla \times (\nabla \times \mathbf{A}) = \nabla(\nabla \cdot \mathbf{A}) - \nabla^2\mathbf{A} \qquad (9.27)$$

$\mathbf{A}$ can always be defined to be *solenoidal*; i.e., such that

$$\nabla \cdot \mathbf{A} = 0 \qquad (9.28)$$

Hence

$$\Omega = -\nabla^2\mathbf{A} \qquad (9.29)$$

The representation of three-dimensional fluid motion in terms of the vorticity and the vector potential is discussed in detail by Hirasaki and Hellums [7], who proposed a procedure for defining unique and appropriate boundary conditions for $\mathbf{A}$ that imply the specified conditions on the velocity.

Hirasaki and Hellums [8] subsequently proposed for *incompressible, three-dimensional flows*, the use of a combined *velocity potential* ϕ (a scalar) and *vector potential* $\mathbf{A}$, with $\mathbf{A}$ defined by

$$\mathbf{v} = \nabla\phi + \nabla \times \mathbf{A} \qquad (9.30)$$

rather than by Equation 9.26. Since $\mathbf{A}$ can be forced to satisfy Equation 9.28, it follows that

$$\nabla \cdot \mathbf{v} = -\nabla^2 \phi \qquad (9.31)$$

Taking the *curl* of Equation 9.30 and applying 9.28 again leads to Equation 9.29. Hirasaki and Hellums [8] demonstrated that the boundary conditions for the two potentials are somewhat simpler than those for **A** alone, and suggest their use even for *compressible* three-dimensional flows. This formulation has not been used for any of the fluid motions considered in this book, but its application for natural convection (which requires solution of a coupled energy balance as well) is described by Aziz and Hellums [9] and Ozoe et al. [10].

The Circulation

The *circulation* is defined as the line integral of the tangential component of the velocity about a closed contour line C wholly immersed in a fluid field in a *simply connected* region. (A region of fluid within a torus or outside a cylinder is not simply connected.) This definition can be expressed as

$$\Gamma = \int_C \mathbf{v} \, dl \qquad (9.32)$$

where Γ = circulation, m^2/s
l = distance along contour line C

Stokes integral theorem[1] *on the circulation* states that

$$\int_C \mathbf{v} \, dl = \int\int_{S_C} (\mathbf{n} \cdot curl \, \mathbf{v}) \, dS \qquad (9.33)$$

where S_C is any area for which C is the contour. Then from Equation 9.13,

$$\int_C \mathbf{v} \, dl = \int\int_{S_C} (\mathbf{n} \cdot \boldsymbol{\Omega}) \, dS \qquad (9.34)$$

The total rate of exchange of the circulation with time, in the absence of field forces producing circulation, can be shown (see, for example, Goldstein [12], p. 97) to be

$$\frac{D\Gamma}{Dt} = \nu \int_C (\mathbf{n} \cdot \nabla^2 \mathbf{v}) \, dl \qquad (9.35)$$

In general,

$$\nabla^2 \mathbf{v} = \nabla(\nabla \cdot \mathbf{v}) - \nabla \times (\nabla \times \mathbf{v})$$
$$= \nabla(\nabla \cdot \mathbf{v}) - \nabla \times (curl \, \mathbf{v}) \qquad (9.36)$$

---

[1] A derivation of this theorem can be found, for example, in Kaufmann [3], p. 152, and Streeter [11], p. 47f.

Hence for an incompressible fluid,

$$\frac{D\Gamma}{Dt} = -\nu \int_C \mathbf{n} \cdot (\nabla \times \Omega) \, dl \tag{9.37}$$

Although the *circulation* is not used to simplify or modify the equations of motion, it does serve as a useful criterion for identifying the regime of flow.

SPECIAL FORMULATIONS FOR CREEPING FLOW

Models for *creeping flow* can be obtained simply by dropping the inertial terms in the previous equations. A formal method of carrying out this reduction is to delete terms in which the density is a multiplier of products of velocity components or their derivatives. This procedure yields a valid model for *steady, fully developed flow* in channels of constant cross section. All of the flows of Chapters 3–7 are of that class. A valid model for the asymptotic condition of Re → 0 is obtained for some, but not all, cases of unconfined flow, flow through channels of varying cross sections, and flow with rotation of one of the boundaries. The possible application of the concept of creeping flow to obtain solutions for small Re is examined in most of the following chapters.

The Stream-Function Formulation

Dropping the inertial terms reduces Equation 9.5 to

$$\frac{\partial(\nabla^2\psi)}{\partial t} = \nu\nabla^4\psi \tag{9.38}$$

For the steady state, Equations 9.38 and 9.5 reduce to

$$\nabla^4\psi = 0 \tag{9.39}$$

Equations 9.38 and 9.39 can be rewritten as

$$\frac{\partial f}{\partial t} = \nu\nabla^2 f \tag{9.40}$$

and

$$\nabla^2 f = 0 \tag{9.41}$$

respectively, where here the function f is defined as

$$f = \nabla^2\psi \tag{9.42}$$

Thus determining the motion is reduced to the consecutive solution of Equation 9.40, the fluid-mechanical analog of *Fourier's equation for thermal conduction*,[1] or Equation 9.41, the *Laplace equation*,[2] plus Equation 9.42, the *Poisson equation*,[3] and Equations 9.1 and 9.2. Most of the repertoire of classical applied analysis is focused on solving these three equations, and hence is available for fluid-flow problems described by Equation 9.38.

It follows from Equation 9.39 that the stream function, and hence the location of the streaklines and streamlines, is independent of the magnitude of the viscosity in steady, creeping flow of a fluid with constant density and viscosity. It further follows from Equations 9.1 and 9.2 that the velocity field is also independent of the viscosity. However, the drag on any surface is directly proportional to the viscosity.

The failure of the postulate of creeping flow in some geometries arises from the impossibility of satisfying the boundary conditions rather than from Equation 9.38 itself. Difficulties may also be encountered in numerical solutions in formulating boundary conditions for f.

The Vorticity Formulation

For creeping flow, Equation 9.15 reduces to

$$\frac{\partial \Omega_i}{\partial t} = \nu \nabla^2 \Omega_i \tag{9.43}$$

For a steady state it reduces to

$$\nabla^2 \Omega_i = 0 \tag{9.44}$$

For two-dimensional motion in x and y, the three equations symbolized by Equation 9.43 reduce to

$$\frac{\partial \Omega_z}{\partial t} = \nu \nabla^2 \Omega_z \tag{9.45}$$

For a steady state they reduce to

$$\nabla^2 \Omega_z = 0 \tag{9.46}$$

Comparison of Equations 9.45 and 9.46 with 9.40 and 9.41, respectively, indicates that $f = \Omega_z$. Hence, the previous remarks concerning solution of Equations 9.40–9.42 are applicable to the formulations in terms of the vorticity.

---

[1] Named after Jean-Baptiste-Joseph Fourier (1768–1830), a French mathematician and physicist who developed the first model for thermal conduction.
[2] See footnote, p. 142.
[3] Named after Siméon-Denis Poisson (1781–1840), another French mathematician and physicist of the same era as Fourier.

SPECIAL FORMULATIONS FOR INVISCID FLOW

Although all fluids have a finite viscosity, the idealized case of zero viscosity has some practical importance. As discussed by Birkhoff [13], inviscid flow does not correspond to the limiting case of viscosity approaching zero (Reynolds number approaching infinity), because in ordinary fluids the velocity goes to zero at all bounding, solid surfaces, thereby producing a region in which the velocity gradient is steep, and hence in which viscous effects (momentum transfer due to molecular motion) cannot be neglected.

At high Reynolds numbers this *boundary layer* may occupy only a small fraction of the fluid stream. Outside the boundary layer, inertial effects predominate and viscous effect may be negligible. In some applications, for example in flow over an immersed solid, an acceptable solution may be developed by patching together the separate, simpler solutions for these two regions. This procedure is illustrated in Part III for flow over plates, wedges, cylinders, and spheres.

In other applications, including those where the fluid is bounded by another fluid rather than by a solid surface, the behavior of primary interest may be characterized by inertial, pressure and field effects only, the region of the boundary layer being ignored. In many practical cases the flow is or can also be treated as one-dimensional. Such applications include free jets, shock waves, detonation waves, surface waves, and flow through nozzles and orifices. They provide the subject matter of a preceding book of this series [14].

Inviscid flow has received attention out of proportion to its practical importance, owing to its mathematical tractability. In most cases the further restriction of incompressibility has been made. Fluids without viscosity or compressibility are known as *ideal* in the field of hydrodynamics. Rayleigh [15] remarked concerning the concept of fluids without viscosity that, "On this principle the screw of a submerged boat would be useless, but, on the other hand, its services would not be needed. It is little wonder that practical men should declare that theoretical hydrodynamics has nothing to do with real fluids." Also, according to Feynman et al. [16], p. 40–3, John von Neumann characterized the theorist who analyzed an *ideal fluid* as "a man who studied dry water." Despite these remarks, solutions for ideal fluids do provide a good approximation for nonturbulent motion at a reasonable distance from surfaces and, as such, an essential expression for the outer flow for boundary-layer models. Indeed Rayleigh demonstrated a number of important applications of the theory of ideal fluids in the article that contains the above quotation.

The general modified formulations are reduced for inviscid flow simply by dropping all terms multiplied by the viscosity.

Compressible Flow

Equation 8.58 is applicable as a starting point. Analytical solution of this equation is difficult because of the nonlinearity of the inertial terms. However, for *one-dimensional, steady flow* Equation 8.58 can be integrated directly to obtain

$$\frac{u_x^2}{2} - g_x x + \int \frac{dp}{\varrho} = \text{constant} \tag{9.47}$$

which is the corresponding reduced form of the *Bernoulli equation*, as derived later for more general conditions. None of the modified earlier formulations appear to have any special utility for this case.

Incompressible Flow

The special formulations obtained by setting $v = 0$ in Equations 9.3–9.5, 9.15, and 9.25 are applicable, but the persistence of nonlinearity, even for steady-state flows, remains a severe restriction. The formulation in Chapter 10 for a *free vortex* is an example of an analytical solution for these conditions.

Irrotational, Incompressible Flow

Consider a small packet of fluid within a large volume of fluid. If the viscosity is zero, the fluid outside the packet cannot induce rotation in the packet; the vorticity is therefore constant unless a nonuniform field force is applied. In the absence of such a force, the vorticity will remain zero if it is initially zero.[1]

Then, from Equation 9.14,

$$\frac{\partial u_i}{\partial x_j} = \frac{\partial u_j}{\partial x_i} \tag{9.48}$$

The vorticity is never exactly zero in a real fluid motion, but it may be negligible far from surfaces. Most commonly, the idealized, hypothetical flows corresponding to this idealization are simply called *irrotational*—inviscidity and incompressibility being implied.

The *circulation*, according to Equation 9.34, is seen to be zero everywhere in an irrotational flow in a simply connected region. This latter restriction allows a finite circulation in irrotational flow across a cylinder.

The Stream-Function Formulation

For irrotational flow, Equation 9.23 reduces to

$$\nabla^2 \psi = 0 \tag{9.49}$$

which provides a greatly simplified basis for obtaining solutions for two-dimensional flows. A further simplification is noted below.

The Vector-Potential Formulation

For irrotational flow, Equation 9.29 reduces to

$$\nabla^2 \mathbf{A} = 0 \tag{9.50}$$

---

[1] For a mathematical proof and generalization of these two statements, see, for example, Kaufmann [3], p. 147f.

which provides a three-dimensional counterpart to Equation 9.49, with **v** in turn determined from Equation 9.26.

The Velocity-Potential Formulation

The velocity can be postulated to be represented by the gradient of a *velocity potential* defined[1] by

$$\mathbf{v} \equiv -\nabla\phi \tag{9.51}$$

where ϕ is the velocity potential in square meters per second. This function, as implied by the symbol, is the same as that used in Equation 9.30. The conditions under which such a potential exists can be determined as follows. From Equation 9.51,

$$u_i = -\frac{\partial\phi}{\partial x_i} \tag{9.52}$$

Hence

$$\frac{\partial u_i}{\partial x_j} = -\frac{\partial^2\phi}{\partial x_j \partial x_i} = -\frac{\partial^2\phi}{\partial x_i \partial x_j} = \frac{\partial u_j}{\partial x_i} \tag{9.53}$$

which is the condition for zero vorticity (i.e., for irrotational flow). The existence of a velocity potential thus implies irrotational flow. The converse is also true.

For an incompressible fluid, substitution from Equation 9.51 in Equation 8.12 indicates that

$$\nabla^2\phi = 0 \tag{9.54}$$

The equation resulting from equating the Laplacian (Equation 8.50) of some quantity, here the velocity potential, to zero is called the *Laplace equation.*[2] Equations 9.41, 9.44, 9.46, 9.49, and 9.50, each of which arose from reduction of some more general equation, have the identical form.

The mathematics dealing with the Laplace equation, which also arises in thermal conduction, electrical conduction, the diffusion of species, and other areas, is called *potential theory*.

For *two-dimensional irrotational* flow in x and y it follows from Equation 9.52 that

$$u_x = -\frac{\partial\phi}{\partial x} \tag{9.55}$$

---

[1] Again, the opposite sign is used in the German literature. See, for example, Kaufmann [3], p. 147, and Schlichting [4], p. 72.
[2] See footnote, p. 163.

and

$$u_y = -\frac{\partial \phi}{\partial y} \qquad (9.56)$$

Then from Equations 9.55 and 9.56,

$$\frac{\partial u_x}{\partial y} = \frac{\partial u_y}{\partial x} \qquad (9.57)$$

and

$$\frac{\partial^2 \phi}{\partial x^2} + \frac{\partial^2 \phi}{\partial y^2} = 0 \qquad (9.58)$$

Also, substituting for u_x and u_y in Equation 9.57 from Equations 9.1 and 9.2, respectively, gives, after rearranging,

$$\frac{\partial^2 \psi}{\partial x^2} + \frac{\partial^2 \psi}{\partial y^2} = 0 \qquad (9.59)$$

This result could also have been obtained directly from Equation 9.49.

Comparison of Equations 9.55 and 9.56 with 9.1 and 9.2, respectively, indicates that

$$\frac{\partial \phi}{\partial x} = \frac{\partial \psi}{\partial y} \qquad (9.60)$$

and

$$\frac{\partial \phi}{\partial y} = -\frac{\partial \psi}{\partial x} \qquad (9.61)$$

Equations 9.60 and 9.61 are the renowned *Cauchy–Reimann* conditions that ensure that the complex velocity potential

$$w\{z\} = \phi + i\psi \qquad (9.62)$$

is an *analytic* function of $z = x + iy$. Here i indicates the *imaginary* part of z and w.

It follows that the entire mathematical repertory of potential theory, including complex-variable theory is available for the solution of problems of plane potential flow. For example, $\psi\{x, y\} = A$ and $\phi\{x, y\} = B$, where A and B are constants, can readily be shown to constitute a set of orthogonal lines, representing the streamlines and equipotential lines. Also, *conformal mapping* can be used to transform these lines from one geometry to another while maintaining the orthogonality; the equipotential lines and streamlines can be interchanged; analogs for flow can be developed using other potentials such as voltage or temperature.

Solutions for planar potential flows that are important as limiting cases or as an outer flow for boundary-layer flows are given in Chapter 10. For details, such as proofs, for planar potential flow, see books on classical hydrodynamics, such as Lamb [6], Milne-Thompson [17], Streeter [11], and Kaufmann [3], and books on complex-variable and potential theory, such as Churchill [18].

Three-Dimensional Potential Flow

A general mathematical structure comparable to that for two dimensions does not exist for potential flow in three dimensions. However, formulations have been devised for some three-dimensional flows that have symmetry in one coordinate. For example, for flows symmetrical about the z-axis (i.e., flows that are the same in all planes intersecting on this axis) the Stokes stream function, as defined by Equations 9.8 and 9.9, is applicable. This stream function can readily be shown to satisfy the equation of continuity in cylindrical coordinates. A corresponding potential function $\tilde{\phi}$ can be defined by the equations

$$u_r = \frac{\partial \tilde{\phi}}{\partial r} \tag{9.63}$$

and

$$u_z = -\frac{\partial \tilde{\phi}}{\partial z} \tag{9.64}$$

which clearly satisfy the requirement

$$\frac{\partial u_r}{\partial z} = \frac{\partial u_z}{\partial r} \tag{9.65}$$

for irrotational flow. Substitution in the equation of continuity in cylindrical coordinates gives

$$\frac{\partial^2 \tilde{\phi}}{\partial r^2} + \frac{1}{r}\frac{\partial \tilde{\phi}}{\partial r} + \frac{\partial^2 \tilde{\phi}}{\partial z^2} = 0 \tag{9.66}$$

Finally substitution of the expressions for the stream function in Equation 9.65 gives, after simplification,

$$\frac{\partial^2 \tilde{\psi}}{\partial r^2} - \frac{1}{r}\frac{\partial \tilde{\psi}}{\partial r} + \frac{\partial^2 \tilde{\psi}}{\partial z^2} = 0 \tag{9.67}$$

Equations 9.66 and 9.67 are not identical, and $\tilde{\psi}$ and $\tilde{\phi}$ are not interchangeable as were ψ and ϕ, yet $\tilde{\psi}\{r, z\} = A$ and $\tilde{\phi}\{r, z\} = B$, where A and B are constants, constitute a set of orthogonal streamlines and equipotential lines in any plane in which θ is constant. The potential function for axially symmetric potential flow can be expressed in spherical coordinates as

$$u_R = -\frac{\partial \tilde{\phi}}{\partial R} \tag{9.68}$$

and

$$u_\theta = -\frac{1}{R}\frac{\partial \tilde{\phi}}{\partial \theta} \tag{9.69}$$

The balance of the derivations corresponding to Equations 9.65–9.67 are left to the problem set.

The Bernoulli Equation

If the field forces in the Euler equation 8.58 are each replaced by the gradient of a potential function, the resulting expression can be integrated for irrotational flow. As an example, for the acceleration due to the gravity let (see Equation 8.52)

$$\mathbf{g} = -g\nabla h \tag{9.70}$$

where g = magnitude of the acceleration due to gravity, m/s^2 or N/kg
        h = elevation, i.e., distance in the direction to opposite the gravitational force, m

Substituting the x-component of Equation 9.70 from Equation 8.52 in Equation 8.58, eliminating y and z through the conditions for irrotational flow provided by Equation 9.14, and substituting for u_x in the non-steady-state term through Equation 9.55 yields

$$-\frac{\partial^2 \phi}{\partial t\, \partial x} + u_x \frac{\partial u_x}{\partial x} + u_y \frac{\partial u_y}{\partial x} + u_z \frac{\partial u_z}{\partial x} = -g\frac{\partial h}{\partial x} - \frac{1}{\varrho}\frac{\partial p}{\partial x} + \cdots \tag{9.71}$$

This equation can be rewritten for constant g as

$$\frac{\partial}{\partial x}\left[\left(\frac{u_x^2 + u_y^2 + u_z^2}{2}\right) - \frac{\partial \phi}{\partial t} + gh\right] = -\frac{1}{\varrho}\frac{\partial p}{\partial x} + \cdots \tag{9.72}$$

which can be integrated with respect to x to give

$$\frac{v^2}{2} - \frac{\partial \phi}{\partial t} + gh + \int \frac{dp}{\varrho} = f(y, z, t) \tag{9.73}$$

Carrying out the same process for the y- and z-components of the Euler equation gives identical expressions on the left side. Hence the constant of integration must be the same for each, and therefore a function of time only. Thus

$$-\frac{\partial \phi}{\partial t} + \frac{v^2}{2} + gh + \int \frac{dp}{\varrho} = f\{t\} \tag{9.74}$$

This is the *Bernoulli*[1] *equation* for *unsteady, irrotational flow*. For the *steady state*, Equation 9.74 reduces to

$$\frac{v^2}{2} + gh + \int \frac{dp}{\varrho} = A \qquad (9.75)$$

where A is a constant in square meters per second per second. The differential form of Equation 9.75 is

$$v \, dv + g \, dh + \frac{dp}{\varrho} = 0 \qquad (9.76)$$

For *incompressible flow*

$$\frac{v^2}{2} + gh + \frac{p}{\varrho} = A \qquad (9.77)$$

Equation 9.75 can be derived (see, for example, Kaufmann [3], p. 155) for *inviscid rotational flow* by integrating the vector form of the Euler equation along a streamline. However, for that case, A is constant only for a particular streamline.

SUMMARY

The equations of motion were reexpressed in terms of the stream function and vorticity, and for irrotational flow in terms of a velocity potential. Expressions were also presented for the *circulation*, and various forms of the Bernoulli equation were derived. These forms provide a convenient starting point for the development of solutions throughout the balance of this book. They also provide some physical insight about the flows.

PROBLEMS

1. **a.** Prove that the average velocity between two streamlines is equal to the difference of the corresponding stream functions.
    b. Develop a corresponding relationship for axially symmetric flow and the Stokes stream function.
2. Prove that the stream function for compressible flow as defined by Equation 9.12 satisfies the continuity equation.
3. Derive the equivalent of Equations 9.3–9.5 for compressible flow, using Equation 9.12.
4. Show that the maximum value of the stream function characterizes the volumetric rate of flow.

---

[1] Named after Daniel Bernoulli (1700–1782), a member of a large family of Swiss mathematicians. Because he first formulated the kinetic theory of gases, he has been called the father of mathematical physics.

5. Derive the equivalent of Equations 9.3–9.5 in polar coordinates.
6. Derive the equivalent of Equations 9.3–9.5 in terms of the Stokes stream function for polar coordinates.
7. Derive the equivalent of Equations 9.3–9.5 in terms of the Stokes stream function for spherical coordinates.
8. Carry out the detailed derivation of Equation 9.15.
9. Derive the equivalent of Equation 9.15 for a compressible fluid.
10. Expand Equation 9.15 in

   a. Cartesian coordinates
   b. polar coordinates
   c. spherical coordinates

11. Derive the equivalent of Equation 9.24 for

   a. polar coordinates in r and θ
   b. polar coordinates for axially symmetric flow
   c. spherical coordinates for axially symmetric flow

12. Derive a vorticity transfer equation equivalent to Equation 9.25 for a compressible fluid with constant viscosity.
13. Outline a process for solving a three-dimensional problem in fluid mechanics using the vector potential. You may postulate that differential equations in a single dependent variable can be solved numerically.
14. Repeat problem 13 for the joint use of the vector and velocity potentials.
15. Confirm the independence of the velocity, the stream function, and the vorticity fields in Chapters 3–7 from the viscosity. (*Hint*: Choose examples and then generalize.)
16. Repeat problem 13 for irrotational flow.
17. Repeat problem 14 for irrotational flow.
18. Derive expressions in polar coordinates analogous to Equations 9.55 and 9.56.
19. Show that Equations 9.68 and 9.69 satisfy the condition of irrotationality.
20. Derive the analog of Equations 9.66 and 9.67 in spherical coordinates.
21. Derive expressions in r and θ analogous to Equations 9.55–9.61.
22. Express the continuity equation for compressible flow in terms of the velocity potential.
23. a. Derive expressions for the vorticity and stream-function distributions for fully developed laminar flow in a pipe.
   b. Relate the friction factor to the vorticity and to the stream function.
   c. Relate the mean velocity to the vorticity and to the stream function.
24. What is the physical interpretation of the first term on the right side of Equation 9.15?

REFERENCES

1. J.-L. Lagrange, "Mémoire sur la théorie du mouvement des fluides," *Nouv. Mém. Acad. Berlin, Oeuvres*, iv., 720 (1981).

2. W. I. M. Rankine, "On Plane Waterlines in Two Dimensions," *Phil. Trans. Roy. Soc. (London)*, *A154*, (1864) 369 (*Miscellaneous Scientific Papers*, Charles Griffin, London (1881), p. 495).
3. W. Kaufmann, *Fluid Mechanics*, English transl. by E. G. Chilton, McGraw-Hill, New York (1963).
4. H. Schlichting, *Boundary Layer Theory*, 7th ed., English transl. by J. Kestin, McGraw-Hill, New York (1979).
5. G. G. Stokes, "On the Steady Motion of Incompressible Fluids," *Trans. Camb. Phil. Soc.*, 7 (1842) 439 (*Mathematical and Physical Papers*, Vol. I, Cambridge University Press (1880), p. 1).
6. H. Lamb, *Hydrodynamics*, 6th ed., Dover, New York (1945).
7. G. J. Hirasaki and J. D. Hellums, "A General Formulation of the Boundary Conditions on the Vector Potential in Three-Dimensional Hydrodynamics," *Quart. Appl. Math.*, *26* 331 (1968).
8. G. J. Hirasaki and J. D. Hellums, "Boundary Conditions on the Vector and Scalar Potentials in Viscous Three-Dimensional Hydrodynamics," *Quart. Appl. Math.*, *28* (1970) 293.
9. K. Aziz and J. D. Hellums, "Numerical Solution of the Three-Dimensional Equations of Motion for Laminar Natural Convection," *Phys. Fluids*, *10* (1967) 314.
10. H. Ozoe, K. Yamamoto, S. W. Churchill, and H. Sayama, "Three-Dimensional Numerical Analysis of Natural Convection in a Confined Fluid Heated from Below," *J. Heat Transfer*, *98C*, (1976) 202, 519.
11. V. L. Streeter, *Fluid Dynamics*, McGraw-Hill, New York (1948).
12. S. Goldstein, Ed., *Modern Developments in Fluid Dynamics*, Vol. I, Oxford University Press, Clarendon (1938).
13. G. Birkhoff, *Hydrodynamics*, Dover, New York (1955).
14. S. W. Churchill, *The Practical Use of Theory in Fluid Flow. Book I. Inertial Flows*, Etaner Press, Thornton, PA (1980).
15. Lord Rayleigh (J. W. Strutt), "Fluid Motions," *Proc. Roy. Inst.*, *21* (1914) 70, (*Scientific Papers*, Vol. VI, Cambridge University Press (1920), p. 237).
16. R. P. Feynman, R. B. Leighton, and Matthew Sands, *The Feynman Lectures on Physics*, Vol. II, Addison-Wesley, Reading, MA (1964).
17. L. M. Milne-Thompson, *Theoretical Hydrodynamics*, 4th ed., Macmillan, New York (1966).
18. R. V. Churchill, *Complex Variables and Applications*, McGraw-Hill, New York (1960).

Chapter 10

Exact, Closed-Form Solutions of the Equations of Motion

INTRODUCTION

The term *exact solution*, when applied to fluid mechanics, has historically had two primary implications:

1. *Exact models* that do not invoke any approximations in the general mass and force-momentum balances
2. Solutions in *closed form*

Exact Models

The term *exact model* is generally interpreted broadly to allow for limiting or asymptotic conditions—for example, constant density and viscosity, Newtonian or idealized non-Newtonian behavior (such as for a power-law fluid), *noninertial* (*creeping or parallel, fully developed*) *flow*, and *inviscid flow* (far from surfaces). The distinction between these models and those in which some terms of the fundamental equations are dropped (as in the *thin-boundary-layer model*) or approximated (as in the *linearized model* of Oseen for slightly inertial flow) is subjective, since these latter simplifications and approximations may be quite justifiable for some range of conditions. The solutions in this chapter, however, are arbitrarily confined to "exact" models in the former sense. "Approximate" models are deferred to Parts III and IV, where they are utilized pervasively.

The solutions of "exact" models are of value in some cases as bounds and as benchmarks to test the validity or range of validity of solutions of approximate models. The solutions for inviscid flow fail in this regard, but provide an essential input for the still approximate but more realistic solutions of *thin-boundary-layer theory* as described in Part III; they also provide a good approximation for the high-velocity, effectively unconfined flows described in a companion volume [1].

Solution in Closed Form

The second restriction, of *exact solutions* to those in *closed form*, even when extended to include solutions in the form of integrals or infinite series, has been rendered moot by history. For example, the term *exact solution* has been interpreted by Schlichting [2], Chapter V, and others to encompass numerical solutions of one or more ordinary differential equations in those particular cases for which the general equations of motion can be reduced to that form. The development of high-speed computing machinery and of efficient algorithms for finite-difference, finite-element, and related methods of solution of the general partial differential equations is now rapidly expanding the class of flows for which numerical results can be obtained. These results can be constructed as *exact* insofar as care is taken to obtain convergent results (see, for example, Chu and Churchill [3] and Churchill et al. [4]), and insofar as the results can be shown to be physically valid by comparison with experimental data. This latter restriction applies to closed-form solutions as well; for example, Equations 4.6–4.12 are not physically unique for Re $\gg$ 2100. Some flows (for example, turbulent ones) are not and may never be computed from the equations of motion without the introduction of some empiricism, owing to their chaotic character, and others (for example, transitions from one mode of motion to another) require intolerable amounts of computing by present standards because of physical and numerical instabilities.

In this chapter, principal attention is given to closed-form solutions (including series and integrals), not because they are more "exact" than numerical solutions, but because their structural continuity provides a basis for physical and mathematical interpretation and understanding. Even in the form of series or integrals they may suggest groupings of variables and/or forms for the correlation of experimental data and of numerical values for more general conditions. As mentioned, solutions for inviscid flow are a required input to the model for thin boundary layers. In that application a closed form, or at least a series form, is essential. A few examples of "exact" results obtained by numerical integration of ordinary or partial differential equations are provided for the sake of completeness.

Conditions Permitting Exact Solutions

Analytical solutions for fluid flow are very limited in category and number because of the inherent complexity of the force-momentum balances, particularly their inhomogeneity and, even more, the nonlinearity introduced by the inertial terms. At sufficiently high Reynolds numbers, the nonlinear terms generate transient and chaotic oscillations that even preclude numerical solutions.

Exact, closed-form solutions of the general equations of motion are therefore necessarily restricted, with only a few exceptions, to flows that involve only one component of the velocity or to conditions such that the nonlinear, inertial terms are identically zero or can be neglected in the asymptotic sense. The viscous terms are identically zero for only a few particular flows and do not drop out asymptotically. Arbitrarily setting the *time-dependent* as well as the viscous terms to zero yields "exact" solutions for a hypothetical behavior called *inviscid flow*, which has restricted but nonetheless surprising utility. Weinbaum and O'Brien [5] provide an excellent discussion of the *mathematical structures* which permit solutions of the equations of motion in various categories, with particular emphasis on those flows for which the inertial terms are identically zero or can be expressed in some simplifying form. Exact solutions (as limited to ordinary differential equations) have been summarized by Berker [6] and Whitman [7], and the results of greatest practical importance by Schlichting [2], Chapter V, and Landau and Lifshitz [8], Chapter II.

Noninertial Flows

The inertial terms of the equations of motion may be negligible by virtue of a small Reynolds number (*creeping flow*), or they may be eliminated by virtue of geometry (*parallel flow*). The absence of the inertial terms usually results in a significant simplification since they are the source of nonlinearity. Also, since the higher-order derivatives arise from the viscous terms, dropping the inertial terms does not preclude satisfaction of all of the boundary conditions.

Unconfined creeping flow is included in the array of solutions of Part III for various immersed shapes. The one-dimensional flows of Chapters 3–7 and the fully developed rectilinear, two-dimensional flows of a companion volume [9] fall in the *parallel* category. Closed-form solutions for a variety of other creeping and parallel flows are included in the examples given here.

Inviscid Flows

Letting the viscosity approach zero in the general, non-steady-state form of the equations of motion results in an increasing Reynolds number and ultimately in turbulent oscillations. On the other hand, setting the viscosity equal to zero in the steady-state form of the equations of motion leads to solutions that do not have a complete physical counterpart. For example, the solution for inviscid flow about a cylinder predicts a symmetrical distribution of pressure on the surface, fore and aft, whereas the pressure difference calculated for a finite viscosity or measured actually provides the main contribution to the drag force at large velocities. This attainment of erroneous solutions by setting the viscosity equal to zero is known as *D'Alembert's paradox*.[1] Such difficulties might be anticipated on mathematical grounds alone, since dropping the viscous terms eliminates the highest-order derivatives in the equations of motion and thereby does not permit satisfying all of the boundary conditions, particularly that of zero velocity on surfaces. Despite such fundamental discrepancies, the solutions for inviscid flow have great practical value in that they closely approximate the physical behavior of unconfined, high-velocity flows, such as in shock, detonation, and gravitational waves, and of effectively unconfined, high-velocity flows through nozzles and orifices. Solutions for those flows that can be approximated as inviscid by virtue of the absence or effective absence of a bounding surface are discussed in a companion volume [1]. The solutions for inviscid flow also provide reasonable approximations for the flow field far from, but effected by, immersed surfaces, and as such are utilized in thin-boundary-layer theory as a boundary condition. In radial flow across an annulus, the absence of a bounding surface (except

---

[1] Birkhoff [10] provides an excellent discussion of this and other paradoxes that arise in the asymptotic theories of fluid flow.

normal to the flow) results in elimination of the Laplacian of the velocity and, thereby, the viscous terms.

EXACT SOLUTIONS FOR GENERAL FLOWS

"Exact" solutions are first illustrated for "general" flows in which neither the viscous nor the inertial flows have been set asymptotically or identically to zero. These solutions are subdivided as (1) closed form, (2) numerical integrations of ordinary differential equations, and (3) numerical integrations of partial differential equations, with principal attention given to the former. The characteristic of the model that permits closed-form solutions or reduction to ordinary differential equations is noted in each case.

E1.   Suddenly Accelerated Plate

The motion generated in a fluid by a wall started impulsively from rest and thereafter moved uniformly in its own plane, as illustrated in Figure 10–1, can be represented by

$$\varrho\frac{\partial u}{\partial t} = \mu\frac{\partial^2 u}{\partial y^2}$$

(10E1.1)

with boundary conditions

$$u = 0 \qquad \text{for } t \leq 0, \quad y \geq 0$$

(10E1.2)

$$u \to 0 \qquad \text{for } y \to \infty$$

(10E1.3)

and

$$u = u_w \qquad \text{for } y = 0, \quad t > 0$$

(10E1.4)

where u = velocity in direction of flow (normal to y), m/s
   u_w = velocity of wall, m/s

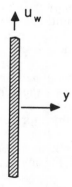

FIGURE 10–1   *Notation for fluid motion generated by the impulsive movement of an infinite plate.*

The infinite extent of the fluid and wall and the confinement of the motion to one direction eliminate all of the steady (nonlinear) inertial terms. The resulting model, although a partial differential equation, is linear and homogeneous in u. This model is analogous to that for thermal conduction into a semi-infinite region due to a step function in temperature at the surface. The well-known solution, which can be obtained by separation of variables, the Laplace transformation, or even elementary means (see problem 2), is

$$\frac{u}{u_w} = \text{erfc}\left\{\frac{y}{2\sqrt{vt}}\right\} \tag{10E1.5}$$

where

$$\text{erfc}\{\eta\} = \frac{2}{\sqrt{\pi}}\int_\eta^\infty e^{-\eta^2}\, d\eta \tag{10E1.6}$$

is the complementary error function and $v = \mu/\varrho$ is the kinematic viscosity in square meters per second. The drag on the plate is

$$\tau_w = \mu\left(\frac{du}{dy}\right)_{y=0} = \mu\left(\frac{2}{\sqrt{\pi}}\right)\left(\frac{u_w}{2\sqrt{vt}}\right) = \frac{\mu u_w}{\sqrt{\pi vt}} \tag{10E1.7}$$

This is known in fluid mechanics as *Stokes' first problem*, and was included in his celebrated memoir on pendulums [11]. The presence of v in the solution implies that both viscous and inertial forces are involved.

Although Equation 10E1.5 is an exact solution of the equations of motion for the cited boundary conditions, the infinite extent of the fluid and wall, and the impulsive start can only be approximated physically.

E2. Developing Couette Flow

The initiation of planar Couette flow by impulsively starting the wall with the fluid initially at rest is represented by the preceding model if the second boundary condition is replaced by $u = 0$ at $y = d$. The well-known solution for the analogous problem in thermal conduction (see, for example, Carslaw and Jaeger [12], p. 310) can be rewritten as

$$\frac{u}{u_w} = \sum_{n=0}^\infty \left(\text{erfc}\left\{\frac{2nd + y}{2\sqrt{vt}}\right\} - \text{erfc}\left\{\frac{(2n + 1)d - y}{2\sqrt{vt}}\right\}\right) \tag{10E2.1}$$

For short times only the first term on the right side for $n = 0$ is significant, which reduces 10E2.1 to 10E1.5. A more rapidly convergent solution can be derived for long times (see problem 5). For very, very long times Equation 10E2.1 approaches Equation 7.8.

Again, the presence of v in the solution implies that both viscous and inertial forces are operative.

E3. Transient Flow in a Pipe

A solution for the motion of initially motionless fluid in a pipe due to the sudden imposition of a pressure gradient was derived by Szymanski [13] in terms of

Bessel functions. Curves representing the resulting velocity profiles at various dimensionless times are plotted in Figure 10–2.

The behavior in which the radial velocity profile varies with time but is the same at all axial positions should not be confused with that for *steady*, two-dimensional developing flow in the *inlet* of a pipe.

Analogous solutions have been developed for many related situations (see Schlichting [2], p. 93).

E4.   Harmonically Oscillating Plate

The fluid motion generated by a plate oscillating harmonically in its own plane is known as *Stokes' second problem* [11]. This behavior is again analogous to thermal conduction in a semi-infinite region with an oscillating surface temperature. The solution (adapted from Carslaw and Jaeger [12], p. 65) for a fluid initially at rest and for a surface velocity

$$u = u_{max} \sin\{\omega t\} \qquad \text{at } y = 0 \tag{10E4.1}$$

is

$$\frac{u}{u_{max}} = \frac{2}{\sqrt{\pi}} \int_{y/2\sqrt{vt}}^{\infty} \sin\left\{\omega\left(t - \frac{y^2}{4v\eta^2}\right)\right\} e^{-\eta^2} \, d\eta \tag{10E4.2}$$

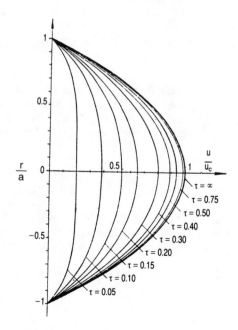

FIGURE 10–2   *Velocity profiles for sudden imposition and maintenance of a pressure gradient on a liquid in a pipe ($\tau = vt/a^2$). (From Szymanski [13].)*

Alternative forms are given by Carslaw and Jaeger and by Schlichting [2], p. 94, who also gives a solution for a fixed plate parallel to the oscillating one. For long times the fluid attains a stationary behavior represented by

$$\frac{u}{u_{max}} = \sin\left\{\omega t - y\sqrt{\frac{\omega}{2\nu}}\right\}\exp\left\{-y\sqrt{\frac{\omega}{2\nu}}\right\} \tag{10E4.3}$$

The period of oscillation of a cylindrical wire in a fluid has been used to determine its viscosity. This and related solutions for various objects have been reviewed by Kestin and Persen [14]. Also, see Kestin and Wang [15], Langlois [16], pp. 100–101, Denn [17], pp. 358–362, and Landau and Lifshitz [8], pp. 88–98.

E5.   Decay of a Free Vortex

The decay of a vortex provides another example of an exact solution in which the nonlinear inertial forces are absent but in which both viscous stresses and transient inertia are operative.

The velocity field generated by a cylinder rotating in an infinite expanse of fluid, as obtained by letting $a_2 \to \infty$ in Equation 7.51, is

$$u = \frac{2\pi a_1^2 \Omega}{r} \tag{10E5.1}$$

where Ω is the rate of rotation in units of 1/seconds. Equation 10E5.1, just as Equation 7.51, is an exact solution. Inertial forces do not affect such velocity fields, but they do influence the corresponding pressure distribution.

The solution for a "free" vortex in inviscid flow (see, for example, Streeter [18], p. 50) is

$$u = \frac{\Gamma}{2\pi r} \tag{10E5.2}$$

where Γ is the strength of the vortex in square meters per second. These solutions (Equations 10E5.1 and 10E5.2) are identical for

$$\Gamma = 4\pi^2 a_1^2 \Omega \tag{10E5.3}$$

Hence Equation 10E5.2 is also an exact solution for the equations of motion, even though it was derived for inviscid flow.

An exact solution for the transient decay of such a line vortex due to the effect of viscosity is

$$u = \frac{\Gamma_0}{2\pi r}(1 - e^{-r^2/4\nu t}) \tag{10E5.4}$$

where Γ_0 is the initial strength of the vortex. The derivation, which was originally carried out independently by Oseen [19] and Hamel [20], is posed as problem 13. A plot of u/u_∞ versus r/r_0 with $\nu t/r_0^2$ as a parameter is shown in Figure 10–3. Here r_0 is arbitrary and u_∞ is the corresponding arbitrary velocity at $r = r_0$ and $t = 0$. The presence of ν in Equation 10E5.4 confirms the role of both inertial and viscous forces in this process.

E6.   *Free Surface of Liquid in a Rotating Bucket*

When a partially filled cylindrical bucket is rotated the surface of the liquid attains a shape such as that sketched in Figure 10–4. For this motion the force-momentum balances *in cylindrical coordinates* reduce to

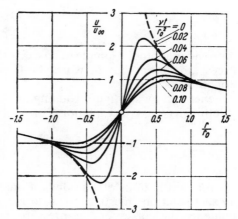

FIGURE 10–3   *Velocity profiles in a line vortex decaying with time. (From Schlichting [2], p. 89.)*

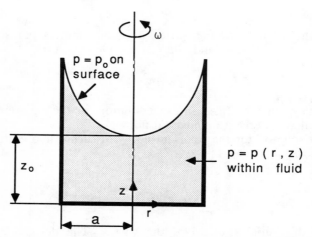

FIGURE 10–4   *Analysis of shape of free surface of water in a rotating bucket.*

$$\frac{\varrho u_\theta^2}{r} = \frac{\partial p}{\partial r} \tag{10E6.1}$$

$$0 = \frac{\partial}{\partial r}\left(\frac{1}{r}\frac{\partial}{\partial r}(ru_\theta)\right) \tag{10E6.2}$$

and

$$0 = -\frac{\partial p}{\partial z} - \varrho g \tag{10E6.3}$$

The boundary conditions, in the absence of surface tension, are

$$u_\theta = \omega a \qquad \text{at } r = a \tag{10E6.4}$$

$$u_\theta < M \qquad \text{as } r \to 0 \tag{10E6.5}$$

and

$$p = p_0 \qquad \text{at free surface} \tag{10E6.6}$$

where ω is the angular velocity of the bucket in radians per second.
Integration of Equation 10E6.2 gives

$$u_\theta = \frac{Ar}{2} + \frac{B}{r} \tag{10E6.7}$$

Boundary conditions 10E6.5 and 10E6.4 give $B = 0$ and $A = 2\omega$, respectively.
Hence

$$u_\theta = \omega r \tag{10E6.8}$$

which indicates that the fluid rotates as if it were a solid. The total derivative of
the pressure is

$$dp = \frac{\partial p}{\partial r}\,dr + \frac{\partial p}{\partial z}\,dz \tag{10E6.9}$$

Hence, from Equations 10E6.1 and 10E6.3, and then 10E6.8,

$$dp = \frac{\varrho u_\theta^2}{r}\,dr - \varrho g\,dz = \varrho\omega^2 r\,dr - pg\,dz \tag{10E6.10}$$

Integrating from $p = p_0$ and $z = z_0$ at $r = 0$ gives

$$p - p_0 = \frac{\varrho\omega^2 r^2}{2} - pg(z - z_0) \tag{10E6.11}$$

The equation for the free surface (where $p = p_0$) is then

$$0 = \frac{\varrho\omega^2 r^2}{2} - \varrho g(z - z_0) \qquad (10\text{E}6.12)$$

or

$$z = z_0 + \frac{\omega^2 r^2}{2g} \qquad (10\text{E}6.12\text{A})$$

The free surface is thus a paraboloid of revolution.

This solution for the shape of the free surface is independent of both the viscosity and the density, although no assumptions were made in this respect. The independence from viscosity is due to the "solid-body" rotation and hence the absence of velocity gradients. The independence from density is due to the "parallel" character of the motion. The pressure distribution (Equation 10E6.11) depends on the density but not on the viscosity.

E7.   Cylindrical Radial Flow

Consider radial flow outward from a vertical porous cylinder, as sketched in Figure 10–5. The continuity equation and r-momentum equation reduce to

$$\frac{1}{r}\frac{\partial}{\partial r}(ru) = 0 \qquad (10\text{E}7.1)$$

and

$$\varrho u \frac{\partial u}{\partial r} = -\frac{\partial \mathscr{P}}{\partial r} + \mu \frac{\partial}{\partial r}\left(\frac{1}{r}\frac{\partial}{\partial r}(ur)\right) \qquad (10\text{E}7.2)$$

The boundary conditions can be expressed as

$$u = u_a, \quad \mathscr{P} = \mathscr{P}_a \qquad \text{at } r = a \qquad (10\text{E}7.3)$$

and

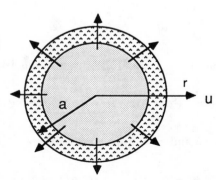

FIGURE 10–5   *Analysis of outward flow from a vertical porous cylinder.*

$$u \to 0 \qquad \text{as } r \to \infty \qquad \text{(10E7.4)}$$

where u = radial component of velocity, m/s
        u_a = radial velocity at $r = a$, m/s
        $\mathscr{P}_a$ = dynamic pressure at $r = a$, Pa
        a = outer radius of porous cylinder, m

Integrating Equation 10E7.1 gives

$$ru = c = \text{a constant} = u_a a \qquad \text{(10E7.5)}$$

Substituting u from Equation 10E7.5 into 10E7.2 gives

$$\frac{\varrho u_a^2 a^2}{-r^3} = -\frac{\partial \mathscr{P}}{\partial r} \qquad \text{(10E7.6)}$$

Integration then gives

$$\mathscr{P}_a - \mathscr{P} = \frac{\varrho u_a^2}{2}\left(1 - \left(\frac{a^2}{r}\right)\right) \qquad \text{(10E7.7)}$$

This solution for the radial distribution of the velocity and pressure is independent of the viscosity because there are no shear stresses.
     The analogous solution for spherical flow is posed as problem 18.

E8. Forced and Longitudinal-Couette Flow in a Rotating Annulus

Consider fully developed flow between an inner cylinder of radius a_1 rotating at a rate Ω_1 and an outer concentric cylinder of radius a_2 rotating at rate Ω_2 and translating at uniform velocity u_w relative to the inner cylinder with a fixed longitudinal pressure gradient, as sketched in Figure 10–6.
     It is apparent that

$$u_r = 0 \qquad \text{(10E8.1)}$$

$$u_\theta = f_1\{r\} \qquad \text{(10E8.2)}$$

$$u_z = f_2\{r\} \qquad \text{(10E8.3)}$$

and

$$\mathscr{P} = f_3\{r\} + f_4\{z\} \qquad \text{(10E8.4)}$$

The dependence on r, represented by f_3, arises from the centrifugal force, and that on z, represented by f_4, from the forced flow.
     The equation of continuity in cylindrical coordinates is automatically satisfied by Equations 10E8.1–10E8.3. The three force-momentum balances reduce to

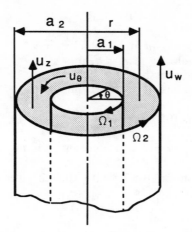

FIGURE 10–6   *Analysis of forced and longitudinal Couette flow in a rotating annulus.*

$$-\frac{\varrho u_\theta^2}{r} = -\frac{d\mathscr{P}}{dr} \tag{10E8.5}$$

$$0 = \frac{1}{r}\frac{d}{dr}\left(r\frac{du_\theta}{dr}\right) - \frac{u_\theta}{r^2} \tag{10E8.6}$$

and

$$0 = -\frac{d\mathscr{P}}{dz} + \frac{\mu}{r}\frac{d}{dr}\left(r\frac{du_z}{dr}\right) \tag{10E8.7}$$

The boundary conditions are

$$u_\theta = 2\pi a_1\Omega_1, \quad u_z = 0 \qquad \text{at } r = a_1 \tag{10E8.8}$$

$$u_\theta = 2\pi a_2\Omega_2, \quad u_z = u_w \qquad \text{at } r = a_2 \tag{10E8.9}$$

Integration of Equation 10E8.6 and satisfaction of the boundary conditions gives

$$u_\theta = \frac{2\pi(a_2^2\Omega_2 - a_1^2\Omega_1)r}{a_2^2 - a_1^2} + \frac{2\pi(\Omega_1 - \Omega_2)a_1^2a_2^2}{(a_1^2 - a_1^2)r} \tag{10E8.10}$$

(Compare with Equations 7.51 and 7.52.)

Recognizing that $d\mathscr{P}/dz$ is constant and integrating Equation 10E8.7 give

$$u_z = \frac{1}{4\mu}\left(-\frac{d\mathscr{P}}{dz}\right)\left[a_1^2 - r^2 + \frac{(a_2^2 - a_1^2)\ln\{r/a_1\}}{\ln\{a_2/a_1\}}\right]$$

$$+ u_w\frac{\ln\{r/a_1\}}{\ln\{a_2/a_1\}} \tag{10E8.11}$$

For $u_w = 0$, Equation 10E8.11 is equivalent to 4.30; for $-d\mathscr{P}/dz = 0$, to 7.34. Integrating Equation 10E8.5 using 10E8.10 for u_θ is left to problem 27.

This solution indicates that the rotary and axial components of the flow are uncoupled. Inertial forces are not operative for the corresponding individual *parallel-type* flows and hence not for this combination. The radial pressure variation defined by Equations 10E8.5 and 10E8.10 arises from the centrifugal (inertial) effect.

E9.   Radial Flow between Parallel Disks

A fluid is presumed to enter through a hole at the center of one or both of a pair of parallel disks and to flow radially outward between the circular plates, as shown in Figure 10–7. Inspection of the figure, neglecting entrance effects, shows that in cylindrical coordinates

$$u_z = u_\theta = 0 \tag{10E9.1}$$

$$u_r = f_1\{r, z\} \tag{10E9.2}$$

and

$$\mathscr{P} = f_2\{r\} \tag{10E9.3}$$

where z is the distance in meters from the central plane between disks. The equation of continuity then reduces to

$$\frac{d(u_r r)}{dr} = 0 \tag{10E9.4}$$

which has the solution

$$u_r r = \phi\{z\} \tag{10E9.5}$$

where $\phi\{z\}$ is an unknown function, independent of r. The force-momentum balance for the radial direction can thereby be reduced to

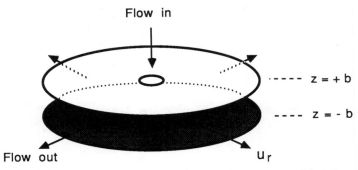

Flow in

$z = + b$

$z = - b$

Flow out u_r

FIGURE 10–7   *Analysis of radial flow between parallel disks.*

$$\frac{-\varrho\phi^2}{r^3} = -\frac{d\mathscr{P}}{dr} + \frac{\mu}{r}\frac{d^2\phi}{dz^2} \qquad (10E9.6)$$

Equation 10E9.3 can be integrated once with respect to r from $\mathscr{P}_1$ at a_1 to obtain

$$\frac{\varrho\phi^2}{2}\left(\frac{1}{r^2} - \frac{1}{a_1^2}\right) = \mathscr{P}_1 - \mathscr{P} + \mu\ln\left\{\frac{r}{a_1}\right\}\frac{d^2\phi}{dz^2} \qquad (10E9.7)$$

Then specifying $\mathscr{P}_2$ at a_2 gives

$$\frac{\varrho\phi^2}{2}\left(\frac{1}{a_2^2} - \frac{1}{a_1^2}\right) = \mathscr{P}_1 - \mathscr{P}_2 + \mu\ln\left\{\frac{a_2}{a_1}\right\}\frac{d^2\phi}{dz^2} \qquad (10E9.8)$$

The total volumetric rate of flow is

$$V = \int_0^b 4\pi r u_r \, dz = 4\pi \int_0^b \phi \, dz \qquad (10E9.9)$$

where b is the half-width between the plates.

Equation 10E9.8 is an ordinary differential equation that can be integrated from $d\phi/dz = 0$ at $z = 0$ to $\phi = 0$ at $z = b$ for chosen

$$\frac{\varrho(1/a_1^2 - 1/a_2^2)}{2\mu\ln\{a_2/a_1\}} \quad \text{and} \quad \frac{\mathscr{P}_1 - \mathscr{P}_2}{\mu\ln\{a_2/a_1\}}$$

to obtain $\phi\{z\}$. The volumetric rate of flow then follows from Equation 10E9.9, and the radial variation in pressure from Equation 10E9.7.

Bird et al. [21], p. 114, assert that an analytical solution of Equation 10E9.8 is possible, but not in closed form (see problem 31). Numerical integration also appears to be feasible (see problem 32). The approximate model for creeping flow is obtained by dropping the left side of Equation 10E9.6, as illustrated in Section EC3.

E10.　Converging and Diverging (Jeffery–Hamel) Flows

The converging (or diverging) flow between two plates that form a small, acute angle, 2α, as sketched in Figure 10–8, can be postulated to be radial only; that is, in cylindrical coordinates

$$u_z = u_\theta = 0 \qquad (10E10.1)$$

The equation of continuity then reduces to

$$\frac{\partial}{\partial r}(u_r r) = 0 \qquad (10E10.2)$$

It follows that

$$u_r r = v\phi\{\theta\} \qquad (10E10.3)$$

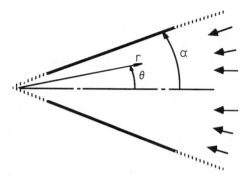

FIGURE 10–8   *Analysis of converging flow between inclined plates.*

That is, the function ϕ defined by Equation 10E10.3 is independent of r. The kinematic viscosity has here been arbitrarily included as a factor in Equation 10E10.3 to make ϕ dimensionless.

The force-momentum equations for the r and θ directions now reduce to

$$-\frac{\phi^2}{r^3} = -\frac{1}{\varrho v^2}\frac{\partial \mathscr{P}}{\partial r} + \frac{1}{r^3}\frac{d^2\phi}{d\theta^2} \tag{10E10.4}$$

and

$$0 = \frac{1}{\varrho v^2}\frac{\partial \mathscr{P}}{\partial \theta} + \frac{2}{r^2}\frac{d\phi}{d\theta} \tag{10E10.5}$$

Differentiating Equations 10E10.4 and 10E10.5 with respect to θ and r, respectively, and subtracting to eliminate the pressure, gives, after multiplying through by r^3,

$$2\phi\phi' + \phi''' + 4\phi' = 0 \tag{10E10.6}$$

Equation 10E10.6 can be integrated once to obtain

$$\phi'' + \phi^2 + 4\phi + C_1 \tag{10E10.7}$$

where C_1 is a constant. Multiplying by ϕ' and integrating again give

$$\frac{(\phi')^2}{2} + \frac{\phi^3}{3} + 2\phi^2 = C_1\phi + C_2 \tag{10E10.8}$$

Designating $\phi\{0\}$ as ϕ_o and noting that owing to symmetry $\phi'\{0\} = 0$ allows C_2 to be eliminated, yielding

$$\frac{(\phi')^2}{2} + \frac{\phi^3 - \phi_0^3}{3} + 2(\phi^2 - \phi_o^2) = C_1(\phi - \phi_o) \tag{10E10.9}$$

Further rearrangement and integration from $\phi = \phi_o$ at $\theta = 0$ gives

$$\theta = \int_{\phi}^{\phi_0} \frac{d\phi}{\sqrt{\frac{2}{3}(\phi^3 - \phi_o^3) + 4(\phi^2 - \phi_o^2) - 2C_1(\phi - \phi_o)}} \qquad (10E10.10)$$

At the surface, as defined by the half-angle α, $\phi = 0$. Hence

$$\alpha = \int_{0}^{\phi_0} \frac{d\phi}{\sqrt{\frac{2}{3}(\phi^3 - \phi_o^3) + 4(\phi^2 - \phi_o^2) - 2C_1(\phi - \phi_o)}} \qquad (10E10.11)$$

Equation 10E10.11 together with

$$\tilde{v} = \int_{-\alpha}^{\alpha} u_r r \, d\theta \qquad (10E10.12)$$

allow evaluation of C_1 and ϕ_0 for a specified $\tilde{v}$. The corresponding pressure field can be obtained from Equations 10E10.4 and 10E10.5. The problem is thus reduced to the evaluation of integrals. This process was first carried out independently by Jeffery [22] and Hamel [20].

Numerical values that were subsequently obtained by Millsaps and Pohlhausen [23] for the velocity distribution are plotted in Figure 10–9 versus θ with ϕ_0 as a parameter. Curve $\bar{A}$, for diverging flow with $\phi_0 = 684$, indicates backflow near the wall, but such *separation* is unstable and would produce turbulence. The model is also obviously invalid for either diverging or converging flows in the limit as $r \to 0$.

Denn [17], p. 217f, discusses approximate solutions for Equation 10E10.7 based on the simplifications of *creeping flow* (described in EC4), *thin-boundary-layer theory* (see Part III), and *lubrication theory*. He also compares (pp. 336–337) the numerical solution of Equation 10E10.7 with a numerical solution of the partial differential equations describing discrete contractions (see his Example XXIII). A detailed discussion of the problem is also provided by Goldstein [24], pp. 105–110.

E11.   Flow Generated by a Rotating Disk

For the velocity field near an isolated, thin rotating disk, as sketched in Figure 10–10, von Kármán [25] postulated that in cylindrical coordinates

$$u_r = r\omega U\{Z\} \qquad (10E11.1)$$

$$u_\theta = r\omega V\{Z\} \qquad (10E11.2)$$

$$u_z = \sqrt{v\omega}\, W\{Z\} \qquad (10E11.3)$$

and

$$\mathscr{P} = \varrho v\omega P\{Z\} \qquad (10E11.4)$$

where

$$z = \sqrt{\frac{v}{\omega}}\, Z \qquad (10E11.5)$$

A

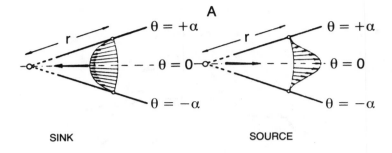

SINK   SOURCE

B

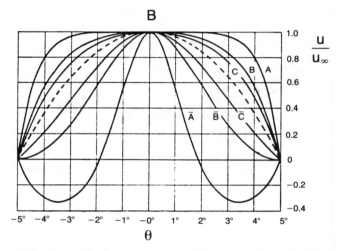

FIGURE 10–9 *Velocity profiles in converging and diverging channels with a total angle of 10°: (A) sketch of velocity profiles; (B) numerical values of velocities.*

ϕ_0	Converging	Diverging
5000	A	$\bar{C}$
1342	B	$\bar{B}$
684	C	$\bar{A}$

The dashed line represents Poiseuille flow. (From Millsaps and Pohlhausen [23].)

u_r = radial component of velocity, m/s
u_θ = angular component of velocity, m/s
u_z = vertical component of velocity, m/s
ω = angular velocity, rad/s

This change of variables reduces the equations of motion to

$$2U + W' = 0 \tag{10E11.6}$$

$$U^2 - V^2 + U'W = U'' \tag{10E11.7}$$

$$2UV + WV' = V'' \tag{10E11.8}$$

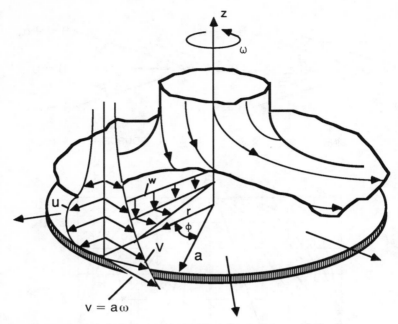

FIGURE 10–10   *Components of velocity near a thin rotating disk of radius a.*

and

$$WW' = -P' + W''$$ (10E11.9)

The boundary conditions are

$$U = 0, \quad W \to 0 \quad \text{as } Z \to \infty$$
$$U = V = 0, \quad W = 1, \quad P = 0 \quad \text{at } Z = 0$$ (10E11.10)

This set of equations has been solved approximately by von Kármán and exactly (by numerical integration) by Cochran [26] (see Goldstein [24], pp. 110–113). The results are summarized in Table 10.1, and the components of the velocity are plotted in Figure 10–11. The dimensionless radial component of the velocity goes through a maximum and then decreases to zero with height, the dimensionless angular component decreases monotonically with height, and the dimensionless normal component increases monotonically to an asymptotic value. The dimensionless pressure also increases to an asymptotic value with height.

The torque on a disk of finite radius a due to motion on both sides is

$$\mathcal{T} = 4\pi \int_0^a r^2 \mu \left(\frac{-du_\theta}{dz}\right)_{z=0} dr$$
$$\mathcal{T} = -\pi \sqrt{\varrho\mu\omega^3}\, a^4 V'\{0\} = 1.935 a^4 \sqrt{\varrho\mu\omega^3}$$ (10E11.11)

Table 10.1
Computed Velocity and Pressure Functions for Flow near a Thin Rotating Disk
(from Cochran [26])

Z	U	V	W	U'	V'	P
0	0	1.000	0	0.510	−0.616	0
0.1	0.046	0.939	−0.005	0.416	−0.611	0.092
0.2	0.084	0.878	−0.018	0.334	−0.599	0.167
0.3	0.114	0.819	−0.038	0.262	−0.580	0.228
0.4	0.136	0.762	−0.063	0.200	−0.558	0.275
0.5	0.154	0.708	−0.092	0.147	−0.532	0.312
0.6	0.166	0.656	−0.124	0.102	−0.505	0.340
0.7	0.174	0.607	−0.158	0.063	−0.476	0.361
0.8	0.179	0.561.	−0.193	0.032	−0.448	0.377
0.9	0.181	0.517	−0.230	0.006	−0.419	0.388
1.0	0.180	0.468	−0.266	−0.016	−0.391	0.395
1.1	0.177	0.439	−0.301	−0.033	−0.364	0.400
1.2	0.173	0.404	−0.336	−0.046	−0.338	0.403
1.3	0.168	0.371	−0.371	−0.057	−0.313	0.405
1.4	0.162	0.341	−0.404	−0.064	−0.290	0.406
1.5	0.156	0.313	−0.435	−0.070	−0.268	0.406
1.6	0.148	0.288	−0.466	−0.073	−0.247	0.405
1.7	0.141	0.264	−0.495	−0.075	−0.228	0.404
1.8	0.133	0.242	−0.522	−0.076	−0.210	0.403
1.9	0.126	0.222	−0.548	−0.075	−0.193	0.402
2.0	0.118	0.203	−0.572	−0.074	−0.177	0.401
2.1	0.111	0.186	−0.596	−0.072	−0.163	0.399
2.2	0.104	0.171	−0.617	−0.070	−0.150	0.398
2.3	0.097	0.156	−0.637	−0.067	−0.137	0.397
2.4	0.091	0.143	−0.656	−0.065	−0.126	0.396
2.5	0.084	0.131	−0.674	−0.061	−0.116	0.395
2.6	0.078	0.120	−0.690	−0.058	−0.106	0.395
2.8	0.068	0.101	−0.721	−0.052	−0.089	0.395
3.0	0.058	0.083	−0.746	−0.046	−0.075	0.395
3.2	0.050	0.071	−0.768	−0.040	−0.063	0.395
3.4	0.042	0.059	−0.786	−0.035	−0.053	0.394
3.6	0.036	0.050	−0.802	−0.030	−0.044	0.394
3.8	0.031	0.042	−0.815	−0.025	−0.037	0.393
4.0	0.026	0.035	−0.826	−0.022	−0.031	0.393
4.2	0.022	0.029	−0.836	−0.019	−0.026	0.393
4.4	0.018	0.024	−0.844	−0.016	−0.022	0.393

A dimensionless coefficient for the torque can be defined and evaluated as

$$C_{\mathcal{T}} = \frac{\mathcal{T}}{\varrho\omega^2 a^5} = \frac{1.935}{a}\sqrt{\frac{\nu}{\omega}} = \frac{2.737}{\text{Re}_\omega^{1/2}} \qquad (10\text{E}11.12)$$

where $\text{Re}_\omega \equiv 2a^2\omega/\nu$.

Experimental data are compared with Equation 10E11.12 in Figure 10–12. The theoretical solution is seen to be valid up to the onset of transition to turbulence.

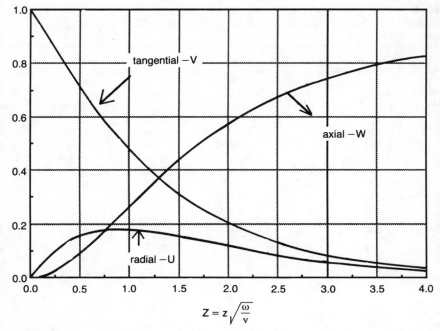

FIGURE 10–11   *Dimensionless components of velocity near a thin rotating disk. (Based on numerical solution of Cochran [26].)*

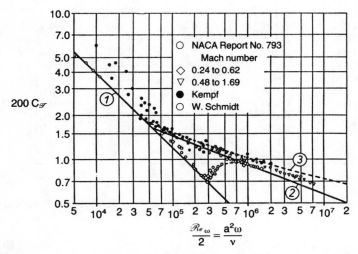

FIGURE 10–12   *Comparison of experimental data for the torque on a thin rotating disk with predictions. (From Schlichting [2], p. 106.) Line 1 corresponds to Equation 10E11.12; curves 2 and 3 represent empirical expressions for the turbulent regime.*

The effect of the finite thickness of the disk is neglected in the foregoing analysis.

E12. Stagnation Flows

In Chapter 14 the equations of motion are reduced to a pair of ordinary differential equations for a fluid stream impacting two-dimensionally on a flat plate (Heimenz flow) and three-dimensionally, with axial symmetry, on a disk (Homann flow), based on the simplifications of boundary-layer theory. However, these simplifications are unnecessary. The indicated pairs of ordinary differential equations can be derived without such idealizations (see problems 37 and 38).

EXACT SOLUTIONS FOR CREEPING FLOW

These solutions all invoke the assumption of sufficiently low Reynolds numbers such that the inertial terms in the force-momentum equations, although finite, can be neglected to yield an asymptotic solution.

EC1. Creeping Flow over Spheres and Other Objects

Solid Spheres. The solution derived by Stokes [27] for axisymmetric flow over a solid sphere was apparently the first for the creeping regime. In spherical coordinates the equation of continuity and the equations for force and momentum (after dropping the inertial terms and, because of symmetry, the φ-terms) can be rearranged as

$$\frac{1}{R^2}\frac{\partial}{\partial R}(R^2 u_R) + \frac{1}{R\sin\{\theta\}}\frac{\partial}{\partial \theta}(u_\theta \sin\{\theta\}) = 0 \qquad (10\text{EC1.1})$$

$$0 = -\frac{\partial \mathscr{P}}{\partial R} + \mu\left[\frac{1}{R^2}\frac{\partial}{\partial R}\left(R^2 \frac{\partial u_R}{\partial R}\right) + \frac{1}{R^2\sin\{\theta\}}\frac{\partial}{\partial \theta}\left(\frac{\partial u_R}{\partial \theta}\sin\{\theta\}\right)\right.$$
$$\left. - \frac{2u_R}{R^2} - \frac{2}{R^2}\frac{\partial u_\theta}{\partial \theta} - \frac{2u_\theta}{R^2}\cot\{\theta\}\right] \qquad (10\text{EC1.2})$$

and

$$0 = -\frac{1}{r}\frac{\partial \mathscr{P}}{\partial \theta} + \mu\left[\frac{1}{R^2}\frac{\partial}{\partial R}\left(R^2 \frac{\partial u_\theta}{\partial r}\right) + \frac{1}{R^2\sin^2\{\theta\}}\frac{\partial}{\partial \theta}\left(\frac{\partial u_\theta}{\partial \theta}\sin\{\theta\}\right)\right.$$
$$\left. + \frac{2}{R^2}\frac{\partial u_R}{\partial \theta} - \frac{u_\theta}{R^2\sin^2\{\theta\}}\right] \qquad (10\text{EC1.3})$$

with boundary conditions

$$u_R = u_\theta = 0 \qquad \text{at } R = a \qquad (10\text{EC1.4})$$

$$u_R \rightarrow u_\infty \cos\{\theta\} \qquad \text{as } R \rightarrow \infty \qquad (10\text{EC1.5})$$

and

$$u_\theta \to -u_\infty \sin\{\theta\} \qquad \text{as } R \to \infty \qquad (10EC1.6)$$

Here θ is measured from the rear of the sphere.

The details of the solution are given by Denn [17], pp. 250–225, in terms of these variables, and by Langlois [16], pp. 133–139, in terms of the stream function. The solution itself can be expressed as

$$\tilde{\psi} = -\frac{u_\infty}{4}R^2\left[\left(\frac{a}{R}\right)^3 - 3\left(\frac{a}{R}\right) + 2\right]\sin^2\{\theta\} \qquad (10EC1.7)$$

$$u_R = \frac{u_\infty}{2}\left[\left(\frac{a}{R}\right)^3 - 3\left(\frac{a}{R}\right) + 2\right]\cos\{\theta\} \qquad (10EC1.8)$$

$$u_\theta = \frac{u_\infty}{4}\left[\left(\frac{a}{R}\right)^3 + 3\left(\frac{a}{R}\right) - 4\right]\sin\{\theta\} \qquad (10EC1.9)$$

$$\mathcal{P} = \mathcal{P}_\infty - \frac{3\mu u_\infty}{2a}\left(\frac{a}{R}\right)^2\cos\{\theta\} \qquad (10EC1.10)$$

It follows that the *viscous drag* is

$$F_f = 4\pi\mu a u_\infty \qquad (10EC1.11)$$

and that the *form drag* (due to pressure) is

$$F_p = 2\pi\mu a u_\infty \qquad (10EC1.12)$$

The total drag coefficient is then

$$\bar{C}_t \equiv \frac{F_f + F_p}{\varrho u_\infty^2 \pi a^2} = \frac{6\mu}{\varrho u_\infty a} = \frac{12}{\text{Re}_D} \qquad (10EC1.13)$$

where $\text{Re}_D = 2au_\infty\varrho/\mu$.

Equation 10EC1.13, which is known as *Stokes' law*, has been found to provide a useful approximation for the drag for Re < 0.1 (see Figures 16–21 to 16–24). The streamlines provided by Equation 10EC1.7 are examined in Chapter 16.

Other Shapes.   A similar solution has been derived for flow over an ellipsoid (see, for example, Lamb [28], pp. 604–605, and Roscoe [29]). This solution can be degenerated to that for an elliptical disk perpendicular to or parallel to the flow, including the limiting case of a long strip, as well as to the preceding solution for a sphere. These results can be summarized for future reference as follows:

1.   Circular disk normal to flow:

$$\bar{C}_t \equiv \frac{F}{\pi a^2 \varrho u_\infty^2} = \frac{16\mu}{\pi a u_\infty \varrho} = \frac{32}{\pi\text{Re}_D} \qquad (10EC1.14)$$

2. Thin, circular disk parallel to flow:

$$\bar{C}_t = \bar{C}_f \equiv \frac{F}{\pi a^2 \varrho u_\infty^2} = \frac{32\mu}{3\pi a u_\infty \varrho} = \frac{64}{3\pi \text{Re}_D} \qquad (10\text{EC}1.15)$$

3. Long, thin strip (ellipse) of height **2b** and length **2a**:

$$\bar{C}_t \equiv \frac{F}{4ab\varrho u_\infty^2} = \frac{2\pi\mu}{b\varrho u_\infty \ln\{4a/b\}} = \frac{8\pi\mu}{\text{Re} \ln\{4a/b\}} \qquad (10\text{EC}1.16)$$

Stokes [27] proved that a solution for *creeping flow* over a cylinder was impossible.

Fluid Spheres. Hadamard [30] and Rybczynski [31] independently derived a solution for creeping flow over a fluid sphere analogous to that of Stokes for a solid sphere (see, e.g., Levich [32], p. 395f). The Stokes stream function in spherical coordinates for the external fluid is

$$\tilde{\psi} = -\frac{u_\infty R^2}{2}\left[1 - \frac{1}{2}\left(\frac{2+3\zeta}{1+\zeta}\right)\frac{a}{R} + \frac{1}{2}\left(\frac{\zeta}{1+\zeta}\right)\left(\frac{a}{R}\right)^3\right]\sin^2\{\theta\} \qquad (10\text{EC}1.17)$$

where $\zeta = \mu_d/\mu_c$
μ_d = viscosity of discontinuous (bubble) phase, Pa·s
μ_c = viscosity of continuous phase, Pa·s

while that inside the sphere is

$$\tilde{\psi} = \frac{u_\infty[1 - (R/a)^2]R^2 \sin^2\{\theta\}}{4(1+\zeta)} \qquad (10\text{EC}1.18)$$

The corresponding velocity components outside the fluid particle are

$$u_R = u_\infty\left[1 - \frac{1}{2}\left(\frac{2+3\zeta}{1+\zeta}\right)\frac{a}{R} + \frac{1}{2}\left(\frac{\zeta}{1+\zeta}\right)\left(\frac{a}{R}\right)^3\right]\cos\{\theta\} \qquad (10\text{EC}1.19)$$

and

$$u_\theta = -u_\infty\left[1 - \frac{1}{4}\left(\frac{2+3\zeta}{1+\zeta}\right)\frac{a}{R} - \frac{1}{4}\left(\frac{\zeta}{1+\zeta}\right)\left(\frac{a}{R}\right)^3\right]\sin\{\theta\} \qquad (10\text{EC}1.20)$$

and within

$$u_R = \frac{-u_\infty[1 - (R/a)^2]\cos\{\theta\}}{2(1+\zeta)} \qquad (10\text{EC}1.21)$$

$$u_\theta = \frac{u_\infty[1 - 2(R/a)^2]\sin\{\theta\}}{2(1+\zeta)} \qquad (10\text{EC}1.22)$$

The total drag coefficient is

$$\bar{C}_t \equiv \frac{F_f + F_p}{\varrho u_\infty^2 \pi a^2} = \frac{2\mu_c}{\varrho u_\infty a}\left(\frac{2 + 3\zeta}{1 + \zeta}\right) = \frac{4}{\text{Re}_D}\left(\frac{2 + 3\zeta}{1 + \zeta}\right) \qquad \text{(10EC1.23)}$$

Equation 10EC1.23 includes Stokes' law (Equation 10EC1.13) as a special case for $\zeta \to \infty$ and provides a solution for bubbles in the asymptotic limit of $\zeta \to 0$. Equation 10EC1.23 is shown in Chapter 17 to be confirmed experimentally for a complete range of ζ.

EC2.   Hele-Shaw Flow

Consider *creeping flow* about a cylinder of arbitrary shape extending between two closely spaced parallel plates as sketched in Figure 10–13. The equations to be satisfied by this three-dimensional flow are

$$\nabla \cdot \mathbf{v} = 0 \qquad \text{(10EC2.1)}$$

$$\frac{\partial \mathscr{P}}{\partial x} = \mu \nabla^2 u \qquad \text{(10EC2.2)}$$

$$\frac{\partial \mathscr{P}}{\partial y} = \mu \nabla^2 v \qquad \text{(10EC2.3)}$$

and

$$\frac{\partial \mathscr{P}}{\partial z} = \mu \nabla^2 w \qquad \text{(10EC2.4)}$$

where $u = u_x, v = u_y,$ and $w = u_z$.

A general solution of these equations, satisfying the boundary condition of no-slip on the parallel plates **but not on the cylindrical surface** is

$$u = \left(1 - \left(\frac{z}{b}\right)^2\right) u_o\{x, y\} \qquad \text{(10EC2.5)}$$

$$v = \left(1 - \left(\frac{z}{b}\right)^2\right) v_o(x, y) \qquad \text{(10EC2.6)}$$

$$w = 0 \qquad \text{(10EC2.7)}$$

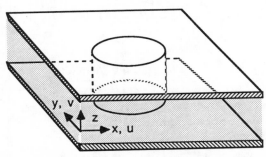

FIGURE 10–13   *Analysis of Hele-Shaw flow about a cylinder of arbitrary cross section.*

and

$$\mathscr{P} = -\frac{2\mu}{b^2} \int_{x_0}^{x} u_o\{x, y\} \, dx = -\frac{2\mu}{b^2} \int_{x_0}^{x} v_o\{x, y\} \, dy \qquad (10EC2.8)$$

where $u_o\{x, y\}$ and $v_o\{x, y\}$ represent the solution for *potential (inviscid, irrotational) flow* over the cylinder—that is, of

$$\frac{\partial u_o}{\partial x} + \frac{\partial v_o}{\partial y} = 0 \qquad (10EC2.9)$$

$$u_o \frac{\partial u_o}{\partial x} + v_o \frac{\partial u_o}{\partial y} = -\frac{1}{\varrho} \frac{\partial \mathscr{P}_0}{\partial x} \qquad (10EC2.10)$$

and

$$u_o \frac{\partial v_o}{\partial x} + v_o \frac{\partial v_o}{\partial y} = -\frac{1}{\varrho} \frac{\partial \mathscr{P}_0}{\partial y} \qquad (10EC2.11)$$

The streamlines corresponding to Equations 10EC2.5 and 10EC2.6 at any z are obviously congruent to those of the *potential flow* represented by $u_o\{x, y\}$ and $v_o\{x, y\}$. This analogy was first proposed and used by Hele-Shaw [33] to obtain photographs of potential flow about various cylinders. An example is provided in Figure 10–14A from Van Dyke [34], p. 9, for flow over an inclined plate (a cylinder of rectangular cross section). These streaklines closely approximate the theoretical solution for potential flow over a thin plate, which is plotted in Figure 10–14B. Note that the streaklines of Figure 10–14A are symmetrical fore and aft, which does not occur for creeping flow.

Schlichting [2], p. 124, asserts that $Re^* \equiv 2u_m b^2 / Dv$ must be less than unity to provide a good approximation of creeping flow in this application. Also, Streeter [18], p. 229, suggests the error in the streamlines due to zero-slip on the cylindrical surface extends outward a distance approximately equal to the spacing between the plates; hence the accuracy near the cylinder can be improved by decreasing the spacing.

EC3. Creeping Radial Flow between Parallel Disks

For *creeping flow*, Equation (10E9.6) can be reduced to

$$\frac{d\mathscr{P}}{dr} = \frac{\mu}{r} \frac{d^2(u_r r)}{dz^2} \qquad (10EC3.1)$$

and integrated once with respect to r and twice with respect to z to obtain

$$u_r = \frac{b^2(\mathscr{P}_2 - \mathscr{P}_1)}{2\mu r \ln\{a_2/a_1\}} \left(1 - \left(\frac{z}{b}\right)^2\right) \qquad (10EC3.2)$$

EC4. Creeping Converging-Diverging Flows between Inclined Plates

For *creeping flow*, Equation 10E10.7 reduces to

$$\phi'' + 4\phi = C_1 \qquad (10EC4.1)$$

A

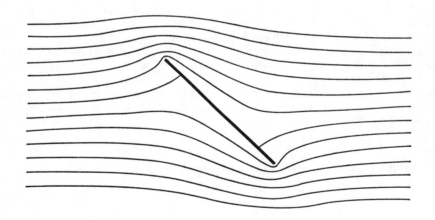

B

FIGURE 10–14   *Streaklines for potential flow about an inclined strip (a cylinder of rectangular cross section): (A) Hele-Shaw flow as photographed by D. H. Peregrine (from Van Dyke [34], p. 9); (B) theoretical values.*

which has the general solution

$$\phi = A \sin\{2\theta\} + B \cos\{2\theta\} + \frac{C_1}{4} \qquad \text{(10EC4.2)}$$

The conditions $\phi'\{0\} = 0$ and $\phi\{\alpha\} = 0$ require

$$\phi = B[\cos\{2\theta\} - \cos\{2\alpha\}] \qquad \text{(10EC4.3)}$$

Then from Equations 10E10.3 and 10E10.12,

$$\tilde{v} = v \int_{-\alpha}^{\alpha} B[\cos\{2\theta\} - \cos\{2\alpha\}] \, d\theta$$

$$= Bv[\sin\{2\alpha\} - 2\alpha \cos\{2\alpha\}] \qquad \text{(10EC4.4)}$$

Hence the final solution for the velocity is

$$u_r = \frac{\tilde{v}[\cos\{2\theta\} - \cos\{2\alpha\}]}{r[\sin\{2a\} - 2\alpha \cos\{2\alpha\}]} \qquad \text{(10EC4.5)}$$

EC5. *Creeping Flow between Fixed, Concentric Spheres*[1]

For the motion sketched in Figure 10–15, symmetry and the neglect of inertial and end effects suggest the postulate that $u_R = u_\phi = 0$. The equation of continuity in spherical coordinates then reduces to

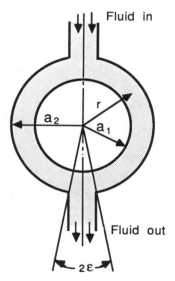

FIGURE 10–15 *Analysis of creeping flow in the annulus between two concentric spheres.*

---

[1] Adapted from Bird et al. [21], pp. 117–118.

$$\frac{d}{d\theta}(u_\theta \sin\{\theta\}) = 0 \tag{10EC5.1}$$

which implies that

$$u_\theta \sin\{\theta\} = f\{R\} \tag{10EC5.2}$$

Also, the force-momentum balance in the θ-direction becomes

$$\frac{1}{R}\frac{d\mathscr{P}}{d\theta} = \mu\left[\frac{1}{R^2 \sin\{\theta\}}\frac{d}{dR}\left(R^2\frac{df}{dR}\right)\right] \tag{10EC5.3}$$

which can be rearranged as

$$\sin\{\theta\}\frac{d\mathscr{P}}{d\theta} = \frac{\mu}{R}\frac{d}{dR}\left(R^2\frac{df}{dR}\right) \tag{10EC5.4}$$

Since the left side is a function of θ only and the right side of R only, they can both be equated to a constant α.

Integration of the resulting expression for $\mathscr{P}$ gives

$$\mathscr{P}_2 - \mathscr{P}_1 = \alpha \ln\left\{\frac{1 + \cos\{\varepsilon\}}{1 - \cos\{\varepsilon\}}\right\} \tag{10EC5.5}$$

where ε is the angle projected by the opening as shown in Figure 10–15. Integration of the corresponding expression for f yields

$$f\{R\} = -\frac{\alpha a_2}{2\mu}\left[1 - \frac{R}{a_2} - \frac{a_1}{a_2}\left(\frac{a_2}{R} - 1\right)\right] \tag{10EC5.6}$$

Combining Equations 10EC5.2, 10EC5.5, and 10EC5.6 yields

$$u_\theta = \frac{(\mathscr{P}_2 - \mathscr{P}_2)a_2}{2\mu \sin\{\theta\} \ln\left\{\dfrac{1 + \cos\{\varepsilon\}}{1 - \cos\{\varepsilon\}}\right\}}\left[1 - \frac{R}{a_2} - \frac{a_1}{a_2}\left(\frac{a_2}{R} - 1\right)\right] \tag{10EC5.7}$$

EC6.   Creeping Flow between Rotating Disks

For *creeping flow* between a fixed disk and a parallel, rotating one, as sketched in Figure 10–16, the equations of motion in cylindrical coordinates reduce to

$$\frac{1}{r}\frac{\partial}{\partial r}(ru_r) + \frac{\partial u_z}{\partial z} = 0 \tag{10EC6.1}$$

$$0 = -\frac{\partial\mathscr{P}}{\partial r} + \mu\left[\frac{\partial}{\partial r}\left(\frac{1}{r}\frac{\partial}{\partial r}(ru_r)\right) + \frac{\partial^2 u_r}{\partial z^2}\right] \tag{10EC6.2}$$

$$0 = \frac{\partial}{\partial r}\left(\frac{1}{r}\frac{\partial}{\partial r}(ru_\theta)\right) + \frac{\partial^2 u_\theta}{\partial z^2} \tag{10EC6.3}$$

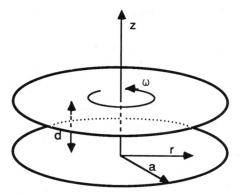

FIGURE 10–16 *Analysis of creeping flow between a fixed and a rotating disk, each of radius a.*

and

$$0 = -\frac{\partial \mathcal{P}}{\partial z} + \mu \left[\frac{1}{r} \frac{\partial}{\partial r} \left(r \frac{\partial u_z}{\partial r} \right) + \frac{\partial^2 u_z}{\partial z^2} \right] \qquad (10\text{EC}6.4)$$

The boundary conditions are

$$u_\theta = u_z = 0, \quad u_\theta = r\omega \qquad \text{at } z = 0 \qquad (10\text{EC}6.5)$$

$$u_r = u_z = u_\theta = 0 \qquad \text{at } z = d \qquad (10\text{EC}6.6)$$

The postulate that

$$u_\theta = rf\{z\} \qquad (10\text{EC}6.7)$$

reduces Equation 10EC6.3 to

$$f'' = 0 \qquad (10\text{EC}6.8)$$

Integration and satisfaction of the boundary conditions gives

$$u_\theta = \omega r \left(1 - \frac{z}{d} \right) \qquad (10\text{EC}6.9)$$

Equations 10EC6.1, 10EC6.2, and 10EC6.4, and the corresponding boundary conditions are satisfied by $u_r = u_z = 0$, and $\mathcal{P} = $ constant.

Equation 10EC6.9 indicates that each fluid particle follows a circular path. Denn [17], pp. 247–248, indicates that the criterion for the applicability of creeping flow in this situation is $a \gg d$. This solution was derived in Chapter 7 (Equation 7.39) by analogy to Couette flow without noting the restriction to the creeping regime. Greater flows would produce a variation of pressure in both radial and axial directions and hence a secondary motion.

*EC7.   Creeping Flow between Concentric Spheres
Rotating about the Same Axis*

For the coordinate system of Figure 10–17 only the φ-component of the velocity is finite. Also, from symmetry, the derivatives with respect to φ are zero. Hence for *creeping flow*

$$\frac{1}{R^2}\frac{\partial}{\partial R}\left(R^2\frac{\partial u_\phi}{\partial R}\right) + \frac{1}{R^2\sin\{\theta\}}\frac{\partial}{\partial\theta}\left(\sin\{\theta\}\frac{\partial u_\phi}{\partial\theta}\right) - \frac{u_\phi}{R^2\sin^2\{\theta\}} = 0 \quad (10EC7.1)$$

The boundary conditions are

$$u_\phi = \omega_1 a_1 \sin\{\theta\} \qquad \text{at } R = a_1 \tag{10EC7.2}$$

$$u_\phi = \omega_2 a_2 \sin\{\theta\} \qquad \text{at } R = a_2 \tag{10EC7.3}$$

The trial solution

$$u_\phi = f\{R\}\sin\{\theta\} \tag{10EC7.4}$$

is suggested by these boundary conditions. Substitution of Equation 10EC7.4 in Equation 10EC7.1 results in

$$R^2 f'' + 2Rf' - 2f = 0 \tag{10EC7.5}$$

A solution satisfying the two boundary conditions is

$$u_\phi = \left[\frac{(a_2^3\omega_2 - a_1^3\omega_1)R}{a_2^3 - a_1^3} + \frac{a_1^3 a_2^3(\omega_1 - \omega_2)}{R^2(a_2^3 - a_1^3)}\right]\sin\{\theta\} \tag{10EC7.6}$$

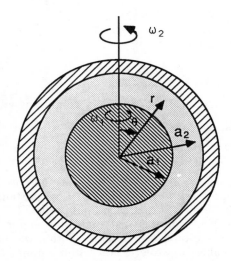

FIGURE 10–17   *Analysis of creeping flow between concentric spheres rotating about the same axis.*

If the inner sphere is absent ($a_1 \to 0$),

$$u_\phi = R\omega_2 \sin\{\theta\} \qquad (10EC7.7)$$

which corresponds to solid-body rotation. The same solution is obtained for equal angular velocities ($\omega_1 = \omega_2$). If the outer sphere is absent ($a_2 \to \infty$),

$$u_\phi \to \frac{a_1^3 \omega_1 \sin\{\theta\}}{R^2} \qquad (10EC7.8)$$

which is the motion produced by a rotating sphere in an infinite medium.

The effect of inertia is to produce secondary motion in the fluid between the spheres (see Langlois [16], pp. 196–200).

EC8. Creeping Flow between Rotating Coaxial Cones

Creeping flow between rotating coaxial cones, as sketched in Figure 10–18, can be represented in spherical coordinates by Equation 10EC7.1 with boundary conditions

$$u_\phi = R\omega_1 \sin\{\alpha_1\} \qquad \text{at } \theta = \alpha_1 \qquad (10EC8.1)$$

and

$$u_\phi = R\omega_2 \sin\{\alpha_2\} \qquad \text{at } \theta = \alpha_2 \qquad (10EC8.2)$$

where α_1 and α_2 are the polar angles of the inner and outer cones, respectively. The trial function

$$u_\phi = R \sin\{\theta\} f\{\theta\} \qquad (10EC8.3)$$

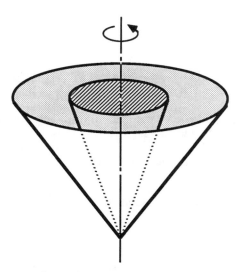

FIGURE 10–18 *Analysis of creeping flow between rotating coaxial cones.*

as suggested by the boundary conditions, reduces Equation 10EC7.1 to

$$\frac{d}{d\theta^2}(f\sin\{\theta\}) + \frac{\cos\{\theta\}}{\sin\{\theta\}}\frac{d}{d\theta}(f\sin\{\theta\}) + \left(\frac{2\sin^2\{\theta\} - 1}{\sin\{\theta\}}\right)f = 0 \quad (10EC8.4)$$

which can be rewritten as

$$\sin\{\theta\}\frac{d^2f}{d\theta^2} + 3\cos\{\theta\}\frac{df}{d\theta} = 0 \qquad (10EC8.5)$$

and integrated to obtain

$$\frac{df}{d\theta} = \frac{A}{\sin^3\{\theta\}} \qquad (10EC8.6)$$

Equation 10EC8.6 can be integrated again to obtain

$$f = B + A\left(\ln\left\{\tan\left\{\frac{\theta}{2}\right\}\right\} - \frac{\cos\{\theta\}}{\sin\{\theta\}}\right) \qquad (10EC8.7)$$

From the boundary conditions

$$A \cdot C = \omega_1\left(\ln\left\{\tan\left\{\frac{\alpha_2}{2}\right\}\right\} - \frac{\cos\{\alpha_2\}}{\sin^2\{\alpha_3\}}\right)$$
$$- \omega_2\left(\ln\left\{\tan\left\{\frac{\alpha_1}{2}\right\}\right\} - \frac{\cos\{\alpha_1\}}{\sin^2\{\alpha_2\}}\right) \qquad (10EC8.8)$$

and

$$B \cdot C = \omega_2 - \omega_1 \qquad (10EC8.9)$$

where

$$C = \ln\left\{\frac{\tan\{\alpha_2/2\}}{\tan\{\alpha_1/2\}}\right\} - \frac{\cos\{\alpha_2\}}{\sin^2\{\alpha_2\}} + \frac{\cos\{\alpha_1\}}{\sin^2\{\alpha_1\}} \qquad (10EC8.10)$$

This solution implies that the cones are infinitely long or that the free surface of the fluid is spherical and stress free. A nonspherical surface, as well as appreciable inertial forces, would result in secondary motion.

Devices with this geometrical form are widely used as viscometers (see problem 59).

*EC9. Creeping Flow between a Rotating Cone and
        Fixed Plate*

The creeping regime of flow induced by a cone rotating above a stationary plate, as sketched in Figure 10–19, is also used to measure the viscosity of liquids. A purely tangential flow may be assumed; that is, $u_R = u_\theta = 0$. The force-momentum balance in the ϕ-direction again reduces to Equation 10EC7.1. The

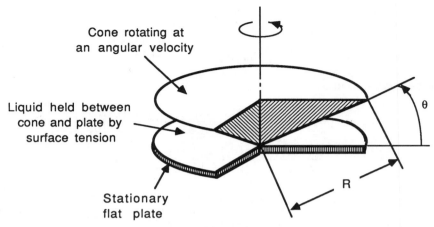

FIGURE 10–19 *Analysis of creeping flow between a rotating cone and a fixed plate.*

boundary conditions for this flow are

$$u_\phi = 0 \quad \text{at } \theta = \frac{\pi}{2} \tag{10EC9.1}$$

$$u_\phi = R\omega \sin\{\theta_1\} \quad \text{at } \theta = \theta_1 \tag{10EC9.2}$$

and

$$u_\phi = 0 \quad \text{at } R = 0 \tag{10EC9.3}$$

Based on the boundary conditions, the following test solution may be postulated:

$$u_\phi = \omega R f\{\theta\} \tag{10EC9.4}$$

Substitution of this expression for u_ϕ in Equation 10EC7.1 gives

$$f'' + \frac{\cos\{\theta\}}{\sin\{\theta\}} f' + \left(2 - \frac{1}{\sin^2\{\theta\}}\right) f = 0 \tag{10EC9.5}$$

which can be reexpressed as

$$\frac{d}{d\theta}\left[\sin^3\{\theta\} \frac{d}{d\theta}\left(\frac{f}{\sin\{\theta\}}\right)\right] = 0 \tag{10EC9.6}$$

One integration gives

$$\sin^3\{\theta\} \frac{d}{d\theta}\left(\frac{f}{\sin\{\theta\}}\right) = A \tag{10EC9.7}$$

Another integration gives

$$\frac{f}{\sin\{\theta\}} = A\left[-\frac{\cos\{\theta\}}{\sin^2\{\theta\}} + \frac{1}{2}\ln\left\{\tan\left\{\frac{\theta}{2}\right\}\right\}\right] + B \qquad (10EC9.8)$$

The boundary conditions for $\theta = \theta_1$ and $\pi/2$ yield

$$u_\phi = \omega R \sin\{\theta_1\}\left[\frac{\cos\{\theta\}}{\sin\{\theta\}} + \frac{\sin\{\theta\}}{2}\ln\left\{\frac{1+\cos\{\theta\}}{1-\cos\{\theta\}}\right\}\right]$$

$$\cdot \left[\frac{\cos\{\theta_1\}}{\sin\{\theta_1\}} + \frac{\sin\{\theta_1\}}{2}\ln\left\{\frac{1+\cos\{\theta_1\}}{1-\cos\{\theta_1\}}\right\}\right]^{-1} \qquad (10EC9.9)$$

Bird et al. [21], pp. 98–101, derive Equation 10EC9.9 in terms of the components of the shear stress and suggest (p. 119) a derivation by separation of variables leading to *Cauchy's linear equation* and *Legendre's equation*.

EC10. Creeping Squeeze-Flows

Consider the motion produced by moving one disk toward another at a uniform velocity, as sketched in Figure 10–20. Symmetry indicates that $u_\theta = \partial\mathscr{P}/\partial\theta = 0$. For the regime of creeping flow and such a low rate of movement of the disk that a pseudo-steady state can be assumed, the equation of continuity and the equations of force-momentum in cylindrical coordinates can be reduced to

$$\frac{1}{r}\frac{\partial}{\partial r}(ru_r) + \frac{\partial u_z}{\partial z} = 0 \qquad (10EC10.1)$$

$$\frac{\partial\mathscr{P}}{\partial r} = \mu\left[\frac{\partial}{\partial r}\left(\frac{1}{r}\frac{\partial}{\partial r}(ru_r)\right) + \frac{\partial^2 u_r}{\partial z^2}\right] \qquad (10EC10.2)$$

and

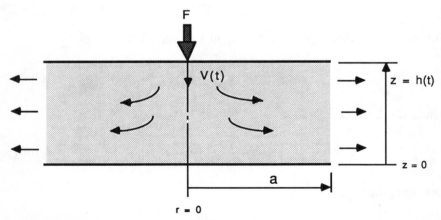

FIGURE 10–20 *Analysis of creeping flow produced by one disk moving toward another, both of radius a.*

$$\frac{\partial \mathscr{P}}{\partial z} = \mu \left[\frac{1}{r} \frac{\partial}{\partial r} \left(r \frac{\partial u_z}{\partial r} \right) + \frac{\partial^2 u_z}{\partial z^2} \right] \tag{10EC10.3}$$

The boundary conditions are

$$u_r = u_z = 0 \qquad \text{at } z = 0 \tag{10EC10.4}$$

$$u_r = 0, \quad u_z = -u_w\{t\} \qquad \text{at } z = h\{t\} \tag{10EC10.5}$$

The conditions on z suggest the postulate

$$u_z = f\{z, t\} \tag{10EC10.6}$$

Substituting this expression into Equation 10EC10.1 gives

$$\frac{1}{r} \frac{\partial}{\partial r}(ru_r) = -\frac{\partial f}{\partial z} \tag{10EC10.7}$$

Integrating with respect to r then gives

$$u_r = -\frac{r}{2} \frac{\partial f}{\partial z} + \frac{A\{z, t\}}{r} \tag{10EC10.8}$$

Since u_r must be finite for all r, $A\{z, t\} = 0$. Substituting for u_r and u_z in Equations 10EC10.2 and 10EC10.3 gives

$$\frac{\partial \mathscr{P}}{\partial r} = -\frac{\mu r f'''}{2} \tag{(10EC10.9)}$$

and

$$\frac{\partial \mathscr{P}}{\partial z} = \mu f'' \tag{10EC10.10}$$

The derivatives on the right side of Equations 10EC10.9 and 10EC10.10 are written as ordinary derivatives, although they are functions of time, since derivatives of time are absent from the force-momentum balances due to the assumption of pseudo-steady flow. Differentiating Equation 10EC10.9 with respect to z and Equation 10EC10.10 with respect to r to eliminate the pressure gives

$$\frac{\partial^2 \mathscr{P}}{\partial z \, \partial r} = -\frac{\mu r f''''}{2} \tag{10EC10.11}$$

and

$$\frac{\partial^2 \mathscr{P}}{\partial r \, dz} = 0 \tag{10EC10.12}$$

Hence

$$f'''' = 0 \qquad (10EC10.13)$$

Integrating Equations 10EC10.13 four times gives

$$f = A + Bz + Cz^2 + Dz^3 \qquad (10EC10.14)$$

The four boundary conditions can then be used to obtain

$$u_z = -3u_w\{t\}\left(\frac{z}{h\{t\}}\right)^2\left[1 - \frac{2}{3}\left(\frac{z}{h\{t\}}\right)\right] \qquad (10EC10.15)$$

and

$$u_r = \frac{3rzu_w\{t\}}{h^2\{t\}}\left[1 - \frac{z}{h\{t\}}\right] \qquad (10EC10.16)$$

Since

$$u_w = -\frac{dh}{dt} \qquad (10EC10.17)$$

either u_w can be eliminated in Equations 10EC10.15 and 10EC10.16 for a specified $h\{t\}$, or conversely.

Integrating Equation 10EC10.9 with respect to r and 10EC10.10 with respect to z then gives

$$\mathscr{P} = -\frac{3\mu r^2 u_w\{t\}}{h^3\{t\}} + f_1\{z\} \qquad (10EC10.18)$$

and

$$\mathscr{P} = -\frac{6\mu z u_w\{t\}}{h^2\{t\}}\left(1 - \frac{z}{h\{t\}}\right) + f_2\{r\} \qquad (10EC10.19)$$

Hence

$$\mathscr{P} = -\frac{3\mu u_w\{t\}}{h\{t\}}\left[\frac{2z}{h\{t\}}\left(1 - \frac{z}{h\{t\}}\right) + \left(\frac{r}{h\{t\}}\right)^2\right] + E \qquad (10EC10.20)$$

where E is a constant. Letting $\mathscr{P} = \mathscr{P}_0$ at $r = a$ and $z = h\{t\}$ then gives

$$\mathscr{P} - \mathscr{P}_0 = \frac{3\mu u_w\{t\}}{h\{t\}}\left[\frac{a^2 - r^2}{h^2\{t\}} - \frac{2z}{h\{t\}}\left(1 - \frac{z}{h\{t\}}\right)\right] \qquad (10EC10.21)$$

and

$$F = \int_0^a 2\pi r (\mathscr{P} - \mathscr{P}_0)_{z=h}\, dr = \frac{3\pi\mu a^4 u_w\{t\}}{2h^3\{t\}} \qquad (10\text{EC}10.22)$$

To obtain the plate spacing with time, as produced by a fixed force, combine Equations 10EC10.22 and 10EC10.16 as follows:

$$u_w = -\frac{dh}{dt} = \frac{2Fh^3}{3\pi\mu a^4} \qquad (10\text{EC}10.23)$$

Integration from h_o at $t = 0$ then gives

$$h\{t\} = \left(\frac{1}{h_o^2} + \frac{4Ftz}{3\pi\mu a^4}\right)^{-1/2} \qquad (10\text{EC}10.24)$$

Denn [17], p. 260, demonstrates good agreement between experimental data and Equation EC10.24. He concludes from analysis that the above solution for squeeze flow is a good approximation for $f \ll (\mu^2/\varrho)(a/h)^4$.

EC11.   Creeping Flow through Orifices

By a similar method to that used by Stokes for the solutions of EC1, Roscoe [29] derived the following exact solution for the volumetric rate of flow through an elliptical orifice:

$$V = \frac{2A^2(-\Delta\mathscr{P})}{3\pi\mu P_w} \qquad (10\text{EC}11.1)$$

where $A = \text{area} = \pi ab,\ \text{m}^2$
  $P_w = \text{perimeter} = 4aE\{\varepsilon\},\ \text{m}$
  $a = \text{half of the major axis, m}$
  $b = \text{half of the minor axis, m}$
  $\varepsilon = \text{eccentricity of ellipse} = \left(1 - \dfrac{b^2}{a^2}\right)^{1/2}$
  $E\{\varepsilon\} = \text{complete elliptic integral of the second kind (see Abramowitz}$
     and Stegun [35], p. 589f)

For a circular orifice ($b \to a$, $\varepsilon \to 0$, $E\{\varepsilon\} \to \pi/2$):

$$V = \frac{2(\pi a^2)^2(-\Delta\mathscr{P})}{3\pi\mu(2\pi a)} = \frac{a^3(-\Delta\mathscr{P})}{3\mu} \qquad (10\text{EC}11.2)$$

and

$$C_o \equiv u_o \sqrt{\frac{\varrho}{2(-\Delta\mathscr{P})}} = \sqrt{\frac{\text{Re}_o}{12\pi}} \qquad (10\text{EC}11.3)$$

Equation 10EC11.3 was shown by Churchill [1], pp. 17–18, to be in good accord with experimental data for $\text{Re}_o = 2au_o/\varrho < 5$.
    For a long narrow ellipse ($b/a \to 0$) Roscoe [29] reduced Equation 10EC11.1 to

$$\tilde{v} = \frac{\pi b^2 (-\Delta \mathscr{P})}{8\mu}$$ (10EC11.4)

where $\tilde{v}$ is the volumetric rate of flow per unit length in square meters per second. Therefore

$$C_o \equiv \left(\frac{\tilde{v}}{2b}\right)\sqrt{\frac{-\Delta \mathscr{P}}{\varrho}} = \sqrt{\frac{\pi \text{Re}}{128}}$$ (10EC11.5)

EC12.   Creeping Flow about Rotating Plates

Roscoe [29] also used a similar method to that of EC1 and EC11 to derive solutions for the torque on rotating plates. For an elliptical plate with a major axis a rotated about its minor axis b,

$$\mathscr{T} = \frac{8\pi\mu\omega a^3 \varepsilon^2}{3(F\{\varepsilon\} - E\{\varepsilon\})}$$ (10EC12.1)

where $F\{\varepsilon\}$ and $E\{\varepsilon\}$ are the complete elliptical integrals of the first and second kind, respectively, and, as before, ω is the angular velocity and $\varepsilon = (a^2 - b^2)^{1/2}/a$. (Again, see Abramowitz and Stegun [35], p. 589f.)

For a circular plate ($b \to a$, $\varepsilon \to 0$) rotating about its diameter, Equation 10EC12.1 reduces to

$$\mathscr{T} \to \frac{32\omega a^3}{3}$$ (10EC12.2)

A solution for a long elliptical plate rotating about its minor axis b is obtained by letting $b/a \to 0$ ($\varepsilon \to 1$):

$$\mathscr{T} \to \frac{8\pi\mu a^3 \omega}{3(\ln\{4a/b\} - 1)}$$ (10EC12.3)

Finally, Roscoe showed that the solution for an elliptical plate rotating about the tangent at one end of its major axis is

$$\mathscr{T} = \left(\frac{a^3 \varepsilon^2}{3[F\{\varepsilon\} - E\{\varepsilon\}]} + \frac{a^3}{F\{\varepsilon\}}\right) 8\pi\mu\omega$$ (10EC12.4)

He pointed out that these solutions provide possible geometrical approximations for viscometers that utilize agitators with complete or staggered blades, as shown in Figure 10–21. The average of the values given by Equation 10EC12.3 as an upper bound and by Equation 10EC12.4 as a lower bound are seen in Figure 10–21B to provide a good asymptote for experimental measurements with a Geddes and Dawson viscometer.

EC13.   Numerical Solutions for Converging Flows

Black and Denn [36] solved by finite differences the equations of motion for creeping flow through a planar 5:1 contraction, at angles of $\pi/4$ and $\pi/2$ rad.

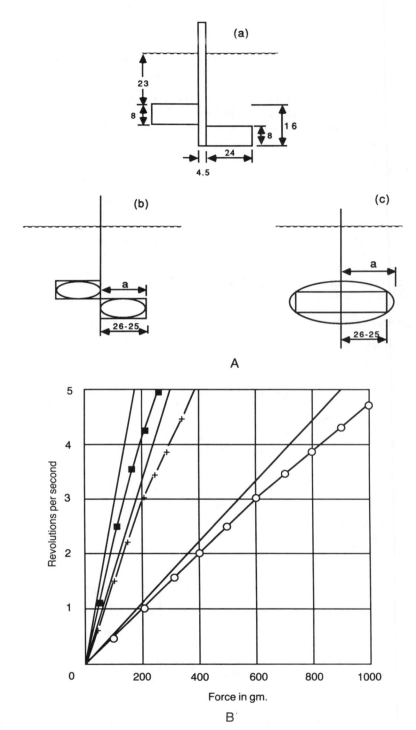

FIGURE 10–21 *(A) Approximation of agitator blades by ellipses. (Numerals refer to dimensions of experimental blades of Figure 10–21B in mm.) (After Roscoe [29].) (B) Comparison of predicted and measured force on blades of a viscometer.* ■:μ = 6.03 cp; +:μ = 10.59 cp; ○:μ = 33.7 cp. *(After Roscoe [29].)*

Their results are compared with the exact solution for infinite inclined plates (Jeffery–Hamel flow, E10) in Figures 10–22 and 10–23.

The exact solution for the idealized geometry is seen in Figure 10–22 to provide a reasonable prediction for the $\pi/4$-rad angle in the region of contraction, but to deviate significantly in Figure 10–23 for the $\pi/2$-rad angle due to the formation of a secondary flow (recirculating eddy) in the corner.

EXACT SOLUTIONS FOR INVISCID FLOW

Attention in this section is confined, except as noted, to *irrotational flow*, as defined and discussed in Chapter 9. Primary attention is given to results, that is, the solution itself, rather than derivations, since the latter virtually constitute a classical and well-documented branch of mathematics (potential theory and the theory of a complex variable) and theoretical mechanics (idealized hydromechanics and aerodynamics). The illustrative solutions were chosen primarily because of their subsequent applicability. Additional solutions may be found in Milne-Thompson [37], Streeter [18], and others.

S1.   Uniform Flow

The solutions for the potential and stream functions for a uniform velocity u_∞ in the *positive* x-direction are readily shown to be

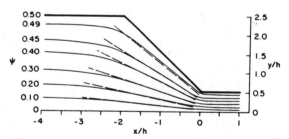

FIGURE 10–22   *Comparison of numerical solution for streamlines in creeping flow through a planar 5:1 contraction at an angle of $\pi/4$ with the Jeffrey-Hamel approximation. (From Black and Denn [36].)*

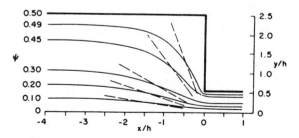

FIGURE 10–23   *Comparison of numerical solution for streamlines in creeping flow through a planar 5:1 contraction at an angle of $\pi/2$ with the Jeffrey-Hamel approximation. (From Black and Denn [36].)*

$$\phi = -u_\infty x \qquad (10S1.1)$$

and

$$\psi = -u_\infty y \qquad (10S1.2)$$

These expressions are seen to conform to Equations 9.60–9.63 and to yield $u_x = -u_\infty$ and $u_y = 0$ from Equations 9.16 and 9.17, and 9.58 and 9.59.

The *equipotential lines* and *streamlines* obviously consist of a grid of lines parallel to the y- and x-axes, respectively. The foregoing solution for this trivial flow proves useful later.

S2. Line Sources and Sinks

The velocity potential of a line source in the z-direction is (see, for example, Streeter [18], pp. 105–106)

$$\phi = -c \ln\{r\} = -c \ln\{\sqrt{x^2 + y^2}\} \qquad (10S2.1)$$

where c is a constant related to the strength of the source. The corresponding stream function (according to Equations 9.62 or 9.63) is

$$\psi = -c\theta = -c \tan^{-1}\left\{\frac{y}{x}\right\} \qquad (10S2.2)$$

Equipotential lines (ϕ = constant) and streamlines (ψ = constant) for this hypothetical flow are plotted in Figure 10–24. The corresponding velocity field is

$$u_x = -\frac{\partial \psi}{\partial y} = -\frac{\partial \phi}{\partial x} = \frac{cx}{x^2 + y^2} = c\frac{\cos\{\theta\}}{r} \qquad (10S2.3)$$

$$u_y = \frac{\partial \psi}{\partial x} = -\frac{\partial \phi}{\partial y} = \frac{cy}{x^2 + y^2} = c\frac{\sin\{\theta\}}{r} \qquad (10S2.4)$$

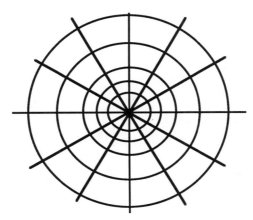

FIGURE 10–24 *Equipotential lines and streamlines for inviscid flow from a line source.*

or, for the velocity components in cylindrical coordinates,

$$u_r = -\frac{1}{r}\frac{\partial \psi}{\partial \theta} = \frac{c}{r} \tag{10S2.5}$$

and

$$u_\theta = \frac{\partial \psi}{\partial r} = 0 \tag{10S2.6}$$

The strength of this line source per unit length in the z-direction is equal to the total outward flow in cubic meters per meter seconds at any radius; that is,

$$S_l = 2\pi r u_r = 2\pi c \tag{10S2.7}$$

A solution for a sink is obtained simply by substituting $-c$ for c in Equations 10.S2.3 and 10S1.7.

This solution could have been derived from that for a uniform planar flow by the proper transformation of coordinates (*conformal mapping*); see, for example, Kaufmann [38], p. 65.

S3. Free Vortex

The solutions for the potential and stream functions of a stationary vortex about a straight axis are (see, for example, Streeter [18], p. 106)

$$\phi = -c\theta \tag{10S3.1}$$

and

$$\psi = c \ln \{r\} \tag{10S3.2}$$

Hence Figure 10–24 represents the equipotential lines and streamlines (inversely) for this flow as well. The velocity field, according to Equations 9.24 and 9.25, is

$$u_r = 0 \tag{10S3.3}$$

and

$$u_\theta = \frac{c}{r} \tag{10S3.4}$$

The *strength of the free vortex*, or *circulation* ($m^3/m \cdot s$), about a closed path, is

$$\Gamma = 2\pi r \left(\frac{c}{r}\right) = 2\pi c \tag{10S3.5}$$

Hence ϕ, ψ, and u_0 can be written in terms of this strength as

$$\phi = -\frac{\Gamma\theta}{2\pi} \tag{10S3.6}$$

$$\psi = \frac{\Gamma}{2\pi}\ln\{r\} \tag{10S3.7}$$

and

$$u_\theta = \frac{\Gamma}{2\pi r} \tag{10S3.8}$$

S4. Multiple Line Sources and Sinks

Equal Source and Sink. Solutions for potential flow can be constructed by superimposing the potential and stream functions for simpler flows. Thus for a line source at $x = a$, $y = 0$ and a sink of equal strength at $x = -a$, $y = 0$,

$$\phi = \frac{S_l}{2\pi}\ln\left\{\frac{r_2}{r_1}\right\} = \frac{S_l}{2\pi}\ln\left\{\frac{(x + a)^2 + y^2}{(x - a)^2 + y^2}\right\} \tag{10S4.1}$$

and

$$\psi = \frac{S_l}{2\pi}(\theta_2 - \theta_1) = \frac{S_l}{2\pi}\left(\tan^{-1}\left\{\frac{y}{x + a}\right\} - \tan^{-1}\left\{\frac{y}{x - a}\right\}\right) \tag{10S4.2}$$

Equations 10S4.1 and 10S4.2 can be rewritten (see problem 81) as

$$\left(x - a\coth\left\{\frac{2\pi\phi}{S_l}\right\}\right)^2 + y^2 = \left(a\operatorname{csch}\left\{\frac{2\pi\phi}{S_l}\right\}\right)^2 \tag{10S4.3}$$

and

$$x^2 + \left(y + a\cot\left\{\frac{2\pi\phi}{S_l}\right\}\right)^2 = \left(a\csc\left\{\frac{2\pi\psi}{S_l}\right\}\right)^2 \tag{10S4.4}$$

The equipotential lines and streamlines are thus, as illustrated in Figure 10–25, seen to be circles of radii $a\operatorname{csch}\{2\pi\phi/S_l\}$ and $a\csc\{2\pi\psi/S_l\}$ with centers at $x = \pm a|\cot\{2\pi\psi/S_l\}|$, respectively.

Method of Images. A solution for a line source (or sink) in a region with a single flat bounding surface can be constructed by superimposing the solutions for two equal sources (or sinks) at equal distances from the boundary (see problem 82). This is sometimes called the *method of images*.

Solutions for a line source between two or four bounding surfaces and for rows and arrays of line sources or sinks can be constructed by superposition of infinite series of line sources and/or sinks.

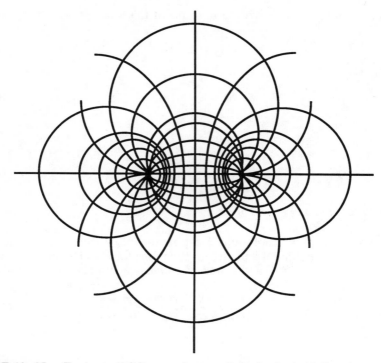

FIGURE 10–25   *Equipotential lines and streamlines for inviscid flow between a line source and a parallel line sink of equal strength.*

Solutions for single or multiple sources and sinks are useful as approximations for flow to or between wells in porous media (see Chapter 19).

Two-Dimensional Doublet.   The limiting solution as a source and a sink of equal strength approach one another (see Figure 10–25) while the product of their strengths and the distance between them remains constant is called a *doublet*. The potential and stream functions for a doublet at the origin are

$$\phi = \frac{M \cos\{\theta\}}{2\pi r} = \frac{Mx}{2\pi(x^2 + y^2)} \tag{10S4.5}$$

and

$$\psi = \frac{M \sin\{\theta\}}{2\pi r} = -\frac{My}{2\pi(x^2 + y^2)} \tag{10S4.6}$$

where the *moment of the doublet* is

$$M = S_l\, dx \tag{10S4.7}$$

From Figure 10–25 it can be inferred that the equipotential lines and streamlines of the doublet are circles tangent to the y-axis and x-axis, respectively.

Flow over Objects. Any closed streamline can be taken as the surface of a solid immersed in the fluid. Hence a solution for flow about such a body can be constructed by superimposing the stream and potential functions for a uniform flow on those for some other condition. This procedure is illustrated in S5 and S6.

S5. Superposition of a Uniform Flow and a Source

Superposition of a line source of strength S_l at the origin and a uniform velocity u_∞ in the positive x-direction gives

$$\psi = \frac{-S_l}{2\pi} \tan^{-1}\left\{\frac{y}{x}\right\} - u_\infty y = -\frac{S_l \theta}{2\pi} - u_\infty r \sin\{\theta\} \tag{10S5.1}$$

and

$$\phi = -\frac{S_l}{4\pi} \ln\{x^2 + y^2\} - u_\infty x \tag{10S5.2}$$

Hence

$$u_x = \frac{S_l x}{2\pi(x^2 + y^2)} + u_\infty = \frac{S_l \cos\{\theta\}}{2\pi r} + u_\infty \tag{10S5.3}$$

and

$$u_y = \frac{S_l y}{2\pi(x^2 + y^2)} = \frac{S_l \sin\{\theta\}}{2\pi r} \tag{10S5.4}$$

The streamlines corresponding to a stream function such as Equation 10S5.1, which consists of two additive terms ψ_1 and ψ_2, can be constructed graphically by drawing curves for $\psi_1 = nc$ and $\psi_2 = nc$ (c a constant; $n = 1, 2, 3, \ldots$) and connecting the corners of the resulting mesh as illustrated in Figure 10–26. This procedure is called the *Rankine method*. The streamlines of Figure 10–26 are symmetrical about the x-axis; hence the values for $y < 0$ are not drawn.

The point of stagnation, which is on the streamline separating the source and uniform fluids, is given by

$$u_x = u_y = 0 \tag{10S5.5}$$

Hence by

$$y = 0, \qquad x = -\frac{S_l}{2\pi u_\infty} \tag{10S5.6}$$

From Figure 10–26, $\theta = \pi$ at this point. The balance of the dividing streamline is

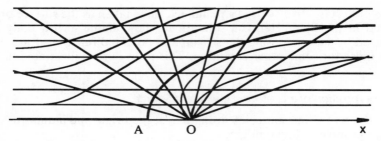

FIGURE 10–26   *Streamlines for inviscid flow due to the superposition of a uniform flow and a line source. A designates a point of stagnation and O shows the location of the line source.*

$$\psi^* = 0 = -\frac{S_l}{2\pi}\tan^{-1}\left\{\frac{y}{x}\right\} - u_\infty y \tag{10S5.7}$$

Hence

$$x = \frac{-y}{\tan\{2\pi u_\infty y/S_l\}} \tag{10S5.8}$$

Equation 10S5.8 is represented by the bold line in Figure 10–26. For $x \to \infty$ ($\theta \to 0$), y approaches an asymptotic value of $S_l/2\pi u_\infty$ along this streamline.

Any of the streamlines for Figure 10–26 can be interpreted as the surface of a solid. Hence the flow about the bold line can be interpreted as the flow of a river about a long island of that shape in a wide river. This shape, which is defined by Equation 10S5.8, is called a two-dimensional *Rankine half-body*. The point $x = -S_l/2u_\infty$, $y = 0$ is obviously the stagnation point for this flow. A photograph of the streaklines for Hele-Shaw flow about a two-dimensional Rankine half-body is shown in Figure 10–27. Note the close agreement with the streamlines of Figure 10.26.

A higher streamline could similarly be used to represent a more gradual turn in the terrain. Hence Equations 10S5.1–10S5.4 provide a representation for the flow field for a stream of water or a wind flowing along this surface.

S6.   Inviscid Flow around a Circular Cylinder

Solutions for inviscid flow about a circular cylinder can similarly be constructed by superimposing the potential and stream functions for a uniform flow on the appropriate source flow.

Fixed Cylinder.   The solution for inviscid flow in the positive x-direction around a fixed cylinder centered at the origin can be written as

$$\phi = -u_\infty\left(r + \frac{a^2}{r}\right)\cos\theta = -u_\infty x\left(1 + \frac{a^2}{x^2 + y^2}\right) \tag{10S6.1}$$

$$\psi = -u_\infty\left(r - \frac{a^2}{r}\right)\sin\theta = -u_\infty y\left(1 + \frac{a^2}{x^2 + y^2}\right) \tag{10S6.2}$$

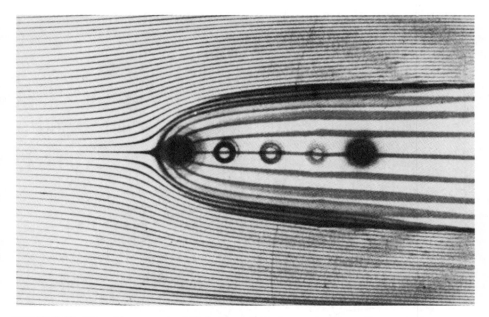

FIGURE 10–27 *Photograph of Hele-Shaw flow due to a uniform flow and a line source by G. I. Taylor. (From Van Dyke [34], p. 9.)*

where here u_∞ is the imposed velocity in the positive x-direction and θ is measured from the rear stagnation point of the cylinder. Equation 10S6.1 is the sum of Equation 10S1.1 and Equation 10S4.5 with $M = 2\pi u_\infty a^2$, that is the sum of the solutions for a uniform flow and a doublet. Equation 10S6.2 similarly follows from Equations 10S1.2 and 10S4.6.

The equipotential lines and streamlines corresponding to Equation 10S6.1 and 10S6.2 are plotted in Figure 10–28. Note that the streamline, $\psi = 0$, consists of the x-axis as well as the surface of the cylinder. A Hele-Shaw simulation for this flow is shown in Figure 10–29.

The components of the velocity in cylindrical coordinates are

$$u_r = -\frac{1}{r}\frac{\partial \psi}{\partial \theta} = u_\infty\left(1 - \frac{a^2}{r^2}\right)\cos\{\theta\} \tag{10S6.3}$$

and

$$u_\theta = \frac{\partial \psi}{\partial r} = -u_\infty\left(1 + \frac{a^2}{r^2}\right)\sin\{\theta\} \tag{10S6.4}$$

The velocity at the surface ($r = a$) is then

$$u_{\theta,a} = -2u_\infty \sin\{\theta\}, \qquad u_r = 0 \tag{10S6.5}$$

The velocity is seen to be zero at the forward ($\theta = \pi$) and rear ($\theta = 0$) *points of stagnation.*

The pressure distribution follows from Equation 9.80 and is

$$\mathscr{P} - \mathscr{P}_0 = -\frac{\varrho v^2}{2}$$

$$= -\frac{\varrho}{2}(u_r^2 + u_\theta^2)$$

$$= -\frac{u_\infty^2 \varrho}{2}\left(\left[1 - \left(\frac{a}{r}\right)^2\right]^2 + 4\left(\frac{a}{r}\right)\sin^2\{\theta\}\right) \qquad (10S6.6)$$

FIGURE 10–28   *Equipotential lines and streamlines for inviscid flow around a cylinder.*

FIGURE 10–29   *Photograph of Hele-Shaw flow around a circular cylinder by D. H. Peregrine. (From Van Dyke [34], p. 8.)*

On the surface $(r = a)$,

$$\mathscr{P}_a = \mathscr{P}_0 - 2u_\infty^2 \varrho \sin^2\{\theta\}$$

$$= \mathscr{P}_\infty + \frac{\varrho u_\infty^2}{2}[1 - 4\sin^2\{\theta\}] \qquad (10S6.7)$$

As shown in Figure 10–30, the dimensionless pressure $2(\mathscr{P} - \mathscr{P}_\infty)/\varrho u_\infty^2$ on the surface decreases from $\mathscr{P}_\infty + \varrho u_\infty^2/2$ at $\theta = \pi$ to -3 at $\theta = \pi/2$ and then increases back to unity at $\theta = 0$.

Moving Cylinder. If the potential and stream functions for the uniform flow are subtracted from Equations 10S6.1 and 10S6.2, one obtains

$$\phi = -\frac{u_\infty a^2 \cos\{\theta\}}{r} \qquad (10S6.8)$$

and

$$\psi = \frac{u_\infty a^2 \sin\{\theta\}}{r} \qquad (10S6.9)$$

which are the same as Equations 10S4.5 and 10S4.6 for $M = -2\pi u_\infty a^2$. The corresponding velocities are

$$u_r = -\frac{u_\infty a^2}{r^2}\cos\{\theta\} \qquad (10S6.10)$$

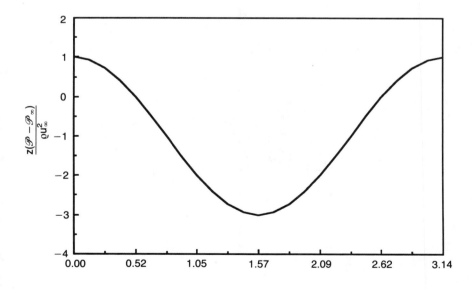

FIGURE 10–30 *Pressure distribution on the surface of a circular cylinder in inviscid flow.*

and

$$u_\theta = -\frac{u_\infty a^2}{r^2} \sin\{\theta\} \qquad (10S6.11)$$

This is the solution for the instantaneous fluid motion generated by the movement of a cylinder in the *x*-direction through a stagnant fluid, as seen by an observer above the cylinder but fixed with respect to the distant fluid. The corresponding equipotential lines and streamlines are plotted in Figure 10–31. They correspond to the doublet of Section S4.

S7.   *Inviscid Flow between and over Inclined Pairs of Walls*

The potential and stream functions for inviscid flow between two walls with an inclosed total angle α, as previously sketched in Figure 10–8, but for a half angle α, are

$$\phi = -\frac{ar^n}{n} \cos\{n\theta\} \qquad (10S7.1)$$

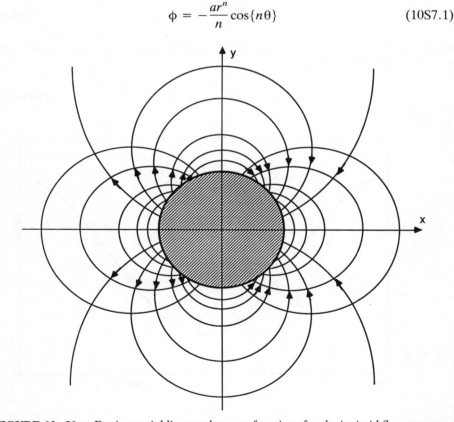

FIGURE 10–31   *Equipotential lines and stream functions for the inviscid flow generated by a cylinder moving from right to left as seen by an observer fixed with respect to the distant fluid.*

and

$$\psi = -\frac{ar^n}{n} \sin\{n\theta\} \tag{10S7.2}$$

The corresponding components of the velocity are

$$u_r = ar^{n-1} \cos\{n\theta\} \tag{10S7.3}$$

and

$$u_\theta = -ar^{n-1} \sin\{n\theta\} \tag{10S7.4}$$

Also,

$$\mathscr{P} = \mathscr{P}_0 - \frac{\varrho(u_r^2 + u_\theta^2)}{2} = \mathscr{P}_0 - \frac{a^2\varrho r^{2n-2}}{2} \tag{10S7.5}$$

where $\mathscr{P} = \mathscr{P}_0$ at $r = 0$ $(n > 1)$
         $n = \pi/\alpha$, rad^{-1}
         α = total angle in fluid between the plates, rad

This solution encompasses a complete range of angles α from 0 to 2π and, hence, a range of n from ∞ to 1/2. Also, this solution is applicable by symmetry, as indicated in Figure 10.32, for flow over wedges and into (as opposed to around) corners with total solid angles

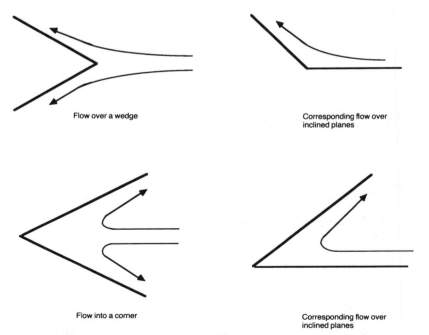

Flow over a wedge

Corresponding flow over inclined planes

Flow into a corner

Corresponding flow over inclined planes

FIGURE 10–32   *Identical divided and undivided inviscid flows.*

$$\beta = 2(\pi - \alpha) = 2\pi\left(\frac{n-1}{n}\right) \qquad (10S7.6)$$

from $\beta = 0$, corresponding to $\alpha = \pi$ and $n = 1$, up to $\beta = 2\pi$, corresponding to $\alpha = 0$ and $n \to \infty$. These flows do not occur physically for $\beta < 0$, corresponding to $\alpha > \pi$ and $n < 1$.

Figures 10–33A and B correspond to $n = 3$ or $\alpha = \pi/3$ and $\beta = 4\pi/3$, respectively; Figure 10.33C to $n = 2$ or $\alpha = \pi/2$ and $\beta = \pi$, respectively; Figures 10–33D and E to $n = 3/2$ or $\alpha = 2\pi/3$ and $\beta = 2\pi/3$, respectively; Figures 10–33F and G to $n = 1$ or $\alpha = \pi$ and $\beta = 0$, respectively; Figure 10–33G to $n = 3/4$ or $\alpha = 4\pi/3$; Figure 10–33H to $n = 2/3$ or $\alpha = 3\pi/2$; and Figure 10–33I to $n = 1/2$ or $\alpha = 2\pi$.

In the special case of $n = 2$ or $\alpha = \pi/2$ and $\beta = \pi$, corresponding to walls and lines of symmetry, respectively, at right angles

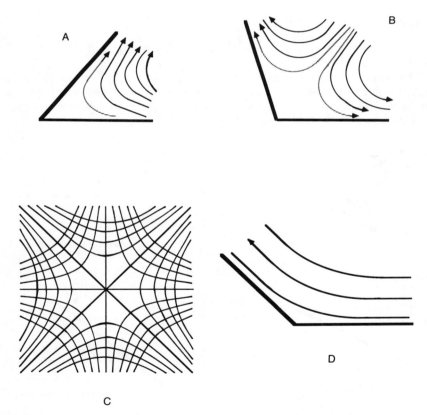

FIGURE 10–33   *Equipotential lines and streamlines for inviscid flow along walls separated by an angle α or impinging on corners or wedges with a total angle β: (A) n = 3, α = π/3; (B) n = 3, β = 4π/3; (C) n = 2, α = π/2 and β = π; (D) n = 3/2, α = 2π/3; (E) n = 3/2, β = 2π/3; (F) n = 1, α = π and β = 2π; (G) n = 3/4, α = 4π/3; (H) n = 2/3, α = 3π/2; (I) n = 1/2, α = 2π.*

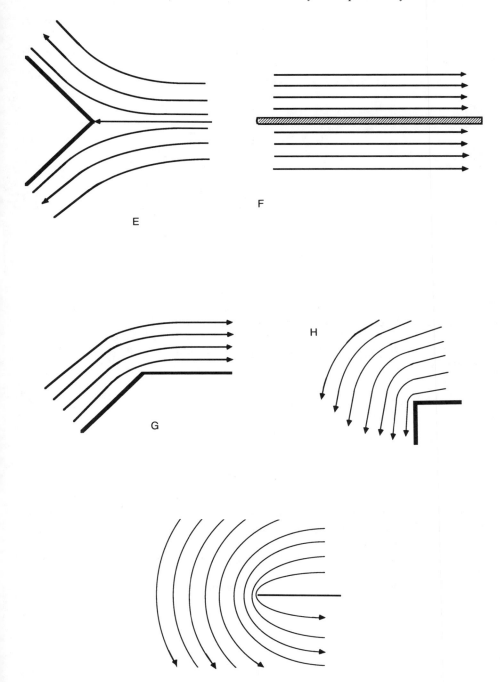

$$\phi = cr^2 \cos\{\theta\} = c(x^2 - y^2) \tag{10S7.7}$$

$$\psi = cr^2 \sin\{\theta\} = cxy \tag{10S7.8}$$

$$u_x = cr \cos\{\theta\} = 2cx \tag{10S7.9}$$

and

$$u_y = -2cr \sin\{\theta\} = -2cy \tag{10S7.10}$$

Hence

$$v = \sqrt{u_x^2 + u_y^2} = 2cr = 2c\sqrt{x^2 + y^2} \tag{10S7.11}$$

and

$$\mathscr{P} = \mathscr{P}_0 - 2c^2\varrho r^2 = \mathscr{P}_0 - 2c^2\varrho(x^2 + y^2) \tag{10S7.12}$$

In Figure 10–33C the streamlines are equilateral hyperbolas about $y = \pm x$, and the equipotential lines are equilateral hyperbolas about $y = 0$ for $|y| < |x|$ and about $x = 0$ for $|y| > |x|$. This solution can be interpreted either for flow around an internal right angle or as a uniform flow impacting on a flat plate.

Letting $x' = x + a$ and $y' = y$ in Equation 10S6.2 gives the following expression for the stream function with the origin shifted to the forward point of stagnation:

$$\psi = u_\infty y'\left(1 - \frac{a^2}{(x' - a)^2 + (y')^2}\right) \tag{10S7.13}$$

For $x' \to 0$ and $y' \to 0$,

$$\psi \to -2\frac{u_\infty y'x'}{a} \tag{10S7.14}$$

Comparison of Equations 10S7.14 with 10S7.8 suggests that flow near the forward point of a cylinder is congruent with impacting flow near the point of stagnation on a plate.

For $\alpha = \pi$, Equations 10S7.1–10S7.5 are seen to reduce to those for uniform flow in the *positive* x-direction.

The solution for flow over wedges is usually expressed as

$$u_x = A(x')^m \tag{10S7.15}$$

and

$$u_y = -Am(x')^{m-1}y' \tag{10S7.16}$$

where x' is parallel to and y' is perpendicular to the surface. The equivalence of Equations 10S7.14 and 10S7.15 to 10S7.1–10S7.5 is examined in problem 88.

S8. Impinging Uniform Flow on a Thin Strip

The solution for impinging flow on a strip of width $4a$ can be expressed in terms of the solution of Section S6 for a fixed cylinder as follows. The transformation

$$\zeta = z - \frac{a^2}{z} \tag{10S8.1}$$

where ζ and z are complex (i.e., $z = x + iy$ and $\zeta = \xi + i\eta$) maps the circle of radius a into a flat stripe of width $4a$ normal to the direction of the undisturbed flow. The coordinates η and ξ are parallel and normal, respectively, to the planar strip. Substituting for ζ and z in Equation 10S8.1 gives

$$\xi + i\eta = x\left(1 - \frac{a^2}{x^2 + y^2}\right) + iy\left(1 + \frac{a^2}{x^2 + y^2}\right) \tag{10S8.2}$$

Hence,

$$\xi = x\left(1 - \frac{a^2}{x^2 + y^2}\right) = r\cos\{\theta\}\left(1 - \frac{a^2}{r^2}\right) \tag{10S8.3}$$

and

$$\eta = y\left(1 + \frac{a^2}{x^2 + y^2}\right) = r\sin\{\theta\}\left(1 + \frac{a^2}{r^2}\right) \tag{10S8.4}$$

ϕ and ψ can be evaluated from Equations 10S6.1 and 10S6.2 for any chosen x and y or r and θ. The corresponding location in the η–ξ plane can be determined from Equations 10S8.3 and 10S8.4. A plot of the streamlines in the η–ξ plane is shown in Figure 10–34.

S9. Flow into a Planar Channel

The fluid motion entering a channel formed by two parallel plates at distance $2b$ apart, immersed in an infinite extent of fluid, can be represented by (see, e.g., Lamb [28], pp. 74–75 or Bird et al. [21], pp. 138–139)

$$x = \frac{\phi}{u_\infty} + \frac{b}{\pi} e^{\pi\phi/bu_\infty} \cos\left\{\frac{\pi\psi}{bu_\infty}\right\} \tag{10S9.1}$$

and

$$y = \frac{\psi}{u_\infty} + \frac{b}{\pi} e^{\pi\phi/bu_\infty} \sin\left\{\frac{\pi\psi}{bu_\infty}\right\} \tag{10S9.2}$$

where u_∞ is the velocity far down the channel in meters per second.

The streamlines can be plotted by locating x for discrete values of ψ and varying values of ϕ. For example, for $\psi = 0$,

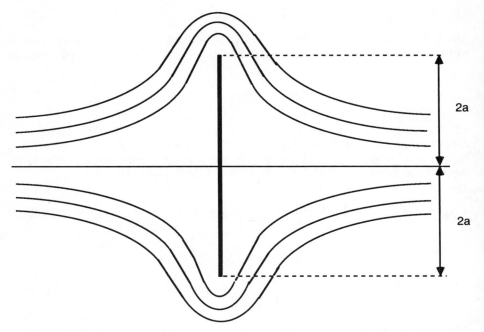

FIGURE 10–34   *Streamlines for inviscid flow impinging on a strip of height 4a.*

$$x = \frac{\phi}{u_\infty} + \frac{b}{\pi}e^{\pi\phi/bu_\infty} \tag{10S9.3}$$

and

$$y = 0 \tag{10S9.4}$$

This streamline consists of the x-axis, with x varying from $-\infty$ to ∞. Also for $\psi = \pm bu_\infty$,

$$x = \frac{\phi}{u_\infty} - \frac{b}{\pi}e^{\pi\phi/bu_\infty} \tag{10S9.5}$$

and

$$y = \pm b \tag{10S9.6}$$

As ϕ varies from $-\infty$ to ∞, x now varies from $-\infty$ to -1 (at $\phi = -1$) and back to -1. Hence the limiting streamlines are described by $y = \pm b$ and $-\infty \leq x < -1$. The streamlines for $\psi = \pm bu_\infty(1 - \delta)$, where δ is very small, thus follow closely along the two outside surfaces of the channel, turn 2π rad at x slightly greater than -1, and then follow closely the two inside walls of the channel. Representative streamlines for intermediate values of ψ are plotted in Figure 10–35.

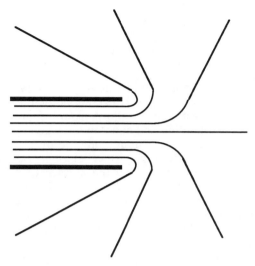

FIGURE 10–35   *Streamlines for inviscid flow entering or leaving a planar channel.*

The components of the velocity are given by

$$u_x = -\frac{v^2}{u_\infty}\left(1 - e^{\pi\phi/u_\infty b}\cos\left\{\frac{\pi\psi}{u_\infty b}\right\}\right) \tag{10S9.7}$$

and

$$u_y = -\frac{v^2}{u_\infty}e^{\pi\phi/u_\infty b}\sin\left\{\frac{\pi\psi}{u_\infty b}\right\} \tag{10S9.8}$$

where $v^2 = u_x^2 + u_y^2$.

This solution is applicable for flow *out* of the channel simply by reversing the signs of u_x, u_y, ψ, and ϕ.

S10.   Free Streamlines

A streamline separates the flowing fluid into two regions. In these regions

$$\frac{\mathcal{P}_a}{\varrho_a} + \frac{v_a^2}{2} = k_a \tag{10S10.1}$$

and

$$\frac{\mathcal{P}_b}{\varrho_b} + \frac{v_b^2}{2} = k_b \tag{10S10.2}$$

Since the pressure is continuous,

$$\mathcal{P}_a = \mathcal{P}_b \tag{10S10.3}$$

Therefore

$$\frac{\varrho_a v_a^2}{2} - \frac{\varrho_b v_b^2}{2} = k_a \varrho_a - k_b \varrho_b = k_c \qquad (10\text{S}10.4)$$

where k_a, k_b, and k_c are constants. If one fluid is at rest ($v_b = 0$) or of relatively low density ($\varrho_b/\varrho_a \ll 1$), then v_a is constant along the streamline. Such a boundary is called a *free streamline*. It follows from Equation 10S10.1 that $\mathscr{P}_a$ is also constant along the streamline and equal to the value in the fluid at rest. The stream function is, by definition, constant along the streamline. This situation is closely approximated by the boundary of a jet of liquid passing through an orifice in a pipe as in Figure 10–36, by the surface of a fluid jet entering a channel abruptly as in Figure 10–37, and by a fluid flowing around a thin plate as in Figure 10–38. The fluid beyond the bounding streamline will in each of these cases have some motion as indicated. However, this induced motion will be very small compared to the main stream and hence can be neglected to a first approximation. The derivation of expressions for the free streamlines is possible and useful in some instances, including those of Figures 10–36 to 10–38. Several such solutions are illustrated next.

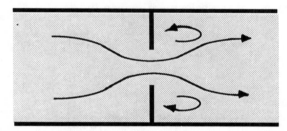

FIGURE 10–36   *Streamlines for flow through an orifice in a pipe at high velocity.*

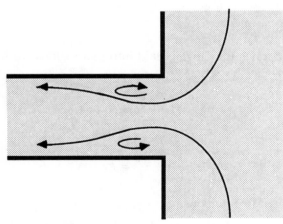

FIGURE 10–37   *Streamlines for abrupt entrance into a channel at high velocity.*

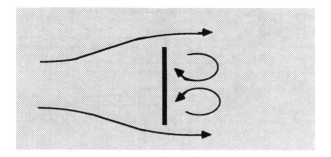

FIGURE 10–38   *Streamlines for flow impinging on a strip at high velocity.*

Borda Mouthpiece.   A *Borda mouthpiece* is an entrance for fluids that extends into a vessel, as shown in Figure 10–39. The expressions defining the location of the free streamline can be shown (see, for example, Lamb [28], pp. 96–98, or Streeter [18], pp. 169–174) to be

$$x_f = \frac{2b}{\pi}\left(\sin^2\left\{\frac{\theta}{2}\right\} - \ln\left\{\sec\left\{\frac{\theta}{2}\right\}\right\}\right) \qquad (10S10.5)$$

and

$$y_f = \frac{b}{\pi}(\theta - \sin\{\theta\}) \qquad (10S10.6)$$

Hence x_f and y_f designate the coordinates of the free streamline as measured outward and downward from the edge of the upper lip, respectively, and b is the final half-width of the jet, as indicated in Figure 10–40. Equation 10S10.6 indicates that the asymptotic value of yf as $\theta \to \pi$ is b. The fractional contraction, which equals the orifice coefficient, is thus

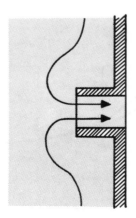

FIGURE 10–39   *Borda mouthpiece.*

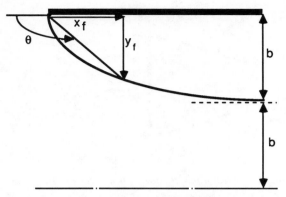

FIGURE 10–40   *Construction of a free streamline for a Borda mouthpiece.*

$$C_b = \frac{2b}{b + 2b + b} = \frac{1}{2} \tag{10S10.7}$$

The coefficient for a circular Borda mouthpiece is also 0.5; that is, the displacement of the free streamline from the wall is not affected by this change in geometry.

Planar Abrupt Entrance (Orifice).   The coordinates of the free streamline for a planar, abrupt entrance producing a jet of width $2b$, as sketched in Figure 10–41, are (see, for example, Streeter [18], pp. 175–177, or Lamb [28], pp. 98–99)

$$x_f = \frac{4b}{\pi} \sin^2\left\{\frac{\theta}{2}\right\} \tag{10S10.8}$$

and

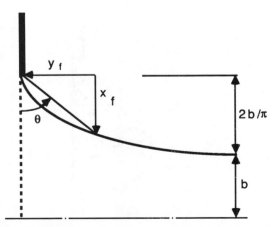

FIGURE 10–41   *Construction of a free streamline for a planar abrupt entrance.*

$$y_f = \frac{2b}{\pi}\left(\ln\left\{\tan\left\{\frac{\pi}{4} + \frac{\theta}{2}\right\}\right\} - \sin\{\theta\}\right) \qquad (10S10.9)$$

The angle θ is seen to be zero at the origin. The asymptotic value of x_f for $\theta \to \pi/2$ is $2b/\pi$. Hence the coefficient of contraction is

$$C_o = \frac{2b}{2b/\pi + 2b + 2b/\pi} = \frac{\pi}{\pi + 2} = 0.6110\ldots$$

This value is also applicable for a circular orifice and has been well confirmed experimentally for $\mathrm{Re}_0 > 10^5$ (see, for example, Churchill [1], p. 17).

The preceding solutions for a Borda mouthpiece and an abrupt entrance are remarkable in that they produce two-dimensional, physically valid results for large Re. This success is a consequence of the negligible role of the confining surfaces.

Drag Force on a Thin Strip Normal to the Flow. The free streamlines for flow about a thin strip of width d normal to the flow, as sketched in Figure 10–42, are (see, for example, Streeter [18], pp. 177–181, or Lamb [28], pp. 99–102)

$$x_f = \frac{2d}{\pi + 4}\left(\sec\{\theta\} + \frac{\pi}{4}\right) \qquad (10S10.10)$$

and

$$y_f = \frac{d}{\pi + 4}\left(\sec\{\theta\}\tan\{\theta\} - \ln\left\{\tan\left\{\frac{\pi}{4} + \frac{\theta}{2}\right\}\right\}\right) \qquad (10S10.11)$$

Here x_f is parallel to the plate and y_f is normal to the plate, with the origin on the centerline of the plate. θ varies from 0 to $-\pi/2$.

The corresponding form drag (net pressure force) on a unit length of the strip is

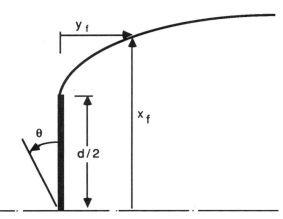

FIGURE 10–42 *Construction of a free streamline for flow impinging on a strip.*

$$F' = \frac{2\pi\rho d u_\infty^2}{\pi + 4} \tag{10S10.12}$$

Hence

$$\bar{C}_t \equiv \frac{F'}{2\rho u_\infty^2 d} = \frac{\pi}{\pi + 4} = 0.4399\ldots \tag{10S10.13}$$

which is less than half the observed experimental value, undoubtedly due to the neglected skin friction on the frontal surface.

Three-Dimensional (Axisymmetric) Flows

The theory of a complex variable, including conformal mapping, is not applicable to three-dimensional flow. However, axisymmetric flows bear some analogy to two-dimensional flows in that a stream function can be defined (Equations 9.26 and 9.27 or 9.28 and 9.29), and when the flow is irrotational a velocity potential exists (Equations 9.66 and 9.67 or 9.70 and 9.71). Solutions prove to be possible for a limited number of important cases. A few of these are illustrated next. For others, see for example, Milne-Thompson [37], Chapter 16.

S11.   Uniform Flow

For a uniform flow u_∞ in the positive z-direction, from Equations 9.66 and 9.67,

$$-\frac{\partial \tilde{\phi}}{\partial r} = u_r = 0 \tag{10S11.1}$$

and

$$-\frac{\partial \tilde{\phi}}{\partial z} = u_z = u_\infty \tag{10S11.2}$$

Hence

$$\tilde{\phi} = -u_\infty z \tag{10S11.3}$$

and similarly

$$\tilde{\psi} = \frac{u_\infty r^2}{2} \tag{10S11.4}$$

Hence, in spherical coordinates with axial symmetry

$$\tilde{\phi} = -u_\infty R \cos\{\theta\} \tag{10S11.5}$$

and

$$\tilde{\psi} = -\frac{u_\infty R^2}{2} \sin^2\{\theta\} \tag{10S11.6}$$

S12. Point Sources and Sinks

Flow outward from a point source of strength S_p in cubic meters per second in an infinite region is obviously radial. Then from continuity

$$u_R = \frac{S_p}{4\pi R^2} \tag{10S12.1}$$

Further, from Equations 9.70 and 9.71,

$$-\frac{\partial\tilde{\phi}}{\partial R} = u_R = \frac{S_p}{4\pi R^2} \tag{10S12.2}$$

and

$$-\frac{1}{R}\frac{\partial\tilde{\phi}}{\partial\theta} = u_\theta = 0 \tag{10S12.3}$$

Hence, assuming $\tilde{\phi} \to 0$ as $R \to \infty$ gives

$$\tilde{\phi} = -\frac{S_p}{4\pi R} \tag{10S12.4}$$

Also, from Equations 9.28 and 9.29,

$$-\frac{1}{R^2 \sin\{\theta\}}\frac{\partial\tilde{\psi}}{\partial\theta} = \frac{S_p}{4\pi R^2} \tag{10S12.5}$$

and

$$\frac{1}{R \sin\{\theta\}}\frac{\partial\tilde{\psi}}{\partial R} = 0 \tag{10S12.6}$$

Then

$$\tilde{\psi} = \frac{S_p}{4\pi} \cos\{\theta\} \tag{10S12.7}$$

The plot of equipotential lines and streamlines for a point source is shown in Figure 10–43. This plot is similar to that of Figure 10–24, but the spacing of the equipotential lines for equal increments of $\tilde{\phi}$ differs.

Equations 10S12.1–10S12.7 are obviously applicable to a point sink simply by changing the signs of u_R, u_θ, $\tilde{\phi}$, and $\tilde{\psi}$.

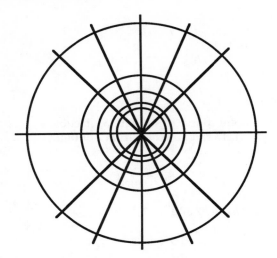

FIGURE 10–43   *Equipotential lines and streamlines for a point source or sink.*

S13.   Multiple Point Sources and Sinks

An Equal Point Source and Sink.   For an equal point source and sink at $z = a$, $r = 0$ and $z = -a$, $r = 0$, respectively,

$$\tilde{\phi} = \frac{S_p}{4\pi}\left(\frac{1}{r_1} - \frac{1}{r_2}\right) = \frac{S_p}{4\pi}\left(\frac{1}{\sqrt{(z-a)^2 + r^2}} - \frac{1}{\sqrt{(z+a)^2 - r^2}}\right) \quad (10S13.1)$$

and

$$\tilde{\psi} = \frac{S_p}{4\pi}(\cos\{\theta_1\} - \cos\{\theta_2\}) \quad (10S13.2)$$

where r_1, θ_1, r_2, and θ_2 are located as indicated in Figure 10–44A. The streamlines and equipotential lines are shown in Figure 10–44B.

Three-Dimensional Doublet.   As $a \to 0$, the potential and stream functions for the equal source and strength become

$$\tilde{\phi} = \frac{aS_p}{4\pi R^2}\cos\{\theta\} \quad (10S13.3)$$

and

$$\tilde{\psi} = \frac{-aS_p}{4\pi R^2}\sin^2\{\theta\} \quad (10S13.4)$$

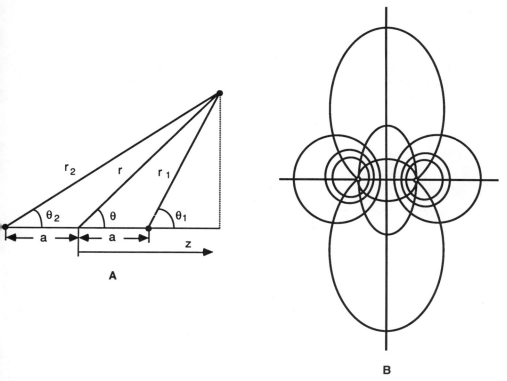

FIGURE 10–44 *Inviscid flow between a point source and a point sink of equal strength separated by a distance 2a: (A) coordinate system; (B) equipotential lines and steamlines.*

S14. Superposition of a Uniform Flow on a Point Source

Combining Equations 10S11.5 and 10S11.6, for uniform flow in the positive *x*-direction, with Equations 10S12.4 and 10S12.7 gives

$$\tilde{\phi} = -u_\infty R \cos\{\theta\} + \frac{S_p}{4\pi R} \tag{10S14.1}$$

and

$$\tilde{\psi} = -\frac{u_\infty R^2}{2}\sin^2\{\theta\} + \frac{S_p}{4\pi}\cos\{\theta\} \tag{10S14.2}$$

Also,

$$u_R = u_\infty \cos\{\theta\} + \frac{S_p}{4\pi R^2} \tag{10S14.3}$$

and

$$u_\theta = -u_\infty \sin\{\theta\} \qquad (10\text{S}14.4)$$

The streamlines and equipotential lines for this flow are plotted in Figure 10–45. The location of the stagnation point ($u_r = u_\theta = 0$) is defined by

$$u_\infty \cos\{\theta\} + \frac{S_p}{4\pi R^2} = 0 \qquad (10\text{S}14.5)$$

and

$$u_\infty \sin\{\theta\} = 0 \qquad (10\text{S}14.6)$$

Therefore, the location of the point of stagnation is given by

$$R = \sqrt{\frac{S_p}{4\pi u_\infty}} \qquad (10\text{S}14.7)$$

and

$$\theta = \pi \qquad (10\text{S}14.8)$$

The streamline through this point is

$$\tilde{\psi}^* = -\frac{S_p}{4\pi} = -\frac{u_\infty R^2}{2}\sin^2\{\theta\} + \frac{S_p}{4\pi}\cos\{\theta\} \qquad (10\text{S}14.9)$$

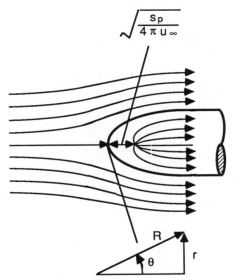

FIGURE 10–45   *Streamlines for inviscid flow due to superposition of a uniform flow and a point source.*

which can be reexpressed as

$$R^2 = \frac{S_p}{2\pi u_\infty \sin^2\theta}(1 + \cos\{\theta\}) \tag{10S14.10}$$

or

$$r^2 = \frac{S_p}{2\pi u_\infty (1 + \cos\{\theta\})} \tag{10S14.11}$$

Thus,

$$r = \sqrt{\frac{S_p}{2\pi u_\infty}} \cos\left\{\frac{\theta}{2}\right\} \tag{10S14.12}$$

This streamline separates the fluids from the source and the uniform stream.

The preceding solution can therefore also be interpreted as that for flow about a cylinder whose blunt nose is defined by Equation 10S14.12. As shown in Figure 10–46, that body has a maximum diameter

$$D_{\max} = 2(r)_{\theta=\pi} = \sqrt{\frac{2S_p}{\pi u_\infty}} \tag{10S14.13}$$

which is equal to twice the distance from the source to the point of stagnation in Figure 10–45.

S15. Inviscid Flow around a Sphere

Fixed Sphere. The potential and stream functions for inviscid flow around a sphere in the positive *z*-direction are

$$\tilde{\phi} = -u_\infty R \cos\{\theta\}\left[1 + \frac{1}{2}\left(\frac{a}{R}\right)^3\right] \tag{10S15.1}$$

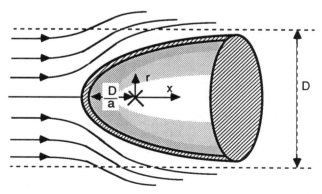

FIGURE 10–46 *Interpretation of solution for a point source superposed on a uniform stream as a solution for inviscid flow over a blunt-nosed cylinder.*

and

$$\tilde{\psi} = -\frac{u_\infty R^2}{2}\left[1 - \left(\frac{a}{R}\right)^3\right]\sin^2\{\theta\}$$
(10S15.2)

Hence

$$u_R = u_\infty\left[1 - \left(\frac{a}{R}\right)^3\right]\cos\{\theta\}$$
(10S15.3)

and

$$u_\theta = -\frac{u_\infty}{2}\left[2 + \left(\frac{a}{R}\right)^3\right]\sin\{\theta\}$$
(10S15.4)

Equations 10S15.2–10S15.4 follow from setting $\zeta = -2/3$ in Equations 10EC1.17, 10EC1.19 and 10EC1.20. The components of the velocity on the surface ($R = a$) are then

$$u_R = 0$$
(10S15.5)

and

$$u_\theta = -\frac{3u_\infty\sin\{\theta\}}{2}$$
(10S15.6)

and

$$v^2 = u_\theta^2 + u_r^2 = \frac{9}{4}u_\infty^2\sin^2\{\theta\}$$
(10S15.7)

It follows that

$$\frac{\mathscr{P}}{\varrho} + \frac{9}{8}u_\infty^2\sin^2\{\theta\} = \frac{\mathscr{P}_\infty}{\varrho} + \frac{u_\infty^2}{2}$$
(10S15.8)

The point of stagnation ($u_R = u_\theta = 0$) occurs at $R = a$, $\theta = 0$. Hence the pressure at that point is

$$\mathscr{P}_0 = \mathscr{P}_\infty + \frac{\varrho u_\infty^2}{2}$$
(10S15.9)

The minimum pressure occurs at $\theta = \pi/2$:

$$\mathscr{P}_{\min} = \mathscr{P}_\infty - \frac{5}{8}\varrho u_\infty^2 = \mathscr{P}_0 - \frac{9}{8}\varrho u_\infty^2$$
(10S15.10)

Equations 10S15.1 and 10S15.2 can be seen to result from the superposition of the functions for a uniform flow (Equations 10S11.5 and 10S11.6) on a doublet of strength $-2\pi u_\infty a^2$ (Equations 10S13.3 and 10S13.4).

The streamlines and equipotential lines are plotted in Figure 10–47 and the elevation of the pressure on the surface above the free-stream pressure in Figure 10–48.

If the contribution of the uniform flow is deleted, the potential and stream functions are

$$\tilde{\phi} = -\frac{u_\infty a^3}{2R^2}\cos\{\theta\} \tag{10S15.11}$$

and

$$\tilde{\psi} = \frac{u_\infty a^3 \sin^2\{\theta\}}{2R} \tag{10S15.12}$$

which are also, as noted earlier, those of a doublet of strength $S_p = -2\pi u_\infty a^2$. These equipotential lines and stream functions are plotted in Figure 10–49. They correspond to those seen by an observer above the sphere and fixed with respect to the distant fluid.

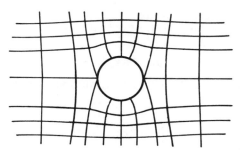

FIGURE 10–47 *Equipotential lines and streamlines for inviscid flow about a fixed sphere.*

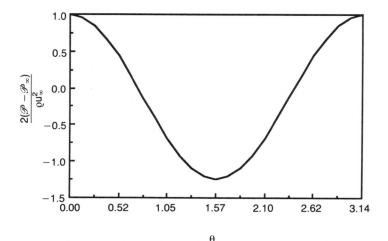

FIGURE 10–48 *Pressure distribution on the surface of a sphere in inviscid flow.*

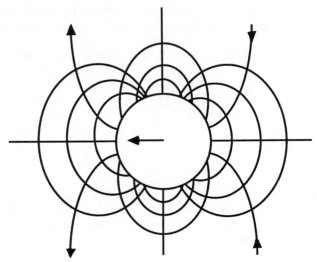

FIGURE 10–49   *Equipotential lines and stream functions for inviscid flow generated by a moving sphere as seen by an observer fixed with respect to the distant fluid.*

S16.   Axially Symmetric Impact on a Disk

For a fluid impacting on a disk such that the motion is axisymmetric, as sketched in Figure 10–50,

$$\tilde{\phi} = \frac{\alpha}{2}(2z^2 - r^2) \tag{10S16.1}$$

$$\tilde{\psi} = \alpha z r^2 \tag{10S16.2}$$

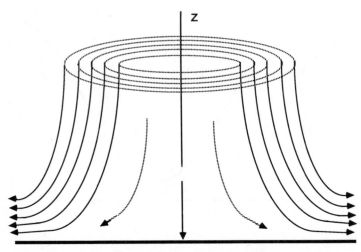

FIGURE 10–50   *Streamlines for inviscid flow impinging on a disk.*

$$u_r = \alpha r \tag{10S16.3}$$

$$u_z = -2\alpha z \tag{10S16.4}$$

and

$$\mathscr{P} = \mathscr{P}_0 - \frac{\alpha^2 \varrho}{2}(r^2 + 4z^2) \tag{10S16.5}$$

where $\mathscr{P}_0$ = pressure at the stagnation point ($u_r = u_z = z = r = 0$)
α = constant, s^{-1}

The solution for inviscid flow may be compared with the general solution in Chapter 14 (Homann flow).

SUMMARY

Solutions in closed form were illustrated for a wide variety of conditions. Similar solutions are possible for many analogous problems, as indicated in the problem set. However, solutions for more complicated problems are generally possible only by numerical methods.

Many of the illustrative solutions are for ideal (completely nonviscous) flows. Such solutions are of value only if one is not concerned with (1) the velocity field near surfaces, (2) the viscous drag force on surfaces, and (3) the exchange of heat or mass with the surface.

The illustrative solutions for purely viscous (noninertial) flows in Chapter 10 as well as in Chapters 3–7 are exact for a wide range of velocities up to some limiting value.

Chapters 11–12 are primarily concerned with geometries for which neither of these limiting conditions is applicable.

PROBLEMS

1. Reduce the equations of continuity and force-momentum in three dimensions to Equation 10E1.1 and justify each simplification.
2. Reduce Equation 10E1.1 to an ordinary differential equation using the method of Hellums and Churchill [39] and solve to obtain Equations 10E1.4 and 10E1.7.
3. Develop approximations in closed form for the behavior of the accelerated plate for long and short times. Compare with the complete solution and indicate the limits of reliability of these approximations.
4. Derive an expression for the "boundary-layer thickness" on a suddenly accelerated plate, corresponding to $u/u_w = 0.01$.
5. Derive a solution (or adapt a thermal solution) that converges more rapidly than Equation 10E1.5 for long times.
6. Prepare a plot of u/u_w versus y/d with vt/d^2 as a parameter for a suddenly accelerated plate.
7. What is the thermal analog of the transient acceleration of fluid in a pipe?

8. Derive a solution for the impulsive start of fluid motion in a pipe.
9. Derive Equation 10E4.2 or the equivalent.
10. What is the amplitude of the oscillation given by Equation 10E4.3? What is the phase lag as a function of depth? What is the "wavelength" of oscillation in the fluid?
11. Prepare a plot of u/u_{max} versus $y\sqrt{\omega/2\nu}$ for several values of ωt for an oscillated plate.
12. Derive the similarity transform indicated by Equation 10E5.4 starting with the appropriately reduced equations of continuity and force-momentum and using the method of Hellums and Churchill [39]. See Equations 11.11–11.36 as an example.
13. Derive Equation 10E5.4 starting with the equations of continuity and force-momentum.
14. Reduce the equations of continuity and force-momentum in cylindrical co-ordinates to obtain Equations 10E6.1–10E6.2. Justify each simplification.
15. Derive a solution analogous to Equation 10E6.12A for the free surface of a fluid between two concentric cylinders of radii a_1 and a_2, rotated about their axis.
16. Reduce the equations of continuity and force-momentum to Equations 10E7.1 and 10E7.2. Justify each simplification.
17. What is the volumetric rate of flow corresponding to Equation 10E7.7 for a finite outer radius of $3a$?
18. Derive a solution analogous to Equation 10E7.7 for spherical radial flow.
19. Justify Equations 10E8.1–10E8.4.
20. Could Equation 10E6.9 have been used instead of Equation 10E8.4? What would be the consequence?
21. Derive Equations 10E8.5–10E8.7 from the general equations of force-momentum. Justify each simplification. Why is $-d\mathscr{P}/dx$ constant?
22. Carry out the detailed derivations of Equations 10E8.10 and 10E8.11 from 10E8.6 and 10E8.7.
23. Show that Equation 10E8.10 reduces to Equations 7.51 and 7.53 for $\Omega_2 = 0$ and $\Omega_1 = 0$, respectively.
24. Using Equation 10E8.10, derive expressions for the torque on each cylinder.
25. Derive an expression for the maximum value of u_z corresponding to Equation 10E8.11, and derive another for its location.
26. Derive an expression for the volumetric rate of flow corresponding to Equation 10E8.11 and for the drag force necessary to produce u_w.
27. Derive an expression for the radial pressure distribution corresponding to Equation 10E8.5.
28. The solution represented by Equations 10E8.10 and 10E8.11 is labeled "helical flow." Derive an expression for the path followed by a particle of fluid and describe the motion.

    a. for $u_w = 0$
    b. for $-d\mathscr{P}/dx = 0$
    c. for finite u_w and $-d\mathscr{P}/dx$

29. Repeat the derivation of Equation 10E8.11 for translation of the inner rather than the outer cylinder.

30. Justify in detail the reduction in the equations of motion leading to Equations 10E9.1 and 10E9.2.

31. Derive a series solution for Equation 10E9.8, assuming $\phi = \sum_{n=1}^{\infty} a_n z^n$. (*Hint*: See Bird et al. [21], p. 122.)

32. Integrate Equation 10E9.8 numerically or iteratively for

$$\frac{\varrho(1/a_1^2 - 1/a_2^2)}{2\mu \ln\{a_2/a_1\}} = \frac{\mathscr{P}_1 - \mathscr{P}_2}{\mu \ln\{a_2/a_1\}}$$

33. Justify all steps in the reduction to Equations 10E10.4 and 10E10.5.

34. Derive the similarity transformation resulting in Equations 10E10.3–10E10.5 using the method of Hellums and Churchill [39].

35. Carry out the complete reduction of the equations of motion to Equation 10E10.6.

36. Derive asymptotic solutions for Equation 10E10.6 for large and small values of ϕ.

37. Derive a set of ordinary differential equations for Heimenz flow (see Schlichting [2], pp. 95–98) using the method of Hellums and Churchill [39] (see Chapter 11).

38. Repeat problem 37 for Homann flow (see Schlichting [2], pp. 100–101).

39. Show that Equations 10EC1.7–10EC1.10 satisfy the equations of motion.

40. Derive Equations 10EC1.11 and 10EC1.12 from Equations 10EC1.8–10EC1.10.

41. Derive an expression for the vorticity corresponding to Equation 10EC1.7. Reduce for $r = a$.

42. Prove that Equations 10EC2.5–10EC2.8 satisfy Equations 10EC2.1–10EC2.4.

43. Show that Equations 10EC1.17–10EC1.22 satisfy the equations of motion and the boundary conditions.

44. Derive F_f and F_p corresponding to Equation 10EC1.17.

45. Show that in Section 10EC1, *Fluid Spheres*, u_θ and u_R match for the fluid sphere and the external fluid for all values of ζ.

46. Plot $\mathscr{P}$ versus r in dimensionless form for the flow corresponding to Equation 10EC3.2.

47. Derive an expression for the volumetric rate of flow corresponding to Equation 10EC3.2.

48. Denn [17], p. 262, suggests that the solution for creeping flow between parallel disks can be adapted to estimate the time required to fill a disk mold, assuming quasi-steady-state flow. Derive such a solution.

49. Determine the pressure field corresponding to Equation 10EC4.5.

50. Derive an expression for the limiting velocity field as $\alpha \to 0$ in Equation 10EC4.5. (*Hint*: Introduce $\phi = \theta/\alpha$ for θ, and apply L'Hôpital's rule.)

51. Prepare a dimensionless plot of u_r versus θ for $\alpha = \pi/4$ and $\pi/2$ corresponding to Equation 10EC4.5. Add a curve for $\alpha \to 0$ (see problem 49).

52. Compare Equation 10EC4.5 with the exact solution plotted in Figure 10–9.

53. Reduce (as far as possible) the equations of E11 for creeping flow.

54. Derive an expression for the volumetric rate of flow corresponding to Equation 10EC6.9.

55. Derive Equation 10EC5.3 and prove that it is equivalent to Equation 10EC5.4.
56. Carry out the detailed derivation of Equations 10EC5.5 and 10EC5.7.
57. Derive a solution for creeping flow outward between a rotating and a fixed disk.
58. Confirm Equations 10EC7.1 and 10EC7.5.
59. Show that Equation 10EC9.6 satisfies Equations 10EC7.1 and 10EC7.5.
60. Calculate the torque on the inner and outer spheres corresponding to Equation 10EC7.6.
61. What is the variation of pressure between the spheres corresponding to Equation 10EC7.6?
62. Reduce Equation 10EC7.1 to a pair of ordinary differential equations by assuming $u\phi = F_1\{\theta\} \cdot F_2\{R\}$. Solve these equations for the velocity distribution between concentric rotating spheres.
63. Confirm that Equation 10EC8.7 is a solution of Equation 10EC8.4 and that 10EC8.7 plus 10EC8.3 is a solution of Equation 10EC7.1.
64. Derive expressions for the torque on the inner and outer cones of EC8 for a finite length corresponding to $R_2 = a_2$. Indicate how such an expression can be used to determine the viscosity.
65. Derive an expression for the torque on a plate-and-cone viscometer of radius a. Ellenberger and Fortuin [40] found this solution to be applicable for $\Omega a^2\theta^2/\nu$ up to about 5.0.
66. Reduce Equation 10EC7.1 to a pair of ordinary differential equations by assuming $u_\phi = F_1\{\theta\} \cdot F_2\{R\}$. Solve these equations for the velocity distribution between a rotating cone and a fixed plate. [See Bird et al. [21], p. 119.]
67. Bird et al. [21], p. 101, assert that the right-hand side of Equation 10EC9.9 reduces to $\omega R(\pi/2 - \theta)/(\pi/2 - \theta_1)$ as θ and θ_1 approach $\pi/2$. Confirm this assertion analytically and evaluate its accuracy numerically. Derive the corresponding expression for the torque.
68. Derive an asymptotic form for Equation 10EC9.9 for θ and $\theta_1 \to 0$. Derive the corresponding expression for the torque.
69. A plate-and-cone viscometer with a radius of 120 mm and an angle of $2\pi/180$ rad registers a torque of 2 mN $\cdot$ m when rotated at 9 rad/min. What is the viscosity? What percentage error results from using the approximation of problem 67?
70. Confirm that Equations 10EC10.15, 10EC10.16, and 10EC10.20 constitute a solution of Equations 10EC10.1–10EC10.9.
71. Derive an expression for the volumetric rate of flow at $r = a$, corresponding to Equation 10EC10.16. Check for constant u_w.
72. Specialize Equations 10EC10.15, 10EC10.16, 10EC10.21, and 10EC10.22 for a uniform plate velocity with $h\{0\} = h_o$.
73. Determine $u_w\{t\}$ corresponding to Equation 10EC10.23.
74. Denn [17], pp. 262–263, suggests that the derivation for squeeze flow can readily be modified for compression molding of a fluid with an initial radius a_o less than a. In this application both plates can be presumed to move at the same velocity. Derive a solution for the outer radius of the fluid as a function of time. (*Hint*: Assume that the curvature of the air–fluid interface can be neglected.)
75. Show how Equation 10EC11.4 is obtained from 10EC11.1.

76. Derive C_o for an elliptical orifice.
77. Show how Equation 10EC12.1 reduces to 10EC12.2 and 10EC12.3.
78. Specialize Equation 10EC12.4 for a circle and for a long thin plate. Compare the latter result with Equation 10EC12.3.
79. Prove that Equations 10S3.1 and 10S3.2 satisfy the equations of motion for inviscid flow.
80. Prove that Equations 10S2.3 and 10S2.4 satisfy the equations of motion for inviscid flow.
81. Show how Equations 10S4.1 and 10S4.2 can be rewritten as 10S4.3 and 10S4.4.
82. Derive a solution for two equal sources a distance $2b$ apart. Plot the equipotential lines and streamlines.
83. Show how a solution can be constructed for a line of equally spaced sources and for a rectangular array.
84. Construct the streamlines corresponding to Equation 10S4.2 by the Rankine method.
85. Use Equation 10S5.6 to find x for which $y = 0.99y_\infty$.
86. Use the Rankine method to construct the streamlines corresponding to Equation 10S6.2.
87. Find u_x and u_y corresponding to Equations 10S6.1 and 10S6.2.
88. Demonstrate the equivalence of Equations 10S7.14 and 10S7.2.
89. Construct the streamlines and equipotential lines for solution S8, using the Rankine method.
90. Prove that Equations 10S8.3 and 10S8.4 satisfy the equations of motion.
91. Calculate the velocity on the plate for S8.
92. Use Equations 10S9.7 and 10S9.8 to determine the components of the velocity and **v** at

   a. $x = 0, y = 0$
   b. $x = y = b/\pi$

93. Explain the differences in the location of the streamlines for S9 and the second subsection of S10.
94. Show that Equation 10S12.7 satisfies the equations of motion.
95. Calculate the pressure distribution for S12.
96. Calculate **v** and $\mathscr{P}$ along $\tilde{\psi}^*$ in S14.
97. Construct the streamlines of Figure 10–45 using the Rankine method.
98. Can the solution corresponding to Equation 10S14.9 be applied to a closed body and a blunt-nosed cylinder? Explain.
99. Plot the streamlines for inviscid flow around a sphere using the Rankine method.
100. Show that $u_z \to u_\infty$ for the solution of S15. (See Bird et al. [21], p. 149).
101. Determine the value(s) of α such that Equation 10EC4.5 is singular. Explain this behavior physically, (*Hint*: See Moffatt and Duffy [41].)
102. Derive a solution for purely viscous flow of a thin film over a conical plate of half-angle θ and radius a at an angular velocity ω. Neglect surface tension and entrance and edge effects. (See Inuzuka et al. [42] for an evaluation of the limit of applicability of this solution.)
103. Repeat problem 103 for a rotating hemisphere.
104. Derive a solution for the fluid motion inside an inviscid sphere moving

through an immiscible, inviscid fluid.
105.   Derive a solution for the fluid motion inside an inviscid sphere within a
       moving, immiscible, inviscid fluid.

REFERENCES

1.   S. W. Churchill, *The Practical Use of Theory in Fluid Flow. Book I. Inertial Flows*,
     Etaner Press, Thornton, PA (1980).
2.   H. Schlichting, *Boundary Layer Theory*, 7th ed., English transl. by J. Kestin,
     McGraw-Hill, New York (1979).
3.   H. H.-S. Chu and S. W. Churchill, "The Development and Testing of a Numerical
     Method for Computation of Laminar Natural Convection in Enclosures," *Computers and Chem. Engng.*, *1* (1977) 103.
4.   S. W. Churchill, P. K.-B. Chao, and Hiroyuki Ozoe, "Extrapolation of Finite-
     Difference Calculations for Laminar Natural Convection in Enclosures to Zero Grid
     Size," *Num. Heat Transfer*, *4* (1981) 39.
5.   Sheldon Weinbaum and Vivien O'Brien, "Exact Navier–Stokes Solutions Including
     Swirl and Cross Flow," *Phys. Fluids*, *10* (1967) 1438.
6.   A. R. Berker, "Intégration des équations du mouvement d'un fluide visqueux
     incompressible," p. 1 in *Encyclopedia of Physics*, S. Flügge, Ed., Vol. 8, part 2,
     Springer-Verlag, Berlin (1963).
7.   A. B. Whitman, "The Navier Stokes Equations," in Chap. III, p. 114, in *Laminar
     Boundary Layers*, L. Rosenhead, Ed., Oxford University Press (1963).
8.   L. D. Landau and E. M. Lifshitz, *Fluid Mechanics*, English translation by J. B.
     Sykes and W. N. Reid, Pergamon, New York (1959).
9.   S. W. Churchill, *The Practical Use of Theory in Fluid Flow. Book III. Laminar,
     Multidimensional Flows in Channels*, Notes, The University of Pennsylvania (1979).
10. Garrett Birkhoff, *Hydrodynamics, A Study in Logic, Fact and Similitude*, Dover,
     New York (1955).
11. G. G. Stokes, "On the Effect of the Internal Friction of Fluids on the Motion of
     Pendulums," *Trans. Camb. Phil. Soc.*, *9* (1851) 8 (*Mathematical and Physical
     Papers, Vol. III*, Cambridge University Press (1901), p. 1).
12. H. S. Carslaw and J. C. Jaeger, *Conduction of Heat in Solids*, Oxford University
     Press, Clarendon (1959).
13. F. Szymanski, "Quelques solutions exactes des équations de l'hydrodynamique de
     fluide visqueux dans le cas d'un tube cylindrique," *J. Math. Pures Appl.*, Ser. 9,
     11 (1932) 67; *Proc. Int. Congr. Appl. Mech., Stockholm*, *1* (1930) 249.
14. J. Kestin and L. N. Persen, "Slow Oscillations of Bodies of Revolution in a Viscous
     Fluid," *Proc. 9th Int. Congr. Appl. Mech., Brussels*, *3* (1957) 326.
15. J. Kestin and H. E. Wang, "Corrections for the Oscillating-Disk Viscometer,"
     J. Appl. Mech., *24* (1957) 197.
16. W. E. Langlois, *Slow Viscous Flow*, Macmillan, New York (1964).
17. M. M. Denn, *Process Fluid Mechanics*, Prentice-Hall, Englewood Cliffs, NJ (1980).
18. V. L. Streeter, *Fluid Dynamics*, McGraw-Hill, New York (1948).
19. C. W. Oseen, "Über die Stokes'sche Formel, und über eine verwandte Aufgabe in
     der Hydrodynamik," *Arkiv Math., Astronom. Fys*, *6* (1910) 75; *Hydromechanik*,
     Academische Verlagsgesellschaft, Leipzig (1927), p. 82.
20. G. Hamel, "Spiralförmige Bewegung zäher Flussigkeiten," *Jahresber. Deutschen
     Math.-Vereinigung*, *34* (1916).
21. R. B. Bird, W. E. Stewart, and E. N. Lightfoot, *Transport Phenomena*, John Wiley,
     New York (1960).
22. G. B. Jeffery, "Steady Motions of a Viscous Fluid," *Phil. Mag.*, *29* (1915) 455.
23. K. Millsaps and K. Pohlhausen, "Thermal Distribution in Jeffery–Hamel Flows
     between Non-Parallel Plane Walls," *J. Aero. Sci.*, *20* (1953) 187.

24. S. Goldstein, Ed., *Modern Developments in Fluid Dynamics*, Oxford University Press, Clarendon (1938).
25. Th. von Kármán, "Laminare und Turbulente Reibung," *Z. Angew. Math. Mech.*, *1* (1921) 233; English transl., "Laminar and Turbulent Friction," NACA TM1092, Washington, D.C. (1946).
26. W. G. Cochran, "The Flow Due to a Rotating Disc," *Proc. Camb. Phil. Soc.*, *30* (1934) 365.
27. G. G. Stokes, "On the Effect of the Internal Friction of Fluids on the Motion of Pendulums," *Trans. Camb. Phil. Soc.*, *9* (1851) 8 (*Mathematical and Physical Papers*, *Vol. III*, Cambridge University Press (1901), p. 55).
28. Horace Lamb, *Hydrodynamics*, Dover, New York (1945).
29. R. Roscoe, "The Flow of Viscous Fluids Round Plane Obstacles," *Phil. Mag.*, *40* (1951), 338.
30. J. Hadamard, "Mouvement permanent lent d'une sphère liquide visqueuse dans un liquid visqueux," *Compt. Rend. Acad. Sci., Paris*, *152* (1911) 1735.
31. W. Rybczynski, "Über die fortschreitende Bewegung einer flüssigen Kugel in einem zähen Medium," *Bull. Acad. Sci., Cracovie*, Ser. A, *1* (1911) 40.
32. V. G. Levich, *Physicochemical Hydrodynamics*, English transl. by Scripta Technica, Prentice-Hall, Englewood Cliffs, NJ (1962).
33. H. S. Hele-Shaw, "Investigation of the Nature of Surface Resistance of Water and Stream Motion under Certain Experimental Conditions," *Trans. Inst. Nav. Arch.*, *40* (1898) 25.
34. Milton Van Dyke, *An Album of Fluid Motion*, Parabolic Press, Stanford, CA (1982).
35. M. Abramowitz and I. A. Stegun, Eds., *Handbook of Mathematical Functions with Formulas, Graphs, and Mathematical Tables*, Nat. Bur. Stds., Appl. Math. Series 55, U.S. Govt. Printing Office, Washington, D.C. (1964).
36. J. R. Black and M. M. Denn, "Converging Flow of a Viscoelastic Fluid," *J. Non-Newtonian Mech.*, *1* (1976) 83.
37. L. M. Milne-Thomson, *Theoretical Hydrodynamics*, Macmillan, New York (1968).
38. Walter Kaufmann, *Fluid Mechanics*, English transl. by E. G. Chilton, McGraw-Hill, New York (1963).
39. J. D. Hellums and S. W. Churchill, "Simplification of the Mathematical Description of Boundary and Initial Value Problems", *AIChE J.*, *10* (1964) 110.
40. J. Ellenberger and J. M. N. Fontuin, "A Criterion for Purely Tangential Laminar Flow in the Cone-and-Plate Rheometer and the Parallel-Plate Rheometer," *Chem. Eng. Sci.*, *40* (1985) 111.
41. H. K. Moffatt and B. R. Duffy, "Local Similarity Solutions and Their Limitations," *J. Fluid Mech.*, *96* (1980) 299.
42. M. Inuzuka, T. Moriya, K. Miwa, I. Yamada, and S. Hiraoka, "Flow Characteristics of a Liquid Film on an Inclined Rotating Surface," *Kagaku Kogaku Ronbunshu*, *5* (1979) 444; English transl., *Int. Chem. Eng.*, *21* (1981) 213.

PART III

Unconfined, Multidimensional, Laminar Flows

Solutions of the equations of motion in good agreement with experiments have been developed for laminar flow over many different immersed surfaces. Some of these solutions, as well as combinations thereof, are examined in this section.

Flow along a flat surface is examined in Chapters 11–13, the closely related problems of flow over oblique or normal flat surfaces in Chapter 14, flow normal to a circular cylinder in Chapter 15, flow relative to solid spheres in Chapter 16, and relative to fluid spheres in Chapter 17. Generalized methods and flow over objects of miscellaneous shapes are examined in Chapter 18. Flow through dispersed solids, including packed beds, is deferred to Part IV.

Fully developed, *confined* flows, as described in Chapters 3–7 (and in two dimensions in a companion volume [1],[1] are wholly laminar for Reynolds numbers below some characteristic, critical value. A singular theoretical treatment for all rates of flow, and, with some minor exceptions, for all geometries is thereby possible.

On the other hand, *unconfined* flows over a surface undergo a succession of regimes as the Reynold number increases, even though the motion remains essentially laminar, at least over the forward portion. More important, two or more of these regimes may exist concurrently at different locations along the surface.

Most theoretical treatments of unconfined flow over surfaces restrict their attention to only one of these regimes and, thereby, to only part of the surface. Examples are Milne-Thompson [2] for inviscid flow, Happel and Brenner [3] for creeping flow, and Rosenhead [4] for laminar boundary layers. By contrast, this book attempts to collate the entire gamut of behavior, invoking the useful aspects of the various theories, identifying their limits, and interpolating in between.

For multidimensional flows, particularly in curvilinear coordinates, the appropriate differential model can usually be derived more easily by simplification of the general equations derived in Chapters 8 and 9 than by direct construction. The classical solutions for multidimensional flow depend upon a

---

[1] References in this introduction are included with those at the end of Chapter 11.

clever choice of idealizations and approximations. These simplifications are easier to identify and rationalize in the process of reduction than in the process of construction.

Models and solutions for flow over immersed surfaces can be categorized as *inviscid (or potential) flow, creeping (or Stokes) flow, slightly inertial (or Oseen) flow, thick-boundary-layer (or Imai) flow, thin-boundary-layer (or Prandtl) flow, stationary-wake flow, periodic-wake flow, irregular-wake flow* and *post-critical flow*. Strictly speaking, irregular wakes and post-critical flow involve turbulent motion, but since a laminar boundary layer still exists on the forward part of the surface, these regimes will be considered briefly herein.

As illustrated in Chapter 10, *potential theory* has been utilized extensively to derive solutions for *inviscid flow* over immersed surfaces. Such solutions are not of much direct practical value because they do not provide any information on the velocity field adjacent to the surface or on the drag. They are, however, a necessary adjunct for boundary-layer theory and will be utilized in this section in that context. Many books on hydrodynamics, such as those by Milne-Thompson [2], Streeter [5], and Lamb [6], are primarily devoted to inviscid flow over immersed objects, and are suggested as sources of solutions not found in Chapter 10 and for further study on that subject. Books on *complex variables* and on potential theory itself may also be relevant sources of such solutions and methods of solutions.

Creeping flow refers to the regime of low Reynolds numbers such that the inertial terms in the force-momentum balances are completely negligible. Stokes [7] first utilized this concept in 1851 to derive a closed-form solution for flow over a sphere [see the first subsection of 10EC.1], and this regime is sometimes called *Stokes flow*. The range of applicability of the approximation of creeping flow is very limited for immersed objects, for example, for a sphere to $Du_\infty \varrho/\mu < \mathcal{O}\{1\}$.

In 1910 Oseen [8] utilized a linear approximation for the inertial terms to derive an improved solution for flow over a sphere, and accordingly, this regime is sometimes called *Oseen flow*. Unfortunately Oseen's solution and subsequent improvements are valid to only slightly higher Reynolds numbers than the Stokes solution. This short, extended regime is herein called *slightly inertial flow*.

The books of Happel and Brenner [3], Langlois [9], and Ladyzhenskaya [10] emphasize creeping and slight inertial flows. Detailed treatments of these topics are also provided by Lamb [6], Goldstein [11], Illingworth [12], and Van Dyke [13]. A number of illustrative solutions were presented in Chapter 10.

Thin-boundary-layer theory is based on the recognition by Prandtl [14] in 1904 that the significant velocity gradients in flow over an immersed surface are confined to a narrow film of fluid, adjacent to the surface called the boundary layer. This concept led him to conceive of some significant simplifications in the partial differential equations of conservation. These simplifications in turn made possible the derivation of relatively simple, highly accurate solutions. This combination of qualitative and quantitative concepts represents one of the milestones in the history of fluid mechanics. The history of boundary-layer theory is discussed by Flügge-Lotz and Flügge [15], Tani [16], and others. Its mathematical character is discussed by Nickel [17].

The primary assumption of thin-boundary-layer theory is that viscosity affects the flow only in a thin layer adjacent to the surface, and hence that inviscid flow can be used as an acceptable approximation outside that layer. In addition, the diffusion of momentum in the direction of flow along the surface is presumed to be negligible relative to diffusion normal to the surface, and the pressure gradient normal to the surface is assumed to be negligible *within* the boundary layer. These latter two asumptions permit reduction of the partial differential equations to ordinary differential equations in many cases.

The change of variables that permits reduction of a set of partial differential equations to a set of ordinary differential equation is called a *similarity transformation*. Historically such transformations have been discovered heuristically, case by case, as by Prandtl [14] for flow over a flat plate. Hellums and Churchill [18] developed a technique that automatically produces, in most cases, the necessary change of variables if a similarity transformation is possible. This technique is described in Chapter 11.

The reduced equations for a thin boundary layer are generally nonlinear, necessitating series or numerical solutions. The results are therefore ordinarily presented in graphical or tabular form. Methods of constructing convenient closed-form **correlating equations** for the solution for a thin boundary layer are accordingly examined here. A new method is formulated in which asymptotic expressions can be derived from the boundary conditions of the differential model and the limiting numerical values of the solution. In some instances these asymptotic solutions provide a sufficient representation in themselves. In others they may be combined in terms of the *Churchill–Usagi model* [19] to give an overall correlating equation. The Churchill–Usagi model is also applicable for interpolation between the solutions for different regimes such as creeping and boundary-layer flow.

The use of *integral boundary-layer theory* to develop approximate solutions is also examined. This method, which was originated in 1921 by von Kármán [20] and Pohlhausen [21], depends on the postulate of a *similar* velocity distribution in the boundary layer. The equations of conservation are then satisfied only on the mean rather than at every point. This procedure is now recognized as a special and heuristic application of the *method of weighted residuals* (see, for example, Finlayson [22]). The achievement of closed-form expressions for the influence of parameters does have some advantages relative to exact numerical solutions. Also, integral boundary-layer theory has conceptual value as an example of the utility of heuristic methods in advance of the development of more sophisticated and exact methods. The primary weakness of integral boundary-layer theory is that the error is unknown, perhaps large, and cannot be reduced systematically. Accordingly, this method has been largely supplanted by more powerful and exact methods. An understanding is, however, essential for reading and interpreting past, and even present, literature on fluid mechanics.

A similarity transformation does not exist for boundary-layer flow along a flat plate with uniform suction or blowing, or for flow over a circular cylinder or sphere. Blasius [23] in 1908 developed a series expansion for some such flows, but the rate of convergence is poor. Dewey and Gross [24] have reviewed methods devised to improve the rate of convergence, some of which are examined in Chapter 15.

Thin-laminar-boundary-layer theory itself has several limitations. It is inapplicable at very low Reynolds numbers such that the boundary-layer thickness is appreciable relative to the distance from the leading edge or point of incidence, or relative to the radius of curvature of a nonflat surface. In flow along flat surfaces, both the leading and trailing edges cause deviations, particularly at low Reynolds number. The boundary layer on a flat surface becomes turbulent at some distance, thus producing a maximum Re_x for applicability of the laminar theory. In flow over curved surfaces, separation occurs at some point, and the boundary layer is replaced by a wake. The thin-boundary-layer solution is good only for the forward portion of such a surface. Furthermore, the wake perturbs the outer flow over the boundary layer itself.

More detailed treatments of various aspects of boundary-layer theory are provided by Goldstein [25], Schlichting [26], Rosenhead [4], Evans [27], Walz [28], Meksyn [29], Van Dyke [13], and others.

Solutions for the regime of *thick boundary layers*, between the regimes of slightly inertial flow and thin boundary layers, have been developed only for a few conditions (see, for example, Van Dyke [13]), and must generally be sought by numerical methods.

In principal, *numerical methods* of integration are applicable to the general equations without any of the idealizations associated with creeping, slightly inertial, and thin-boundary-layer flow. However, in practice, success for flow over bodies is limited primarily to Reynolds numbers below that at which the wake become unsteady. A few results have been obtained for nonsteady periodic wakes, but the computational requirements are very great.

Numerical solutions have been obtained by finite-difference, finite-element, boundary-element, and weighted-residual methods. The details of these methods are beyond the scope of this book, but some of the results are examined. An extensive survey of such work is provided by reference [30].

Knudsen and Katz [31] and Goldstein [11] give particular attention to the experimental foundations of the material in this section. Shapiro [32] provides an excellent physical interpretation.

A review of the early, and apparently independent, Russian work on boundary layers is provided by Loitsianskii [33].

Photographs of the flow patterns for many of the geometries and conditions covered in this section have been anthologized by Van Dyke [34].

Chapter 11

The Blasius Solution for Laminar Flow along a Flat Plate

The pattern of flow over the wing of an airplane is affected by the width, thickness, length, and shape of the wing as well as by the rest of the plane. Flow over the roof of a building is affected by the configuration of the roof, its roughness, the shape of the rest of the building, and perhaps by the trees and ground. Flow over the surface of the inlet of a channel is affected by the inlet configuration and the shape of the channel. Nevertheless, these flows, as well as many others, are approximated in the limit by the flow over a thin isolated plate. This idealized problem serves as the base for all boundary-layer treatments, and hence will be examined here in considerable detail.

DIFFERENTIAL BOUNDARY-LAYER MODEL

Consider a semi-infinite thin plate with a sharp leading edge immersed in a uniform horizontal flow in the x-direction with a velocity u_∞, as illustrated in Figure 11–1. Symmetry in the z-direction (parallel to the plate and perpendicular to the direction of flow) and a steady state are postulated. Negligible changes in density and viscosity are assumed. The general equations for the conservation of mass and momentum then reduce (see problem 1) to

$$\frac{\partial u_x}{\partial x} + \frac{\partial u_y}{\partial y} = 0 \tag{11.1}$$

$$u_x \frac{\partial u_x}{\partial x} + u_y \frac{\partial u_x}{\partial y} = -\frac{1}{\varrho} \frac{\partial p}{\partial x} + v \left(\frac{\partial^2 u_x}{\partial x^2} + \frac{\partial^2 u_x}{\partial y^2} \right) \tag{11.2}$$

and

$$u_x \frac{\partial u_y}{\partial x} + u_y \frac{\partial u_y}{\partial y} = -g - \frac{1}{\varrho} \frac{\partial p}{\partial y} + v \left(\frac{\partial^2 u_y}{\partial x^2} + \frac{\partial^2 u_y}{\partial y^2} \right) \tag{11.3}$$

Far from the plate, potential flow is presumed to exist with uniform velocity and hence with $\partial p/\partial x = 0$ and $\partial p/\partial y = -g\varrho$. The latter condition is assumed

255

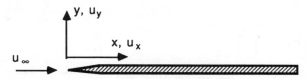

FIGURE 11–1   *Coordinates and variables for analysis of flow along a flat plate.*

to be a good approximation across the thin boundary layer next to the plate as well. The former condition must then also exist within the boundary layer. With the elimination of pressure as a dependent variable, the problem is over-specified; since u_y is expected to be very small with respect to u_x, Equation 11.3 may be dropped. Also $\partial^2 u_x / \partial x^2$ is expected to be much smaller than $\partial^2 u_x / \partial y^2$ and can be dropped. Then only

$$u_x \frac{\partial u_x}{\partial x} + u_y \frac{\partial u_x}{\partial y} = \nu \frac{\partial^2 u_x}{\partial y^2} \tag{11.4}$$

remains of Equations 11.2 and 11.3.

The simplifications described in the preceding paragraph are known as "the usual simplifying assumptions of thin-boundary-layer theory." They were originated by Prandtl [14] and can be justified more formally by an order-of-magnitude analysis (see, for example, Schlichting [26], p. 128). A modern view is presented by Evans [27], p. 5f. The real test is a comparison of the subsequent solution of this reduced model with experimental data or with numerical solutions of the more general representation. As noted later, good agreement is obtained between the solutions of Equations 11.1 and 11.4 and experimental results for $u_x\{y\}$ and the shear stress on the plate over a reasonable range of conditions. However, the available data are not sufficiently precise to evaluate critically the individual effects of the neglected terms, nor does this evaluation seem to have been accomplished with numerical solutions.

A sufficient set of boundary conditions for this model is

$$u_x = u_y = 0 \qquad \text{at } y = 0, \, x > 0 \tag{11.5}$$

$$u_x \to u_\infty \qquad \text{as } y \to \infty, \quad \text{and also for } x < 0 \tag{11.6}$$

The latter condition implies that the plate does not perturb the flow upstream. This postulate will later be shown to be a significant source of error in the solution.

STREAM-FUNCTION MODEL

Introducing the Lagrange stream function defined by Equations 9.1 and 9.2 satisfies Equation 11.1 and transforms Equation 11.4 to

$$-\frac{\partial \psi}{\partial y} \frac{\partial^2 \psi}{\partial x \, \partial y} + \frac{\partial \psi}{\partial x} \frac{\partial^2 \psi}{\partial y^2} = \nu \frac{\partial^3 \psi}{\partial y^3} \tag{11.7}$$

The boundary conditions corresponding to Equations 11.5 and 11.6 are

$$\frac{\partial \psi}{\partial y} = \frac{\partial \psi}{\partial x} = 0 \qquad \text{at } y = 0, \quad x > 0 \tag{11.8}$$

and

$$\frac{\partial \psi}{\partial y} \rightarrow -u_\infty \qquad \text{as } y \rightarrow \infty, \quad \text{and also for } x < 0 \tag{11.9}$$

The arbitrary condition

$$\psi = 0 \qquad \text{at } y = 0, \quad x > 0 \tag{11.10}$$

can be added to complete the description.

REDUCTION OF MODEL

Dedimensionalization, identification of a possible similarity transformation, and minimization of the number of parameters can be accomplished by the technique of Hellums and Churchill [18]. This technique is outlined and illustrated here, but reference [18] is suggested for further study. Reduction of the number of parameters is in itself very valuable when the problem must be solved numerically. Alternative techniques with the same objective are compared by Ames [35].

With the *Hellums–Churchill method* each dependent and independent variable is replaced by a reference quantity times a dimensionless variable.[1] Thus in this case let

$$\Psi \equiv \psi/\psi_A, \quad X \equiv x/x_A, \quad \text{and} \quad Y \equiv y/y_A$$

where ψ_A = reference stream function, m²/s
x_A = reference distance, m
y_A = reference distance, m

The model can be rewritten in terms of these new dimensionless variables and reference quantities as

$$-\frac{\partial \Psi}{\partial Y} \frac{\partial^2 \Psi}{\partial X \partial Y} + \frac{\partial \Psi}{\partial X} \frac{\partial^2 \Psi}{\partial Y^2} = \frac{\nu x_A}{y_A \psi_A} \frac{\partial^3 \Psi}{\partial Y^3} \tag{11.11}$$

$$\frac{\partial \Psi}{\partial Y} = \frac{\partial \Psi}{\partial X} = 0 \qquad \text{at } Y = 0, \quad X > 0 \tag{11.12}$$

$$\frac{\partial \Psi}{\partial Y} \rightarrow -\frac{u_\infty y_A}{\psi_A} \qquad \text{as } Y \rightarrow \infty \quad \text{and for } X < 0 \tag{11.13}$$

---

[1] As explained by Hellums and Churchill [18], a reference quantity should also be added to a dependent variable for which conditions are specified at two finite boundaries.

The boundary-value problem described by Equations 11.11–11.13 can be expressed in functional form as

$$\frac{\psi}{\psi_A} = \phi \left\{ \frac{x}{x_A}, \frac{y}{y_A}, \frac{\nu x_A}{y_A \psi_A}, \frac{u_\infty y_A}{\psi_A} \right\} \tag{11.14}$$

The parameters in Equation 11.14 can both be made unity and therefore eliminated as parameters by choosing

$$\psi_A = u_\infty y_A \tag{11.15}$$

and

$$x_A = \frac{y_A \psi_A}{\nu} = \frac{u_\infty y_A^2}{\nu} \tag{11.16}$$

The reduced functional dependence in terms of the original variables is then

$$\frac{\psi}{u_\infty y_A} = \phi \left\{ \frac{x\nu}{u_\infty y_A^2}, \frac{y}{y_A} \right\} \tag{11.17}$$

The reference quantity y_A is arbitrary and cannot exist in the final description of the problem. It can be eliminated by combining the groups in Equation 11.17 to give

$$\frac{\psi}{\sqrt{u_\infty \nu x}} = \phi \left\{ y \sqrt{\frac{u_\infty}{x\nu}} \right\} \tag{11.18}$$

The groups in this relationship constitute a *similarity transformation*; that is, the partial differential equation in ψ, x, and y can be reduced to an ordinary differential equation in terms of these compound variables. This assertion can be proven, and the ordinary differential equation derived as follows. Let[1]

$$\Psi \equiv \frac{\psi}{\sqrt{u_\infty \nu x}} \tag{11.19}$$

and

$$\xi \equiv y \sqrt{\frac{u_\infty}{\nu x}} \tag{11.20}$$

The derivatives in Equation 11.7 can be transformed into these new variables as follows:

---

[1] This transformation was discovered heuristically by Prandtl [14] in 1904, but a complete solution was first developed by Blasius [23] in 1908.

$$\frac{\partial \psi}{\partial y} = \frac{\partial (\sqrt{u_\infty vx}\,\Psi)}{\partial y} = \sqrt{u_\infty vx}\,\frac{\partial \Psi}{\partial \xi}\frac{\partial \xi}{\partial y}$$

$$= \sqrt{u_\infty vx}\,\sqrt{\frac{u_\infty}{xv}}\frac{\partial \Psi}{\partial \xi} = u_\infty \frac{\partial \Psi}{\partial \xi} \tag{11.21}$$

$$\frac{\partial^2 \psi}{\partial y^2} = \frac{\partial}{\partial y}\left(\frac{\partial \psi}{\partial y}\right) = \frac{\partial}{\partial y}\left(u_\infty \frac{\partial \Psi}{\partial \xi}\right)$$

$$= u_\infty \frac{\partial^2 \Psi}{\partial \xi^2}\frac{\partial \xi}{\partial y} = \sqrt{\frac{u_\infty^3}{xv}}\frac{\partial^2 \Psi}{\partial \xi^2} \tag{11.22}$$

$$\frac{\partial^3 \psi}{\partial y^3} = \frac{\partial}{\partial y}\left(\frac{\partial^2 \psi}{\partial y^2}\right) = \frac{\partial}{\partial y}\left(\sqrt{\frac{u_\infty^3}{xv}}\frac{\partial^2 \Psi}{\partial \xi^2}\right)$$

$$= \sqrt{\frac{u_\infty^3}{xv}}\frac{\partial^3 \Psi}{\partial \xi^3}\frac{\partial \xi}{\partial y} = \frac{u_\infty^2}{xv}\frac{\partial^3 \Psi}{\partial \xi^3} \tag{11.23}$$

$$\frac{\partial^2 \psi}{\partial x \partial y} = \frac{\partial}{\partial x}\left(u_\infty \frac{\partial \Psi}{\partial \xi}\right) = u_\infty \frac{\partial^2 \Psi}{\partial \xi^2}\frac{\partial \xi}{\partial x}$$

$$= -\frac{yu_\infty}{2}\sqrt{\frac{u_\infty}{vx^3}}\frac{\partial^2 \Psi}{\partial \xi^2} = -\frac{y}{2}\sqrt{\frac{u_\infty^3}{vx^3}}\frac{\partial^2 \Psi}{\partial \xi^2} \tag{11.24}$$

and

$$\frac{\partial \psi}{\partial x} = \frac{\partial}{\partial x}(\sqrt{u_\infty vx}\,\Psi) = \frac{1}{2}\sqrt{\frac{u_\infty v}{x}}\,\Psi - \sqrt{u_\infty vx}\left(\frac{y}{2}\sqrt{\frac{u_\infty}{vx^3}}\right)\frac{\partial \Psi}{\partial \xi}$$

$$= \frac{1}{2}\sqrt{\frac{u_\infty v}{x}}\left(\Psi - y\sqrt{\frac{u_\infty}{xv}}\frac{\partial \Psi}{\partial \xi}\right) \tag{11.25}$$

Substituting these expressions in Equation 11.7 and simplifying the resulting equations yield the ordinary differential equation

$$2\Psi''' - \Psi\Psi'' = 0 \tag{11.26}$$

where the primes indicate differentiation with respect to ξ. The boundary conditions resulting from this transformation are

$$\Psi = \Psi' = 0 \quad \text{at } \xi = 0 \tag{11.27}$$

and

$$\Psi' \to -1 \quad \text{as } \xi \to \infty \tag{11.28}$$

Thus the asserted reduction has been accomplished.

The factor of 2 in Equation 11.26 and the negative sign in Equations 11.26 and 11.28 can be removed as follows. Introducing $f = a\Psi$ and $\eta = b\xi$ produces

$$\frac{2b^3}{a}f''' - \frac{b^2}{a^2}ff'' = 0 \tag{11.29}$$

with

$$\frac{f}{a} = \frac{b}{a}f' = 0 \qquad \text{at } \eta = 0 \tag{11.30}$$

and

$$\frac{bf'}{a} \rightarrow -1 \qquad \text{as } \eta \rightarrow \infty \tag{11.31}$$

Choosing $-a = b = 1/\sqrt{2}$ then produces

$$f''' + ff'' = 0 \tag{11.32}$$

$$f = f' = 0 \qquad \text{at } \eta = 0 \tag{11.33}$$

and

$$f' \rightarrow 1 \qquad \text{as } \eta \rightarrow \infty \tag{11.34}$$

where

$$f = -\frac{\psi}{\sqrt{2u_\infty \nu x}} \tag{11.35}$$

and

$$\eta = y\sqrt{\frac{u_\infty}{2\nu x}} \tag{11.36}$$

Choosing $a = -b = 1/\sqrt{2}$ would yield the same model but with the signs of f and η reversed. Equations 11.32–11.36 represent the simplest, although not the most common, form of the Blasius problem in terms of an ordinary differential equation. The representation by Equations 11.19, 11.20, and 11.26–11.28 can just as readily be solved. The change was made here to illustrate how a canonical form of an equation with the simplest possible coefficients can be obtained. This procedure can also be used to determine whether an equation can be transformed to another that differs only in the sign and magnitude of the coefficients. Such an example is provided later.

Series Solution

Blasius [23] obtained the following series solution for the foregoing boundary-value problem for small η:

$$f\{\eta\} = \frac{A\eta^2}{2!} - \frac{A^2\eta^5}{5!} + \frac{11A^3\eta^8}{8!} - \frac{375A^4\eta^{11}}{11!} + \cdots$$

$$= \sum_{n=0}^{\infty} (-1)^n \frac{A^{n+1}C_n\eta^{3n+2}}{(3n+2)!} \tag{11.37}$$

where $C_n = 1$; $1,11$; 375; $27,897$; $3,817,137$; $856,874,115$; $298,013,975$; and so on. Weyl [36] has shown that these coefficients are given indefinitely by

$$C_n = (3n-1)! \sum_{i=0}^{n-1} \frac{C_i C_{n-1-i}}{(3i)!(3n-1-3i)!} \tag{11.38}$$

but he pointed out that the range of convergence of the Blasius series is finite. Blasius also derived the equivalent of the following asymptote for large η:

$$f\{\eta\} = \eta - B + \frac{\gamma}{2}e^{-(\eta-B)^2/2}\gamma\sqrt{\pi}(\eta - B)\,\text{erfc}\left\{\frac{\eta - B}{\sqrt{2}}\right\} + \cdots \tag{11.39}$$

which has the limiting behavior

$$f\{\eta\} = \eta - B \tag{11.40}$$

Modern values of $A = f''\{0\}$, B, and γ are 0.4695999, 1.216768, and 2.34, respectively. Values of $f\{\eta\}$, $f'\{\eta\}$, and $f''\{\eta\}$ computed by Howarth [37] are given in Table 11.1. The factors of $\sqrt{2}$ in this table occur because Howarth developed his solution in terms of $\eta/\sqrt{2}$ and $-\sqrt{2}f$.

The velocity field provided by this solution is described by the expressions

$$u_x = u_\infty f'\{\eta\} \tag{11.41}$$

and

$$u_y = \sqrt{\frac{u_\infty\nu}{2x}}(\eta f'\{\eta\} - f\{\eta\}) \tag{11.42}$$

Note from Equations 11.42 and 11.40 that for $\eta \to \infty$,

$$u_y \to 0.860\sqrt{\frac{u_\infty\nu}{x}} \tag{11.43}$$

This asymptotic upward flow is a consequence of the increasing thickness of the boundary layer.

The corresponding *local drag coefficient* for the plate is

$$C_f \equiv \frac{(j_{yx})_{y=0}}{\varrho u_\infty^2} = \frac{\mu}{\varrho u_\infty^2}\left(\frac{\partial u_x}{\partial y}\right)_{y=0} = \sqrt{\frac{\nu}{u_\infty x}}\frac{f''\{0\}}{\sqrt{2}} = \frac{0.332}{\sqrt{Re_x}} \tag{11.44}$$

This coefficient is analogous to the friction factor for flow in channels, differing only in the use of the free-stream velocity rather than the mean velocity. (The free-stream velocity could logically be considered to be the mean velocity.) In flow over curved or oblique surfaces the drag force may arise from the pressure distribution as well as from friction, hence the different name. The *overall* or *mean drag coefficient* obtained by integrating the local coefficient from $x = 0$ to x is

$$\bar{C}_f \equiv \frac{1}{x} \int_0^x C_f \, dx = 2C_f = \frac{0.664}{\sqrt{Re_x}} \tag{11.45}$$

In these equations $Re_x \equiv u_\infty x / \nu$. This definition of the Reynolds number differs fundamentally from that used for flow in channels in that the distance in the direction of flow rather than some dimension perpendicular to the flow is used as the characteristic dimension. Local and mean drag coefficients equal to twice those defined by Equations 11.44 and 11.45 also appear in the literature. These other values arise from dividing the shear stress by the kinetic energy per unit volume, $\varrho u_\infty^2 / 2$, rather than by the momentum per unit area, ϱu_∞^2.

DERIVATION OF ASYMPTOTIC SOLUTIONS

Asymptotic solutions for large and small values of the independent variable are often more useful than tabulated values in succeeding derivations and may be necessary for extrapolation (for example, for large η in Table 11.1). The Blasius solution itself provides such asymptotic expressions. However, they will be rederived directly from the boundary conditions and limiting values of the numerical solution as an illustration of a procedure that will subsequently prove useful.

Equation 11.33 specifies that $f\{0\} = f'\{0\} = 0$. Table 11.1 further indicates that $f''\{0\} = 0.4696$. These three conditions are sufficient to derive an asymptote for $\eta \to 0$. It will be assumed that

$$f''\{\eta\} \cong f''\{0\} \tag{11.46}$$

Then integration gives

$$f'\{\eta\} = f''\{0\}\eta + a \tag{11.47}$$

Since $f'\{0\} = 0$, $a = 0$. A second integration gives

$$f\{\eta\} = f''\{0\}\frac{\eta^2}{2} + b \tag{11.48}$$

Since $f\{0\} = 0$, $b = 0$. An asymptotic expression for $\eta \to 0$ is therefore

$$f\{\eta\} = \frac{f''\{0\}}{2}\eta^2 = 0.2348\eta^2 \tag{11.49}$$

which agrees exactly with the first term of Equation 11.37

Table 11.1
Numerical Solution for Laminar Thin-Boundary-Layer Flow along a Flat Plate
(from Howarth [37])

$\dfrac{\eta}{\sqrt{2}}$	$\sqrt{2}\,f$	$2f'$	$2\sqrt{2}\,f''$
0.0	0.00000	0.00000	1.32824
0.1	0.00664	0.13282	1.32795
0.2	0.02656	0.26553	1.32589
0.3	0.05974	0.39788	1.32033
0.4	0.10611	0.52942	1.30957
0.5	0.16557	0.65957	1.29204
0.6	0.23795	0.78756	1.26637
0.7	0.32298	0.91253	1.23147
0.8	0.42032	1.03352	1.18666
0.9	0.52952	1.14953	1.13173
1.0	0.65003	1.25954	1.06701
1.1	0.78120	1.36263	0.99341
1.2	0.92230	1.45798	0.91237
1.3	1.07252	1.54492	0.82582
1.4	1.23099	1.62303	0.73603
1.5	1.39682	1.69210	0.64544
1.6	1.56911	1.75218	0.55651
1.7	1.74696	1.80354	0.47151
1.8	1.92954	1.84666	0.39234
1.9	2.11605	1.88224	0.32050
2.0	2.30576	1.91104	0.25694
2.1	2.49806	1.93392	0.20208
2.2	2.69238	1.95174	0.15589
2.3	2.88826	1.96537	0.11793
2.4	3.08534	1.97558	0.08748
2.5	3.28329	1.98309	0.06363
2.6	3.48189	1.98849	0.04537
2.7	3.68094	1.99231	0.03171
2.8	3.88031	1.99496	0.02173
2.9	4.07990	1.99675	0.01459
3.0	4.27964	1.99795	0.00961
3.1	4.47948	1.99873	0.00620
3.2	4.67938	1.99922	0.00392
3.3	4.87931	1.99954	0.00243
3.4	5.07928	1.99973	0.00148
3.5	5.27926	1.99984	0.00088
3.6	5.47925	1.99991	0.00051
3.7	5.67924	1.99995	0.00029
3.8	5.87924	1.99997	0.00017
3.9	6.07923	1.99999	0.00009
4.0	6.27923	1.99999	0.00005
4.1	6.47923	2.00000	0.00003
4.2	6.67923	2.00000	0.00001
4.3	6.87923	2.00000	0.00001
4.4	7.07923	2.00000	0.00000

The only boundary condition for large η is $f'\{\infty\} = 1$. However, from Table 11.1 it is evident that $f''\{\infty\} = 0$. Thus it may be assumed that for large η,

$$f''\{\eta\} \cong 0 \tag{11.50}$$

Integration gives

$$f'\{\eta\} = a \tag{11.51}$$

From $f'\{\infty\} = 1$, $a = 1$. A second integration then gives

$$f\{\eta\} = \eta + b \tag{11.52}$$

From the value of $f\{\eta\}$ for the largest value of η in Table 11.1,

$$b = f\{\eta\} - \eta = 5.005\,77 - 6.222\,54 = -1.21\,677$$

The asymptotic solution for large η is therefore

$$f\{\eta\} = \eta - 1.2168 \tag{11.53}$$

which agrees with Equation 11.40.

DERIVATION OF A CORRELATING EQUATION

Correlating equations are more convenient for interpolation and succeeding computations than are tabulated values. The series solution (Equation 11.37) and the asymptotic solution (Equation 11.39) together are sufficient for calculations for all η. Hence the following construction of a correlating equation is for illustrative purposes only. Using the Churchill–Usagi [19] method based on the limiting asymptotes, try

$$f^p\{\eta\} = (0.2348\eta^2)^p + (\eta)^p \tag{11.54}$$

where p is an arbitrary constant. Evaluating p at $\eta = 2\sqrt{2}$, which is roughly the lower limit of validity of Equation 11.40, gives

$$\left(\frac{2.305\,76}{\sqrt{2}}\right)^p = (0.234\,799\,9 \times 8)^p + (2\sqrt{2})^p \tag{11.55}$$

By trial and error $p \simeq -2.315$. The correlating equation, in more convenient form is

$$f = \eta\left[1 + \left(\frac{4.240\,88}{\eta}\right)^{2.315}\right]^{-1/2.315} \tag{11.56}$$

The exponent and coefficient in Equation 11.56 can be rounded off with small loss to yield the simpler equation

$$f = \eta \left[1 + \left(\frac{4.25}{\eta} \right)^{7/3} \right]^{-3/7} \tag{11.57}$$

Either Equation 11.56 or Equation 11.57 provides an approximation within about 1% of the values in table 11.1 for $\eta < 2\sqrt{2}$. For higher η, Equation 11.40 can be used. Churchill and Char [38] developed analogous correlating equations for the equivalent of f', f'', u_y, and $\int_0^\eta f \, d\eta$.

ALTERNATIVE METHODS OF SOLUTION

The boundary-value problem represented by Equations 11.29–11.34 has been solved by many methods, including numerical integration with a digital computer (see problem 3). As a further alternative, the following simple iterative method proposed by Piercy and Preston [39] can be used. For purposes of iteration Equation 11.32 can be written as

$$f'''_{p+1} + f_p f''_{p+1} = 0 \tag{11.58}$$

where the subscript p indicates the order of iteration. Integrating formally from 0 to η gives

$$\ln\{f''_{p+1}\} = -\int_0^\eta f_p \, d\eta + \ln\{f''_{p+1}\{0\}\} \tag{11.59}$$

which can be arranged as

$$f''_{p+1} = f''_{p+1}\{0\} \exp\left(-\int_0^\eta f_p \, d\eta \right) \tag{11.59A}$$

Integrating again and employing $f'\{0\} = 0$ give

$$f'_{p+1} = f''_{p+1}\{0\} \int_0^\eta \exp\left(-\int_0^\eta f_p \, d\eta \right) d\eta \tag{11.60}$$

From the condition $f'\{\infty\} = 1$,

$$f''_{p+1}\{0\} = 1 \bigg/ \int_0^\infty \exp\left(-\int_0^\eta f_p \, d\eta \right) d\eta \tag{11.61}$$

Integrating Equation 11.60 with $f\{0\} = 0$ finally gives

$$f_{p+1} = f''_{p+1}\{0\} \int_0^\eta \left(\int_0^\eta \exp\left(-\int_0^\eta f_p \, d\eta \right) d\eta \right) d\eta \tag{11.62}$$

The postulate of $f_0\{\eta\}$ allows the calculation of $f''_1\{0\}$ from Equation 11.61, and $f''_1\{\eta\}$, $f'_1\{\eta\}$, and $f_1\{\eta\}$ from Equations 11.59A, 11.60, and 11.62. $f_1\{\eta\}$ can be used to calculate $f''_2\{0\}$, and so on. This procedure has been proven by Weyl [4] to converge to the exact solution. (Also see problem 8.)

SUMMARY

The general differential model for flow over a thin, horizontal, semi-infinite plate was simplified according to thin-boundary-layer theory, and the resulting simplified model was reduced to a single ordinary differential by the introduction of the Lagrange stream function and the use of a similarity transformation. A method of identifying the appropriate similarity transformation was demonstrated. The Blasius solution of the simplified model, which consists of one series for small values of the dimensionless coordinate and another for large, was described. Tabulated values for this solution for the velocity field were provided as well as a simple correlating equation. Expressions for the local and integrated mean drag coefficient were also obtained.

The Blasius solution is compared with experimental data and with numerical solutions of the more general equations of motion in Chapter 13 in order to appraise the limits of validity of the various simplifications that were introduced.

PROBLEMS

1. Reduce Equations 8.12 and 8.57 to Equations 11.1–11.3. Identify and justify each simplification.
2. Compare the series solution (Equation 11.37) and the asymptotic solution (Equation 11.39) with the values in table 11.1.

   a. What are the limits of the series and the asymptotic solution in terms of η for a 5% error in u_x and C_f?
   b. What are the corresponding ranges of validity of the first term in the series?
   c. How much do the coefficients 27,897 and 3,817,137 extend the range of validity of the series?

3. Write a computer program to solve the Blasius problem by a finite-difference method. Compute and tabulate $\Psi\{\xi\}$ and $\Psi'\{\xi\}$ from $\xi = 0$ to $\xi = 10$ at intervals of $\xi = 0.2$. Compare with the values in Table 11.1. Compute $A = \Psi''\{0\}$ and $B = \lim_{\eta \to \infty}(2\eta - \Psi\{\eta\})$ and compare with the values given below Equation 11.40.
4. Prepare dimensionless plots of u_x, u_y, and ψ for the Blasius solution.
5. a. Reduce the model represented by Equations 11.1, 11.4–11.6 to a set of ordinary differential equations and boundary conditions in u_x and u_y without using the stream function.
   b. Combine the two differential equations into a single differential equation in u_y. Specify the boundary conditions.
   c. Show that this result is equivalent to Equations 11.26–11.28.
6. For air at 15°C flowing at 3 m/s over a flat plate, and for $\mathrm{Re}_x = 300,000$, calculate

   a. C_f, $\bar{C}_f$ and the total drag force on both sides of the plate
   b. the boundary-layer thickness
   c. the limiting velocity u_y as $y \to \infty$

7. Rephrase the Blasius model and solution for

   a. horizontal flow along a vertical plate
   b. vertical flow along a vertical plate

8. Carry out the iteration suggested by Piercy and Preston. Compare the successive values of $f''\{0\}$ with the exact value. Begin with (a) $f_0 = \eta$, and (b) $f_0 = \eta^2$.

9. Assuming that u_x, u_y, x, and y can be scaled by (are of the order of) u_∞, $u_\infty \delta/L$, L, and δ, respectively, determine the order of the terms in Equations 11.1 and 11.2. (*Hint*: Note that $\delta^2 = \mathcal{O}\{Lv/u_\infty\}$, and assume $\mathcal{P} = p + \varrho gh - p_0 = \mathcal{O}\{L\mu u_\infty/\delta^2\}$.)

10. A steady west wind is blowing at 30 mi/h. Calculate the local shear stress and the total drag on the flat roof of a building 200 ft long (in the east–west direction) and 50 ft wide.

11. Equation 11.43 implies an upward flow that attains a finite asymptotic value at large distances from the plate. Is this result physically valid? Explain.

12. Develop a model for the velocity field resulting from a sheet issuing from a slit with velocity u_w into parallel flow of an otherwise unbounded stream with a velocity u_∞. Compare with the Blasius model and conjecture on the behavior of the solution for $u_w \ll y_\infty$ and $u_w \gg u_\infty$. Sketch the corresponding velocity fields and the relationship between C_f and $|u_w - u_\infty|$.

13. Determine the minimum set of dimensionless groups required to represent the degenerate motion described by the following model.

$$\frac{\partial u}{\partial t} + u\frac{\partial u}{\partial x} = 0$$

$$u = 0 \qquad \text{at } t = 0$$

$$u = U \qquad \text{at } x = 0, \quad t > 0$$

$$u \to 0 \qquad \text{as } x \to 0$$

14. Use the Blasius solution to estimate the relative magnitude of the various velocity terms in the equations of conservation, including the terms that were dropped to obtain the thin-boundary-layer model.

REFERENCES

1. S. W. Churchill, *The Practical Use of Theory in Fluid Flow. Book III. Laminar, Multidimensional Flows in Channels*, Notes, The University of Pennsylvania (1979).
2. L. M. Milne-Thompson, *Theoretical Hydrodynamics*, 5th ed., Macmillan, New York (1968).
3. J. Happel and H. Brenner, *Low Reynolds Number Hydrodynamics*, Martinus Nijhoff, Dordrecht (1986).
4. L. Rosenhead, Ed., *Laminar Boundary Layers*, Oxford University Press (1963).
5. V. L. Streeter, *Fluid Dynamics*, McGraw-Hill, New York (1948).

6. H. Lamb, *Hydrodynamics*, Dover, New York (1945).
7. G. C. Stokes, "On the Effect of the Internal Friction of Fluids on the Motion of Pendulums," *Trans. Camb. Phil. Soc. 9* (1851) 8 (*Mathematical and Physical Papers*, Vol. III, Cambridge University Press (1901), p. 1).
8. C. W. Oseen, "Über die Stokes'sche Formel und über die verwandte Aufgabe in der Hydrodynamik," *Arkiv Math., Astronom. Fys.*, 6, (1910) 75.
9. W. E. Langlois, *Slow Viscous Flow*, Macmillan, New York (1964).
10. O. A. Ladyzhenskaya, *The Mathematical Theory of Viscous Incompressible Flow*, Gordon and Breach, New York (1964).
11. S. Goldstein, Ed., *Modern Developments in Fluid Dynamics*, Vol. 2, Oxford University Press, Clarendon (1938).
12. C. R. Illingworth, "Flow at Small Reynolds Number," Chapter IV; p. 163, in *Laminar Boundary Layers*, L. Rosenhead, Ed., Oxford University Press (1963).
13. M. Van Dyke, *Perturbation Methods in Fluid Mechanics*, annotated ed., Parabolic Press, Stanford (1975).
14. L. Prandtl, "Über Flüssigkeitsbewegung bei sehr kleiner Reibung," *Verhandl. 3rd Int. Math. Kongr., Heidelberg* (1904), Leipzig (1905), p. 484; English transl., "Motion of Fluids with Very Little Viscosity," *NACA* TM 452, Washington, D.C. (1928).
15. I. Flügge-Lotz and W. Flügge, "Ludwig Prandtl in the Nineteen Thirties: Reminiscences," *Ann. Rev. Fluid Mech.*, 5 (1973) 1.
16. I. Tani, "History of Boundary-Layer Theory," *Ann. Rev. Fluid Mech.*, 9 (1977) 87.
17. K. Nickel, "Prandtl's Boundary Layer Theory from the Viewpoint of a Mathematician," *Ann. Rev. Fluid Mech.*, 5 (1973) 405.
18. J. D. Hellums and S. W. Churchill, "Mathematical Simplification of Boundary Value Problems," *AIChE. J.*, 10, (1964) 110.
19. S. W. Churchill and R. Usagi, "A General Expression for the Correlation Rates of Transfer and Other Phenomena," *AIChE J.*, 18 (1972) 1121.
20. Th. von Kármán, "Über laminare und turbulente Reibung," *Z. Angew. Math. Mech.*, 1 (1921) 233; English transl., "On Laminar and Turbulent Friction," *NACA* TM 1092, Washington, D.C. (1947).
21. K. Pohlhausen, "Zur näherungsweisen Integration der Differentialgleichung der laminaren Grenzschicht," *Z. Angew. Math. Mech.*, 1 (1921) 252.
22. B. A. Finlayson, *The Method of Weighted Residuals and Variational Principles*, Academic Press, New York (1977).
23. H. Blasius, "Grenzschichten in Flüssigkeiten mit kleiner Reibung," *Z. Math. Phys.*, 56 (1908) 1; English transl., "The Boundary Layers in Fluids with Little Friction," *NACA* TM 1256, Washington, D.C. (1950).
24. C. F. Dewey, Jr. and J. F. Gross, "Exact Similar Solutions of the Laminar Boundary Layer Equations," *Advances in Heat Transfer*, Vol. 4, Academic Press, New York (1967), p. 317.
25. S. Goldstein, "Concerning some Solutions of the Boundary Layer Equations in Hydrodynamics," *Proc. Camb. Phil. Soc.*, 26 (1930) 1.
26. H. Schlichting, *Boundary Layer Theory*, 7th ed., English transl. by J. Kestin, McGraw-Hill, New York (1979).
27. N. L. Evans, *Laminar Boundary Layer Theory*, Addison-Wesley, Reading, MA (1968).
28. A. Walz, *Boundary Layers of Flow and Temperature*, English transl. by H. J. Oser, MIT Press, Cambridge, MA (1969).
29. D. Meksyn, *New Methods in Laminar Boundary Layer Theory*, Pergamon, Oxford (1961).
30. "Computational Methods for Inviscid and Viscous Two- and Three-Dimensional Flow Fields," *AGARD Lecture Series*, No. 73 (1975).
31. H. G. Knudsen and D. L. Katz, *Fluid Dynamics and Heat Transfer*, McGraw-Hill, New York (1958).

32. A. H. Shapiro, *Shape and Flow*, Doubleday, New York (1961).
33. L. G. Loitsianskii, "Pogranichnyi Sloi," p. 300, in *Mechanics in the USSR over Thirty Years, 1917–1947*; English transl., "Boundary Layers," *NACA* TM 1400, Washington, D.C. (1956).
34. M. Van Dyke, *An Album of Fluid Motion*, Parabolic Press, Stanford, CA (1982).
35. W. F. Ames, "Recent Developments in the Nonlinear Equations of Transport Processes," *Ind. Eng. Chem. Fundam.*, 8 (1969) 522.
36. H. Weyl, "Concerning the Differential Equations of Some Boundary Value Problems," *Proc. Nat. Acad. Sci.*, 27 (1941) 578.
37. L. Howarth, "On the Solution of the Laminar Boundary Layer Equations," *Proc. Roy. Soc. (London)*, *A164* (1938) 547.
38. S. W. Churchill and H. W. Char, "Correlating Equations for the Recovery Factor and the Blasius Functions for Flow along a Flat Plate," *Chem. Eng. Commum.*, 20 (1983) 355.
39. N. A. V. Piercy and G. N. Preston, "A Simple Solution of the Flat Plate Problem of Skin Friction and Heat Transfer," *Phil. Mag.*, Ser. 7, *21* (1936) 996; also, see N. A. V. Piercy, *Aerodynamics*, Van Nostrand, New York (1937), p. 299.

Chapter 12

Integral Boundary-Layer Solution for Laminar Flow along a Flat Plate

In 1921 von Kármán [1] developed an integral model for flow along a flat plate. At the same time Pohlhausen [2] devised a method of obtaining approximate solutions from that model. This model and method are simple and reasonably accurate and have been applied to a variety of problems for which solutions are difficult or impossible by classical methods, including other geometries, heat and component transfer, variable physical properties, power-law fluids, and even turbulent transfer. A whole literature exists on the method and results. However, with the development of modern computing machinery and techniques, the integral method has been supplanted by numerical methods that give essentially exact results. The integral method is now only of historical interest and is described here in that context for incompressible laminar flow along a flat horizontal plate.

HEURISTIC DERIVATION OF THE MODEL

The distance from the plate at which u_x closely approaches u_∞ is sketched in Figure 12–1. As indicated by the dashed line in Figure 12–1A, the *boundary-layer thickness*, δ, continually increases from zero at the leading edge as the drag of the plate slows down more and more of the moving stream.

A material balance can be written over a finite element of length Δx, height h greater than δ, and width H, as indicated in Figure 12–2, as follows:

$$H \int_0^h u \, dy = H \int_0^h u \, dy + H \left(\frac{\partial}{\partial x} \int_0^\eta u \, dy \right) \Delta x + H v_h \Delta x \qquad (12.1)$$

where to avoid double subscripts,

u = component of velocity in x-direction, m/s
v = component of velocity in y-direction, m/s

Equation 12.1 can be simplified to give the following expression for the normal velocity at $y = h$:

271

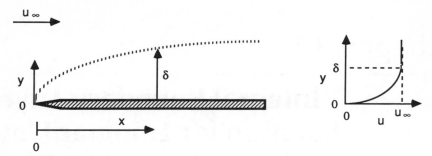

FIGURE 12–1   *Integral boundary-layer model for flow along a flat plate.*

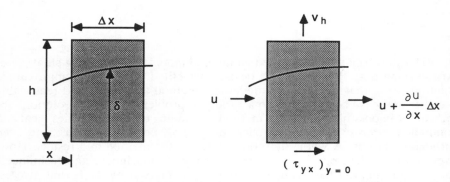

FIGURE 12–2   *Control element for integral mass and momentum balances for flow along a flat plate.*

$$v_h = -\frac{\partial}{\partial x} \int_0^h u\, dy \qquad (12.1A)$$

A force-momentum balance over the same element takes the form

$$(j_{yx})_{y=0} H\, dx = \frac{\partial}{\partial x} \left[H\varrho \int_0^h u^2\, dy \right] \Delta x + v_h u_h \varrho H\, \Delta x \qquad (12.2)$$

where $u = u_h$ at $y = h$. The term on the left side, representing the influx of x-momentum at $y = 0$, is equal to the negative of the drag of the wall on the fluid. The first term on the right side is the momentum leaving the right side of the element minus that entering the left side. The second term on the right side is the x-momentum leaving the element with the outgoing fluid at $y = h$. Since $h > \delta$, $u_h \simeq u_\infty$. The pressure is assumed to be uniform outside and to vary negligibly with x inside the boundary layer just as in the differential model. Substituting for v_h from Equation 12.1A and for $(j_{yx})_{y=0}$ from Equation 1.1, and rearranging give

$$\nu \left(\frac{\partial u}{\partial y} \right)_{y=0} = \frac{\partial}{\partial x} \int_0^\delta u(u_\infty - u)\, dy \qquad (12.3)$$

where μ/ϱ has been replaced by v. Also h has been replaced by δ since $u_\infty - u \cong 0$ for $y > \delta$. Equation 12.3 is a form of the *integral momentum-equation for the boundary layer* derived by von Kármán [1].

DERIVATION FROM THE DIFFERENTIAL MODEL

Equation 12.3 can also be derived directly from the differential equations of conservation as follows. Equation 11.1 is integrated across the boundary layer from $y = 0$, where $v = 0$ to $y = \delta$ where $v \simeq v_\infty$, and then the order of integration and differentiation are reversed to give

$$v_\infty = -\int_0^\delta \frac{\partial u}{\partial x}\, dy = -\frac{\partial}{\partial x}\int_0^\delta u\, dy \tag{12.4}$$

Equation 11.4 can be expanded as

$$u\frac{\partial u}{\partial x} + \frac{\partial (uv)}{\partial y} - u\frac{\partial v}{\partial y} = v\frac{\partial^2 u}{\partial y^2} \tag{12.5}$$

Substituting for $\partial v/\partial y$ from Equation 11.1 and rearranging give

$$\frac{\partial u^2}{\partial x} + \frac{\partial (uv)}{\partial y} = v\frac{\partial^2 u}{\partial y^2} \tag{12.6}$$

Integrating Equation 12.6 across the boundary layer from $y = 0$ where $u = v = 0$ to $y = \delta$ where $u \cong u_\infty$, $v \cong v_\infty$, and $\partial u/\partial y = a$, then inverting the order of integration and differentiation of the term in u^2 produce

$$\frac{\partial}{\partial x}\int_0^\delta u^2\, dy + u_\infty v_\infty = -v\left(\frac{\partial u}{\partial y}\right)_{y=0} \tag{12.7}$$

Eliminating v_∞ between Equations 12.4 and 12.7 then gives

$$-v\left(\frac{\partial u}{\partial y}\right)_{y=0} = \frac{\partial}{\partial x}\int_0^\delta u^2\, dy - u_\infty\frac{\partial}{\partial x}\int_0^\delta u\, dy$$

$$= \frac{\partial}{\partial x}\int_0^\delta u(u - u_\infty)\, dy \tag{12.8}$$

which is equivalent to Equation 12.3. It is apparent from this second derivation that no serious errors beyond those already made to obtain 11.1 and 11.4 have been made in reducing the general equations of conservation to Equation 12.8. However, a solution of Equation 12.8 satisfies 11.1 and 11.4 only on-the-mean with respect to y.

POHLHAUSEN METHOD OF SOLUTION

Equation 12.8 can be rearranged in the form

$$\frac{v}{u_\infty \delta}\left(\frac{\alpha(u/u_\infty)}{\alpha(y/\delta)}\right)_{y/\delta=0} = \frac{\partial}{\partial x}\left[\delta \int_0^1 \frac{u}{u_\infty}\left(1 - \frac{u}{u_\infty}\right)d\left(\frac{y}{\delta}\right)\right]$$ (12.9)

Now, following the suggestion of Pohlhausen [2], u/u_∞ is postulated to be a function of y/δ only. The derivative at $u/\delta = 0$ and the integral in Equation 12.9 are then constants and may be designated as

$$\psi_0 = \left(\frac{d(u/u_\infty)}{d(y/\delta)}\right)_{y/\delta=0}$$ (12.10)

and

$$\phi_0 = \int_0^1 \frac{u}{u_\infty}\left(1 - \frac{u}{u_\infty}\right)d\left(\frac{y}{\delta}\right)$$ (12.11)

Substituting from Equations 12.10 and 12.11 in 12.9 and rearranging gives

$$\delta\frac{d\delta}{dx} = \frac{v\psi_o}{u_\infty\phi_o}$$ (12.12)

Equation 12.12 is an ordinary differential equation in x with no remaining dependence on y and can be integrated directly from $\delta = 0$ at $x = 0$ to give

$$\frac{\delta^2}{2} = \frac{\psi_o vx}{\phi_o u_\infty}$$ (12.13)

Hence

$$\frac{\delta}{x} = \left(\frac{2\psi_o v}{\phi_o u_\infty x}\right)^{1/2} = \left(\frac{2\psi_o}{\phi_o \mathrm{Re}_x}\right)^{1/2}$$ (12.14)

Also

$$\tau_w \equiv (-j_{yx})_{y=0} = \mu\left(\frac{du}{dy}\right)_{y=0} = \left(\frac{\psi_o\phi_o\mu u_\infty^3 \varrho}{2x}\right)^{1/2}$$ (12.15)

The local drag coefficient, again defined as $\tau_w/\varrho u_\infty^2$, is

$$C_f = \left(\frac{\psi_o\phi_o v}{2u_\infty x}\right)^{1/2} = \left(\frac{\psi_o\phi_o}{2\mathrm{Re}_x}\right)^{1/2}$$ (12.16)

The corresponding integrated-mean drag coefficient is

$$\bar{C}_f \equiv \frac{1}{x}\int_0^x C_f\, dx = \left(\frac{\psi_o\phi_o}{\mathrm{Re}_x}\right)^{1/2}$$ (12.17)

It is now necessary to postulate a functional dependence for u/u_∞. After the example of Pohlhausen, a power series is chosen:

$$\frac{u}{u_\infty} = a_o + a_1\left(\frac{y}{\delta}\right) + a_2\left(\frac{y}{\delta}\right)^2 + a_3\left(\frac{y}{\delta}\right)^3 + \cdots \qquad (12.18)$$

As a first try, only the first two terms are used. Since $u = 0$ at $y = 0$, and $u = u_\infty$ at $y = \delta$, it follows that $a_o = 0$ and $a_1 = 1$. Then

$$\psi_o = 1$$

$$\phi_o = \int_0^1 \frac{y}{\delta}\left(1 - \frac{y}{\delta}\right) d\left(\frac{y}{\delta}\right) = \frac{1}{6}$$

and

$$\left(\frac{\psi_o\phi_o}{2}\right)^{1/2} = \left(\frac{1}{12}\right)^{1/2} = 0.289$$

As a second try, using three terms and the additional condition that $d(u/u_\infty)/d(y/\delta) = 0$ at $y = \delta$ gives $a_o = 0$, $a_1 = 2$, and $a_2 = -1$. Then

$$\psi_o = 2$$

$$\phi_o = \int_0^1 (2z - z^2)(1 - 2z + z^2)\, dz = \frac{2}{15}$$

and

$$\left(\frac{\psi_o\phi_o}{2}\right)^{1/2} = 0.365$$

As a third try, using four terms and the additional condition that $d^2(u/u_\infty)/d(y/\delta)^2 = 0$ at $y = 0$ (which may be rationalized on the basis that inertial effects should be negligible and hence $d(u/u_\infty)/d(y/\delta) = a$ constant near the wall) gives $a_0 = 0$, $a_1 = 3/2$, $a_2 = 0$, and $a_3 = -1/2$. Hence

$$\psi_o = \frac{3}{2}$$

$$\phi_o = \int_0^1 \left(\frac{3}{2}z - \frac{1}{2}z^3\right)\left(1 - \frac{3}{2}z + \frac{1}{2}z^3\right) dz = \frac{39}{280}$$

and

$$\left(\frac{\psi_o\phi_o}{2}\right)^{1/2} = 0.323$$

These several values of $\sqrt{\psi_o\phi_o/2}$ are compared in Table 12.1 with the exact value of 0.332 derived by Blasius. The agreement is reasonably good for all of the values, and the error appears to decrease as the number of terms in the series increases. However, the good agreement is largely due to the insensitivity of the integral to the kernel. Furthermore, there is no assurance that the series solution would converge to the exact solution even if more conditions could be contrived to evaluate further coefficients in the series.

The velocity distributions corresponding to the several solutions given here are compared with the exact solution in Figure 12–3. The agreement is not as

Table 12.1
Illustrative Solutions for Integral Boundary-Layer Model for Laminar Flow along a Flat Plate

u/u_∞	ψ_0	ϕ_0	$C_f Re_x^{1/2}$	$\delta\sqrt{u_\infty/x\nu}$
$\dfrac{y}{\delta}$	1	1/6	0.289	3.46
$2\dfrac{y}{\delta} - \left(\dfrac{y}{\delta}\right)^2$	2	2/15	0.365	5.48
$\dfrac{3}{2}\dfrac{y}{\delta} - \dfrac{1}{2}\left(\dfrac{y}{\delta}\right)^3$	3/2	39/280	0.323	4.64
(Exact)	—	—	0.332	~5.00

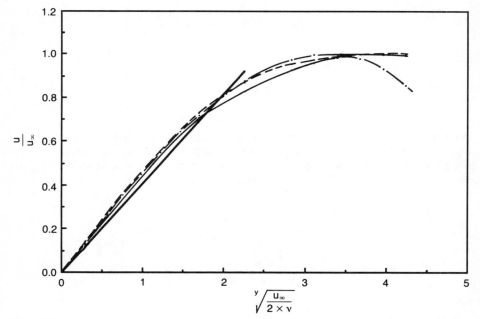

FIGURE 12–3   *Comparison of integral boundary-layer solutions with exact solution for laminar flow along a flat plate.* ———— = *first order solution;* ———— = *second order solution;* —·— = *third order solution;* ———— = *exact (Blasius) solution.*

impressive as for the drag, but naturally improves with the number of terms. Additional postulated distributions are examined in problem 2.

Integral boundary-layer theory is now recognized as an application of the method of weighted residuals. As discussed by Finlayson [3], such solutions can be proven to converge only under special conditions, although the error can be bounded under some circumstances.

The error and uncertainty in integral boundary-layer solutions arise from two sources: (1) the satisfaction of the equations of conservation on-the-mean in y rather than at each value of y, and (2) the arbitrary postulate of a velocity field. The error cannot be reduced indefinitely and, in the absence of an exact solution, cannot be evaluated. This is a decisive handicap relative to numerical solutions, whose convergence can usually be forced.

Integral boundary-layer theory can also be seen to produce the similarity transformation derived in the previous chapter, that is, the postulate that $u/u_\infty = f\{y/\delta\}$ leads directly to $u/u_\infty = f\{y(u_\infty/xv)^{1/2}\}$.

SUMMARY

This chapter illustrated the use of integral boundary-layer theory to derive a solution for the Blasius model for flow along a thin flat plate. This technique produced the similarity transformation that was derived in Chapter 11. The solution itself has the merit of being in closed form and appears to be in fair numerical agreement with the exact (Blasius) solution, depending on the choice of the form of the velocity distribution.

Unfortunately this technique cannot always be depended on to produce similarity transformations, as can the one described in Chapter 11. The choice of the form of the velocity distribution is highly arbitrary and cannot be generalized or extended indefinitely. The fair agreement of the drag coefficient with the exact value is largely fortuitous and results from the well-known insensitivity of the value of integrals to the integrand. The particular functional dependence on $y(u_\infty/xv)^{1/2}$ provides no insight since it was chosen a priori. Finally, the integral boundary-layer solution also incorporates the errors arising from the simplifications leading to the Blasius model.

PROBLEMS

1. Derive Equation 12.5 by integrating Equation 11.4 by parts.
2. Derive expressions for ϕ_o, ϕ_o, and $C_f\sqrt{\mathrm{Re}_x}$ based on the distributions:

    a. $u/u_\infty = \sin\{\pi y/2\delta\}$
    b. $u/u_\infty = \mathrm{erf}\{y/\delta\}$
    c. $u/u_\infty = 1 - e^{-y/\delta}$
    d. a fourth-order polynominal with the additional condition $d^3(u/u_\infty)/d(y/\delta)^3 = 0$ at $y = \delta$
    e. $u/u_\infty = \tanh\{y/\delta\}$

3. Derive expressions for u/u_∞ as a function of $\eta = y\sqrt{u_\infty/2xv}$ for the postulated functions in Table 12.1 and problem 2. Compare.

4. Derive an expression for v_∞ in terms of ψ_o and ϕ_o. Compute the coefficient in this expression for the several velocity distributions in Table 12.1 and problem 2. compare numerically with Equation 11.43.

5. Develop a solution for flow along a flat plate using the method of collocation. That is, postulate a velocity distribution $u = b_1 + b_2 y + \cdots$ and determine the coefficients that make the residual of the differential equations zero at a finite number of points in the boundary layer.

REFERENCES

1. Th. von Kármán, "Über laminare und turbulente Reibung," *Z. Angew. Math. Mech.*, *1* (1921) 233; English transl., "On Laminar and Turbulent Friction," *NACA TM 1092*, Washington, D.C. (1947).

2. K. Pohlhausen, "Zur näherungsweisen Integration der Differentialgleichung der laminaren Grenzschicht," *Z. Angew. Math. Mech.*, *1* (1921) 252.

3. B. A. Finlayson, *The Method of Weighted Residuals and Variational Principles*, Academic Press, New York (1977).

Chapter 13

Experimental Results and Extended Solutions for Laminar Flow along a Flat Plate

Analytical solutions of limited scope have been developed for *slow* flow over a flat plate, for which the postulates of a thin boundary-layer are known to be invalid, and also for a plate of finite length. These solutions are described in this chapter and, together with the solutions in Chapters 11 and 12, are compared with experimental data and numerical solutions of the unsimplified equations of motion.

The effects of blowing and sucking on flow along a porous plate, and of non-Newtonian behavior are also examined.

SLIGHTLY INERTIAL (TOMOTIKA–AOI) FLOW ALONG A FLAT PLATE

Oseen [1] derived the first few terms of a series solution for flow over a sphere at very low Reynolds numbers by utilizing a linear approximation for the inertial terms in the force-momentum balance. That solution is discussed in Chapter 16. The regime for which such a linearization is valid for a sphere is called *slightly inertial* or *Oseen* flow. Tomotika and Aoi [2] used this concept to develop the following approximation for flow along a plate at small Re_x:

$$
\bar{C}_f = \frac{2\pi}{Re_x S}\left[1 - \frac{1}{S}\left(S^2 - S - \frac{5}{12}\right)\frac{Re_x}{128} \right.
$$
$$
\left. - \frac{1}{S^2}\left(S^4 - \frac{S^3}{12} + \frac{23S^2}{24} - \frac{133S}{360} - \frac{25}{144}\right)\left(\frac{Re_x^2}{128}\right)^2 \right] \tag{13.1}
$$

where $S = 1 - \gamma - \ln\{Re_x/16\}$
γ = Euler's constant = $0.577\,22\ldots$

The term in brackets differs negligibly from unity for $Re_x < 1$. However, for $Re_x > 1$ the error due to the Oseen approximation is greater than the correction

279

due to the terms in brackets. Hence, the principal value of Equation 13.1 is as an approximation for $Re_x \ll 1$ in the reduced form

$$\bar{C}_f = \frac{2\pi}{Re_x[1 - \gamma - \ln\{Re_x/16\}]} \tag{13.3}$$

Equation 13.3 was derived much earlier by Harrison [3].

SUB-BOUNDARY-LAYER (IMAI) FLOW ALONG A FLAT PLATE

The boundary layer itself necessarily perturbs the flow ahead of the leading edge as well as displacing the streamlines along the plate as shown (in exaggerated form) in Figure 13–1. Using the Blasius solution within the boundary layer and an outer flow corresponding to the resulting displacement, Imai [4] derived the following corrected boundary-layer solution for the local and mean drag coefficients for short distances from the leading edge of a semi-infinite plate:

$$C_f = \frac{0.332}{Re_x^{1/2}} + 0.2755\frac{\ln\{Re_x\}}{Re_x^{3/2}} + \frac{A - 1}{2Re_x^{3/2}} + \cdots \tag{13.4}$$

and

$$\bar{C}_f = \frac{0.664}{Re_x^{1/2}} + \frac{1.163}{Re_x} - 0.551\frac{\ln\{Re_x\}}{Re_x^{3/2}} - \frac{0.102 + A}{Re_x^{3/2}} + \cdots \tag{13.5}$$

where A is an unknown constant. The second-order term in Equation 13.5 is due to the two-dimensionality of the free-stream velocity at the leading edge. Van Dyke [5], p. 138, says "It is remarkable that 50 years were required to discover that one term of the boundary layer solution provides two terms of the friction drag." The higher-order terms in Equation 13.5 and the corresponding terms in Equation 13.4 are discussed in detail by Van Dyke, p. 131f, Goldstein [6] and Rosenhead [7].

Botta and Dijkstra [8] carried out finite-difference calculations and thereby evaluated the coefficients in the *Stokes expansion*:

$$C_f = \frac{A}{Re_x^{1/2}} + B\,Re_x^{1/2} - \frac{\pi A^2}{32}Re_x + \cdots \tag{13.6}$$

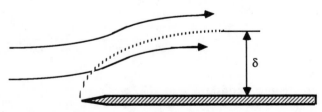

FIGURE 13–1   *Streamlines for laminar flow along a flat plate.* ——— = *actual for low Re;* ...... = *thin-boundary-layer theory.*

They obtained $A = 0.37737$ and $B = 0.0205$. This value of A indicates a 14% higher local skin friction near the leading edge than that given by the Blasius solution.

EFFECT OF FINITE LENGTH OF PLATE

Kuo [9] assumed that the displacement due to the boundary layer was constant in the trailing edge, as sketched in Figure 13–2, and derived the following expression for a finite plate of length l:

$$\bar{C}_f = \frac{0.664}{\text{Re}_l^{1/2}} + \frac{2.06}{\text{Re}_l} + \cdots \tag{13.7}$$

Van Dyke [5], p. 139, asserts that the coefficient of the second term should be "something like" 2.65 rather than 2.06.

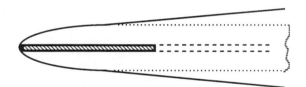

FIGURE 13–2 *Displacement thickness over a flat plate of finite length in the laminar boundary-layer regime.* ——— = *actual;* ...... = *postulated by Kuo [9].*

Stewartson [10] and Messiter [11] used a "triple-deck" model to derive for a finite plate for large Re an expression of the form

$$\bar{C}_f = \frac{0.664}{\text{Re}_l^{1/2}} + \frac{C}{\text{Re}_l^{7/8}} \tag{13.8}$$

Melnick and Chow [12] obtained $C = 1.33$ using numerical integration. Slightly differing values have been obtained by others (see Van Dyke [5], p. 231).

NUMERICAL SOLUTIONS FOR FLOW ALONG A FLAT PLATE

Janssen [13] in 1957 used an electrical analog to solve the general equations of motion and obtained the values in Table 13.1 for a plate of finite breadth.

The first numerical solution of the unreduced equations was apparently by Dennis and Dunwoody [14], who in 1966 obtained the values in Table 13.2 for a semi-infinite plate. (Their results converge to the Tomotika–Aoi solution for

Table 13.1
Drag Coefficients from Electrical Analog of Janssen [13]

Re_x	0.1	1.0	10
$\bar{C}_f$	11.11	1.40	0.285

Table 13.2
Drag Coefficients from Numerical Solution of Dennis and Dunwoody for a Plate of
Semi-Infinite Length [14]

Re_x	$\bar{C}_f$	Re_x	$\bar{C}_f$	Re_x	$\bar{C}_f$
0.1	11.33	2	1.10	100	0.094
0.2	6.40	4	0.68	500	0.03655
0.4	3.665	10	0.374	1,000	0.0251
0.6	2.67	15	0.2905	2,000	0.01705
1.0	1.82	20	0.2415	5,000	0.0103
		40	0.158	10,000	0.00705

small Re_x and to the Blasius solution for large Re_x.) They also computed the
local skin friction and pressure distribution for a plate of finite length l as shown
in Figures 13–3 and 13–4. As the Reynolds number increases, the pressure
variation appears to become negligible, as postulated in the Blasius solution.

Brauer and Sucker [15] carried out more extensive calculations for a plate of
finite length. Their computed boundary-layer thickness (defined for $u_x =
0.99u_\infty$) and the two components of the velocity field are compared with the
Blasius solution in Figures 13–5 to 13–7, respectively. The corresponding dimensionless local shear stress on the plate is plotted in Figure 13.8. The leading-
and trailing-edge effects are seen to be quite significant. Large deviations from
the Blasius solution would be expected at these low Reynolds numbers due to
the thickness of the boundary layer. Direct comparison of their results with
those of Dennis and Dunwoody for finite lengths is difficult because of the
different dimensionless variables chosen for display, but qualitative agreement
is apparent.

Brauer and Sucker concluded from their studies that laminar boundary-

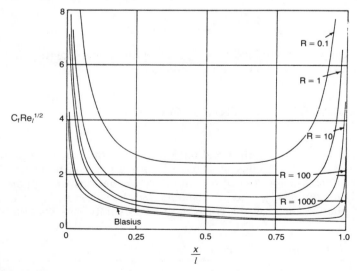

FIGURE 13–3   *Local skin friction on a plate of finite length. Parameter $R = Re_l =
lu_\infty/v$. (From Dennis and Dunwoody [14].)*

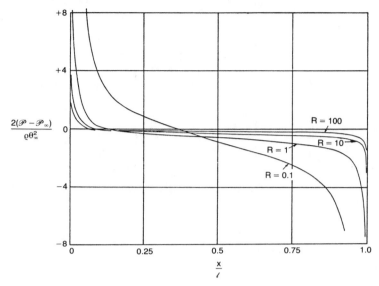

FIGURE 13-4 *Pressure distribution on a plate of finite length. Parameter $R = Re_l = lu_\infty/\nu$. (From Dennis and Dunwoody [14].)*

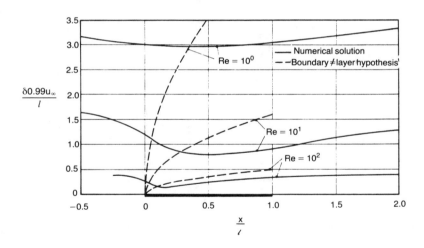

FIGURE 13-5 *Comparison of numerical solution of Brauer and Sucker for the boundary layer thickness along a plate of finite length with the Blasius solution [15]. Parameter: $Re = Re_l = u_\infty l/\nu$.*

layer-theory "must be regarded as out of date." However, to the contrary, boundary-layer solutions still appear to be necessary for Re_x beyond the present capabilities of numerical methods, and still provide asymptotic expressions for correlation. Indeed, Brauer and Sucker, in the same article, used such expressions for the latter purpose, as indicated next.

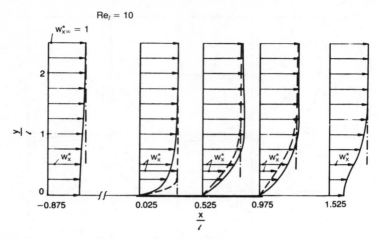

FIGURE 13–6　*Longitudinal velocity profiles computed by Brauer and Sucker for flow along a plate of finite length at $Re_l = 10$. ——— = numerical solution; ——— = Blasius solution. ($w_x^* = u_x/u_\infty$) [15].*

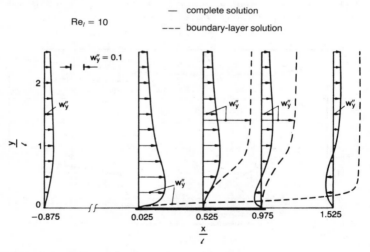

FIGURE 13–7　*Profiles of normal velocity computed by Brauer and Sucker for flow along a plate of finite length at $Re_l = 10$. ——— = numerical solution; ——— = Blasius solution. ($w_y^* = u_y/u_\infty$) [15].*

EXPERIMENTAL RESULTS FOR THE FLAT PLATE

Experimental data for large Re_x are compared with the Blasius solution in Figures 13–9 and 13–10. The agreement between the predicted velocity profile and the measured values of Nikuradse [16] is seen in Figure 13–9 from Schlichting [17], p. 142, to be remarkable. The independence of the plot from the Reynolds number as well as the agreement in absolute value provides strong

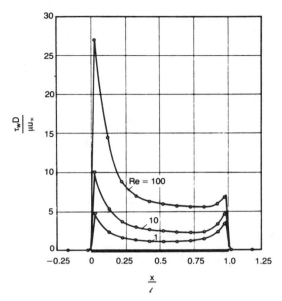

FIGURE 13–8 *Dimensionless drag computed by Brauer and Sucker for flow along a plate of finite length.* $(Re = Re_l = lu_\infty/\nu)$ *[15].*

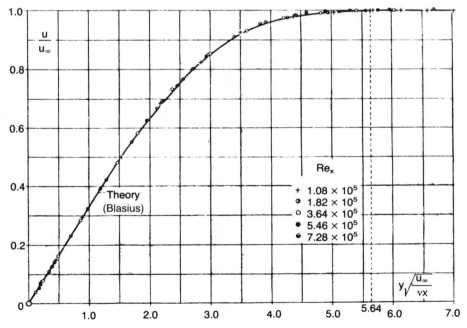

FIGURE 13–9 *Comparison by Schlichting [17, p. 142] of experimental data of Nikuradse [16], with solution of Blasius for laminar flow along a plate.*

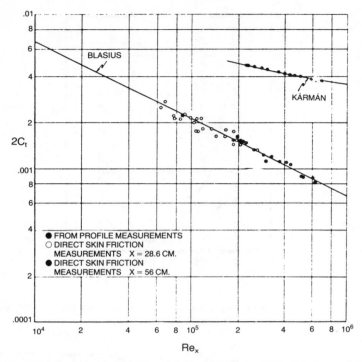

FIGURE 13–10   *Comparison by Liepmann and Dhawan [18] of their experimental data for the local drag coefficient in flow along a plate with the solution of Blasius.*

confirmation for the validity of the boundary-layer model over this range of flow. The dimensionless boundary-layer thickness $y\sqrt{u_\infty/x\nu}$ is observed to be about 5.0, which agrees with the prediction.

The experimental data of Liepmann and Dhawan [18] for the local drag coefficient are seen in Figure 13–10 to agree well with the Blasius solution as long as the boundary layer remains laminar. Transition from laminar to turbulent motion may occur as low as $Re_x = 3 \times 10^5$ or may be deferred to as high as 3×10^6, depending on the free-stream turbulence, the sharpness of the leading edge, and the roughness of the surface. Brauer and Sucker [15] proposed the empirical equation

$$Re_x^* = \frac{3 \times 10^6}{1 + 10^4 \, Tu^{1.7}} \tag{13.9}$$

for the critical Reynolds Re_x^* for transition, where

$$Tu \equiv \sqrt{\frac{(u')^2}{u_\infty^2}}$$

is the *turbulence number* of the free stream. This relationship is limited to $Tu < 0.1$.

The mean drag coefficients measured by Janour [19] are compared with the computed values of Janssen [13] and Dennis and Dunwoody [14] in Figure

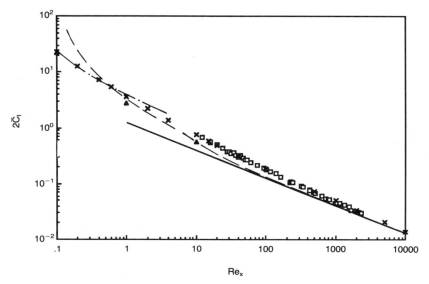

FIGURE 13–11   *Comparison of experimental data, computed values, and theoretical solutions for the mean drag coefficient on a plate. x = Dennis and Dunwoody, computed; △ = Janssen, computed; × = Janour, measured; —·—·— = Tomotika and Aoi/Harrison (Equation 13.3); ———— = Imai (first three terms of Equation 13.5); ———— = Blasius (Equation 11.57).*

13–11. The agreement between the measured and experimental values is seen to be very fair except for the values of Janssen at $Re_x = 1$ and 10. Curves representing the equations of Blasius (11.57), Imai (13.5), and Tomotika–Aoi–Harrison (13.3) are included. The Blasius equation appears to be accurate only for $Re_x > 10^4$, but is a lower bound for all Re_x. The first three terms of the Imai equation provide a good representation for all $Re_x > 1$. The Tomotika–Aoi–Harrison equation provides a better representation for $Re_x < 1$ and appears to be an upper bound for all Re_x.

Brauer and Sucker [15] proposed to encompass all regimes of flow with the two empirical equations

$$\bar{C}_f = \frac{1.325}{Re_x^{7/8}} - \frac{1}{8Re_x + 0.016/Re_x} + \frac{0.664}{Re_x^{1/2}} \qquad \text{for } 10^{-2} < Re_x < Re_x^* \qquad (13.10)$$

and

$$\bar{C}_f = \frac{0.2275}{(\ln\{Re_x\})^{2.58}} - \frac{4.95 \times 10^3/Re_x}{1 + 10^4\,Tu^{1.7}} \qquad \text{for } Re_x > Re_x^* \qquad (13.11)$$

FLOW ALONG A POROUS FLAT PLATE WITH SUCTION OR BLOWING

Solutions can also be derived for suction or blowing at the surface of a plate, as might be made to occur with a porous material. Suction is sometimes applied to aircraft surfaces to prevent separation and to chemical processes to remove reactants. Blowing is used to add reactants, cool the surface, prevent corrosion or scaling, reduce the drag, etc. Equations 11.1, 11.4, and 11.6 are applicable, but the condition for u_y in 11.5 must be replaced by

$$u_y = v_0\{x\} \qquad \text{at } y = 0, \quad x > 0 \tag{13.12}$$

where $v_0\{x\}$, the normal superficial velocity in meters per second, at the surface —positive for blowing and negative for suction, as illustrated in Figure 13–12.

Similarly, Equation 11.7 and boundary conditions 11.8–11.10 are applicable when the condition $\partial\psi/\partial x = 0$ in 11.8 is replaced by

$$\frac{\partial \psi}{\partial x} = v_0\{x\} \qquad \text{at } y = 0, \quad x > 0 \tag{13.13}$$

Dedimensionalizing as before produces the preliminary result

$$\frac{\psi}{\psi_A} = \phi\left\{\frac{x}{x_A}, \frac{y}{y_A}, \frac{vx_A}{y_A\psi_A}, \frac{u_\infty y_A}{\psi_A}, \frac{x_A v_0\{x\}}{\psi_A}\right\} \tag{13.14}$$

The extra group, compared to Equation 11.14, arises from the new boundary condition.

Letting

$$\frac{vx_A}{y_A\psi_A} = 1 \tag{13.15}$$

and

$$\frac{u_\infty y_A}{\psi_A} = 1 \tag{13.16}$$

requires

$$y_A = \sqrt{\frac{vx_A}{u_\infty}} \tag{13.17}$$

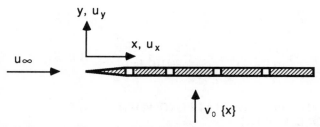

FIGURE 13–12   *Coordinates and variables for analysis of flow along a plate with blowing (or suction).*

and

$$\psi_A = \sqrt{\nu x_A u_\infty} \tag{13.18}$$

and reduces Equation 13.14 to

$$\frac{\psi}{\sqrt{\nu x_A u_\infty}} = \phi\left\{\frac{x}{x_A}, \; y\sqrt{\frac{u_\infty}{\nu x_A}}, \; \sqrt{\frac{x_A}{\nu u_\infty}}\, v_0\{x\}\right\} \tag{13.19}$$

Eliminating x_A gives

$$\frac{\psi}{\sqrt{\nu x u_\infty}} = \phi\left\{y\sqrt{\frac{u_\infty}{\nu x}}, \; \sqrt{\frac{x}{\nu u_\infty}}\, v_0\{x\}\right\} \tag{13.20}$$

A similarity transformation exists only if the rightmost group is a constant, that is, if

$$v_0\{x\} = A\sqrt{\frac{\nu u_\infty}{x}} \tag{13.21}$$

where A is a constant that may be positive (blowing) or negative (suction). This special case will be considered first.

$v_0\{x\} = (A/2)\sqrt{\nu u_\infty/x}$ (Schlichting–Bussmann flows). Letting $f = -\psi/\sqrt{\nu x u_\infty}$ and $\eta = y\sqrt{u_\infty/2\nu x}$ again gives Equations 11.29–11.34, except that $f = 0$ at $\eta = 0$ is replaced by

$$f = -A \qquad \text{at } \eta = 0 \tag{13.22}$$

This boundary-value problem was apparently first solved by Schlichting and Bussmann [20, 21]. Extensive results have been tabulated by Emmons and Leigh [22]. Their computed values of u/u_∞ are plotted versus η in Figure 13–13.

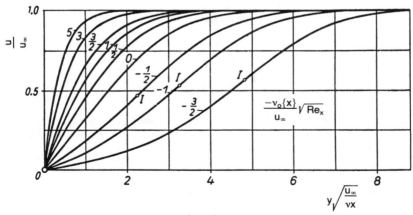

FIGURE 13–13 *Velocity profiles computed by Emmons and Leigh [22] for Schlichting-Bussmann flow [suction and blowing on a plate in proportion to $(\nu u_\infty/x)^{1/2}$]. (From Schlichting [17], p. 390.)*

Asymptotic Solution for Uniform Suction (Meredith–Griffith Flow)

For the more practical case of uniform suction, a similarity transformation does not exist and the partial differential equations must be solved numerically. However, Meredith and Griffith, according to Goldstein [23], p. 534, showed that a closed-form solution can be obtained for uniform suction ($v_0 < 0$) for the asymptotic case of $x \to \infty$. It is first postulated that

$$\frac{\partial u_x}{\partial x} \to 0 \qquad \text{as } x \to \infty \tag{13.23}$$

From Equation 11.1 it follows that

$$u_y\{x, y\} \to v_0 \tag{13.24}$$

Equation 11.4 then reduces to

$$v_0 \frac{du_x}{dy} = v \frac{d^2 u_x}{dy^2} \tag{13.25}$$

with boundary conditions

$$u_x = 0 \qquad \text{at } y = 0 \tag{13.26}$$

and

$$u_x \to u_\infty \qquad \text{as } y \to \infty \tag{13.27}$$

The solution is

$$u_x = u_\infty (1 - e^{v_0 y/v}) \tag{13.28}$$

It follows that

$$C_f = \frac{\tau_w}{\varrho u_\infty^2} = \frac{\mu}{\varrho u_\infty^2} \left(\frac{\partial u_x}{\partial y}\right)_{y=0} = -\frac{v_0}{u_\infty} \tag{13.29}$$

The drag is independent of the viscosity in this limit. This solution satisfies the general equations of conservation and does not depend on the idealizations of boundary-layer theory (see problem 18).

Uniform Suction with a Developing Boundary Layer (Iglisch Flow)

Iglisch [24] presented results obtained by numerical solution of the partial differential equations for a developing boundary layer. These calculations reveal that the foregoing asymptotic solution of Meredith and Griffith is a good approximation for

$$-v_0\sqrt{\frac{x}{vu_\infty}} = C_f\sqrt{Re_x} > 2 \tag{13.30}$$

Dimensional analysis of the differential equations and boundary conditions solved by Iglisch (see problem 10) suggests that the drag coefficient can be expressed functionally as

$$C_f\sqrt{Re_x} = \phi_1\left\{\frac{-v_0}{u_\infty}\sqrt{Re_x}\right\} \tag{13.31}$$

or as

$$C_f\left(\frac{-u_\infty}{v_0}\right) = \phi_2\left\{\frac{-v_0}{u_\infty}\sqrt{Re_x}\right\} \tag{13.32}$$

The former expression is more convenient for small $(-v_0/u_\infty)\sqrt{Re_x}$, since the right side approaches a constant value of 0.332 corresponding to the Blasius solution, the latter expression for large $(-v_0/u_\infty)\sqrt{Re_x}$, since that right side approaches unity in accordance with Equation 13.29. The computed values of Iglisch are converted to the indicated forms in Table 13.3. A correlating equation for the values in Table 13.3, based on Equations 11.44 and 13.29 is suggested in problem 12.

Table 13.3
Functions Based on Computations of Iglisch for Flow along a Flat Plate with Uniform Suction [24]

$-v_0\sqrt{\dfrac{x}{u_\infty v}} = \dfrac{-v_0}{u_\infty}\sqrt{Re_x}$	$\dfrac{-v}{u_\infty v_0}\left(\dfrac{\partial u_x}{\partial y}\right)_{y=0} = \dfrac{-u_\infty}{v_0}C_f$	$\sqrt{\dfrac{xv}{u_\infty^3}}\left(\dfrac{\partial u_x}{\partial y}\right)_{y=0} = C_f\sqrt{Re_x}$
0	∞	0.3321
0.0707	5.322	0.3763
0.1414	2.986	0.4223
0.2121	2.216	0.4701
0.2828	1.835	0.5190
0.3536	1.612	0.5699
0.4243	1.467	0.6224
0.4950	1.366	0.6761
0.5657	1.292	0.7309
0.6364	1.237	0.7872
0.7071	1.194	0.8443
0.8485	1.135	0.9631
0.9899	1.094	1.0830
1.131	1.068	1.2083
1.414	1.036	1.4651
1.697	1.019	1.7293
2.263	1.009	2.283
2.828	1.000	2.828
∞	1.000	∞

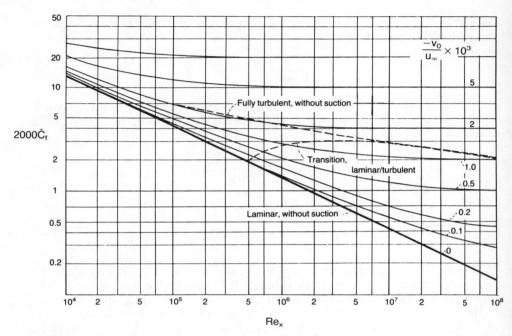

FIGURE 13-14   *The effect of uniform suction on the mean drag coefficient for a flat plate. (From Iglisch [24].)*

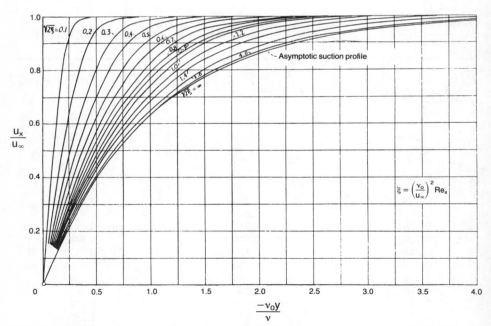

FIGURE 13-15   *Velocity profiles for flow along a flat plate with uniform suction. (From Iglisch [24].)*

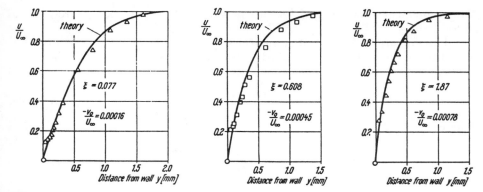

FIGURE 13-16 *Comparison of Iglisch solution with the experimental values of Head [25] for the velocity profile along a flat plate with uniform suction.* $\xi = (-v_0/u_\infty)^2 Re_x$ *(From Schlichting [17], p. 389.)*

Although uniform suction increases the drag above that for laminar flow without suction, it delays the transition to turbulent flow and hence may reduce the overall drag, as indicated in Figure 13-14. The distribution of u_x, as computed by Iglisch, is reproduced in Figure 13-15 and is compared with the experimental data of Head [25] in Figure 13-16. The agreement is excellent. Uniform blowing is considered in Chapter 18 in connection with the methods of local similarity and local nonsimilarity.

POWER-LAW (ACRIVOS) FLOWS

Solutions for laminar boundary-layer flows utilizing the power-law model (Equation 2.10) have been developed by Acrivos and co-workers for many situations. Accordingly they will be designated herein as *Acrivos flows*.

The momentum-force balance for a power-law fluid in the thin-boundary-layer regime becomes

$$u_x \frac{\partial u_x}{\partial x} + u_y \frac{\partial u_x}{\partial y} = M \frac{\partial}{\partial y} \left(\frac{\partial u_x}{\partial y} \right)^\alpha \tag{13.33}$$

The mass balance and the boundary conditions are unchanged. Application of the Hellums–Churchill technique directly to this representation indicates that

$$\frac{u_x}{u_\infty} = \phi \left\{ y \left(\frac{\varrho u_\infty^{2-\alpha}}{Mx} \right)^{1/(1+\alpha)}, \alpha \right\} \tag{13.34}$$

α is a parameter in this expression because it occurs as an exponent of the derivative of u_x on the right side of Equation 13.33. It follows that

$$C_f = \frac{M}{\varrho u_\infty^2} \left(\frac{-\partial u_x}{\partial y} \right) \left| \frac{\partial u_x}{\partial y} \right|^{\alpha-1} = \left(\frac{M}{\varrho x^\alpha u_\infty^{2-\alpha}} \right)^{1/(1+\alpha)} \phi\{\alpha\} \tag{13.35}$$

The equivalent value of Re_x for this power-law flow may be inferred to be $x^\alpha u_\infty^{2-\alpha} \varrho / M$.

The similarity transformation indicated by Equation 13.34 or the equivalent in terms of the Lagrange stream function can be used to reduce the thin-boundary-layer model to

$$\alpha(1 + \alpha)2^{-(\alpha+1)/2} f''' + (f'')^{2-\alpha} f = 0 \tag{13.36}$$

where

$$f\{\eta\} = \psi \left(\frac{\varrho}{2Mxu_\infty^{2-\alpha}} \right)^{1/(1+\alpha)} \tag{13.37}$$

and the primes indicate differentiation with respect to

$$\eta = y \left(\frac{\varrho u_\infty^{2-\alpha}}{2Mx} \right)^{1/(1+\alpha)} \tag{13.38}$$

In terms of these variables

$$C_f = \left(\frac{M}{x^\alpha u_\infty^{2-\alpha} \varrho} \right)^{1/(1+\alpha)} \left(\frac{f''\{0\}}{2^{1/(1+\alpha)}} \right)^\alpha \tag{13.39}$$

Equations 13.36–13.39 reduce to the prior expressions for a Newtonian fluid for $\alpha = 1$ and $M = \mu$.

Acrivos et al. [26] in 1960 solved the equivalent of Equation 13.36 numerically and obtained the values of $(f''\{0\}/2^{1/(1+\alpha)})^\alpha$ (labeled "Exact") in Table 13.4. They also solved this problem using the integral boundary-layer method, with the third-degree polynomial derived in Chapter 12, and obtained

Table 13.4
Solutions of Acrivos et al. [26] for the Drag Coefficient for the Flow of a Power-Law Fluid along a Flat Plate

$$C_f \left(\frac{x^\alpha u_\infty^{2-\alpha} \varrho}{M} \right)^{1/(1+\alpha)}$$

α	Exact	Equation 13.40	% Error
0.05	1.017	0.926	−8.9
0.1	0.969	0.860	−11.2
0.2	0.8725	0.747	−14.3
0.3	0.7325	0.656	−10.4
0.5	0.5755	0.518	−10.0
1.0	0.33206	0.323	−2.7
1.5	0.2189	0.226	+3.2
2.0	0.1612	0.169	+4.8
2.5	0.1226	0.134	+9.3
3.0	0.09706	0.109	+12.3
4.0	0.06777	0.079	+16.6
5.0	0.05111	0.061	+19.4

$$C_f \left(\frac{x^\alpha u_\infty^{2-\alpha} \varrho}{M} \right)^{1/(1+\alpha)} = \left(\frac{117}{560(1 + \alpha)} \right)^{\alpha/(1+\alpha)} \tag{13.40}$$

Values computed from Equation 13.40 are included in Table 13.4. The error is seen to increase rapidly as α departs in either direction from unity.

BOUNDARY-LAYER THICKNESS

Various definitions of the boundary-layer thickness are in vogue. The simplest is perhaps the distance y at which u/u_∞ attains some arbitrary value such as 0.9, 0.98, or 0.99. The latter criterion was, for example, used by Brauer and Sucker [15] in preparing Figure 13–5.

The boundary-layer thickness defined by integral boundary-layer theory can be expressed as

$$\delta = \left[\left(\frac{2x\nu}{u_\infty} \right) \left(\frac{d(u_x/u_\infty)}{d(y/\delta)} \right)_{y/\delta=0} \Big/ \int_0^1 \frac{u_x}{u_\infty} \left(1 - \frac{u_x}{u_\infty} \right) d\left(\frac{y}{\delta} \right) \right]^{1/2} \tag{13.41}$$

A *displacement thickness* can be defined as

$$\delta_D = \int_0^\infty \left(1 - \frac{u_x}{u_\infty} \right) dy \tag{13.42}$$

This quantity can be interpreted, as illustrated in Figure 13–17, as the thickness of a hypothetical stagnant layer with the same *velocity defect* $(u_\infty - u_x)$ as the integrated velocity defect of the actual boundary layer. It follows that the streamline at the outer edge of the boundary layer is displaced a distance δ_D from its location for inviscid flow.

A *momentum thickness* can be defined as

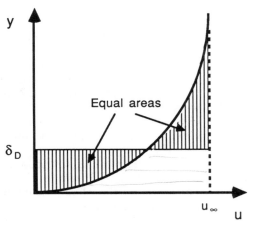

FIGURE 13–17   *Definition of displacement thickness of the boundary layer on a plate.*

$$\delta_M = \int_0^\infty \frac{u_x}{u_\infty}\left(1 - \frac{u_x}{u_\infty}\right) dy \tag{13.43}$$

and an *energy-dissipation thickness* as

$$\delta_E = \int_0^\infty \frac{u_x}{u_\infty}\left(1 - \left(\frac{u_x}{u_\infty}\right)^2\right) dy \tag{13.44}$$

corresponding to the loss of momentum and energy, respectively, relative to inviscid flow. Note from comparison of Equations 12.11 and 13.43 that δ as defined by Equation 12.14 equals δ_M/ϕ_0.

Values of these various boundary-layer thicknesses are compared in problems 22 and 23.

SUMMARY

The analytical solution of Tomotika and Aoi for slightly inertial flow and that of Imai for the sub-boundary-layer regime, as well as the Blasius equation, were compared with experimental data and with values computed numerically from the general equations of motion. These analytical solutions were each found to be in good agreement with the experimental and numerically computed values over a limited range of Re. In particular, the Blasius solution was found to be accurate only for $10^4 < \mathrm{Re}_x < 2 \times 10^5 - 6 \times 10^5$. The Imai solution (Equations 13.4 and 13.5) extends this range downward to $\mathrm{Re}_x \cong 1$, which is the approximate upper limit of the Tomotika–Aoi solution.

Theoretical solutions were also examined for plates of finite length, for blowing and sucking through a porous plate, and for power-law fluids. Blowing and sucking were seen to have significant effects. The numerical solution for uniform suction was shown to be confirmed experimentally.

Different expressions for the boundary-layer thickness were compared.

PROBLEMS

1. Derive an expression for C_f corresponding to Equation 13.8. Compare with the computed values in Figure 13–3.
2. Derive Equation 13.5 by integration of 13.4. Explain any discrepancies.
3. Explain qualitatively the difference between Equations 13.4 and 13.6. Which would be expected to represent experimental data better? Why?
4. Derive C_f from the values in Table 13.2.
5. Compare Equation 13.6 with the data of Figure 13–10. Derive the corresponding expression for $\bar{C}_f$ and compare with the data and equations of Figure 13–11.
6. Derive expressions for C_f corresponding to Equations 13.7 and 13.8. Compare with the computed values of Figure 13–3.
7. Compare the local shear stress on the plate as computed by Dunwoody (Figure 13–3) and Brauer and Sucker (Figure 13–8).
8. Compare Equations 13.10 and 13.11 with the experimental and computed values and estimate their reliability.

9. Explain by physical reasoning why suction increases and blowing decreases the drag in the laminar regime.
10. Derive Equations 13.31 and 13.32 starting from

    a. Equations 11.1, 11.4 and the appropriate boundary conditions
    b. Equation 13.20

11. Derive an expression equivalent to Equation 13.20 for u/u_∞. Reduce for

    a. $v_0\{x\} \to -(A/2)\sqrt{vu_\infty/x}$
    b. $v_0 = $ constant

12. a. Use the Churchill-Usagi model [27] with Equations 11.44 and 13.29 to develop an empirical equation for the dependence of C_f on Re_x and v_0/u_∞ for a porous plate with uniform suction. Evaluate the accuracy of this solution.
    b. Modify the results of part (a) to obtain an expression for $\bar{C}_f$.

13. a. Derive an integral momentum equation for the boundary layer on a porous flat plate over which a fluid is flowing at a uniform velocity u_∞ and through which a fluid is removed at a uniform superficial velocity v_0.
    b. Assuming that u/u_∞ is still a function of only y/δ for small v_0/u_∞, reduce the integral equation to a differential equation and solve the differential equation to obtain an expression for δ.
    c. Prepare a plot relating the local drag coefficient to Re_x and u_∞/v_0 for any ϕ_0 and ψ_0.
    d. Obtain a complete solution for a third-degree polynomial. Compare with the solution of Iglisch. Plot C_f versus Re_x for $v_0/u_\infty = -0.01$.
    e. Derive an expression for the value of $v_0\sqrt{x/vu_\infty}$ for which the boundary-layer thickness will approach a limiting value. Compare this result with Equation 13.30.

14. Develop and solve an integral boundary-layer model for $v_0 = (A/2)\sqrt{vu_\infty/x}$. Compare with the exact solution in Figure 13–13.
15. Derive Equation 13.35 starting from 13.33.
16. Derive an integral boundary-layer solution for flow of a power-law fluid along a flat plate in terms of an arbitrary velocity distribution. Show that for the third-degree polynomial this general solution yields Equation 13.40.
17. Solve Equation 13.36 for $\alpha = 2$.
18. Prove that Equation 13.28 is independent of the simplifications of boundary-layer theory.
19. Derive an expression for $\bar{C}_f$ corresponding to Equation 13.39 for C_f.
20. Use the Churchill–Usagi model to derive a correlating equation as an alternative to Equation 13.10.
21. Prepare sketches indicating the variation with x and y of the velocity component, u_y, normal to a flat plate for laminar flow along the plate for a

    a. nonporous plate
    b. porous plate with uniform suction v_0

Make your coordinates as general and quantitative as possible.

22. Compare $\delta_{u_x} = 0.099u_\infty$, δ_D, δ_M, and δ_E using the Blasius solution.

23. Compare δ, $\delta_{u_x} = 0.99u_\infty$, δ_D, δ_M, and δ_E using the velocity distributions in Table 12.1 and those of problem 2 in Chapter 12.

24. Calculate and plot $\delta_{u_x} = 0.99$ versus x for air at 15°C flowing along a plate at $u_\infty = 3$ m/s.

25. Derive expressions for $\delta_{u_x} = 0.99u_\infty$, δ_D, δ_M, and δ_E corresponding to Equation 13.28.

26. Derive an expression for the velocity for inviscid flow over a cylindrical body whose shape is defined by $y = 4.64(vx/u_\infty)^{1/2}$. Specialize for $y = 0$

   a. numerically or graphically
   b. analytically

27. Derive expressions for the added skin friction and measure drag due to the perturbation in the velocity at the surface as derived in problem 26.

REFERENCES

1. C. W. Oseen, "Über die Stokes'sche Formel und über die verwandte Aufgabe in der Hydrodynamik," *Arkiv Math., Astronom. Fys.*, 6 (1910) 75.
2. S. Tomotika and T. Aoi, "The Steady Flow of Viscous Fluid," *Quart. J. Mech. Appl. Math.*, 6 (1953) 290.
3. W. J. Harrison, "On the Motion of Spheres, Circular Cylinders and Elliptic Cylinders through Viscous Fluid," *Trans. Camb. Phil. Soc.*, 23 (1923) 71.
4. I. Imai, "Second Approximation to the Laminar Boundary Layer Flow over a Flat Plate," *J. Aero. Sci.*, 24 (1957) 155.
5. M. Van Dyke, *Perturbation Methods in Fluid Mechanics*, annotated ed., Parabolic Press, Stanford, CA (1975).
6. S. Goldstein, *Lectures on Fluid Mechanics*, Wiley-Interscience, New York (1960).
7. L. Rosenhead, Ed., *Laminar Boundary Layers*, Oxford University Press (1963).
8. E. F. F. Botta and D. Dijkstra, "An Improved Numerical Solution of the Navier–Stokes Equations for Laminar Flow Past a Semi-Infinite Flat Plate," *Rept. Math. Inst. Univ. Groningen*, No. TW-80 (1970).
9. Y. H. Kuo, "On the Flow of an Incompressible Viscous Fluid Past a Hot Plate at Moderate Reynolds Numbers," *J. Math. Phys.*, 32 (1953) 83.
10. K. Stewartson, "On the Flow Near the Trailing Edge of a Flat Plate. II," *Mathematika*, 16 (1969) 106.
11. A. R. Messiter, "Boundary-Layer Flow near the Trailing Edge of a Flat Plate," *SIAM J. Appl. Math.*, 18 (1970) 241.
12. R. E. Melnick and R. Chow, "Asymptotic Theory of Two-Dimensional Trailing Flows, Aerodynamic Analyses Requiring Advanced Computers," *NASA*, SP No. 347, Washington, D.C. (1975).
13. E. Janssen, "Flow past a Flat Plate at Low Reynolds Numbers," *J Fluid Mech.*, 3 (1958) 329.
14. S. R. C. Dennis and J. Dunwoody, "The Steady Flow of a Viscous Fluid past a Flat Plate," *J. Fluid Mech.*, 24 (1966) 577.
15. H. Brauer and D. Sucker, "Umströmung von Platten, Zylinder und Kugeln," *Chem.-Ing.-Tech.*, 48 (1976) 665; English transl., "Flow about Plates, Cylinders and Spheres," *Int. Chem. Eng.*, 18 (1978) 367.
16. J. Nikuradse, "Laminare Reibungsschichten an der längsangeströmten Platte," Monographe, Zentrale f. wiss. Berichtswesen, Berlin (1942).

17. H. Schlichting, *Boundary Layer Theory*, 7th ed., English transl. by J. Kestin, McGraw-Hill, New York (1979).
18. H. W. Liepmann and S. Dhawan, "Direct Measurements of Local Skin Friction in Low-Speed and High-Speed Flow," *Proc. First U.S. Congr. Appl. Mech*, ASME, New York (1951), p. 869.
19. Z. Janour, "Odpor podelne obtékané desky při malých Reynoldsových cislech," Lecktecký Vyzkumný Ústav, Praha, Rep. 2 (1947); English transl., "Resistance of a Plate in Parallel Flow at Low Reynolds Numbers," *NACA* TM1316, Washington, D.C. (1951).
20. H. Schlichting and K. Bussmann, "Exakte Lösungen für die laminare Reibungsschicht mit Absaugung und Ausblasen," *Schriften Deut. Akad. Luftfahrtforschung*, 7B, No. 2 (1943).
21. H. Schlichting, "Die Beeinflussung der Grenzschicht durch Absaugung und Ausblasen," *Jahrb. Deutsche Akad. Luftfahrtforschung*, (1943/1944) 90.
22. H. W. Emmons and D. C. Leigh, "Tabulation of the Blasius Function with Blowing and Suction," *Aero. Res. Council, Gt. Brit.*, Current Papers, No. 157 (1954).
23. S. Goldstein, Ed., *Modern Developments in Fluid Dynamics*, Vol. 2, Oxford University Press, Clarendon (1938).
24. R. Iglisch, "Exakte Berechnung der laminaren Reibungsschicht an der längsangeströmten ebenen Platte mit homogener Absaugung," *Schriften Deut. Akad. Luftfahrtforschung*, 8B, No. 1 (1944); English transl. "Exact Calculation of Laminar Boundary Layer in Longitudinal Flow over a Flat Plate with Homogeneous Suction," *NACA* TM 1205, Washington, D.C. (1949).
25. M. R. Head, "An Approximate Method of Calculating the Laminar Boundary Layer in Two-Dimensional Incompressible Flow, *Aero Res. Council, Gt. Brit.*, R. & M. 3121 (1957).
26. A. Acrivos, M. J. Shah, and E. E. Petersen, "Momentum and Heat Transfer in Laminar Boundary Layer Flows of Non-Newtonian Fluids past External Surfaces," *AIChE J.*, 6 (1960) 312; also see "On the Solution of the Two Dimensional Boundary-Layer Flow Equations for a Non-Newtonian Power-Law Fluid," *Chem. Eng. Sci.*, 20 (1965) 101.
27. S. W. Churchill and R. Usagi, "A General Expression for the Correlation of Rates of Transfer and Other Phenomena," *AIChE J.*, 18 (1972) 1121.

Chapter 14

Laminar Flow over Wedges and Disks

Flow over wedges has little direct interest but comprises a general class of solutions of which flow along a flat plate and impinging flow on a flat plate are limiting cases. Impinging plane flow on a flat plate is of direct interest, but of even more interest is the limiting case of flow near the forward point of stagnation on a bluff body, such as a circular cylinder, whose generatrix is perpendicular to the direction of flow. Impinging axisymmetric flow on a circular disk plays the same role for bluff bodies of rotation with their axes in the direction of flow, such as spheres, ellipsoids, and paraboloids.

The development here parallels that for flow along a flat plate. First the boundary-layer model is derived. Exact solutions, asymptotic solutions, correlating equations, and approximate solutions are examined in turn or in the problem set.

THIN-BOUNDARY-LAYER FLOW OVER A WEDGE (FALKNER–SKAN FLOW)

Model

A solution for thin-boundary-layer flow over a wedge can be developed by a simple extension of the foregoing method for a flat plate. The geometrical situation is shown in Figure 14–1. Here x is measured from the leading edge *along* the surface, and y is measured *normal* to the surface.

Outside the boundary layer the effect of viscosity is negligible, and the Euler equation 8.58 is applicable in the form

$$-\frac{1}{\varrho}\frac{\partial p}{\partial x} + g_x = U_x \frac{\partial U_x}{\partial x} \tag{14.1}$$

where U_x is the velocity in meters per second just outside the boundary layer. Substituting for $-(1/\varrho)(\partial p/\partial x) + g_x$ in Equation 8.57 from Equation 14.1 and dropping $\partial u_x/\partial t$, $\partial^2 u_x/\partial x^2$, and the terms in the z-direction yield

$$u_x \frac{\partial u_x}{\partial x} + u_y \frac{\partial u_x}{\partial y} = U_x \frac{\partial U_x}{\partial x} + v \frac{\partial^2 u_x}{\partial y^2} \tag{14.2}$$

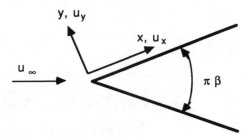

FIGURE 14–1   *Coordinates and variables for analysis of thin-boundary-layer flow along a wedge.*

The x-component of the velocity field for inviscid flow along the surface of a wedge (Section 10.S7) is

$$U_x\{x\} = cx^m \tag{14.3}$$

where c = constant, m^{1-m}/s
   $m = \beta/(2 - \beta)$
   $\pi\beta$ = total angle of wedge, rad

For $\beta = m = 0$, $c = u_\infty$; otherwise c is undefined.
   Substitution of U_x from Equation 14.3 in Equation 14.2 yields

$$u_x\frac{\partial u_x}{\partial x} + u_y\frac{\partial u_x}{\partial y} = c^2mx^{2m-1} + v\frac{\partial^2 u_x}{\partial y^2} \tag{14.4}$$

Reduction

Introducing the Lagrange stream function, as before for the flat plate, then gives

$$-\frac{\partial\psi}{\partial y}\frac{\partial^2\psi}{\partial x\,\partial y} + \frac{\partial\psi}{\partial x}\frac{\partial^2\psi}{\partial y^2} = c^2mx^{2m-1} + v\frac{\partial^3\psi}{\partial y^3} \tag{14.5}$$

The boundary condition for $y \to \infty$ becomes

$$-\frac{\partial\psi}{\partial y} \to U_x = cx^m \tag{14.6}$$

By the same procedure as in Chapter 11 it can be shown that

$$\frac{\psi}{\sqrt{cvx^{m+1}}} = \phi\left\{y\sqrt{\frac{x^{m-1}c}{v}}\right\} \tag{14.7}$$

Therefore the similarity transformation

$$\Psi = \frac{\psi}{\sqrt{cvx^{m+1}}} \tag{14.8}$$

and

$$\xi = y\sqrt{\frac{x^{m-1}c}{v}} \tag{14.9}$$

reduces Equation 14.5 to the ordinary differential equation

$$2\Psi''' - (m + 1)\Psi\Psi'' - 2m(1 - \Psi'^2) = 0 \tag{14.10}$$

where the primes indicate differentiation with respect to ξ. Equation 14.10 can be seen to reduce to 11.26 for $m = \beta = 0$.

The choice of unity for the coefficients in the similarity transformation was arbitrary. The coefficients in the final differential equation can often be simplified by some other choice. Thus let

$$\eta = b\xi \tag{14.11}$$

and

$$f = a\Psi \tag{14.12}$$

Equation 14.10 is thereby transformed to

$$\frac{2b^3 f'''}{a} - (m + 1)\frac{b^2}{a^2}ff'' - 2m\left(1 - \frac{b^2}{a^2}f'^2\right) = 0 \tag{14.13}$$

Choosing $b = -a = \sqrt{(m + 1)/2}$ reduces Equation 14.13 to

$$f''' + ff'' + \beta(1 - f'^2) = 0 \tag{14.14}$$

where

$$f = -\Psi\sqrt{\frac{m + 1}{2}} = -\psi\sqrt{\frac{m + 1}{2cvx^{m+1}}} \tag{14.15}$$

and the primes indicate differentiation with respect to

$$\eta = \xi\sqrt{\frac{m + 1}{2}} = y\sqrt{\frac{(m + 1)x^{m-1}c}{2v}} \tag{14.6}$$

This form is usually called the *Falkner–Skan* equation after the original investigators [1] of 1931. The boundary conditions for Equation 14.14 are

$$f\{0\} = f'\{0\} = 0 \tag{14.17}$$

and

$$f'\{\infty\} = 1 \tag{14.18}$$

Table 14.1
Functions Computed by Hartree for Wedge (Falkner–Skan) Flows (from Knudsen and Katz [3], p. 288)

β	-0.198_8	-0.19	-0.18	-0.16	-0.14	-0.10	0	0.1	0.2	0.3	0.4	0.5	0.6	0.8	1.0	1.2	1.6	2.0	2.4
$f''(0)$	0.0000	0.086	0.128	0.190	0.239	0.319	0.4696	0.5870	0.686	0.774	0.854	0.927	0.996	1.120	1.2326	1.336	1.521	1.687	1.837
η / $f'(\eta)$																			
0.0	0.0000	0.0000	0.0000	0.0000	0.0000	0.0000	0.0000	0.0000	0.0000	0.0000	0.0000	0.0000	0.0000	0.0000	0.0000	0.0000	0.0000	0.0000	0.0000
0.1	0.0010	0.0095	0.0137	0.0198	0.0246	0.0324	0.0469	0.0582	0.0677	0.0760	0.0834	0.0903	0.0966	0.1080	0.1183	0.1276	0.1441	0.1588	0.1720
0.2	0.0040	0.0209	0.0293	0.0413	0.0507	0.0659	0.0939	0.1154	0.1334	0.1490	0.1628	0.1756	0.1872	0.2081	0.2266	0.2433	0.2726	0.2980	0.3206
0.3	0.0089	0.0343	0.0467	0.0643	0.0781	0.1003	0.1408	0.1715	0.1970	0.2189	0.2382	0.2558	0.2719	0.3003	0.3252	0.3475	0.3859	0.4186	0.4472
0.4	0.0158	0.0495	0.0659	0.0889	0.1069	0.1356	0.1876	0.2265	0.2584	0.2858	0.3097	0.3311	0.3506	0.3848	0.4144	0.4405	0.4849	0.5219	0.5537
0.5	0.0248	0.0665	0.0868	0.1151	0.1370	0.1718	0.2342	0.2803	0.3177	0.3495	0.3771	0.4015	0.4235	0.4619	0.4946	0.5231	0.5708	0.6096	0.6424
0.6	0.0358	0.0855	0.1094	0.1427	0.1684	0.2088	0.2806	0.3328	0.3747	0.4100	0.4403	0.4670	0.4907	0.5317	0.5662	0.5959	0.6446	0.6834	0.7155
0.7	0.0487	0.1063	0.1338	0.1719	0.2010	0.2466	0.3266	0.3839	0.4294	0.4672	0.4994	0.5276	0.5524	0.5947	0.6298	0.6596	0.7076	0.7449	0.7752
0.8	0.0636	0.1289	0.1598	0.2023	0.2347	0.2849	0.3720	0.4335	0.4816	0.5212	0.5545	0.5834	0.6086	0.6512	0.6859	0.7150	0.7610	0.7858	0.8235
0.9	0.0803	0.1533	0.1874	0.2341	0.2694	0.3237	0.4167	0.4815	0.5312	0.5718	0.6055	0.6344	0.6596	0.7015	0.7350	0.7629	0.8058	0.8376	0.8624
1.0	0.0991	0.1794	0.2166	0.2671	0.3050	0.3628	0.4606	0.5274	0.5782	0.6190	0.6526	0.6811	0.7056	0.7460	0.7778	0.8037	0.8432	0.8717	0.8934
1.2	0.1423	0.2364	0.2791	0.3362	0.3784	0.4415	0.5453	0.6135	0.6640	0.7033	0.7351	0.7615	0.7837	0.8194	0.8467	0.8682	0.8997	0.9214	0.9373
1.4	0.1927	0.2991	0.3463	0.4083	0.4534	0.5194	0.6244	0.6907	0.7383	0.7743	0.8027	0.8258	0.8449	0.8748	0.8968	0.9137	0.9375	0.9530	0.9640
1.6	0.2498	0.3665	0.4170	0.4820	0.5284	0.5948	0.6967	0.7583	0.8011	0.8326	0.8568	0.8860	0.8917	0.9154	0.9324	0.9450	0.9620	0.9726	0.9799
1.8	0.3126	0.4372	0.4896	0.5555	0.6016	0.6660	0.7610	0.8160	0.8528	0.8791	0.8988	0.9141	0.9264	0.9443	0.9569	0.9658	0.9775	0.9845	0.9892
2.0	0.3802	0.5095	0.5621	0.6269	0.6712	0.7314	0.8167	0.8637	0.8940	0.9151	0.9305	0.9421	0.9514	0.9644	0.9732	0.9793	0.9871	0.9914	0.9944
2.2	0.4509	0.5814	0.6327	0.6944	0.7354	0.7896	0.8633	0.9019	0.9260	0.9421	0.9537	0.9621	0.9689	0.9779	0.9841	0.9879	0.9928	0.9954	0.9970
2.4	0.5230	0.6509	0.6995	0.7561	0.7927	0.8398	0.9011	0.9315	0.9500	0.9617	0.9700	0.9760	0.9807	0.9867	0.9905	0.9931	0.9961	0.9976	0.9985
2.6	0.5946	0.7162	0.7605	0.8107	0.8422	0.8817	0.9306	0.9537	0.9672	0.9754	0.9812	0.9852	0.9884	0.9922	0.9946	0.9962	0.9980	0.9989	0.9993
2.8	0.6635	0.7754	0.8146	0.8574	0.8836	0.9153	0.9529	0.9697	0.9792	0.9847	0.9886	0.9913	0.9933	0.9956	0.9971	0.9980	0.9990	0.9994	0.9996
3.0	0.7278	0.8273	0.8607	0.8959	0.9168	0.9413	0.9691	0.9808	0.9873	0.9908	0.9933	0.9952	0.9962	0.9976	0.9985	0.9989	0.9995	0.9997	0.9998
3.2	0.7858	0.8713	0.8986	0.9265	0.9425	0.9607	0.9804	0.9883	0.9924	0.9946	0.9962	0.9974	0.9979	0.9987	0.9992	0.9995	0.9998	0.9999	0.9999
3.4	0.8364	0.9071	0.9286	0.9499	0.9616	0.9746	0.9880	0.9931	0.9957	0.9970	0.9979	0.9986	0.9989	0.9993	0.9996	0.9997	0.9999	1.0000	1.0000
3.6	0.8789	0.9352	0.9515	0.9669	0.9752	0.9841	0.9929	0.9961	0.9976	0.9984	0.9989	0.9993	0.9995	0.9997	0.9998	0.9999	1.0000		
3.8	0.9132	0.9563	0.9681	0.9789	0.9845	0.9904	0.9959	0.9978	0.9987	0.9991	0.9994	0.9997	0.9997	0.9998	0.9999	1.0000			
4.0	0.9399	0.9716	0.9798	0.9871	0.9907	0.9944	0.9978	0.9988	0.9993	0.9995	0.9997	0.9999	0.9999	0.9999	1.0000				
4.2	0.9598	0.9822	0.9876	0.9924	0.9946	0.9969	0.9988	0.9994	0.9996	0.9997	0.9999	0.9999	0.9999	1.0000					
4.4	0.9741	0.9893	0.9927	0.9957	0.9970	0.9983	0.9994	0.9997	0.9998	0.9999	0.9999	1.0000	1.0000						
4.6	0.9839	0.9938	0.9959	0.9977	0.9984	0.9991	0.9997	0.9998	0.9999	0.9999	1.0000								
4.8	0.9904	0.9965	0.9978	0.9988	0.9992	0.9996	0.9999	0.9999	0.9999	1.0000									
5.0	0.9945	0.9981	0.9988	0.9994	0.9996	0.9998	0.9999	0.9999	1.0000										
5.2	0.9969	0.9990	0.9994	0.9997	0.9998	0.9999	1.0000	1.0000											
5.4	0.9984	0.9995	0.9997	0.9999	0.9999	1.0000													
5.6	0.9992	0.9997	0.9999	0.9999	1.0000														
5.8	0.9996	0.9999	0.9999	1.0000															
6.0	0.9998	0.9999	1.0000																
6.2	0.9999	1.0000																	
6.4	1.0000																		

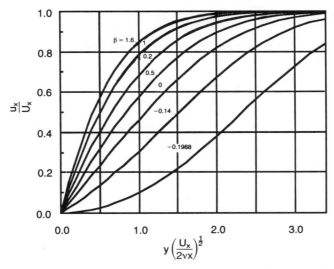

FIGURE 14–2 *Velocity fields for thin-boundary-layer (Falkner-Skan) flow along a wedge. (After Eckert and Drake [5], p. 285).*

General Results

Six years later, in 1937, Hartree [2] solved this boundary-value problem numerically for a number of values of β. Some of his results for $f'\{\eta\}$ and $f''\{0\}$ are reproduced from Knudsen and Katz [3], p. 288, in Table 14.1. Values for a more extended range of β are given by Evans [4]. The velocity distribution (after Eckert and Drake [5], p. 285) is shown in Figure 14–2 for several wedge angles. As β increases above zero, these profiles become steeper at the wall. As β decreases below zero, the profiles become S-shaped. At β = −0.1988 the velocity gradient at the surface is zero. A lower value of β would produce a negative gradient (backflow) and separation. The curves for β < 0 correspond to decelerating flow and therefore to a geometry such as that illustrated in Figure 14–3, in which the boundary layer on the horizontal wall is removed by suction.

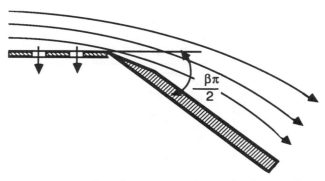

FIGURE 14–3 *A geometrical configuration producing decelerating flow. (After Eckert and Drake [5], p. 289).*

Drag Coefficient

The local drag coefficient for friction for this solution can be defined in terms of the *local* free-stream velocity:

$$C_f \equiv \frac{\mu}{\varrho U_x^2\{x\}} \left(\frac{\partial u_x}{\partial y}\right)_{y=0} = \frac{-\mu}{\varrho U_x^2\{x\}} \left(\frac{\partial^2 \psi}{\partial y^2}\right)_{y=0}$$

$$= f''\{0\} \sqrt{\left(\frac{m+1}{2}\right) \frac{\mu}{x\varrho U_x\{x\}}} = f''\{0\} \sqrt{\left(\frac{m+1}{2}\right)} \frac{1}{\mathrm{Re}_x} \qquad (14.19)$$

where here $\mathrm{Re}_x = x\varrho\, U_x\{x\}/\mu$.

Blasius Flow

The special case of $\beta = m = 0$ and $c = u_\infty$ is the Blasius problem considered in detail in Chapter 11. The special dimensionless variables for that case were chosen in anticipation of the canonical form of the Falkner–Skan equation.

Heimenz Flow

The special case of $\beta = m = 1$ is also of particular interest. It corresponds to a stream impinging normally on a flat plate and flowing away symmetrically about the plane $x = 0$, as sketched in Figure 14–4. This flow is named after Heimenz [6], who in 1911 first solved the Falkner–Skan equation for $\beta = 1$. More detailed results were obtained by Howarth [7] in 1935 and by Yuge [8] in 1956. The latter values are reproduced in Table 14.2. Goldstein [9], p. 140, notes that this

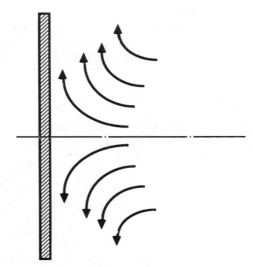

FIGURE 14–4   *Streamlines for planar impinging (Heimenz) flow.*

Table 14.2
Functions Computed by Yuge [8] for Planar Impingement (Heimenz Flow)

$y\sqrt{\dfrac{c}{\nu}}$	f	f'	f''
0	0.0000	0.0000	1.23259
0.1	0.00600	0.11826	1.13283
0.2	0.02332	0.22661	1.03445
0.3	0.05100	0.32524	0.93863
0.4	0.08806	0.41446	0.84633
0.5	0.13359	0.49465	0.75831
0.6	0.18670	0.56628	0.67517
0.7	0.24657	0.62986	0.59735
0.8	0.31242	0.68594	0.52513
0.9	0.38353	0.73508	0.45867
1.0	0.45923	0.77787	0.39801
1.1	0.53891	0.81487	0.34310
1.2	0.62203	0.84667	0.29378
1.3	0.70809	0.87381	0.24984
1.4	0.79665	0.89681	0.21100
1.5	0.88733	0.91617	0.17696
1.6	0.97978	0.93235	0.14735
1.7	1.07371	0.94577	0.12181
1.8	1.16886	0.95683	0.09996
1.9	1.26501	0.96588	0.08143
2.0	1.36198	0.97322	0.06583
2.1	1.45960	0.97913	0.05281
2.2	1.55776	0.98385	0.04204
2.3	1.65634	0.98760	0.03321
2.4	1.75526	0.99055	0.02602
2.5	1.85443	0.99285	0.02023
2.6	1.95381	0.99463	0.01560
2.7	2.05334	0.99600	0.01193
2.8	2.15300	0.99705	0.00905
2.9	2.25274	0.99783	0.00681
3.0	2.35256	0.99842	0.00508
3.1	2.45242	0.99886	0.00376
3.2	2.55233	0.99919	0.00276
3.3	2.65226	0.99942	0.00200
3.4	2.75221	0.99959	0.00144
3.5	2.85218	0.99972	0.00103
3.6	2.95215	0.99980	0.00073
3.7	3.05214	0.99987	0.00051
3.8	3.15212	0.99990	0.00036
3.9	3.25212	0.99994	0.00025
4.0	3.35211	0.99996	0.00017
4.1	3.45211	0.99997	0.00012
4.2	3.55211	0.99998	0.00008
4.3	3.65211	0.99999	0.00005
4.4	3.75210	0.99999	0.00003
4.5	3.85210	1.00000	0.00002
4.6	3.95210	1.00000	0.00002
4.7	4.05210	1.00000	0.00001

solution can be obtained without invoking the simplifications of thin-boundary-layer theory (see problem 27). The application of the Heimenz solution to flow over cylinders is discussed in Chapter 15.

Asymptotic solutions can readily be developed as for a flat plate, this time leading to

$$f\{\eta\} = \frac{f''\{0\}}{2}\eta^2 = 0.6163\eta^2 \qquad \text{for } \eta \to 0 \qquad (14.20)$$

and

$$f\{\eta\} = \eta - 0.6479 \qquad \text{for } \eta \to \infty \qquad (14.21)$$

The following test equation is then suggested for correlation for small and moderate η:

$$f^p\{\eta\} = (0.6163\eta^2)^p + \eta^p \qquad (14.22)$$

A value of $p = -1.442$ fits the value of $f\{2.0\} = 1.362$. However, the rounded-off exponent and coefficient in the following expression predict values within about 1% for $\eta < 2$:

$$f = \eta\left[1 + \left(\frac{1.65}{\eta}\right)^{10/7}\right]^{-7/10} \qquad (14.23)$$

For $\eta > 2$, Equation 14.21 is better.

THIN-BOUNDARY-LAYER FLOW IMPINGING ON A DISK (HOMANN FLOW)

The axisymmetric flow resulting from impingement on a disk is most readily solved in cylindrical coordinates. For thin-boundary-layer flows, Equations 8.13 and 8.59 can be reduced to

$$\frac{1}{r}\frac{\partial(u_r r)}{\partial r} + \frac{\partial u_z}{\partial z} = 0 \qquad (14.24)$$

and

$$u_r\frac{\partial u_r}{\partial r} + u_z\frac{\partial u_r}{\partial z} = g_r - \frac{1}{\varrho}\frac{\partial p}{\partial r} + \nu\frac{\partial^2 u_r}{\partial r^2} \qquad (14.25)$$

where r = radial distance from the axis of the disk, m
z = distance from the surface of the disk, m

The solution for inviscid flow for this geometry (see Section 10.S16) is

$$g_r - \frac{1}{\varrho}\frac{\partial p}{\partial r} = U_r\frac{dU_r}{dr} = \alpha^2 r \qquad (14.26)$$

where α is a constant with units of reciprocal seconds and $U_r\{r\}$ represents the velocity in the outer region. Combining Equations 14.25 and 14.26 gives

$$u_r \frac{\partial u_r}{\partial r} + u_z \frac{\partial u_r}{\partial z} = \alpha^2 r + \nu \frac{\partial^2 u_r}{\partial r^2} \tag{14.27}$$

A solution for Equation 14.24 is provided by the Stokes stream function in cylindrical coordinates (Equations 9.8 and 9.9). Introducing these expressions into Equation 14.27 yields

$$-\frac{1}{r^3}\left(\frac{\partial \tilde{\psi}}{\partial z}\right)^2 + \frac{1}{r^2}\frac{\partial \tilde{\psi}}{\partial z}\frac{\partial^2 \tilde{\psi}}{\partial z \partial r} - \frac{1}{r^2}\frac{\partial \tilde{\psi}}{\partial r}\frac{\partial^2 \tilde{\psi}}{\partial z^2} = \alpha^2 r + \frac{\nu}{r}\frac{\partial^3 \tilde{\psi}}{\partial z^3} \tag{14.28}$$

Dedimensionalization by the method of Hellums and Churchill produces

$$A^2\tilde{\Psi}''' + 2\tilde{\Psi}\tilde{\Psi}'' - \tilde{\Psi}'^2 + 1 = 0 \tag{14.29}$$

where

$$\tilde{\Psi} = \frac{A\tilde{\psi}}{r^2\sqrt{\alpha\nu}} \tag{14.30}$$

and the primes indicate differentiation with respect to

$$\tilde{\xi} = Az\sqrt{\frac{\alpha}{\nu}} \tag{14.31}$$

A is an arbitrary coefficient that can be chosen to produce the desired canonical form. The boundary conditions in terms of the primitive variables are

$$u_r = u_z = 0 \qquad \text{at } z = 0 \tag{14.32}$$

and, from Equation 14.26,

$$u_r \to U_r\{r\} = \alpha r \qquad \text{as } z \to \infty \tag{14.33}$$

Hence, in the transformed variables

$$\tilde{\Psi}\{0\} = \tilde{\Psi}'\{0\} = 0 \tag{14.34}$$

and

$$\tilde{\Psi}'\{\infty\} = 1 \tag{14.35}$$

Equation 14.29 together with 14.34 and 14.35, in the form resulting from letting $A = 1$, was first solved by Homann [10] in 1936. However, note that letting $A = \sqrt{2}$ yields the same mathematical problem as the Falkner–Skan problem in x and y with $\beta = 1/2$ ($m = 1/3$), corresponding to a wedge with an enclosed angle of $\pi/2$.

The values of $\tilde{\Psi}\{\tilde{\xi}\}$, $\tilde{\Psi}'\{\tilde{\xi}\}$, and $\Psi''\{\tilde{\xi}\}$ computed in 1956 by Yuge [8] (who also let $A = 1$) are given in Table 14.3 in both the original form with $A = 1$ and the canonical form with $A = \sqrt{2}$. The canonical form is utilized in the balance of this section. Goldstein [9], p. 143, notes that Equation 14.29 can be derived without invoking the assumptions of thin-boundary-layer theory.

Table 14.3
Functions Computed by Yuge [10] for Axisymmetric Impingement (Homann Flow) [Adjusted to $A = \sqrt{2}$].

for $A = 1$	$\tilde{\xi}$	$\tilde{\Psi}$	$\tilde{\Psi}'$	$\tilde{\Psi}''$
for $A = \sqrt{2}$	$\tilde{\xi}/\sqrt{2}$	$\tilde{\Psi}/\sqrt{2}$	$\tilde{\Psi}'$	$\sqrt{2}\tilde{\Psi}''$
	0	0	0	1.31194
	0.1	0.00639	0.12619	1.21195
	0.2	0.02491	0.24240	0.11211
	0.3	0.05454	0.34863	1.01278
	0.4	0.09430	0.44499	0.91455
	0.5	0.14321	0.53160	0.81816
	0.6	0.20031	0.60871	0.72451
	0.7	0.26465	0.67663	0.63454
	0.8	0.33534	0.73577	0.54921
	0.9	0.41153	0.78666	0.46942
	1.0	0.49242	0.82987	0.39593
	1.1	0.57726	0.86607	0.32933
	1.2	0.66542	0.89598	0.26999
	1.3	0.75628	0.92032	0.21803
	1.4	0.84932	0.93983	0.17336
	1.5	0.94411	0.95522	0.13565
	1.6	1.04025	0.96717	0.10443
	1.7	1.13745	0.97630	0.07906
	1.8	1.18636[a]	0.98316	0.05885
	1.9	1.33402	0.98822	0.04305
	2.0	1.43303	0.99189	0.03095
	2.1	1.53236	0.99451	0.02186
	2.2	1.63191	0.99635	0.01517
	2.3	1.73161	0.99761	0.01034
	2.4	1.83142	0.99846	0.00692
	2.5	1.93129	0.99903	0.00455
	2.6	2.03122	0.99939	0.00293
	2.7	2.13117	0.99963	0.00186
	2.8	2.23114	0.99978	0.00116
	2.9	2.33113	0.99987	0.00071
	3.0	2.43112	0.99993	0.00042
	3.1	2.53112	0.99996	0.00025
	3.2	2.63111	0.99998	0.00014
	3.3	2.73111	0.99999	0.00008
	3.4	2.83111	1.00000	0.00005
	3.5	2.93111	1.00000	0.00002
	3.6	3.03111	1.00000	0.00001
	3.7	3.13111	1.00000	0.00001

[a] This value should presumably be 1.23(636)

The boundary values for the foregoing problem together with the observed values of $\tilde{\Psi}''\{0\} = 1.31194/\sqrt{2} = 0.927682$, $\tilde{\Psi}'''\{\infty\} = 0$, and $\tilde{\Psi}\{3.7\sqrt{2}\} = 3.13111\sqrt{2} = 4.42806$ can be used to derive asymptotic expressions for small and large $\tilde{\xi}$. A slightly modified procedure will be illustrated this time. For small $\tilde{\xi}$ let

$$\tilde{\Psi} = a_0 + a_1\tilde{\xi} + a_2\tilde{\xi}^2 + a_3\tilde{\xi}^3 + \cdots \qquad (14.36)$$

Then $\Psi\{0\} = 0$ requires $a_0 = 0$, $\Psi'\{0\}$ requires $a_1 = 0$, and $\Psi''\{0\} = 0.927682$ requires $a_2 = 0.927682/2 = 0.46384$. Hence the first-order asymptote for $\tilde{\xi} \to 0$ is

$$\tilde{\Psi} = 0.463\,84\tilde{\xi}^2 \qquad (14.37)$$

Equation 14.36 can also be postulated for large $\tilde{\xi}$. $\tilde{\Psi}''\{\infty\} = 0$ requires that $a_2 = a_3 = a_4 = \cdots = 0$, $\tilde{\Psi}'\{\infty\} = 1$ requires $a_1 = 1$, and $\tilde{\Psi}\{5.23259\} = 4.42806$ requires that $a_0 = 4.42806 - 5.23259 = -0.80453$. Hence the asymptote for large $\tilde{\xi}$ is

$$\tilde{\Psi} = \tilde{\xi} - 0.804\,53 \qquad (14.38)$$

Correlation of the values in Table 14.3 using these asymptotes is posed in problem 8.

Equation 14.31 indicates that for this flow the boundary-layer thickness is constant (independent of r). The velocity field itself can be expressed as

$$u_r = \frac{1}{r}\frac{\partial \Psi}{\partial z} = r\alpha\tilde{\Psi}'\{\tilde{\xi}\} \qquad (14.39)$$

and

$$u_z = -\frac{1}{r}\frac{\partial \Psi}{\partial r} = -2\sqrt{\alpha\nu}\,\tilde{\Psi}\{\tilde{\xi}\} \qquad (14.40)$$

The drag coefficient is

$$C_f = \frac{\mu}{\varrho U_r^2\{r\}}\left(\frac{\partial u_r}{\partial z}\right)_{z=0} = \frac{\nu r\alpha}{U_r^2\{r\}}\sqrt{\frac{2\alpha}{\nu}}\,\tilde{\Psi}''\{0\}$$

$$= \sqrt{\frac{2\nu}{rU_r\{r\}}}\,\Psi''\{0\} = \sqrt{\frac{2}{\mathrm{Re}_r}}\,\tilde{\Psi}''\{0\} \qquad (14.41)$$

where $\mathrm{Re}_r \equiv rU_r\{r\}/\nu$.

INTEGRAL BOUNDARY-LAYER METHOD FOR A WEDGE

The von Kármán–Pohlhausen integral method can be adapted to derive solutions for wedge flows. A convenient starting point is Equation 14.2. Expanding the first term on the left side, using Equation 11.1, just as in the derivation of Equation 12.6, and integrating over y give

$$\int_0^\delta \left(\frac{\partial u_x^2}{\partial x} - U_x \frac{dU_x}{dx} \right) dy + U_x U_y = -v \left(\frac{\partial u_x}{\partial y} \right)_{y=0} \tag{14.42}$$

Again, from Equation 11.1,

$$U_y = -\int_0^\delta \frac{\partial u_x}{\partial x} dy \tag{14.43}$$

Combining Equations 14.42 and 14.43 gives

$$\int_0^\delta \left(\frac{\partial u_x^2}{\partial x^2} - U_x \frac{dU_x}{dx} - U_x \frac{\partial u_x}{\partial x} \right) dy = -v \left(\frac{\partial u_x}{\partial y} \right)_{y=0} \tag{14.44}$$

Expanding the third term in the integral and rearranging produces

$$\int_0^\delta \left(\frac{\partial u_x^2}{\partial x} - \frac{\partial (U_x u_x)}{\partial x} + u_x \frac{dU_x}{dx} - U_x \frac{dU_x}{dx} \right) dy = -v \left(\frac{\partial u_x}{\partial y} \right)_{y=0} \tag{14.45}$$

Inverting the order of differentiation and integration for the first two terms on the left side permits rearrangement as

$$\frac{\partial}{\partial x} \int_0^\delta u_x (U_x - u_x) \, dy + \frac{dU_x}{dx} \int_0^\delta (U_x - u_x) \, dy = v \left(\frac{\partial u_x}{\partial y} \right)_{y=0} \tag{14.46}$$

which is a general integral momentum equation for the boundary layer for any potential flow, that is, for any $U\{x\}$, including those of Chapters 15 and 16 for curved surfaces. Equation 14.46 degenerates to Equation 12.3 for the flat plate.
Equation 14.46 can be rewritten as

$$\frac{\partial}{\partial x} \left[\delta U_x \int_0^1 \frac{u_x}{U_x} \left(1 - \frac{u_x}{U_x} \right) d \left(\frac{y}{\delta} \right) \right] + \delta U_x \frac{dU_x}{dx} \int_0^1 \left(1 - \frac{u_x}{U_x} \right) d \left(\frac{y}{\delta} \right)$$
$$= \frac{v U_x}{\delta} \left[\frac{\partial (u_x / U_x)}{\partial (y/\delta)} \right]_{y/\delta = 0} \tag{14.47}$$

If u_x/U_x is postulated to be a function of y/δ only, the two integrals and the derivatives become numbers, and Equation 14.47 becomes an ordinary differential equation. In postulating a velocity distribution $u_x/U_x = f\{y/\delta\}$, the boundary condition

$$v \frac{\partial^2 u_x}{\partial y^2} = -U_x \frac{dU_x}{dx} \qquad \text{at } y = 0 \tag{14.48}$$

which follows from Equation 14.4, together with $u_x = u_y = 0$, must be satisfied, as well as $u_x = U_x$, $\partial u_x / \partial y = 0$, and

$$\frac{\partial^2 u_x}{\partial y^2} = 0 \qquad \text{at } y = \delta$$

Schlichting [11], p. 207, describes a formal structure for the solution of Equation 14.47 for a power-series representation for u_x/U_x.

BOUNDARY-LAYER THICKNESSES

The boundary-layer thicknesses defined by Equations 13.42–13.44 can be generalized for two-dimensional flow over an arbitrary body such as a wedge simply by replacing u_∞ with U_x. Thus

$$\delta_D = \int_0^\infty \left(1 - \frac{u_x}{U_x}\right) dy \qquad (14.49)$$

$$\delta_M = \int_0^\infty \frac{u_x}{U_x}\left(1 - \frac{u_x}{U_x}\right) dy \qquad (14.50)$$

and

$$\delta_E = \int \frac{u_x}{U_x}\left(1 - \left(\frac{u_x}{U_x}\right)^2\right) dy \qquad (14.51)$$

It follows that Equation 14.46 can be rewritten as

$$\frac{d}{dx}\left(U_x^2 \delta_M\right) + \delta_D u_x \frac{dU_x}{dx} = \nu \left(\frac{\delta u_x}{\delta y}\right)_{y=0} \qquad (14.52)$$

SUMMARY

Solutions were developed and numerical results given for thin-boundary-layer flow over wedges, including the limiting cases of parallel (Blasius) and normal (Heimenz) flow. Impinging flow on a disk was also examined. These solutions have little interest in themselves, but provide valuable asymptotes for the important cases of flow over a cylinder and over a sphere, as examined in Chapters 15 and 16.

PROBLEMS

1. a. Determine the minimum set of dimensionless variables required to represent Equations 11.1 and 14.2 and the boundary conditions, all reexpressed in terms of u_x and u_y.
    b. Reduce Equations 11.1 and 14.2 to ordinary differential equations and boundary conditions in terms of dimensionless velocities.
    c. Combine the equations of part b into a single equation.
    d. Show that the solution of part c is equivalent to Equation 14.14.

2.   Derive an expression for the mean drag coefficient for friction corresponding to Equation 14.19 for the local coefficient. Note any restrictions.
3.   Determine the dynamic pressure variation

  a.   along a wedge
  b.   for Homann flow

4.   Derive expressions for the total, local, and mean drag coefficients on a wedge (taking into account the pressure).
5.   Construct generalized asymptotic solutions for wedge flows corresponding to Equations 14.20 and 14.21 for normal impingment.
6.   Derive Equation 14.28 from Equations 8.13 and 8.59, noting all assumptions.
7.   Show that Equation 14.29 is independent of the simplifications of boundary-layer theory.
8.   The values in Table 14.3 indicate that for $A = \sqrt{2}$, $\tilde{\Psi}\{0\} = \tilde{\Psi}'\{0\} = 0$, $\tilde{\Psi}''\{0\} = 0.92768$, $\tilde{\Psi}\{\infty\} = 0$, and $\tilde{\Psi}'\{\infty\} = 1$. Develop asymptotic equations for $\eta \rightarrow 0$ and ∞ and use them to develop a correlating equation for the Homann solution for $\tilde{\Psi}$ for values of $\tilde{\xi}$ for which Equation 14.38 is not adequate. Evaluate this expression.
9.   Reduce Equation 14.14 for the Heimenz problem and solve.
10.   Derive an integral force-momentum balance for the Heimenz problem using an element of finite size. Compare with the solution of problem 9.
11.   Reduce Equation 14.47 for the Falkner–Skan problem.
12.   Does Equation 14.47 hold for the Homann problem? If so, reduce and solve for that condition.
13.   Does Equation 14.48 hold for axisymmetric flows in general? If not, revise as required.
14.   Potential flow in a planar converging channel can be represented by $U_x = -Q/x$, where Q is a constant with dimensions of square meters per second and x is the distance from the intersection of the planes. Derive a differential model for the boundary layer along $y = 0$. Reduce the representation to $f''' - f'^2 + 1 = 0$. Determine the functional dependence of the drag coefficient in terms of $f''\{0\}$. Compare the result with the exact solution (Section 10E10) and the solution for creeping flow (Section 10EC4).
15.   Is the flow of problem 14 a special case of the Falkner–Skan equation? Explain.
16.   Derive a closed-form analytical solution for u_x/U_x for the flow of problem 14. (*Hint:* Multiply the reduced differential equation by the integrating factor f''.)
17.   Explain the behavior of the inviscid flow over a wedge for $x = 0$.
18.   Derive an expression for the boundary-layer thickness for Falkner–Skan flows. Reduce for Heimenz flow.
19.   Construct a thin-boundary-layer model for the flow of a power-law fluid over a wedge. Determine what angles, if any, permit reduction of the model to an ordinary differential equation. Carry out the reduction and compare with the Falkner–Skan problem (see Lee and Ames [12]).
20.   Generalize the Schlichting–Bussmann solution for suction in wedge flow.
21.   Compare the values in Table 14.3 with those for $\beta = 0.5$ in Table 14.1.

22. Derive additional terms in Equation 14.36 using the values in Table 14.3 until 14.37 and 14.38 provide a complete approximation.

23. Determine $\delta_D \mathrm{Re}_x^{1/2}/x$, $\delta_M \mathrm{Re}_x^{1/2}/x$, δ_D/δ, and δ_M/δ for each of the velocity fields of problem 2 in Chapter 12. Here, δ is the value defined by Equation 13.41.

24. Repeat problem 23 for the velocity distributions in Table 12.1.

25. Repeat problem 23 for the Blasius solution for a flat plate and compare with the results of problem 23.

26. Prepare a sketch of the profiles of $u_x\{y\}$ for various x, and of $u_y\{x\}$ for various y for Homann flow.

27. Show that Heimenz solution is independent of the simplifications of thin-boundary-layer theory.

28. Develop a model for two planar impinging flows of the same gas. Reduce this model to a set of ordinary differential equations and boundary conditions.

29. Repeat problem 28 for impinging axisymmetric flows.

REFERENCES

1. V. M. Falkner and S. W. Skan, "Some Approximate Solutions of the Boundary Layer Equations," *Aero. Res. Council, Gt. Brit.*, R. & M. 1314 (1930); *Phil. Mag.*, *12* (1931) 865.

2. D. R. Hartree, "On an Equation Occurring in Falkner and Skan's Approximate Treatment of the Equations of the Boundary Layer," *Proc. Camb. Phil. Soc.*, *33* (1937) 223.

3. H. G. Knudsen and D. L. Katz, *Fluid Dynamics and Heat Transfer*, McGraw-Hill, New York (1958).

4. N. L. Evans, *Laminar Boundary Layer Theory*, Addison-Wesley, Reading, MA (1968).

5. E. R. G. Eckert and R.M. Drake, Jr., *The Analysis of Heat and Mass Transfer*, McGraw-Hill, New York (1972).

6. K. Heimenz, "Die Grenzschicht an einem in den gleichförmigen Flüssigkeitsstrom eigentauchten geraden Kreiszylinder," [Thesis, Göttingen (1911)], *Dingler's Polytech. J.*, *326* (1911) 326.

7. W. Howarth, "On the Calculation of the Steady Flow in the Laminar Boundary Layer near the Surface of a Cylinder in a Stream," *Aero Res. Council, Gt. Brit.*, R. & M. 1632 (1935).

8. T. Yuge, "Theory of Distributions of the Coefficients of Heat Transfer of Spheres," *Repts. Inst. High Speed Mech.*, *Tohoku Univ.*, 6, (1956) 115.

9. S. Goldstein, Ed., *Modern Developments in Fluid Dynamics*, Vol. 2, Oxford University Press, Clarendon (1938).

10. F. Homann, "Der Einfluss grosser Zähigkeit bei der Strömung um den Zylinder und um die Kugel," *Z. Angew. Math. Mech.*, *16* (1936) 153 and *Forsch. Gebiete Ingenieurow.*, *7* (1936) 1; English transl., "The Effect of High Viscosity on the Flow around a Cylinder and around a Sphere," *NACA* TM 1334, Washington, D.C. (1952).

11. H. Schlichting, *Boundary Layer Theory*, 7th ed., English transl. by J. Kestin, McGraw-Hill, New York (1979).

12. S. Y. Lee and W. F. Ames, "Similarity Solutions for Non-Newtonian Fluids," *AIChE J.*, *12* (1966) 700.

Chapter 15

Laminar Flow over a Circular Cylinder

Flow over a cylinder involves many regimes. Therefore a general description is helpful in advance of the derivations and for interpretation of solutions and data for the individual regimes.

If inviscid flow actually occurred, the streamlines would be symmetrical fore and aft, as in Figure 10–29. This photograph is of the streaklines of dye in water flowing at 1 mm/s between glass plates 1 mm apart. These conditions produce a close approximation of the idealized case of *Hele-Shaw flow*, for which the streamlines are the same as the hypothetical ones of inviscid flow.

At very low Reynolds number, $Re_D = Du_\infty/\nu$, where D is the diameter of the cylinder, inertial forces are small but not negligible relative to viscous forces. As illustrated in Figure 15–1 by the stream functions computed by Sucker and Brauer [1] for $Re_D = 1.0$, and in Figure 15–2A by a photograph of Taneda (Van Dyke [2]) of actual streaklines at $Re_D = 1.54$, the effect of this level of inertia is to extend the streamlines slightly outward and to the rear.

As Re_D increases above 6, *separation* occurs at the rear and a circulating *wake* is formed, as illustrated in Figure 15–2B for $Re_D = 9.6$. As the Reynolds number increases, the point of separation gradually moves forward. At $Re_D \cong 1.2 \times 10^4$ the point of separation attains a maximum advancement of about $99\pi/180$ rad ($81\pi/180$ rad from the forward point of incidence) and remains fixed until $Re_D \gtrsim 1.2 \times 10^5$. The exact point of separation depends on secondary variables such as the intensity and scale of free-stream turbulence, surface roughness, and the length of the cylinder. Nevertheless, as illustrated subsequently, a reasonable correlation is attained for this point and for the closely related point of attachment of the wake.

Up to $Re_D \cong 44$, the wake consists of two symmetrical and stationary eddies or vortices, as shown in Figure 15–3 by the stream functions computed by Sucker and Brauer [1] for $Re_D = 30$ and in Figure 15–2B, C, and D by the photographs of actual streaklines for $Re_D = 9.6$, 13.1, and 26. The size of the eddies is seen in these photographs to increase with the Reynolds number. Beginning at $Re_D \cong 44$, the vortices are shed alternately and periodically. For $60 \lesssim Re_D \lesssim 150$ a very structured pattern, called a *Kármán vortex street*, is formed behind the cylinder. For the lower Reynolds numbers in this range (as illustrated in Figure 15–4A for $Re_D = 105$), the vortices retain the same size and spacing after being shed. For slightly larger Re_D, as illustrated in Figure 15–4B for $Re_D = 140$, the vortices grow in size and spacing. As illustrated in Figure

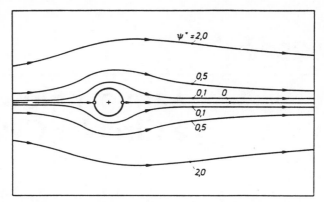

FIGURE 15–1   *Computed stream functions for flow over a cylinder at $Re_D = 1$. $\psi^* = 2\psi/Du_\infty$. (From Sucker and Brauer [1].)*

A

FIGURE 15–2   *Photographs by Taneda of streamlines for flow of water over a cylinder (from right to left): (A) $Re_D = 1.54$; (B) $Re_D = 9.6$; (C) $Re_D = 13.1$; (D) $Re_D = 26$. (From Van Dyke [2], pp. 20, 28.)*

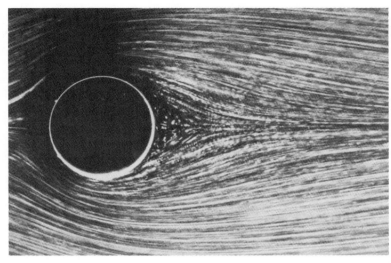

B

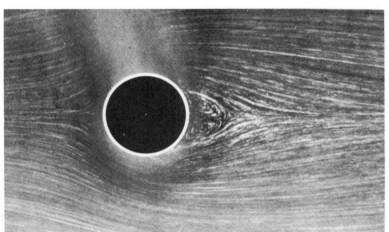

C

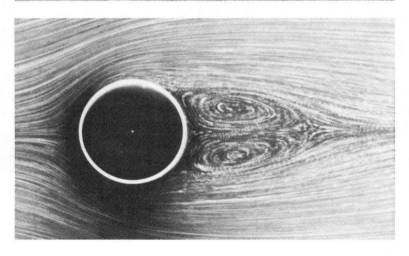

D

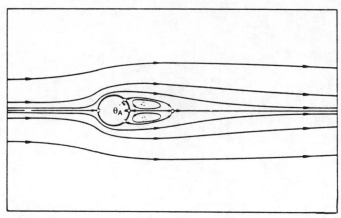

FIGURE 15–3   *Computed stream functions for flow over a cylinder at $Re_D = 30$. (From Sucker and Brauer [1].)*

15–5 by a photograph from Werlé [3], and in Figure 15–6 from Kaufmann [4], p. 220, by the theoretical structure derived by von Kármán [5], the flow pattern within the vortices is very structured. For higher Re_D (up to about 3×10^5), as illustrated in Figure 15–7 for $Re_D = 10^4$, the vortices are shed at a constant frequency but are quickly obliterated by turbulent mixing.

A

FIGURE 15–4   *Photographs by Taneda of von Kármán vortex sheet behind a cylinder: (A) $Re_D = 105$; (B) $Re_D = 150$. (From Van Dyke [2], pp. 57, 56).*

When Re_D attains about 1.2×10^5, the point of separation of the laminar boundary layer starts moving backward. At $\mathrm{Re}_D \cong 3 \times 10^5$ it reaches $110\pi/180$ rad from the front, and the free shear layer behind the point of separation becomes turbulent and reattaches to the surface as a turbulent boundary layer, which in turn separates at about $140\pi/180$ rad. This transition is accompanied by a sharp increase in the frequency at which vortices are shed. At $\mathrm{Re}_D \cong 2 \times 10^6$ the frequency of vortex shedding decreases and becomes irregular.

The total drag on a cylinder is made up of skin friction and the net force due to pressure. The drag due to pressure is equal to that of skin friction at very low Re_D but becomes dominant as Re_D increases. Hence, results for the pressure distribution over the cylinder are necessary to complement boundary-layer solutions.

Theoretical solutions and supplementary empirical equations for these various regimes are examined here.

INVISCID FLOW

The solution for inviscid flow around a cylinder does not provide any direct information on the skin friction. However it does provide an asymptotic expression for the pressure distribution over the forward portion of the cylinder and an approximation for the external velocity field for the boundary-layer solutions.

The solution for the Lagrange stream function of this flow (see Section 10.S6) is

B

FIGURE 15–5   *Photograph by Wille of von Kármán vortices behind a cylinder with the camera fixed with respect to the vortices. (From Werlé [3].)*

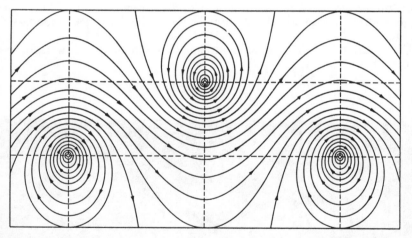

FIGURE 15–6   *Streamlines in a vortex street behind a cylinder as calculated by von Kármán [5]. (From Kaufmann [4], p. 220.)*

$$\psi = -u_\infty \left(r - \frac{a^2}{r} \right) \sin\{\theta\} \tag{15.1}$$

where u_∞ = free-stream velocity, m/s
   r = radial distance from the center of the cylinder, m
   a = radius of cylinder, m
   θ = *angle measured around cylinder from rear, rad*

The components of the velocity are then

$$u_r = -\frac{1}{r}\frac{\partial\psi}{\partial\theta} = u_\infty\left(1 - \frac{a^2}{r^2}\right)\cos\{\theta\} \tag{15.2}$$

and

$$u_\theta = \frac{\partial\psi}{\partial r} = -u_\infty\left(1 + \frac{a^2}{r^2}\right)\sin\{\theta\} \tag{15.3}$$

At the surface $(r = a)$, $u_r\{\theta, a\} = 0$ and

$$u_\theta\{\theta, a\} = -2u_\infty\sin\{\theta\} \tag{15.4}$$

The pressure distribution along the surface is then found from Bernoulli's law and Equation 15.4 to be

$$\mathscr{P}\{\theta, a\} = \mathscr{P}\{\pi, a\} - 2u_\infty^2\varrho\sin^2\{\theta\} \tag{15.5}$$

Since for inviscid flow the stagnation pressure (i.e., the pressure attained by completely stopping the flow) is

$$\mathscr{P}_s = \mathscr{P}\{\pi, a\} = \mathscr{P}_\infty + \frac{u_\infty^2\varrho}{2} \tag{15.6}$$

Equation 15.5 can be rewritten as

$$\mathscr{P}\{\theta, a\} - \mathscr{P}_\infty = \frac{u_\infty^2\varrho}{2}(1 - 4\sin^2\{\theta\}) \tag{15.7}$$

Equations 15.5–15.7 are reproduced from Chapter 10, but are renumbered for convenience.

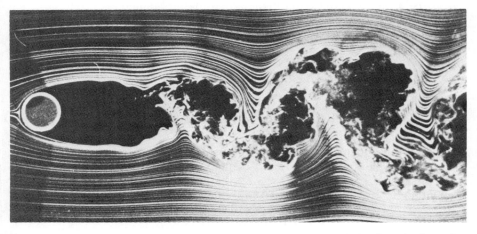

FIGURE 15–7   *Photograph by Corke and Nagib of streaklines behind a cylinder at $Re_D = 10^4$. (From Van Dyke [2], p. 31.)*

CREEPING FLOW

In 1851 Stokes [6] proved that a solution for creeping flow was not possible for a cylinder; that is, inertial effects are appreciable even in the limit as Re_D approaches zero. This is in contrast with the situation for the sphere for which he derived just such a solution.

SLIGHTLY INERTIAL (LAMB) FLOW

Lamb [7], p. 616, used a linear approximation for the inertial terms, analogous to that used by Oseen [8] for spheres, to obtain a first-order solution for slightly inertial flow. From this solution, he noted that $\bar{C}_f = \bar{C}_p = \bar{C}_t/2$, and that the total drag coefficient is

$$\bar{C}_t = \frac{4\pi}{S' Re_D} \tag{15.8}$$

where

$$S' = \frac{1}{2} - \gamma - \ln\left\{\frac{Re_D}{8}\right\} \cong 2.0022 - \ln Re_D \cong \ln\left\{\frac{7.4055}{Re_D}\right\} \tag{15.9}$$

and $\gamma = 0.57722\ldots$ is Euler's constant.

The various drag coefficients conventionally and here are based on the projected area, A_p, rather than on the surface area. Thus, for the total drag on a cylinder

$$\bar{C}_t = \frac{F_t}{A_p \varrho u_\infty^2} = \frac{4F_t}{\pi D^2 \varrho u_\infty^2} \tag{15.10}$$

Higher-order terms have been derived by various investigators (see Van Dyke [9], p. 161). For example Tomotika and Aoi [10] obtained

$$\bar{C}_t = \frac{4\pi}{Re_D S'}\left[1 - \left(S' - \frac{1}{2} + \frac{5}{16S'}\right)\frac{Re_D^2}{32} + \cdots\right] \tag{15.11}$$

whereas Kaplun [11] obtained

$$\bar{C}_t = \frac{4\pi}{Re_D S'}\left(1 - \frac{0.87}{(S')^2} + \mathscr{O}\left\{\frac{1}{(S')^3}\right\}\right) \tag{15.12}$$

As shown subsequently, Equation 15.11 provides an upper bound, and 15.12 a lower bound. Both reduce to Lamb's equation as $Re_D \to 0$.

The pressure $\mathscr{P}\{\pi, a\} = \mathscr{P}_0$ at the forward point of incidence, corresponding to Equation 15.8, is given by the expression

$$\frac{2(\mathscr{P}_0 - \mathscr{P}_\infty)}{\varrho u_\infty^2} = \frac{1 + 8/Re_D}{S'} \tag{15.13}$$

which becomes singular because of S' as Re_D increases.

BOUNDARY-LAYER FLOW

Differential Model

In addition to the usual postulates, the boundary layer will be presumed to be thin relative to the radius of the cylinder, which allows the curvature to be neglected. The mass and force-momentum balances are again represented by Equations 11.1 and 14.2, with y the radial distance from the surface and x the distance along the surface from the forward point of incidence of the flow, i.e., in the negative θ-direction, such that $x = a(\pi - \theta)$.

In this coordinate system the distribution of the velocity, $U\{x\}$, outside the boundary layer can be adapted from Equation 15.4 as

$$U\{x\} = 2u_\infty \sin\left\{\frac{x}{a}\right\} \tag{15.14}$$

Substitution in Equation 14.2 gives

$$u_x \frac{\partial u_x}{\partial x} + u_y \frac{\partial u_x}{\partial y} = \frac{2u_\infty^2}{a} \sin\left\{\frac{2x}{a}\right\} + v \frac{\partial^2 u_x}{\partial y^2} \tag{15.15}$$

Solution for Region near Point of Incidence

Near the forward point of the cylinder, Equation 15.14 can be approximated by

$$U_x = 2u_\infty \left[\frac{x}{a} - \frac{1}{3!}\left(\frac{x}{a}\right)^3 + \frac{1}{5!}\left(\frac{x}{a}\right)^5 + \cdots\right] \Rightarrow \frac{2u_\infty x}{a} \tag{15.16}$$

and, hence, Equation 15.15 by

$$u_x \frac{\partial u_x}{\partial x} + u_y \frac{\partial u_x}{\partial y} = \frac{4u_\infty^2 x}{a^2} + v \frac{\partial^2 u_x}{\partial y^2} \tag{15.17}$$

Equation 15.17 is equivalent to 14.4 with $m = 1$ and $c = 2u_\infty/a$. Hence the solution is given by Table 14.1 for $m = 1$, by Table 14.2, and also by Equations 14.20, 14.21, and 14.23, where

$$\eta = y \sqrt{\frac{2u_\infty}{av}} \tag{15.18}$$

and

$$f\{\eta\} = -\sqrt{\frac{a}{2u_\infty v}} \frac{\psi}{x} \tag{15.19}$$

The components of the velocity become

$$u_x = -u_\theta = -\frac{\partial \psi}{\partial y} = \frac{2u_\infty x}{a} f'\{\eta\} = 2u_\infty (\pi - \theta) f'\{\eta\} \tag{15.20}$$

and

$$u_y = u_r = \frac{\partial \psi}{\partial x} = -\sqrt{\frac{2u_\infty \nu}{a}} f\{\eta\} \tag{15.21}$$

and the local shear stress is

$$\frac{\tau_w}{\varrho u_\infty^2} \sqrt{Re_D} = \frac{4x}{a} f''\{0\} = 4(\pi - \theta) f''\{0\} = 4.9304 (\pi - \theta) \tag{15.22}$$

Following Grove et al. [12], an expression for the pressure at the point of incidence can be derived as follows in x, y coordinates. Along the streamline at $x = 0$, $\partial u_y / \partial x = \partial^2 u_y / \partial x^2 = 0$. Hence,

$$u_y \frac{\partial u_y}{\partial y} = -\frac{1}{\varrho} \frac{\partial \mathscr{P}}{\partial y} + \nu \frac{\partial^2 u_y}{\partial y^2} \tag{15.23}$$

Integrating with respect to y from 0 to ∞ gives

$$\frac{u_\infty^2}{2} = \frac{\mathscr{P}\{0, 0\} - \mathscr{P}_\infty}{\varrho} + \nu \int_0^\infty \frac{\partial^2 u_y}{\partial y^2} dy \tag{15.24}$$

The integrand is negligible outside the edge of the boundary layer at $y = \delta$. Hence

$$\frac{2(\mathscr{P}\{0, 0\} - \mathscr{P}_\infty)}{\varrho u_\infty^2} = 1 - \frac{\nu}{u_\infty^2} \left[\left(\frac{\partial u_y}{\partial y} \right)_{y=\delta} - \left(\frac{\partial u_y}{\partial y} \right)_{y=0} \right] \tag{15.25}$$

The latter term is zero, and from Equations 15.16 and 11.1,

$$\left(\frac{\partial u_y}{\partial y} \right)_{y=\delta} \cong -\frac{2u_\infty}{a}.$$

Therefore

$$\frac{2(\mathscr{P}\{0, 0\} - \mathscr{P}_\infty)}{\varrho u_\infty^2} = 1 + \frac{4\nu}{au_\infty} = 1 + \frac{8}{Re_D} \tag{15.26}$$

This expression indicates that $\mathscr{P}_o \equiv \mathscr{P}\{x = 0, y = 0\} \equiv \mathscr{P}\{\theta = \pi, r = a\}$, which is the dynamic pressure at the point of incidence, approaches $\mathscr{P}_s \equiv \varrho u_\infty^2/2$ as $Re_D \to \infty$.

The accuracy of the foregoing approximations for finite angles can be extended somewhat by reversing the substitution of θ for $\sin \theta$, to give

$$u_x = -u_\theta \cong 2u_\infty \sin\{\theta\} f'\{\eta\} \tag{15.27}$$

$$u_y = u_r = \sqrt{\frac{2u_\infty \nu}{a}} \cos\{\theta\} f\{\eta\} \tag{15.28}$$

and

$$\frac{\tau_w}{\varrho u_\infty^2}\sqrt{\text{Re}_D} \cong 4\sin\{\theta\}f''\{0\} = 4.9304\sin\{\theta\} \tag{15.29}$$

Also, in consideration of Equation 15.7,

$$\frac{2(\mathscr{P}\{\theta, a\} - \mathscr{P}_\infty)}{\varrho u_\infty^2} \cong (1 - 4\sin^2\{\theta\})\left(1 + \frac{8}{\text{Re}_D}\right) \tag{15.30}$$

Homann [13] took the "displacement thickness" on the front of the cylinder into account and derived the following improvement on Equation 15.26:

$$\frac{2(\mathscr{P}_0 - \mathscr{P}_\infty)}{\varrho u_\infty^2} = 1 + \frac{8}{\text{Re}_D + 0.646\sqrt{\text{Re}_D}} \tag{15.31}$$

General Solutions

A similarity transformation is not possible for Equations 11.1 and 15.15. However, Blasius [14] developed a series solution by postulating the following representation for the stream function:

$$\psi = -\sqrt{\frac{v}{a_1}}\Big(a_1 \times f_1\{\eta\} + 4a_3 x^3 f_3\{\eta\} + 6a_5 x^5 f_5\{\eta\}$$

$$+ 8a_7 x^7 f_7\{\eta\} + \cdots\Big) \tag{15.32}$$

Here $\eta = y\sqrt{a_1/v}$, and the coefficients a_i are given by the following series expansion for the velocity just outside the boundary layer:

$$U_x\{x\} = a_1 x + a_3 x^3 + a_5 x^5 + \cdots \tag{15.33}$$

Substituting in Equations 11.1 and 14.2 for u_x and u_y from Equation 15.32, and for $U_x\{x\}$ from Equation 15.33, and collecting terms with the same power of x yield a set of simultaneous differential equations beginning with

$$f_1''' + f_1 f_1'' - f'^2 + 1 = 0 \tag{15.34A}$$

and

$$f_3''' + 3f_3' f_1'' + f_1 f_3'' - 4f_1' f_3' + 1 = 0 \tag{15.34B}$$

The functions $f_1\{\eta\}$ and $f_3\{\eta\}$ are obviously independent of the coefficients a_i, but the higher-order functions do depend on a_i. Howarth [15] succeeded in choosing a linearly related set of functions that are independent of a_i. These relationships are

$$f_5 = g_5 + \frac{a_3^2}{a_1 a_5} h_5 \tag{15.35}$$

$$f_7 = g_7 + \frac{a_3 a_5}{a_1 a_7} h_7 + \frac{a_3^3}{a_1^2 a_7} k_7 \tag{15.36}$$

$$f_9 = g_9 + \frac{a_3 a_7}{a_1 a_9} h_9 + \frac{a_5^2}{a_1 a_9} k_9 + \frac{a_3^2 a_5}{a_1^2 a_9} j_9 + \frac{a_3^4}{a_1^3 a_9} q_9 \tag{15.37}$$

$$f_{11} = g_{11} + \frac{a_3 a_9}{a_1 a_{11}} h_{11} + \frac{a_5 a_7}{a_1 a_{11}} k_{11} + \frac{a_3^2 a_7}{a_1^2 a_{11}} j_{11} + \frac{a_3 a_5^2}{a_1^2 a_{11}} q_{11}$$

$$+ \frac{a_3^3 a_5}{a_1^3 a_{11}} m_{11} + \frac{a_3^5}{a_1^4 a_{11}} n_{11} \tag{15.38}$$

with g_i, h_i, k_i, and so on, the functions independent of a_i. The computed values of Tifford [16] (according to Schlichting [17], p. 150) for the derivatives of these functions, which are used directly to compute u, are given in Table 15.1.

If Equation 15.14 is postulated for the velocity outside the boundary layer,

$$a_1 = \frac{2u_\infty}{a}; \quad a_3 = -\frac{2}{3!} \frac{u_\infty}{a^3}; \quad a_5 = \frac{2}{5!} \frac{u_\infty}{a^5};$$

$$\dots a_n = \frac{2(-1)^{(n+1)/2}}{n!} \frac{u_\infty}{a^n} \tag{15.39}$$

Then,

$$\frac{\tau_w}{\varrho u_\infty^2} \sqrt{\mathrm{Re}_D} = 4 \left[\frac{x}{a} f_1''\{0\} - \frac{4}{3!} \left(\frac{x}{a}\right)^3 f_3''\{0\} \right.$$

$$\left. + \frac{6}{5!} \left(\frac{x}{a}\right)^5 f_5''\{0\} + \cdots \right] \tag{15.40}$$

Using the functions computed by Tifford gives

$$\frac{\tau_w}{\varrho u_\infty^2} \sqrt{\mathrm{Re}_D} = 4.9304 \frac{x}{a} - 1.9317 \left(\frac{x}{a}\right)^3 + 0.2064 \left(\frac{x}{a}\right)^5 - 0.012\,93 \left(\frac{x}{a}\right)^7$$

$$+ 3.04 \times 10^{-5} \left(\frac{x}{a}\right)^9 - 8.13 \times 10^{-5} \left(\frac{x}{a}\right)^{11} + \cdots \tag{15.41}$$

The corresponding point of zero shear stress can be calculated by equating $(\tau_w/\varrho u_\infty^2)\sqrt{\mathrm{Re}_D}$ to zero and solving for x/a. The result is $x/a = 1.8984$ or $\theta = 71.2\pi/180$ rad. If the shear stress in the wake is assumed to be negligible, Equation 15.41 can be used to calculate the mean coefficient for drag due to skin friction based on the projected area:

$$\bar{C}_f \sqrt{\mathrm{Re}_D} = \int_0^{1.8984} \left(\frac{\tau_w}{\varrho u_\infty^2}\right) \sqrt{\mathrm{Re}_D} \sin\left\{\frac{x}{a}\right\} d\left(\frac{x}{a}\right) = 2.9921 \tag{15.42}$$

Table 15.1
Derivatives of Coefficients of the First Six Terms of the Blasius Series for a Cylinder (Equation 15.32) as Computed by Tifford [16] according to Schlichting [17], p. 150

η	f'_1	f'_3	g'_5	h'_5	g'_7	h'_7	k'_7	g'_9	h'_9	k'_9	j'_9	q'_9	g'_{11}	h'_{11}	k'_{11}	j_{11}	q_{11}	m'_{11}	n'_{11}
0	0	0	0	0	0	0	0	0	0	0	0	0	0	0	0	0	0	0	0
0.2	0.2266	0.1251	0.1072	0.0141	0.0962	0.0173	0.0016	0.0884	0.0112	0.0019	0.0125	-0.0062	0.0824	0.0074	-0.0041	0.0165	0.0237	-0.0359	0.0103
0.4	0.4145	0.2129	0.1778	0.0117	0.1563	0.0030	0.0044	0.1413	-0.0079	-0.0112	0.0288	-0.0124	0.1299	-0.0145	-0.0368	0.0371	0.0514	-0.0721	0.0206
0.6	0.5663	0.2688	0.2184	-0.0011	0.1879	-0.0286	0.0096	0.1669	-0.0417	-0.0311	0.0525	-0.0190	0.1511	-0.0489	-0.0799	0.0651	0.0863	-0.1095	0.0307
0.8	0.6859	0.2997	0.2366	-0.0177	0.1994	-0.0637	0.0174	0.1740	-0.0760	-0.0501	0.0833	-0.0262	0.1553	-0.0816	-0.1181	0.0992	0.1267	-0.1489	0.0406
1.0	0.7779	0.3125	0.2399	-0.0331	0.1980	-0.0925	0.0271	0.1700	-0.1019	-0.0639	0.1171	-0.0341	0.1498	-0.1046	-0.1428	0.1342	0.1666	-0.1895	0.0503
1.2	0.8467	0.3133	0.2341	-0.0442	0.1896	-0.1102	0.0369	0.1604	-0.1157	-0.0704	0.1480	-0.0423	0.1397	-0.1152	-0.1520	0.1641	0.1995	-0.2288	0.0595
1.4	0.8968	0.3070	0.2239	-0.0499	0.1782	-0.1159	0.0452	0.1489	-0.1176	-0.0703	0.1710	-0.0502	0.1283	-0.1146	-0.1474	0.1841	0.2202	-0.2627	0.0678
1.6	0.9323	0.2975	0.2123	-0.0504	0.1665	-0.1114	0.0506	0.1375	-0.1101	-0.0649	0.1829	-0.0567	0.1175	-0.1055	-0.1330	0.1919	0.2266	-0.2874	0.0747
1.8	0.9568	0.2871	0.2012	-0.0468	0.1558	-0.0997	0.0525	0.1276	-0.0965	-0.0562	0.1827	-0.0610	0.1083	-0.0911	-0.1131	0.1876	0.2190	-0.2996	0.0796
2.0	0.9732	0.2775	0.1916	-0.0406	0.1469	-0.0839	0.0510	0.1195	-0.0798	-0.0460	0.1718	-0.0625	0.1008	-0.0746	-0.0912	0.1731	0.2001	-0.2982	0.0822
2.2	0.9839	0.2695	0.1839	-0.0332	0.1400	-0.0669	0.0466	0.1132	-0.0628	-0.0359	0.1528	-0.0610	0.0951	-0.0581	-0.0701	0.1515	0.1737	-0.2834	0.0820
2.4	0.9905	0.2632	0.1781	-0.0257	0.1349	-0.0507	0.0402	0.1087	-0.0470	-0.0267	0.1290	-0.0568	0.0910	-0.0432	-0.0516	0.1263	0.1436	-0.2575	0.0790
2.6	0.9946	0.2586	0.1740	-0.0189	0.1313	-0.0367	0.0330	0.1055	-0.0337	-0.0190	0.1037	-0.0504	0.0882	-0.0308	-0.0364	0.1003	0.1133	-0.2237	0.0733
2.8	0.9970	0.2554	0.1712	-0.0133	0.1288	-0.0254	0.0257	0.1033	-0.0231	-0.0129	0.0795	-0.0426	0.0863	-0.0210	-0.0246	0.0761	0.0854	-0.1858	0.0655
3.0	0.9984	0.2532	0.1694	-0.0089	0.1273	-0.0168	0.0191	0.1020	-0.0152	-0.0085	0.0581	-0.0344	0.0851	-0.0138	-0.0160	0.0552	0.0616	-0.1476	0.0563
3.2	0.9992	0.2519	0.1682	-0.0057	0.1263	-0.0107	0.0135	0.1011	-0.0096	-0.0053	0.0406	-0.0265	0.0843	-0.0087	-0.0100	0.0383	0.0425	-0.1121	0.0463
3.4	0.9996	0.2510	0.1675	-0.0035	0.1257	-0.0065	0.0091	0.1006	-0.0058	-0.0032	0.0271	-0.0195	0.0839	-0.0052	-0.0060	0.0255	0.0251	-0.0814	0.0365
3.6	0.9998	0.2506	0.1671	-0.0021	0.1254	-0.0038	0.0059	0.1003	-0.0034	-0.0019	0.0173	-0.0137	0.0836	-0.0030	-0.0036	0.0162	0.0178	-0.0565	0.0275
3.8	0.9999	0.2503	0.1669	-0.0012	0.1252	-0.0021	0.0036	0.1002	-0.0019	-0.0010	0.0106	-0.0092	0.0835	-0.0017	-0.0019	0.0099	0.0108	-0.0375	0.0199
4.0	1.0000	0.2501	0.1668	-0.0006	0.1251	-0.0011	0.0021	0.1001	-0.0010	-0.0006	0.0062	-0.0059	0.0834	-0.0009	-0.0010	0.0057	0.0063	-0.0238	0.0137
η	f''_1	f''_3	g''_5	h''_5	g''_7	h''_7	k''_7	g''_9	h''_9	k''_9	j''_9	q''_9	g''_{11}	h''_{11}	k''_{11}	j''_{11}	q''_{11}	m''_{11}	n''_{11}
0	1.2326	0.7244	0.6347	0.1192	0.5792	0.1829	0.0076	0.5399	0.1520	0.0572	0.0607	-0.0308	0.5100	0.1323	0.0742	0.0806	0.1164	-0.1796	0.0516

Table 15.2
Local Shear Stress on a Cylinder as Computed by Schönauer [18] for the
Thin-Boundary-Layer Regime

x/a	0.3	0.5	0.8	1.0	1.2	1.5	1.6
$(\tau_w/\varrho u_\infty^2)\sqrt{Re_D}$	1.428	2.230	3.019	3.190	3.042	2.204	1.754
x/a	1.7	1.8	1.8234				
$(\tau_w/\varrho u_\infty^2)\sqrt{Re_D}$	1.196	0.418	0				

Reprinted by permission of the publisher.

Table 15.3
Computed Values of Evans [19], p. 186, for the Local Shear Stress on a Circular Cylinder

θ'	0	0.20	0.40	0.60	0.80	0.90	1.00	1.10
$(\tau_w/\varrho u_\infty^2)\sqrt{Re_D}$	0	0.838	1.572	2.277	2.360	2.361	2.284	2.216

Schönauer [18] solved this same model numerically and obtained the values in
Table 15.2 for $(\tau_w/\varrho u_\infty^2)\sqrt{Re_D}$. These values differ only slightly from those of
Equation 15.41.

Evans [19] (see Chapter 18) used the method of Merk together with
the following empirical velocity of Sogin and Subramanian [20] for $Re_D = 1.22 \times 10^5$,

$$\frac{U_{\theta'}}{u_\infty} = 1.82\theta' - 0.4(\theta')^3 \qquad (15.43)$$

to solve Equations 11.1 and 14.2. His results, which are summarized in Table
15.3, differ significantly from those in Table 15.2, presumably due to the
difference in Equations 15.14 and 15.43 rather than to the methodology.

Wake Theory

Attempts have been made to develop a theoretical solution for the boundaries of
the velocity field and, more importantly, the pressure distribution in the wake.
However, the closed-form results obtained to date are either inaccurate or
require the introduction of some empiricism. (See, for example, Goldstein [21].)

Since the thin-boundary-layer solutions do not cover the region of the wake
behind a cylinder, numerical solutions provide the only reliable theoretical
values for that region and, hence, for $\bar{C}_t$ beyond the upper limits of Equations
15.8, 15.11, and 15.12.

NUMERICAL SOLUTIONS

The first successful numerical solution for flow around a cylinder was by Thom
[22] in 1928[1] for $Re_D = 10$ and 20. With the development of computing facilities
and the improvement of numerical techniques, reliable solutions have since been

---

[1] This appears to be the earliest finite-difference solution for any two-dimensional velocity field.

Table 15.4
Numerically Computed Characteristics for Flow over a Circular Cylinder at $Re_D = 40$

Year	Investigator	$\bar{C}_f$	$\bar{C}_p$	$\bar{C}_t$	θ_S (deg)
1953	Kawaguti [23]	0.283	0.526	0.809	52.5
1961	Apelt [24]	0.284	0.464	0.748	50.0
1966	Kawaguti and Jain [25]	0.264	0.501	0.765	53.7
1969	Jain and Rao [26]	0.269	0.526	0.795	54.2
1969	Son and Hanratty [27]	0.257	0.498	0.755	
1969	Takami and Keller [28]			0.768	
1970	Dennis and Chang [29]	0.262	0.499	0.761	53.8
1973	Nieuwstadt and Keller [30]			0.775	
1975	Sucker and Brauer [1]	0.278	0.538	0.817	51.9

obtained for Re_D up to 44. Results of less certainty have been obtained even for Re_D up to 3×10^5.

The results for $Re_D \leq 40$ have generally been computed from a steady-state model, since the shedding of vortices is known to begin at $Re_D \cong 44$. Some steady-state calculations have been carried out for Re_D up to 500. Since the shedding of the vortices can be prevented by the installation of a plate behind the cylinder in the plane formed by the axis of the cylinder and the incident direction of flow, the steady-state calculations have physical meaning in that sense. In any event, the computed velocity field, pressure distribution, and skin friction ahead of the wake would be expected to be relatively independent of the behavior in the wake.

Computed values for $\bar{C}_f$, $\bar{C}_p$, and $\bar{C}_t$ for $Re_D = 40$, using a variety of numerical methods, are compared in Table 15.4. Values of θ_S, the point of separation of the boundary layer from the surface of the cylinder, are included. All of these values are in reasonable agreement, but which are the most accurate is not certain because of the lack of an absolute standard.

Various characteristic quantities, including $\bar{C}_f$, $\bar{C}_p$, $\bar{C}_t$, θ_S, l_w/a (the dimensionless length of the wake), and the dimensionless pressures at the forward point of incidence ($\theta = \pi$) and at the rear ($\theta = 0$) are listed in Tables 15.5–15.9.

The non-steady-state calculations of Thoman and Szewczyk [31], which extend to $Re_D = 3 \times 10^5$, include the prediction of the frequency of eddy shedding and the structure of the vortex trail, but are not tabulated and hence are not reproduced here.

The various computed values are subsequently compared with and shown to be in reasonable agreement with experimental data. Hence the entire gamut of subcritical behavior now appears to be within the grasp of numerical integration.

Table 15.5
Mean Coefficients Computed Numerically by Son and Hanratty [27] for Flow over a Circular Cylinder

Re_D	$\bar{C}_f$	$\bar{C}_p$	$\bar{C}_t$
40	0.257	0.498	0.755
200	0.095	0.367	0.462
500	0.045	0.255	0.300

Table 15.6
Characteristic Functions Computed Numerically by Takami and Keller [28] for Flow over a Circular Cylinder

Re_D	$\dfrac{2(\mathscr{P}_\infty - \mathscr{P}_0)}{\varrho u_\infty^2}$	$\dfrac{2(\mathscr{P}_\pi - \mathscr{P}_\infty)}{\varrho u_\infty^2}$	$\bar{C}_t$	θ_S (deg)	$\dfrac{l_w}{a}$
1	2.719	3.905	5.415		
2	1.652	2.715	3.319		
4	1.057	2.000	2.218		
6	0.848	1.723	1.783		
7	0.783	1.637	1.645	14.5	0.115
10	0.670	1.474	1.377	29.3	0.500
15	0.582	1.336	1.133	38.7	1.178
20	0.537	1.261	1.001	43.65	1.870
30	0.530	1.184	0.858	49.6	3.223
40	0.512	1.141	0.768	53.55	4.650
50	0.499	1.114	0.709	56.6	6.10
60	0.491	1.096	0.662	59.0	7.53

Table 15.7
Characteristic Functions Computed Numerically by Dennis and Chang [29] for Flow over a Circular Cylinder

Re_D	$\dfrac{2(\mathscr{P}_\infty - \mathscr{P}_0)}{\varrho u_\infty^2}$	$\dfrac{2(\mathscr{P}_\pi - \mathscr{P}_\infty)}{\varrho u_\infty^2}$	$\bar{C}_f$	$\bar{C}_p$	$\bar{C}_t$	θ_S (deg)	$\dfrac{l_w}{a}$
5	1.044	1.872	0.9585	1.0995	2.058		
7	0.870	1.660	0.7765	0.934	1.7105	15.9	0.19
10	0.742	1.489	0.623	0.800	1.423	29.6	0.53
20	0.589	1.269	0.406	0.6165	1.0225	43.7	1.88
40	0.509	1.144	0.262	0.499	0.761	53.8	4.69
70	0.439	1.085	0.180	0.426	0.606	61.3	8.67
100	0.393	1.060	0.141	0.387	0.528	66.2	13.11

Table 15.8
Characteristic Functions Computed Numerically by Nieuwstadt and Keller [30] for Flow over a Circular Cylinder

Re_D	$\dfrac{2(\mathscr{P}_\infty - \mathscr{P}_0)}{\varrho u_\infty^2}$	$\dfrac{2(\mathscr{P}_\pi - \mathscr{P}_\infty)}{\varrho u_\infty^2}$	$\bar{C}_t$	θ_S (deg)	$\dfrac{l_w}{a}$
1	2.928	3.73	5.16		
7	0.9316	1.595	1.707		
10	0.6921	1.500	1.414	28.0	0.434
20	0.5817	1.274	1.026	43.3	1.786
30	0.5556	1.176	0.866	49.4	3.086
40	0.5535	1.117	0.775	53.3	4.357

Table 15.9
Characteristic Functions Computed Numerically by Sucker and Brauer [1] for Flow over a Circular Cylinder

Re_D	$2\bar{C}_f$	$2\bar{C}_p$	$2\bar{C}_t$	θ_A	$\dfrac{l_w}{a}$
0.0001	20380.0	20180.0	40560.0		
0.01	211.7	213.0	424.7		
0.1	28.635	28.602	57.237		
1	5.225	5.355	10.570		
4	2.128	2.374	4.502		
6	1.680	1.971	3.651		
6.2	1.646	1.942	3.588		
6.4	1.614	1.914	3.528	3.42	
6.6	1.585	1.881	3.465	8.07	
7	1.529	1.836	3.365	13.21	0.096
8	1.410	1.717	3.127	21.16	0.208
10	1.231	1.600	2.831	28.73	0.483
15	1.016	1.440	2.456	36.06	1.085
20	0.853	1.325	2.178	42.80	1.683
23	0.782	1.256	2.037	45.21	2.075
30	0.664	1.161	1.825	49.07	2.855
40	0.557	1.076	1.633	51.94	4.046
50	0.485	1.035	1.520	56.01	5.321
60	0.434	1.001	1.435	58.40	6.395
70	0.395	0.976	1.371	60.44	7.421
80	0.364	0.957	1.321	62.08	8.642
100	0.318	0.924	1.243	64.76	10.940
120	0.263	0.941	1.204	67.11	13.476

COMPARISON OF SOLUTIONS WITH EXPERIMENTAL MEASUREMENTS

Point of Attachment

Experimental and computed values for the angle of attachment of the wake, θ_A, as measured from the rear, are shown in Figures 15–8 and 15–9. The agreement is seen to be reasonably good. The solid curves in both figures correspond to the empirical equation of Sucker and Brauer [1]:

$$\theta'_A = \left(\frac{\ln\{Re_D\} - 1.83}{3 \times 10^{-4}}\right)^{0.456} \tag{15.44}$$

This expression represents the data well only for the stationary regime.

Length of Recirculation Zone

Figure 15–10 shows the dimensionless length, l_w/a, of the stationary wake beyond the rear point of symmetry. The points for $Re_D > 44$ represent wakes

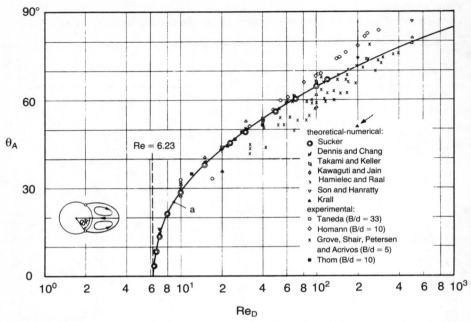

FIGURE 15-8   *Angle of attachment of wake on a cylinder in low range of Re_D. Here* a *designates Equation 15.44, B = width of tunnel and d = diameter of cylinder. (From Sucker and Brauer [1].)*

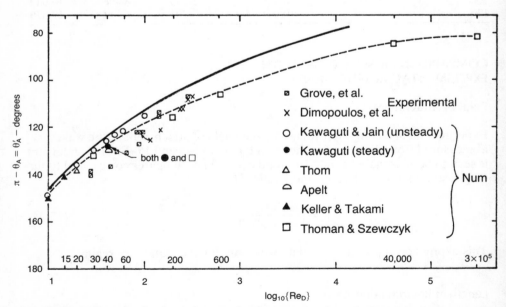

FIGURE 15-9   *Angle of attachment of wake on a circular cylinder in the high range of Re_D. (From Thoman and Szewczyk [31].) The solid curve, representing Equation 15.44, has been added to their original.*

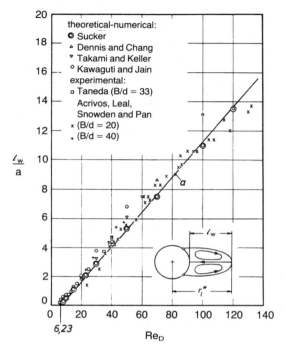

FIGURE 15–10 *Length of stationary wake on a cylinder. Here a designates Equation 15.45, B = width of tunnel, d = diameter of cylinder. (From Sucker and Brauer [1].)*

stabilized by the presence of a *splitter plate* behind the cylinder in the $\theta = 0$ plane. The computed values for $\text{Re}_D > 44$ were obtained by postulating a hypothetical steady state. The straight line represents the empirical equation of Sucker and Brauer [1]:

$$\frac{l_w}{a} = 0.12 \, \text{Re}_D - 0.748 \tag{15.45}$$

Pressure

Measured values of the pressure at the point of incidence are seen in Figure 15–11 to be in excellent agreement with the predictions of Equation 15.26. However, computed values for low Re_D are seen in Figure 15–12 to fall below the predictions of Equations 15.26 and 15.31. These values do not extend to the presumed range of validity of Equation 15.13.

The experimental pressure distributions of Thom [22] at several low values of Re_D are compared with his computed distribution at $\text{Re}_D = 20$ in Figure 15–13. The agreement is very good. The experimental values of Achenbach [32] at $\text{Re}_D = 5 \times 10^4$ are seen in the upper part of Figure 15–14 to be in agreement with the prediction of inviscid flow (Equation 15.7) over the forward half of the cylinder.

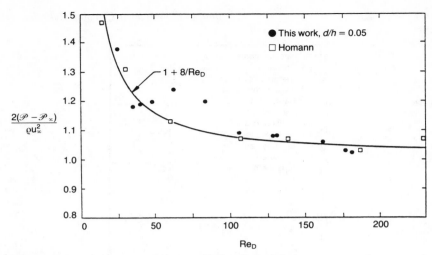

FIGURE 15–11 *Comparison of measured and predicted pressure at the point of incidence on a cylinder. Here d = diameter of cylinder and h = width of tunnel. (From Grove et al. [12].)*

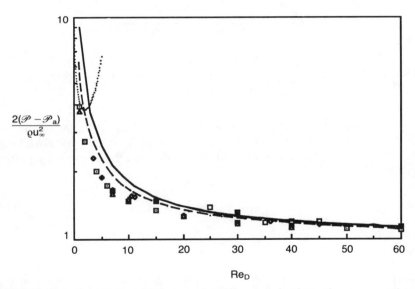

FIGURE 15–12 *Comparison of measured and computed values of pressure at the point of incidence on a circular cylinder at low Re_D with theoretical expressions:*

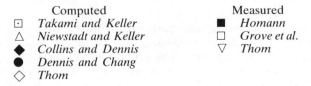

Computed	Measured
⊡ *Takami and Keller*	■ *Homann*
△ *Niewstadt and Keller*	□ *Grove et al.*
◆ *Collins and Dennis*	▽ *Thom*
● *Dennis and Chang*	
◇ *Thom*	

······ *Equation 15.13;* ——— *Equation 15.26;* ——— *Equation 15.31.*

336

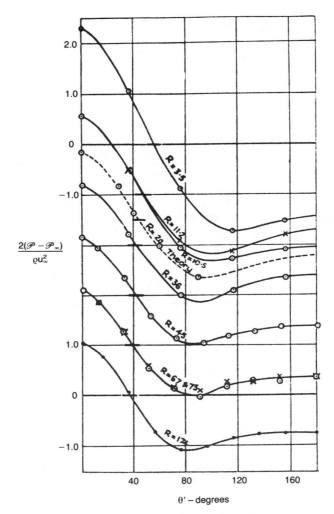

FIGURE 15–13   *Pressure distributions around a circular cylinder as measured in oil and water and computed by Thom [22]. R designates Re_D.*

Velocity Just Outside the Boundary Layer

Separation is known to influence the velocity outside the boundary layer somewhat, even up to the point of incidence. The velocity fields at the edge of the boundary layer can be derived from experimental measurements of the pressure distribution along the surface through

$$U_\theta^2\{\theta\} = u_\infty^2 - \frac{2(\mathscr{P}\{\theta, a\} - \mathscr{P}_\infty)}{\varrho} \qquad (15.46)$$

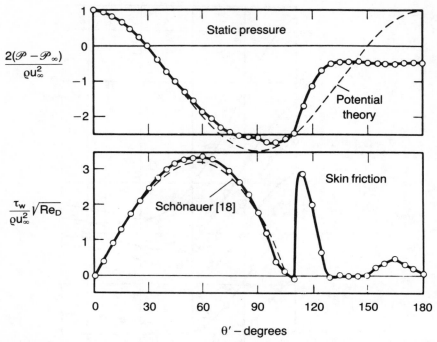

$$\frac{2(\mathscr{P} - \mathscr{P}_\infty)}{\varrho u_\infty^2}$$

Static pressure

Potential
theory

$$\frac{\tau_w}{\varrho u_\infty^2}\sqrt{Re_D}$$

Skin friction

Schönauer [18]

θ' – degrees

FIGURE 15–14   *Distribution of pressure and shear stress (skin friction) on a circular cylinder at* $Re_D = 4 \times 10^5$ *as measured by Achenbach [32].*

Equation 15.43 was so derived. For $Re_D = 1.85 \times 10^4$ Heimenz [33] determined the empirical representation

$$\frac{U_{\theta'}}{u_\infty} = 1.8157\theta' - 0.271\,36(\theta')^3 - 0.047\,325(\theta')^5 \tag{15.47}$$

where $\theta' = \pi - \theta$ is the angle from the point of incidence.

From the experimental measurements of Schmidt and Wenner [34] at $Re_D = 1.7 \times 10^5$, Eckert [35] derived

$$\frac{U_{\theta'}}{u_\infty} = 1.8157\theta' - 0.4094(\theta')^3 - 0.005\,25(\theta')^5 \tag{15.48}$$

Equations 15.43, 15,47, and 15.48 are seen in Figure 15–15 to differ less than 10% from one another for $\theta' < \pi/2$ despite the wide difference in Re_D. They do differ significantly from Equation 15.14. The boundary-layer results for θ near the point of incidence could be corrected to correspond to Equations 15.43, 15.47, or 15.48 merely by multiplying u_x, u_y, and $C_f\sqrt{Re_D}$ by $(1.8157/2)^{3/2} = 0.8650$. The complete Blasius solution for a cylinder could be corrected similarly by revising the expressions for a_i in Equation 15.33 (see problems 5 and 6).

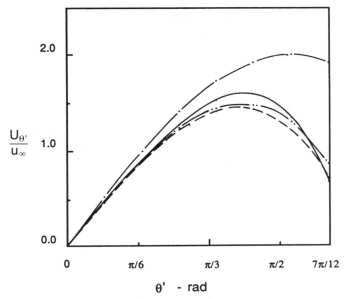

FIGURE 15–15 *Comparison of correlating equations for variation of the free-stream velocity around a circular cylinder as computed from measured pressure distributions.* —·— *Equation 15.14 (potential flow);* —— *Equation 15.47 (Re$_D$ = 1.85 × 10⁴);* —··— *Equation 15.43 (Re$_D$ = 1.22 × 10⁵);* ———— *Equation 15.48 (Re$_D$ = 1.7 × 10⁵).*

Local Shear Stress

The experimental values of Achenbach [32] for the local shear stress at $Re_D = 4 × 10^5$ are included in the lower part of Figure 15–14. The agreement with the predicted values of Schönauer (see Table 15.2) from thin-laminar-boundary-layer theory is very good. As shown in Figure 15–16, the agreement between the experimental values of Achenbach [36] at higher values of Re_D, with the prediction of Schönauer is fair for $θ' < 30π/180$ but only qualitative for higher angles. Similar discrepancies would also be expected for $Re_D < 4 × 10^5$, due to thickening of the boundary layer.

Total Drag Coefficient

The most reliable values of $\bar{C}_t$ for small Re_D appear to be those of Tritton [37], as shown in Figure 15–17. The several solutions for slightly inertial flow are compared with these values in Figure 15–18. Equation 15.12 of Kaplun [11] appears to provide the best representation as $Re_D \to 0$ and is quite satisfactory for $Re_D < 1$. The various numerical solutions for $3 < Re_D < 100$ are compared with the same set of data in Figure 15–19 and seem to be in general agreement.

For $Re_D > 100$, the measurements of Wieselberger [38] are the most complete and precise. His values and others are compared with computed values in Figure 15–20. Good agreement may be noted except for $Re_D > 44$, for which

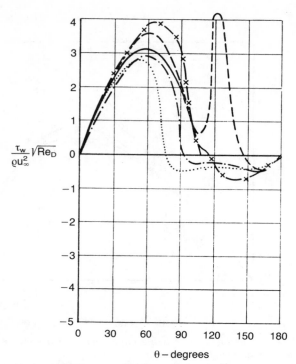

FIGURE 15–16   *Distribution of shear stress (skin friction) on a circular cylinder at high Re_D as measured by Achenbach [36].* ······ $Re_D = 1 \times 10^5$; —·— $Re_D = 2.6 \times 10^5$; ———— $Re_D = 8.5 \times 10^5$; —x— $Re_D = 3.6 \times 10^6$; ———— *Schönauer [18], thin-laminar–boundary-layer theory.*

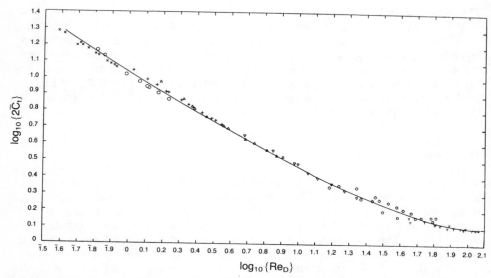

FIGURE 15–17   *Total drag coefficients for a circular cylinder at low Re_D as measured by Tritton [37].*

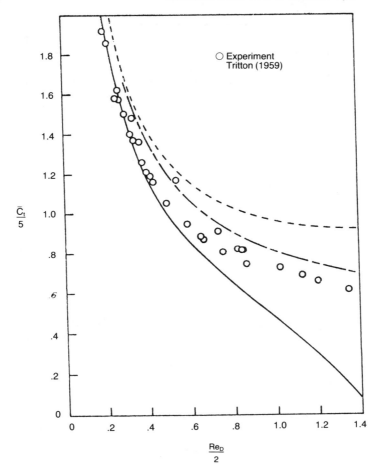

FIGURE 15–18 *Comparison by Van Dyke [9], p. 164, of analytical solutions for the total drag coefficient for slightly inertial flow over a circular cylinder with experimental data of Tritton [37].* ———— *Equation 15.12;* ——— *Equation 15.11;* —---— *Equation 15.8.*

a hypothetical steady state was postulated in the computations. The non-steady-state values of Thoman and Szewczyk [31] for $Re_D = 30, 40, 100, 200, 600, 4 \times 10^4$, and 3×10^5 (not shown) are, however, in good agreement.

Sucker and Brauer [1] proposed the following empirical equation to represent these data for $10^{-4} < Re_D < 2 \times 10^5$:

$$\bar{C}_t = 0.59 + \frac{3.4}{Re_D^{0.89}} + \frac{0.98}{Re_D^{1/2}} - \frac{2 \times 10^{-4} Re_D}{1 + 3.64 \times 10^{-7} Re_D^2} \qquad (15.49)$$

The solid curve labeled *a* in Figure 15–20 corresponds to this expression.

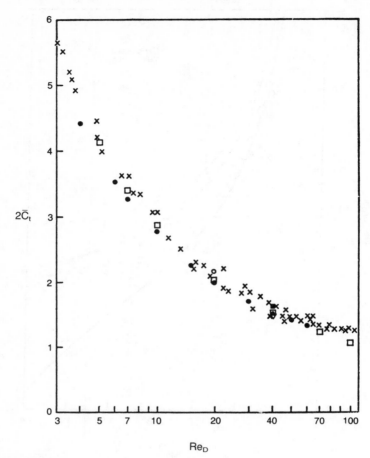

FIGURE 15–19   *Comparison of experimental and computed total drag coefficient for a circular cylinder of low* Re_D. *(From Dennis and Chang [29].)*

	Numerical solutions	Experimental data
□	*Dennis and Chang (1970)*	*x Tritton (1959)*
●	*Takami and Keller (1969)*	
▽	*Apelt (1961)*	
■	*Kawaguti (1953)*	
○	*Thom (1933)*	

Drag Coefficients for Skin Friction and Pressure

The behavior of the drag coefficient is easier to interpret if the coefficients for the contributions of skin friction and pressure are plotted separately, as in Figure 15–21. The two coefficients approach equality for $Re_D < 1$. The computed experimental values of $\bar{C}_f$ and $\bar{C}_p$ scatter somewhat more than those for $\bar{C}_t$ in Figures 15–19 and 15–20, but are reasonably well represented by the (solid) empirical curves. The computed values for $Re_D > 40$ appear to be generally low,

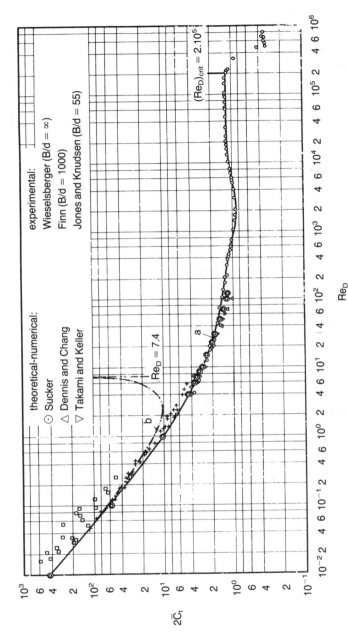

FIGURE 15–20 *Comparison of experimental and computed total drag coefficients for a circular cylinder for all Re_D. Here a designates Equation 15.49, b designates Equation 15.11, B = width of channel, d = diameter of cylinder. (From Sucker and Brauer [1].)*

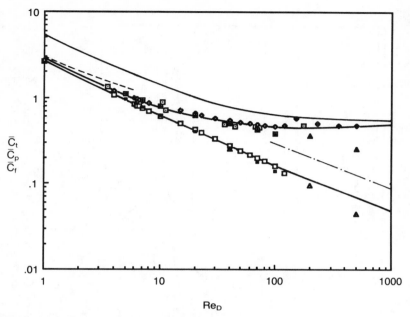

FIGURE 15–21   *Computed and measured coefficients for skin friction and pressure on a circular cylinder.*

Computed			Experimental	
	$\bar{C}_f$	$\bar{C}_f$		$\bar{C}_f$
Sucker and Brauer	□	◇	*Thom*	⊡
Son and Hanratty	△	▲	*Linke*	◆
Dennis and Chang	■	⊞		

——— *Equation 5.11;* ——— *Equation 5.42;* ——— *graphical correlations (upper line is sum of lower ones).*

as might be expected due to the neglect of vortex shedding. The numerical solution of the Oseen approximation by Tomotika and Aoi [10] is an upper bound. Presumably, Equations 15.8 and 15.12 are also satisfactory for $\bar{C}_f = \bar{C}_p = \bar{C}_t/2$ for $\mathrm{Re}_D \lesssim 0.5$. Equation 15.42, from thin-boundary-layer theory, appears to give the correct functional dependence for $\bar{C}_f$ on Re_D for $\mathrm{Re}_D > 100$ but excessive values. For $\mathrm{Re}_D > 100$, $\bar{C}_p \cong 0.45$. $\bar{C}_f$ is negligible with respect to $\bar{C}_p$ for $\mathrm{Re}_D > 1000$.

Behavior of the Wake

When an air stream blows over a wire at certain velocities, the wire "sings." Such *aeolian tones* were presumably observed even in antiquity. In 1878 Strouhal [39] showed that the frequency of the sound depends on the velocity of the air and not on the elastic properties of the wire. Soon afterward, Rayleigh [40] showed that, more generally, the frequency depends on the Reynolds number.

Leonardo da Vinci in the fifteenth century described[1] [41], p. 200, and sketched the vortex trail behind a bluff body, shown in Figure 15–22, which incorrectly indicates symmetrical rather than alternating vortices. Bénard [42] in 1908 recognized the significance of the alternating vortices and von Kármán [4] in 1911 and 1912 first derived a model for them. By making the postulates of (a) an ideal fluid, (b) an infinitely long body (such as a cylinder), and (c) potential vortices with axes parallel to the long dimension of the body, he determined that only an arrangement such as that shown in Figure 15–23, with uniformly staggered and spaced vortices was stable. He furthermore showed that stability requires the axial spacing l between the vortices on one side to be given by

$$l = \frac{\pi h}{\cosh^{-1}\{\sqrt{2}\}} = \frac{h}{0.2806} \tag{15.50}$$

where h is the normal separation in meters of the centers of the vortices. This prediction is in fair accord with the observations of von Kármán and Rubach [43] and subsequent experimenters. Also, see problem 29.

Lamb [7] subsequently completed the derivation outlined by von Kármán [4] and obtained the following expressions for the velocity components within such vortices:

$$u_x = \left(\frac{S_l}{2l}\right)\frac{\sinh\{2\pi y/l\}}{\cosh\{2\pi y/l\} - \cos\{2\pi x/l\}} \tag{15.51}$$

and

$$u_y = \left(\frac{S_l}{2l}\right)\frac{\sin\{2\pi x/l\}}{\cosh\{2\pi y/l\} - \cos\{2\pi x/l\}} \tag{15.52}$$

where the origin for x and y is at the center of the vortex, and

where S_l = strength of line vortex, m²/s
 x = distance in direction of unperturbed motion, m
 y = distance perpendicular to direction of unperturbed motion, m
 l = distance between centers of vortices, m

Figure 15–6 is a plot of the streamlines corresponding to Equations 15.51 and 15.52.

The velocity of the center of the vortices relative to the unperturbed fluid can be shown to be

$$u_c = \frac{S_l}{2l}\tanh\left\{\frac{\pi h}{l}\right\} \tag{15.53}$$

---

[1] "Observe the motion of the surface of the water which resembles that of hair, and has two motions, of which one goes on with the flow of the surface, the other forms the lines of eddies; thus the water forms eddying whirlpools one part of which are due to the impetus of the principle current and the other to the incidental motion and return flow."

FIGURE 15–22 *Sketch by Leonardo da Vinci of an old man studying the vortex trail behind cylindrical objects in a river. (From Richter [41], plate XXV.)*

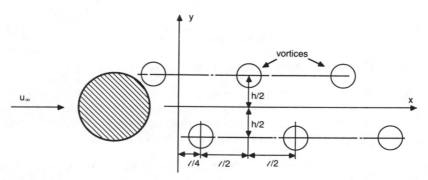

FIGURE 15–23 *Coordinate system for derivation of von Kármán solution for the stable periodic wake behind a cylinder.*

It follows from Equation 15.50 that

$$u_c = \frac{S_l}{l\sqrt{8}} \tag{15.54}$$

Von Kármán [4] calculated the drag force per unit length produced by the vortex street to be

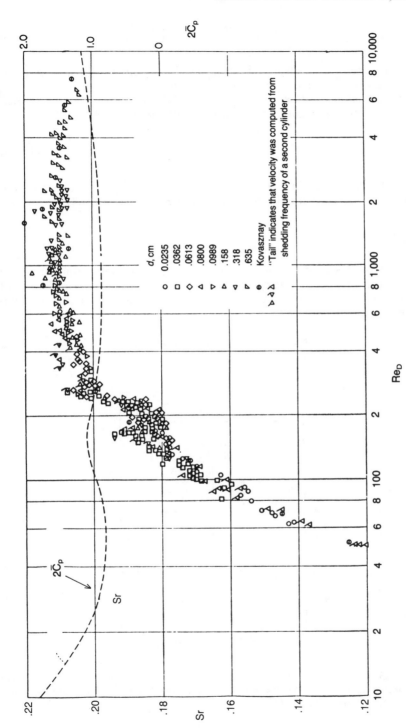

FIGURE 15-24 *Experimental data for the frequency of vortex shedding. (From Roshko [441].)*

$$\frac{F}{L} = \varrho u_\infty^2 l \left[0.794 \frac{u_c}{u_\infty} - 0.314 \left(\frac{u_c}{u_\infty} \right)^2 \right] \tag{15.55}$$

Unfortunately, the normal displacement h, and hence the axial displacement l, is not predicted by the theory nor is S_l and, hence, u_c.

The experimental observations of Roshko [44] for the frequency of vortex shedding are plotted in Figure 15–24 in terms of the *Strouhal number*:

$$Sr = \frac{\omega D}{u_\infty} \tag{15.56}$$

where ω is the frequency in hertz or reciprocal seconds. A relative constant value of 0.21 is observed for the Strouhal number for $Re_D > 300$.

Transition

The critical Reynolds number for the formation of a turbulent boundary layer depends on the free-stream turbulence, as indicated by the correlation of experimental data in Figure 15–25. The curve represents the empirical equation of Sucker and Brauer [1]:

$$Re_D^* = 3.78 \times 10^5 \, e^{-6\sqrt{Tu}} \tag{15.57}$$

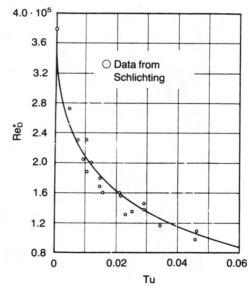

FIGURE 15–25 *Correlation for the effect of free-stream turbulence on the critical Reynolds number for transition to a turbulent boundary layer.* ——— *designates Equation 15.57. (From Sucker and Brauer [1].)*

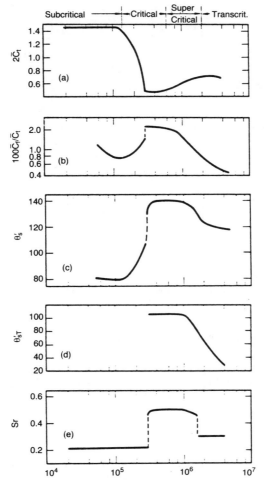

FIGURE 15–26   *Experimentally determined characteristics of flow over a circular cylinder in the subcritical, critical, supercritical, and transcritical regimes. (From Achenbach [32].)*

Critical, Supercritical, and Transcritical Behavior

All of the preceding results are for the subcritical regime. The experimental results of Achenbach [32] for the variation of $\bar{C}_t$, $\bar{C}_f/\bar{C}_t$, θ'_S, θ'_{ST} (the angle of separation of the turbulent boundary layer), and Sr in the transcritical regime are shown in Figure 15–26. Sucker and Brauer [1] proposed the following empirical expression for the drag coefficient for $4 \times 10^5 < \mathrm{Re}_D < 10^7$:

$$\bar{C}_t = 1.75 \times 10^{-4}\,\mathrm{Re}_D^{1/2} + 0.5[1 - (1 + 1.5 \times 10^{-14}\,\mathrm{Re}_D^2)]^{1/4} \quad (15.58)$$

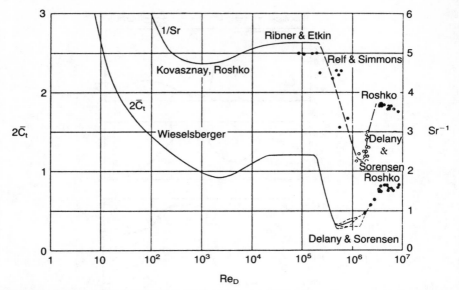

FIGURE 15–27   *Correspondence between the total drag coefficient and the Strouhal number for a circular cylinder. (From Roshko [45].)*

Relationship between Vortex Shedding and Drag

Figure 15–27 of Roshko [45] indicates that the total drag coefficient has the same trends with Reynolds number as the reciprocal of the Strouhal number for all regimes of vortex shedding.

Effect of Cylinder Length

A real cylinder is of finite length although the solutions all imply infinite length. The boundary conditions at the ends of the cylinder, whether walls, free ends, or otherwise, induce flow in the third dimension—i.e., in the direction of the axis of the cylinder. Some discrepancies would therefore be expected with respect to all of the theoretical solutions. Such effects seem to be significant for cylinders of small diameter (wires), even for very long lengths, since the scale of the disturbances is not proportional to the diameter.

SUMMARY

The behavior of the flow around a cylinder has been shown to involve many regimes. Hence, many different theoretical approaches are necessary for prediction, interpretation, and correlation.

Kaplun's extension of Lamb's solution for slightly inertial flow provides sound predictions for $\mathrm{Re}_D < 1$.

Because of the presence of a wake for all $\mathrm{Re}_D < 6.3$, the contribution of boundary-layer theory is far more limited than for the flat plate. Quantitative predictions are accurate only for the region near the point of incidence, but to that extent are valid even for critical, supercritical, and transcritical flows.

Fortunately, numerical solutions provide quantitative results, if not a general structure, for the entire range of subcritical flow.

Except for the laminar boundary layer on the forward portion of the cylinder the descriptions of critical, supercritical, and transcritical flows are almost wholly empirical.

The greatest theoretical need is for complete, closed-form solutions for the region of the wake.

PROBLEMS

1. Show that Equation 15.1 is a solution for inviscid flow around a cylinder.
2. Find expressions in the literature for the following functions in Lamb's solution for slightly inertial flow around a cylinder: u_r, u_θ, $\psi\{r,\theta\}$, $\phi\{r,\theta\}$.
3. Derive an expression for the local coefficient of skin friction in terms of the vorticity.
4. Show from Equation 15.8 that a solution does not exist for creeping flow.
5. Rederive Equation 15.41 using

    a. Equation 15.43
    b. Equation 15.47
    c. Equation 15.48

6. Compare the results obtained from Equation 15.41 with those in Table 15.2.
7. Verify Equation 15.42.
8. Compute $\bar{C}_f\sqrt{\mathrm{Re}_D}$ using the values in Table 15.2 and compare with Equation 15.42.
9. Compute $\bar{C}_f\sqrt{\mathrm{Re}_D}$ using Equations 15.22 and 15.29 and compare with Equation 15.42.
10. Prove that a similarity transformation is not possible for Equations 11.1 and 15.15.
11. Explain the derivation of the Blasius solution in view of the absence of a similarity transformation.
12. Derive expressions for t_5''', f_7''', f_9''', and f_{11}''' corresponding to Equations 15.34A and 15.34B.
13. Calculate $\bar{C}_p$ corresponding to the curves in Figures 15–13 and 15–14 and compare with the values in Figure 15–21.
14. Calculate $\bar{C}_t$ corresponding to the curves in Figure 15–14.
15. Develop a correlation for $\bar{C}_t$ using the Churchill–Usagi model [46] with the first term of Equation 15.11 and $\bar{C}_{t\infty}$ (a pseudolimiting value).
16. Derive an empirical equation for $\bar{C}_t$ for $\mathrm{Re}_D < 1$ in the form A/Re_D^n. What is the lower limit of validity?
17. Derive an empirical expression for $\bar{C}_p$.
18. Derive an empirical expression for $\bar{C}_f$.
19. What wind velocity is required to produce a pure tone of 440 Hz with a 1.0-mm wire? What is the Reynolds number?

20. Construct a dimensionless graph relating velocity explicitly to frequency, thus avoiding the trial and error of problem 19.

21. Repeat problem 20 for ω and D.

22. Prepare a plot of Sr versus $\bar{C}_p$.

23. Derive an empirical expression for $2(\mathscr{P}_0 - \mathscr{P}_\infty)/\varrho u_\infty^2$ using the computed values in Tables 15.6–15.8 as well as the measured and computed values in Figures 15–11 to 15–13.

24. Construct expressions for $2(\mathscr{P}_0 - \mathscr{P}_\infty)/\varrho u_\infty^2$ corresponding to Equations 15.11 and 15.12. Compare with Equation 15.13.

25. From the dimensionless vorticities at the surface in Figure 15–28, determine $\tau_w/\varrho u_\infty^2$ and $\bar{C}_f$.

26. Derive an expression for the velocity distribution in terms of the coefficients in Table 15.1.

27. Show how the coefficients f_1, f_2, ... in Equation 15.32 could be determined from the values in Table 15.1. Why were the derivatives of these coefficients rather than the coefficients themselves tabulated?

28. Determine h/l from Figure 15–4A and compare with Equation 15.50.

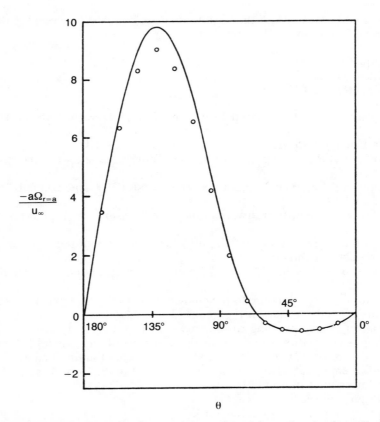

FIGURE 15–28   *Dimensionless vorticity at the surface of a circular cylinder at $Re_D = 100$. (From Collins and Dennis [47].)*

29. Calculate and plot the streamlines for a von Kármán vortex using Equations 15.51 and 15.52. Compare with Figure 15.6.
30. Knudsen and Katz [48], p. 204f, refer to $U\{x\}$ for a cylinder as u_{max}. Is this designation justified? Explain quantitatively.
31. Langmuir [49] derived a correction for the effect of curvature and the finite thickness of the boundary layer on the rate of convective heat transfer from a cylinder by postulating that convection could be represented by conduction across a cylindrical shell of effective thickness δ. Show that for viscous shear this concept leads to

$$\left(\frac{\tau_w'}{\varrho u_\infty^2}\right)\mathrm{Re}_D = \frac{2}{\ln\{1 + 2/(\tau_w/\varrho u_\infty^2)\mathrm{Re}_D\}} \tag{15.59}$$

where τ_w' is the corrected shear stress.
32. Calculate the percentage correction in τ_w due to curvature as a function of Re_D and x/a on the basis of Equations 15.59 and 15.41.
33. Repeat problem 32 for $\bar{C}_f$.
34. Calculate the percentage correction of $\bar{C}_f$ as a function of Re_D based on the results of problem 33.
35. Compute and compare the velocity distributions given by the Blasius solution (Equation 15.32 and Table 15.1) with the computed and experimental values given in Figure 15–29.

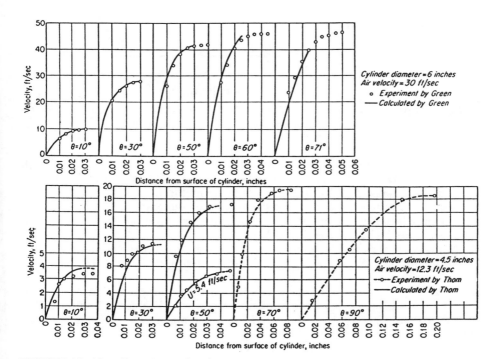

FIGURE 15–29 *Experimental and computed velocity distributions in the boundary layer of a circular cylinder. (From Knudsen and Katz [48], p. 294.)*

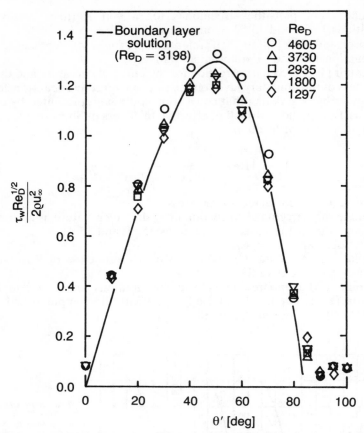

FIGURE 15–30　*Experimental and computed shear stress on the surface of a circular cylinder. (From Nishimura and Kawamura [50], as corrected [51].)*

36. Compare the experimental shear stress distribution of Figure 15–30 with those of Tables 15.2 and 15.3. These data were determined by Nishimura and Kawamura [50] by an electrochemical method. The plot itself is a corrected version provided by Kawamura [51]. The curve labeled boundary-layer solution was computed using the experimental pressure distribution of Figure 15.31 for $\mathrm{Re}_D = 3198$.

37. Calculate the free-stream velocity distribution corresponding to the pressure distribution of Figure 15–31. Compare with the curves of Figure 15–15.

38. Calculate $\bar{C}_f$ based on Figure 15–31 and the assumption that $\tau_w \cong 0$ for $\theta' > 84°$. Compare with the numerically computed values in Figure 15–21.

39. Calculate $\bar{C}_p$ based on Figure 15–31. Compare with the numerically computed values of Figure 15–21.

40. Explain why the values of $C_f\sqrt{\mathrm{Re}_p}$ calculated in problem 38 agree with the numerically-computed values in Figure 15–21, whereas Equation 15.42 does not.

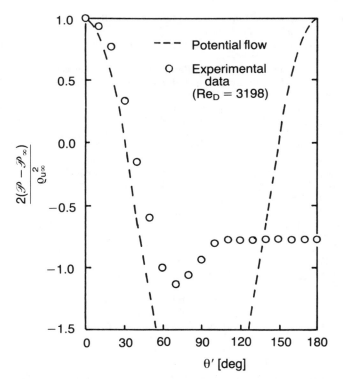

FIGURE 15–31 *Experimental pressure distribution on the surface of a circular cylinder. (From Kawamura [51].)*

REFERENCES

1. D. Sucker and H. Brauer, "Fluiddynamik bei der angeströmten Zylindern," *Wärme- und Stoffübertragung*, 8 (1975) 149.
2. M. Van Dyke, *An Album of Fluid Motion*, Parabolic Press, Stanford, CA (1982).
3. H. Werlé, "Hydrodynamic Flow Visualization," *Ann. Rev. Fluid Mech.*, 5 (1973) 361.
4. Walther Kaufmann, *Fluid Mechanics*, English transl. by E. G. Chilton, McGraw-Hill, New York (1963).
5. Th. von Kármán, "Über den Mechanismus des Widerstandes, den ein bewegter Körper in einer Flüssigkeit erfährt," *Nachr. Wiss. Ges. Göttingen, Math. Phys. Klasse* (1911) 509; (1912) 547.
6. G. C. Stokes, "On the Effect of the Internal Friction of Fluids on the Motion of Pendulums," *Trans. Camb. Phil. Soc. 9* (1851) 8 (*Mathematical and Physical Papers*, Vol. III, Cambridge University Press (1901), p. 55).
7. H. Lamb, *Hydrodynamics*, Dover, New York (1945).
8. C. W. Oseen, "Über die Stokes'sche Formel und über die verwandte Aufgabe in der Hydrodynamik," *Arkiv Math., Astronom. Fys.*, 6, (1910) 75.
9. *M. Van Dyke, Perturbation Methods in Fluid Mechanics*, annotated ed., Parabolic Press, Stanford, CA (1975).
10. S. Tomotika and T. Aoi, "An Expansion Formula for the Drag on a Circular

Cylinder Moving through a Viscous Fluid at Low Reynolds Numbers," *Quart. J. Mech. Appl. Math.*, *3* (1951) 401.

11. S. Kaplun, "Low Reynolds Number Flow past a Circular Cylinder," *J. Math. Mech.* *6* (1957) 595.

12. A. S. Grove, F. H. Shair, E. E. Petersen, and A. Acrivos, "An Experimental Investigation of the Steady Separated Flow past a Circular Cylinder," *J. Fluid Mech.*, *19* (1964) 60.

13. F. Homann, "Der Einfluss grosser Zähigkeit bei der Strömung um den Zylinder und um die Kugel," *Z. Angew. Math. Mech.*, *16* (1936) 153, and *Forsch. Gebiete Ingenieurw.*, *7* (1936) 1; English transl., "The Effect of High Viscosity on the Flow around a Cylinder and around a Sphere," *NACA* TM 1334, Washington, D.C. (1952).

14. H. Blasius, "Grenzschichten in Flüssigkeiten mit kleiner Reibung," *Z. Math. Phys.*, *56* (1908) 1; English transl., "The Boundary Layers in Fluids with Little Friction," *NACA* TM 1256, Washington, D.C. (1950).

15. L. Howarth, "On the Calculation of Steady Flow in the Boundary Layer near the Surface of a Cylinder in a Stream," *Aero. Res. Council, Gt. Brit.*, R. & M. 1932, London (1935).

16. A. N. Tifford, "Heat Transfer and Frictional Effects in Laminar Boundary Layers," *WADC*, Tech. Rept. 53-228, Part 4, Washington, D.C. (1954).

17. H. Schlichting, *Boundary Layer Theory*, 4th ed., English transl. by J. Kestin, McGraw-Hill, New York (1960).

18. W. Schönauer, "Ein Differenzenverfahren zur Lösung der Grenzschichtgleichung für stationäre, laminare, inkompressible Strömung," *Ing.-Arch.*, *33* (1964) 173.

19. N. L. Evans, *Laminar Boundary Layer Theory*, Addison-Wesley, Reading, MA (1968).

20. H. H. Sogin and V. S. Subramanian, "Local Mass Transfer from Cylinders in Crossflow," *J. Heat Transfer*, *83C* (1961) 483.

21. S. Goldstein, Ed., *Modern Developments in Fluid Dynamics*, Vol. 2, Oxford University Press, Clarendon (1938).

22. A. Thom, *Aero. Res. Comm., Gt. Brit.*, R. & M. 1176 (1928) and R. & M. 1194 (1929), and in final form in "The Flow Past Circular Cylinders at Low Speed," *Proc. Roy. Soc. (London)*, A 141 (1933) 651.

23. M. Kawaguti, "Numerical Solution of the Navier–Stokes Equations for the Flow around a Circular Cylinder at a Reynolds Number of 40," *J. Phys. Soc. Japan*, *8* (1953) 747.

24. C. J. Apelt, "The Steady Flow of a Viscous Fluid past a Circular Cylinder at Reynolds Numbers of 40 and 44," *Aero. Res. Comm., Gt. Brit.*, R. & M. 3175 (1961).

25. M. Kawaguti and P. Jain, "Numerical Study of a Viscous Fluid Flow Past a Circular Cylinder," *J. Phys. Soc. Japan*, *21* (1966) 2055.

26. P. C. Jain and K. S. Rao, "Numerical Solution of Unsteady Viscous Incompressible Fluid Flow past a Circular Cylinder," *Phys. Fluids*, *12*, Suppl. II (1969) II-57.

27. J. S. Son and T. J. Hanratty, "Numerical Solution for the Flow around a Cylinder at Reynolds Numbers of 40, 200 and 500," *J. Fluid Mech.*, *35* (1969) 369.

28. H. Takami and H. B. Keller, "Steady Two-Dimensional Viscous Flow of an Incompresible Fluid past a Circular Cylinder," *Phys. Fluids*, *12*, Suppl. II (1969) II-51.

29. S. C. R. Dennis and G.-Z. Chang, "Numerical Solutions for Steady Flow past a Circular Cylinder at Reynolds Numbers up to 100," *J. Fluid Mech.*, *42* (1970) 471.

30. F. Nieuwstadt and H. B. Keller, "Viscous Flow past Circular Cylinders," *Computers and Fluids*, *1* (1973) 59.

31. D. C. Thoman and A. A. Szewczyk, "Time Dependent Viscous Flow over a Circular Cylinder," *Phys. Fluids*, *12*, Suppl. II (1969) II-76.

32. E. Achenbach, "Total and Local Heat Transfer from a Smooth Cylinder in Cross-Flow at High Reynolds Number," *Int. J. Heat Mass Transfer*, *18* (1975) 1387.

33. K. Heimenz, "Die Grenzschicht an einem in den gleichförmigen Flüssigkeitsstrom eigentauchten geraden Kreiszylinder" [Thesis, Göttingen (1911)], *Dingler's Polytech. J.*, *326* (1911) 321.

34. E. Schmidt and K. Wenner, "Wärmeabgabe über den Umfang eines angeblasenen geheizten Zylinders," *Forsch. Gebiete Ingenieurw.*, *12* (1941) 65; English transl., *NACA* TM 1050, Washington, D.C. (1943).

35. E. R. A. Eckert, "Die Berechnung des Wärmeübergangs in der laminaren Grenzeschicht," *Ver. Deut. Ing. Forschungsheft* 416 (1942) 1.

36. E. Achenbach, "Distribution of Local Pressure and Skin Friction around a Circular Cylinder in Cross-Flow up to Re = 5 × 10⁶," *J. Fluid Mech.*, *34* (1968) 625.

37. D. J. Tritton, "Experiments on the Flow past a Circular Cylinder at Low Reynolds Numbers," *J. Fluid Mech.*, *6* (1959) 547.

38. C. Wieselberger, *Ergebnisse der aerodynamischen Versuchsanstalt zu Göttingen*, IId, Lieferung (1923) p. 23.

39. V. Strouhal, "Über eine besondere Art der Tonerregung," *Ann. Phys. und Chemie, Neue Folge*, *5*, (1878) 216.

40. Lord Rayleigh (J. W. Strutt), "Acoustical Observations," *Phil. Mag.*, Ser. 5, *7* (1879) 149; *The Theory of Sound*, Dover, New York (1945).

41. J. P. Richter, Ed., *The Notebooks of Leonardo da Vinci*, Vol. 1, Dover, New York (1970).

42. H. Bénard, "Formation de centres de giration à l'arrière d'un obstacle en mouvement," *Comp. Rend., Acad. Sci., Paris, 147* (1908) 839.

43. Th. von Kármán and H. Rubach, "Über die Mechanismus des Flüssigkeits- und Luftwiderstandes," *Physik. Z.*, *13* (1912) 49.

44. A. Roshko, "On the Development of Turbulent Wakes from Vortex Streets," *NACA* Rept. 1191, Washington, D.C. (1954).

45. A. Roshko, "Experiments on Flow past a Circular Cylinder at High Reynolds Number," *J. Fluid Mech.*, *10* (1961) 345.

46. S. W. Churchill and R. Usagi, "A General Expression for the Correlation of Rates of Transfer and Other Phenomena," *AIChE J.*, *18* (1972) 1121.

47. W. M. Collins and S. C. R. Dennis, "Flow past an Impulsively Started Cylinder," *J. Fluid Mech.*, *150* (1973) 105.

48. H. G. Knudsen and D. L. Katz, *Fluid Dynamics and Heat Transfer*, McGraw-Hill, New York (1958).

49. I. Langmuir, "Convection and Conduction of Heat in Gases," *Phys. Rev.*, *34* (1912) 401.

50. T. Nishimura and Y. Kawamura, "Flow Pattern in the Region of Separation on a Single Cylinder," *Kagaku Kogaku Ronbunshu*, 7 (1981) 120; English transl., *Int. Chem. Eng.*, *23* (1983) 78.

51. Y. Kawamura, private communication, 1986.

Chapter 16

Laminar Flow over a Solid Sphere

Flow over a solid sphere is analogous to that over a cylinder in some respects, but on close examination differs significantly owing to axial rather than plane symmetry. Also, the more important applications arise from the movement of a sphere through a fixed body of fluid rather than from flow over a fixed sphere. A wide range of behavior is observed for the relative motion of a sphere and the surrounding fluid, just as for a cylinder. Hence, a general description in advance of the mathematical treatment is again appropriate.

Inviscid Flow

The streamlines and equipotential lines obtained from the theoretical solution for a stagnant, *inviscid* fluid through which a sphere is moving at uniform velocity were sketched in Figure 10–49. Since the frame of reference is the distant fluid, these curves represent the unsteady values at any instant in time. Figure 10–47 provides the corresponding sketch of the streamlines and equipotential lines for uniform motion of an inviscid fluid past a fixed sphere with the sphere itself as the point of reference. The variation of the dynamic pressure around the perimeter of the sphere in the direction of flow was plotted in Figure 10–48. All of these sketches indicate that inviscid flow is symmetrical fore and aft. This hypothetical flow does not occur under any physical conditions, but provides a useful approximation for flow outside the boundary layer of a sphere, just as it did for a cylinder.

Creeping Flow

As contrasted with flow over a cylinder, a regime of flow exists for spheres for $Re_D < 0.1$ for which inertial forces are truly negligible. The instantaneous streamlines obtained from the exact theoretical solution for a fixed sphere in a uniform velocity field are sketched in Figure 16–1 with the velocity profile at $\theta = \pi/2$ superimposed. The dynamic pressure around a perimeter of the sphere in the direction of flow is plotted in Figure 16–2.

In the regime of creeping flow the streamlines are still seen to be symmetrical fore and aft, although the pressure distribution is antisymmetrical.

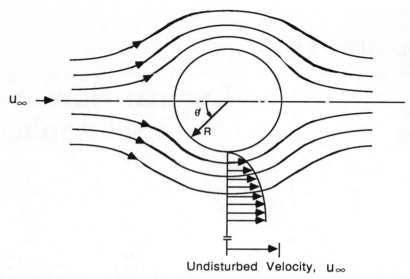

FIGURE 16–1   *Streamlines and velocity distribution at θ' = π/2 for creeping flow about a sphere.*

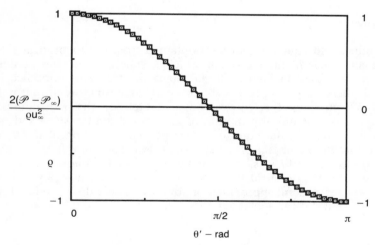

FIGURE 16–2   *Pressure distribution on the surface of a sphere in creeping flow.*

Photographs of particle traces in actual flows are shown in Figure 16–3 for a sphere moving through a fixed fluid (glycerine) in a tube with $D_{\text{tube}}/D_{\text{sphere}} = 2$ and $\text{Re}_D = 0.1$. In Figure 16–3A the camera was fixed with respect to the distant fluid, but in Figure 16–3B it was moving with the sphere. The streaklines are somewhat compressed in Figure 16–3B and are closed in Figure 16–3A due to the confinement, but the theoretical motion is generally confirmed, particularly the symmetry fore and aft.

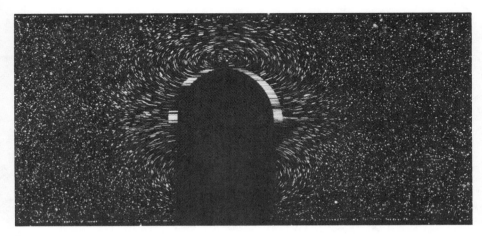

A

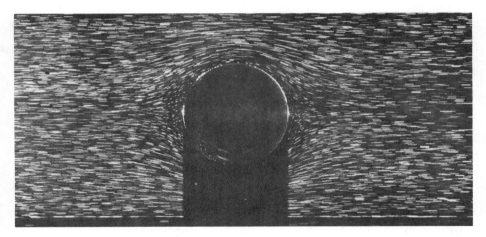

B

FIGURE 16–3   *Photographs by Coutanceau [1] of particle traces for a sphere falling in a tube of glycerine at $Re_D = 0.1$ with $D_{tube}/D = 2$: (A) camera fixed with respect to the distant fluid; (B) camera moving with the sphere.*

Slightly Inertial Flow

For $Re_D \gtrsim 1.0$ inertial effects become significant and the flow becomes asymmetric fore and aft, as shown by the sketch in Figure 16–4 of the instantaneous streaklines predicted by an approximate theoretical solution for a moving sphere and a fixed frame of reference relative to the distant fluid, for $Re_D = 5$. The pressure distribution on the surface of the sphere is plotted in Figure 16–5 for $Re_D = 0$ (creeping flow), $Re_D = 1$ and $Re_D = 5$. The photo-

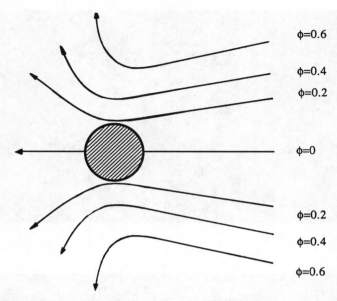

FIGURE 16–4   *Theoretical streaklines for slightly inertial (Oseen) motion of a sphere (to the left) at $Re_D = 5$ with the distant fluid as a frame of reference.*

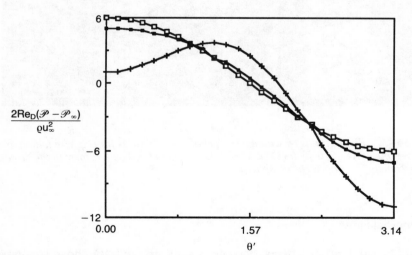

FIGURE 16–5   *Theoretical pressure distribution on the surface of a sphere for creeping and slightly inertial flow.*

☐ $Re_D \to 0$ *(creeping flow)*
■ $Re_D = 1$ ⎱
+ $Re_D = 5$ ⎰ *Slightly inertial (Oseen) flow*

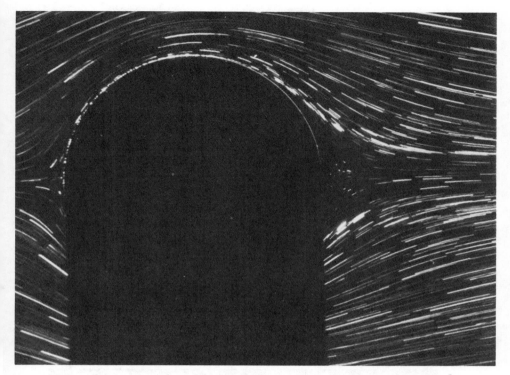

FIGURE 16–6   *Photograph by Coutanceau and Payard of particle traces for a fluid moving past a sphere at $Re_D = 9.8$. (From Van Dyke [2], p. 20.)*

graph in Figure 16–6 of particle traces for the motion of fluid past a sphere at $Re_D = 9.8$ confirms the predicted asymmetry fore and aft for slightly inertial flow, and indicates that separation has not yet occurred.

Flow with a Stationary Wake

At $Re_D \cong 24$, separation occurs at the rear of the sphere and a thin standing vortex ring is formed, as indicated by the particle tracks in Figure 16–7A for $Re_D = 26.8$. The point of separation gradually moves forward and eventually reaches a stationary point about 81 $\pi/180$ rad from the forward point of incidence. The wake itself grows more slowly than for a cylinder but eventually becomes fully developed, as shown by the numerically computed streamlines in Figure 16–8 for $Re_D = 100$ and the photograph of particle tracks in Figure 16–7B for $Re_D = 118$. Although not apparent from these two-dimensional representations, the stationary wake for a sphere consists of a single annular vortex, symmetrical about the axis of the sphere in the direction of flow.

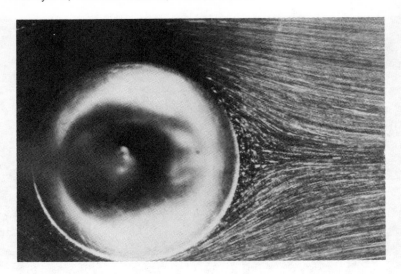

A

B

FIGURE 16–7   *Photographs of particle traces for flow about a sphere: (A) $Re_D = 26.8$; (B) $Re_D = 118$. (From Taneda [3].)*

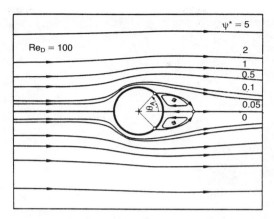

FIGURE 16–8   *Numerical computed profiles of the Stokes stream function for flow about a sphere at* $Re_D = 100$. $\tilde{\psi}^* = 4\tilde{\psi}/u_\infty D^2$. *(From Brauer and Sucker [4].)*

Flow with an Unsteady Wake

Starting at $Re_D \cong 140$ the annular vortex is shed periodically. At $Re_D \cong 450$, which is therefore known as the *lower critical Reynolds number*, a nonstationary spiral-like vortex is formed. This oscillatory behavior extends up to $Re_D \cong 2 \times 10^5$, but the boundary layer becomes increasingly chaotic, as illustrated for $Re_D = 1.5 \times 10^4$ in Figure 16–9. At $Re_D = 2$–3×10^5 the boundary layer becomes turbulent and the point of separation moves backward, decreasing the size of the wake and sharply reducing the drag. The point at which this transition occurs is known as the *higher critical Reynolds number*.

In subsequent paragraphs the various solutions that have been developed to describe these several regimes are examined quantitatively. An excellent historical review of the theoretical and experimental work up to 1959 on flow over a sphere is provided by Torobin and Gauvin [6].

INVISCID FLOW

In spherical coordinates, the Stokes stream function for inviscid flow in the z-direction around a fixed solid sphere can be shown to be (see Section 10S15)

$$\tilde{\psi} = -\frac{u_\infty R^2}{2}\left(1 - \left(\frac{a}{R}\right)^3\right)\sin^2\{\theta\} \tag{16.1}$$

where a = radius of sphere, m
R = radial distance from center of sphere, m
θ = *angle along surface from rear*, rad
u_∞ = free-stream velocity, m/s

Figure 10–47 is a plot of Equation 16.1 and the corresponding equipotential lines.

FIGURE 16–9   *Time-exposure of air bubbles in water flowing about a sphere at Re_D = 1.5 × 10⁴. (From Werlé [5].)*

It follows from Equation 16.1 that

$$u_R = \frac{-1}{R^2 \sin\{\theta\}} \frac{\partial \tilde{\psi}}{\partial \theta} = u_\infty \left(1 - \left(\frac{a}{R}\right)^3\right) \cos\{\theta\} \tag{16.2}$$

and

$$u_\theta = \frac{1}{R \sin\{\theta\}} \frac{\partial \tilde{\psi}}{\partial R} = -u_\infty \left(1 + \frac{1}{2}\left(\frac{a}{R}\right)^3\right) \sin\{\theta\} \tag{16.3}$$

Along the surface $u_r\{\theta, a\} = 0$ and

$$u_\theta\{\theta, a\} = -\frac{3}{2} u_\infty \sin\{\theta\} \tag{16.4}$$

The distribution of the dynamic pressure over the surface of the sphere follows from Equation 16.4 (see problem 19) and is

$$\mathscr{P}\{\theta, a\} = \mathscr{P}_0 - \frac{9}{8} u_\infty^2 \varrho \sin^2\{\theta\} \tag{16.5}$$

$\mathscr{P}_s$ from Equation 15.6 is applicable for $\mathscr{P}_0$; hence

$$2\left(\frac{\mathscr{P}\{\theta, a\} - \mathscr{P}_\infty}{\varrho u_\infty^2}\right) = 1 - \frac{9}{4}\sin^2\{\theta\} \tag{16.6}$$

Figure 10–48 is a plot of Equation 16.6.

CREEPING (STOKES) FLOW

In 1851 Stokes [7] showed that a solution could be obtained for a sphere when the inertial terms in the equations of motion were dropped, reducing the analog of Equation 9.5 in terms of ψ to

$$\nabla^4\tilde{\psi} = 0 \tag{16.7}$$

This regime is called *creeping* or *Stokes flow*. The model in terms of u_R, u_θ, and $\mathscr{P}$ was presented in Section 10EC1. In addition to the references given there, the process of solution is given in detail, for example, by Lamb [8], p. 597, and Streeter [9], p. 235. The resulting expression for the Stokes stream function in spherical coordinates is (also see Section 10EC1)

$$\tilde{\psi} = \frac{-u_\infty R^2}{2}\left(1 - \frac{3}{2}\frac{a}{R} + \frac{1}{2}\left(\frac{a}{R}\right)^3\right)\sin^2\{\theta\} \tag{16.8}$$

For the dynamic pressure the expression is

$$\mathscr{P} - \mathscr{P}_\infty = -\frac{3\mu u_\infty a \cos\{\theta\}}{2R^2} \tag{16.9}$$

At the surface

$$\mathscr{P}\{\theta\} - \mathscr{P}_\infty = -\frac{3\mu u_\infty \cos\{\theta\}}{2a} \tag{16.10}$$

or

$$\frac{2(\mathscr{P}\{\theta\} - \mathscr{P}_\infty)}{\varrho u_\infty^2} = -\frac{6\cos\{\theta\}}{\mathrm{Re}_D} \tag{16.11}$$

In contrast to Equation 16.6 the dimensionless pressure distribution is not symmetrical fore and aft and depends, as indicated, on Re_D. The components of the velocity follow from the stream function (Equation 16.8) by the same procedure as for inviscid flow and therefore are

$$u_r = u_\infty\left(1 - \frac{3}{2}\frac{a}{R} + \frac{1}{2}\left(\frac{a}{R}\right)^3\right)\cos\{\theta\} \tag{16.12}$$

and

$$u_\theta = -u_\infty\left(1 - \frac{3}{4}\frac{a}{R} - \frac{1}{4}\left(\frac{a}{R}\right)^3\right)\sin\{\theta\} \tag{16.13}$$

Figure 16–1 is based on Equations 16.8 and 16.13; Figure 16.2 on Equation 16.11. The velocity field in this regime is independent of the viscosity.

The local shear stress is

$$\frac{\tau_w}{\varrho u_\infty^2} = \frac{\mu}{\varrho u_\infty^2}\left(-\frac{\partial u_\theta}{\partial R}\right)_{R=a} = \frac{3\mu \sin\{\theta\}}{2a\varrho u_\infty} = \frac{3\sin\{\theta\}}{\mathrm{Re}_D} \tag{16.14}$$

The mean coefficient for drag due to skin friction is defined as in Chapter 15 to be the component of the local shear stress in the direction of flow, integrated over the surface of the sphere and divided by the projected area and ϱu_∞^2. Thus

$$\bar{C}_f = \frac{1}{\pi a^2 \varrho u_\infty^2} \int_0^\pi \left(\frac{3\mu u_\infty \sin\{\theta\}}{2a}\right)\sin\{\theta\}(2\pi a\sin\{\theta\})a\,d\theta$$

$$= \frac{4\mu}{u_\infty \varrho a} = \frac{8}{\mathrm{Re}_D} \tag{16.15}$$

Similarly the component of the dynamic pressure in the direction of flow (Equation 16.11) gives, on integration over the sphere,

$$\bar{C}_p = \frac{1}{\pi a^2 \varrho u_\infty^2} \int_0^\pi \left(\frac{3\mu u_\infty \cos\{\theta\}}{2a}\right)\cos\{\theta\}(2\pi a\sin\{\theta\})a\,d\theta$$

$$= \frac{2\mu}{u_\infty \varrho a} = \frac{4}{\mathrm{Re}_D} \tag{16.16}$$

The drag due to friction is thus equal to twice the drag due to pressure, and the total drag is

$$\bar{C}_t = \bar{C}_p + \bar{C}_f = \frac{12}{\mathrm{Re}_D} \tag{16.17}$$

Equation 16.17 is known as *Stokes' law.*

Stokes' solution remains to this day the only successful, completely theoretical derivation for the drag on an immersed body of finite thickness.

SLIGHTLY INERTIAL (OSEEN) FLOW

In 1910 Oseen [10] used a linear approximation for the inertial terms to obtain an extension of the solution of Stokes for higher Reynolds numbers. That solution is

$$\tilde{\psi} = \frac{3va}{2}(1 + \cos\{\theta\})(1 - e^{-u_\infty R(1-\cos\{\theta\})/2v})$$

$$- u_\infty R^2\left(\frac{1}{2} + \frac{1}{4}\left(\frac{a}{R}\right)^3\right)\sin^2\{\theta\} \tag{16.18}$$

and

$$\frac{2(\mathscr{P} - \mathscr{P}_\infty)}{\varrho u_\infty^2} = -\frac{6}{\mathrm{Re}_D}\left(\frac{a}{R}\right)^2 \cos\{\theta\} + \left(\frac{a}{R}\right)^3\left(\frac{1 - 3\cos^2\{\theta\}}{2}\right) \quad (16.19)$$

Figure 16–4B is a plot for $\mathrm{Re}_D = 5$ of Equation 16.18 with $u_\infty R^2 \sin^2\{\theta\}/2$ added to the right side, and Figure 16–5 of Equation 16.19 for $R = a$. It follows from Equations 16.18 and 16.19 that

$$\bar{C}_t = \frac{12}{\mathrm{Re}_D}\left(1 + \frac{3}{16}\mathrm{Re}_D\right) \quad (16.20)$$

The added term with respect to Equation 16.17 is entirely due to skin friction, since the added pressure term is symmetrical with respect to the direction of flow (see Figure 16–5). Unfortunately, as shown subsequently, the added term for the drag does not represent a significant improvement of Stokes' law.

Just as for the cylinder, numerous attempts have been made to obtain higher-order terms. A historical survey of this work and an interpretive discussion of the method is provided by Proudman and Pearson [11], who derived the three-term expansion

$$\bar{C}_t = \frac{12}{\mathrm{Re}_D}\left(1 + \frac{3}{16}\mathrm{Re}_D + \frac{9}{160}\mathrm{Re}_D^2 \ln\left\{\frac{\mathrm{Re}_D}{2}\right\}\right) \quad (16.21)$$

Chester and Breach [12] extended this expansion to

$$\bar{C}_t = \frac{12}{\mathrm{Re}_D}\left[1 + \frac{3}{16}\mathrm{Re}_D + \frac{9}{160}\mathrm{Re}_D^2\left(\ln\{2^{2/3}\mathrm{Re}_D\} + \gamma - \frac{323}{360}\right)\right.$$
$$\left. + \frac{27}{640}\mathrm{Re}_D^3 \ln\left\{\frac{\mathrm{Re}_D}{2}\right\}\right] \quad (16.22)$$

They assert that further terms would not improve the range of approximation ($\mathrm{Re}_D < 1$). Tomotika and Aoi [13] earlier showed that for all orders of expansion the Oseen-type approximation gives

$$\bar{C}_p = \frac{1}{2}\bar{C}_f = \frac{1}{3}\bar{C}_t \quad (16.23)$$

just as for Stokes flow.

Despite the preceding restriction on the range of Re_D with respect to $\bar{C}_t$, Van Dyke [14] used a two-term expansion to derive an expression for the envelope of the stationary wake behind the sphere. He first derived the following approximation for the Stokes stream function in the vicinity of the surface:

$$\tilde{\psi} = -a^2 u_\infty\left(\frac{R}{a} - 1\right)^2 \frac{\sin^2\{\theta\}}{4}\left[\left(1 + \frac{3}{16}\mathrm{Re}_D\right)\left(2 + \frac{a}{R}\right)\right.$$
$$\left. - \frac{3}{16}\mathrm{Re}_D\left(2 + \frac{a}{R} + \left(\frac{a}{R}\right)^2\right)\cos\{\theta\}\right] \quad (16.24)$$

He then noted that $\tilde{\psi}$, as given in Equation 16.24, vanishes not only along the surface of the sphere and along the axis of symmetry but also for

$$\cos\{\theta\} = \left(\frac{16}{3\,\mathrm{Re}_D} + 1\right)\left(2 + \frac{a}{R}\right)\Big/\left(2 + \frac{a}{R} + \left(\frac{a}{R}\right)^2\right) \tag{16.25}$$

This is the envelope of the wake. The length of the standing eddy can be determined by setting $\theta = 0$, which reduces Equation 16.25 to

$$2\left(\frac{R}{a}\right)^2 + \frac{R}{a} - \frac{3\,\mathrm{Re}_D}{16} = 0 \tag{16.26}$$

Equation 16.26 has solutions

$$\frac{R}{a} = \frac{-1 \pm \sqrt{1 + (3/2)\,\mathrm{Re}_D}}{4} \tag{16.27}$$

The negative root has no physical meaning, but the positive one gives

$$\frac{l_w}{D} = \frac{1}{2}\left(\frac{R}{a} - 1\right) = \frac{1}{8}(\sqrt{1 + (3/2)\,\mathrm{Re}_D} - 1) - \frac{1}{2} \tag{16.28}$$

Furthermore, setting $R = a$ in Equation 16.26 yields $\mathrm{Re}_D = 16$ as the value for which separation just begins. Setting $R = a$ in Equation 16.25 likewise gives the following expression for the angle of separation:

$$\theta_S = \pi - \theta_S' = \cos^{-1}\left\{\frac{4}{\mathrm{Re}_D} + \frac{3}{4}\right\} \tag{16.29}$$

BOUNDARY-LAYER FLOW

Equation 14.2 is applicable as an approximation for the thin-boundary-layer regime on a sphere as well, with U_x by Equation 16.4, and x measured in the negative θ-direction, hence by

$$U_x = \frac{3}{2}u_\infty \sin\left\{\frac{x}{a}\right\} \tag{16.30}$$

Solution for Region near Point of Incidence

Near the forward point of the sphere Equation 16.30 can be approximated by

$$U_x = \frac{3}{2}u_\infty\left[\frac{x}{a} - \frac{1}{3!}\left(\frac{x}{a}\right)^3 + \cdots\right] \Rightarrow \frac{3u_\infty x}{2a} \tag{16.31}$$

Hence the force-momentum equation becomes

$$u_x \frac{\partial u_x}{\partial x} + u_y \frac{\partial u_x}{\partial y} = \frac{9u_\infty^2 x}{4a^2} + v \frac{\partial^2 u_x}{\partial y^2} \tag{16.32}$$

Equation 16.32 is equivalent to Equation 14.27 with $\alpha \to 3u_\infty/2a$, $r \to x$ and $z \to y$. Equation 14.24 is applicable in the same terms. It follows that the solution is

$$u_x = -u_\theta = \frac{3u_\infty x \tilde{\psi}'\{\tilde{\xi}\}}{a} = 3u_\infty(\pi - \theta)\tilde{\psi}'\{\tilde{\xi}\} \tag{16.33}$$

and

$$u_y = u_R = -\sqrt{\frac{3vu_\infty}{a\tilde{\psi}\{\tilde{\xi}\}}} \tag{16.34}$$

where

$$\tilde{\xi} = y\sqrt{\frac{3u_\infty}{av}} \tag{16.35}$$

and

$$\tilde{\psi} = \frac{-\tilde{\psi}}{x^2}\sqrt{\frac{4a}{3u_\infty v}} \tag{16.36}$$

$\tilde{\psi}\{\tilde{\xi}\}$, $\tilde{\psi}'\{\tilde{\xi}\}$, $\tilde{\psi}''\{\tilde{\xi}\}$ are given in Table 14.3 (for $A = \sqrt{2}$). If follows that

$$\left(\frac{\tau_w}{\varrho u_\infty^2}\right)\sqrt{\mathrm{Re}_D} = 3\sqrt{\frac{3}{2}}\left(\frac{x}{a}\right)\psi\{0\} = 3.4085\frac{x}{a} = 3.4085\theta' \tag{16.37}$$

A more rigorous procedure than that used for the cylinder will now be illustrated for the derivation of the pressure at the point of incidence. The momentum equation in the y-direction, normal to the surface, corresponding to Equation 16.32 for the x-direction, is

$$u_y \frac{\partial u_y}{\partial y} + u_x \frac{\partial u_y}{\partial x} + u_z \frac{\partial u_z}{\partial z} = -\frac{1}{\varrho}\frac{\partial \mathcal{P}}{\partial y} + v\left(\frac{\partial^2 u_y}{\partial x^2} + \frac{\partial^2 u_y}{\partial y^2} + \frac{\partial^2 u_y}{\partial z^2}\right) \tag{16.38}$$

Introducing the change of variables (after Homann [15])

$$\frac{u_x}{x} = \frac{u_z}{z} = f'\{y\} \tag{16.39}$$

$$u_y = -2f\{y\} \tag{16.40}$$

and

$$\mathcal{P}_0 - \mathcal{P} = \frac{9}{8}\varrho u_\infty^2\left(\left(\frac{x}{a}\right)^2 + \left(\frac{z}{a}\right)^2 + F\{y\}\right) \tag{16.41}$$

where $\mathscr{P}_0$ is the unknown dynamic pressure on the surface at the point of incidence ($\theta = 0$, $y = 0$), gives

$$ff' = \frac{9}{8}u_\infty^2 F' - 2\nu f'' \tag{16.42}$$

The boundary conditions can be seen from Equations 16.29–16.31 to be $f'\{0\} = f\{0\} = F\{0\}$, respectively. Integration of Equation 16.42 with these conditions gives

$$\frac{9u_\infty^2}{8}F = \frac{f^2}{2} + 2\nu f' \tag{16.43}$$

Substitution of this expression for F into Equation 16.41 gives

$$\mathscr{P}_0 - \mathscr{P} = \frac{9u_\infty^2 x^2}{8a^2} + \varrho\left(\frac{f^2}{2} + 2\nu f'\right) \tag{16.44}$$

Specializing Equation 16.44 for the axis of symmetry gives

$$\mathscr{P}_0 - \mathscr{P}\{0, y\} = \varrho\left(\frac{f^2\{y\}}{2} + 2\nu f'\{y\}\right) \tag{16.45}$$

For a point just outside the boundary layer where $\mathscr{P}\{0, y\} = \mathscr{P}_\infty$, $f'\{y\} = U_x/x \cong 3u_\infty/2a$, and $f\{y\} = -U_y = -u_\infty$. Hence

$$\mathscr{P}_0 - \mathscr{P}_\infty = \varrho\left(\frac{u_\infty^2}{2} + \frac{3\nu u_\infty}{a}\right) \tag{16.46}$$

or

$$\frac{2(\mathscr{P}_0 - \mathscr{P}_\infty)}{\varrho u_\infty^2} = 1 + \frac{12}{\text{Re}_D} \tag{16.47}$$

Homann [15] took into account the displacement thickness at the point of incidence and derived the modified expression

$$\frac{2(\mathscr{P}_0 - \mathscr{P}_\infty)}{\varrho u_\infty^2} = 1 + \frac{12}{\text{Re}_D + 0.643\,\text{Re}_D^{1/2}} \tag{16.48}$$

The range of validity of the approximations represented by Equations 16.33, 16.34, 16.37, and 16.47 can be extended to somewhat higher angles by reversing the approximation of $\sin\{\theta\}$ by θ, yielding

$$u_\theta = -3u_\infty \sin\{\theta\}\tilde{\Psi}'\{\xi\} \tag{16.49}$$

$$u_R = \sqrt{\frac{3\nu u_\infty}{a}}\cos\{\theta\}\tilde{\Psi}\{\xi\} \tag{16.50}$$

$$\frac{\tau_w}{\varrho u_\infty^2}\sqrt{\mathrm{Re}_D} = 3.4085 \sin\{\theta'\} \tag{16.51}$$

and

$$\frac{2(\mathscr{P}_0 - \mathscr{P}_\infty)}{\varrho u_\infty^2} = \left(1 + \frac{12}{\mathrm{Re}_D}\right)\left(1 - \frac{9}{4}\sin^2\{\theta'\}\right) \tag{16.52}$$

General Solution

Frössling [16] used the procedures of Blasius [17] to develop a solution for the thin boundary layer over a solid sphere. This solution has the same form as for a cylinder, but the coefficients a_i are multiplied by 3/4, corresponding to the ratio of the inviscid velocity fields, and the functions f_i, g_i, and so on, are different. Some of the functions computed by Frössling are given in Table 16.1. The velocity distribution in the boundary layer is

$$\begin{aligned}
\frac{u_x}{u_\infty} &= \frac{3\theta'}{2}f_1' - \frac{(\theta')^3}{2} - (g_3' + h_3') \\
&+ \frac{3(\theta')^5}{80}\left(g_5' + h_5' + \frac{10}{3}j_5' + \frac{10}{3}k_5' + \frac{10}{3}q_5'\right) \\
&- \frac{(\theta')^7}{840}\left(g_7' + h_7' + \frac{7}{3}j_7' + \frac{7}{3}k_7' + 7l_7' + 7p_7' + \frac{7}{3}q_7'\right. \\
&+ \left.\frac{7}{3}t_7' + 7v_7' + 7z_7'\right)
\end{aligned} \tag{16.53}$$

where $\eta = y\sqrt{3u_\infty/av}$. The corresponding local shear stress can then be shown to be

$$\frac{\tau_w}{\varrho u_\infty^2}\sqrt{\mathrm{Re}_D} = 3.4086\theta' - 1.3370(\theta')^3 + 0.1435(\theta')^5$$
$$- 0.00883(\theta')^7 \tag{16.54}$$

This expression indicates that $\tau_w = 0$ at $\theta' = 1.9136$ rad (109.6°). The corresponding mean drag coefficient due to skin friction and based on the projected area is

$$\bar{C}_f\sqrt{\mathrm{Re}_D} = 2\int_0^{1.9136} \frac{\tau_w}{\varrho u_\infty^2}\sqrt{\mathrm{Re}_D}\sin^2\{\theta'\}\,d\theta' = 3.4640 \tag{16.55}$$

Yuge [18] carried out equivalent calculations but used a slightly different formulation and the free-stream velocity distribution

$$\frac{U_x}{u_\infty} - 1.398095\theta' - 0.189389(\theta')^3 - 0.0419(\theta')^5 \tag{16.56}$$

Table 16.1
Computed Functions for Thin-Boundary-Layer Solution of Frössling for a Sphere [16]

η	f_1	f_1'	f_1''	g_3	g_3'	g_3''	h_3	h_3'	h_3''
0.0	0	0	0.9277	0	0	1.0475	0	0	0.0448
.1	.0046	.0903	.8777	.0051	.0998	.9477	.0002	.0044	.0448
.2	.0179	.1755	.8277	.0196	.1896	.8488	.0009	.0090	.0444
.3	.0395	.2558	.7778	.0427	.2696	.7517	.0020	.0133	.0434
.4	.0689	.3311	.7282	.0732	.3400	.6574	.0036	.0176	.0416
.5	.1056	.4014	.6788	.1104	.4012	.5666	.0055	.0217	.0391
.6	.1490	.4669	.6300	.1532	.4535	.4802	.0079	.0254	.0356
.7	.1988	.5275	.5819	.2008	.4974	.3986	.0106	.0288	.0314
.8	.2544	.5833	.5348	.2524	.5334	.3227	.0136	.0316	.0265
.9	.3153	.6345	.4888	.3072	.5621	.2528	.0169	.0340	.0210
1.0	.3811	.6811	.4443	.3646	.5842	.1895	.0204	.0358	.0152
1.1	.4514	.7234	.4014	.4239	.6002	.1328	.0241	.0370	.0091
1.2	.5256	.7614	.3604	.4845	.6110	.0832	.0278	.0377	.0032
1.3	.6035	.7954	.3215	.5459	.6171	.0403	.0316	.0377	−.0026
1.4	.6846	.8258	.2850	.6078	.6193	.0044	.0353	.0372	−.0080
1.5	.7686	.8526	.2508	.6696	.6182	−.0251	.0390	.0361	−.0127
1.6	.8550	.8761	.2192	.7313	.6144	−.0483	.0425	.0346	−.0168
1.7	.9437	.8966	.1901	.7925	.6087	−.0657	.0459	.0327	−.0202
1.8	1.0342	.9142	.1637	.8530	.6015	−.0780	.0491	.0306	−.0228
1.9	1.1264	.9294	.1398	.9127	.5932	−.0857	.0520	.0282	−.0244
2.0	1.2200	.9422	.1185	.9716	.5845	−.0894	.0547	.0258	−.0254
2.1	1.3148	.9530	.0996	1.0296	.5755	−.0898	.0572	.0233	−.0256
2.2	1.4106	.9622	.0831	1.0867	.5666	−.0876	.0594	.0207	−.0252
2.3	1.5072	.9698	.0688	1.1430	.5580	−.0834	.0613	.0182	−.0243
2.4	1.6045	.9760	.0564	1.1984	.5500	−.0776	.0630	.0158	−.0229
2.5	1.7024	.9811	.0458	1.2530	.5425	−.0709	.0645	.0136	−.0212
2.6	1.8007	.9853	.0370	1.3069	.5358	−.0637	.0657	.0116	−.0193
2.7	1.8994	.9886	.0296	1.3602	.5298	−.0563	.0668	.0097	−.0174
2.8	1.9984	.9912	.0234	1.4129	.5245	−.0490	.0677	.0081	−.0153
2.9	2.0977	.9932	.0184	1.4651	.5200	−.0420	.0684	.0067	−.0133
3.0	2.1971	.9949	.0143	1.5169	.5161	−.0356	.0690	.0054	−.0114
3.1	2.2966	.9962	.0110	1.5683	.5128	−.0297	.0695	.0044	−.0097
3.2	2.3963	.9972	.0085	1.6195	.5102	−.0245	.0699	.0035	−.0082
3.3	2.4961	.9979	.0064	1.6704	.5079	−.0200	.0702	.0028	−.0067
3.4	2.5959	.9985	.0048	1.7211	.5061	−.0161	.0705	.0022	−.0054
3.5	2.6958	.9989	.0036	1.7716	.5047	−.0128	.0706	.0016	−.0044
3.6	2.7957	.9992	.0026	1.8220	.5036	−.0101	.0708	.0013	−.0035
3.7	2.8956	.9995	.0020	1.8723	.5027	−.0078	.0709	.0010	−.0028
3.8	2.9956	.9996	.0014	1.9226	.5020	−.0060	.0710	.0007	−.0021
3.9	3.0955	.9997	.0010	1.9727	.5015	−.0046	.0711	.0006	−.0016
4.0	3.1955	.9998	.0007	2.0229	.5011	−.0034	.0711	.0004	−.0013
4.1	3.2955	.9999	.0005	2.0730	.5008	−.0026	.0711	.0002	−.0009
4.2	3.3955	.9999	.0004	2.1230	.5006	−.0019	.0712	.0002	−.0007
4.3	3.4955	.9999	.0003	2.1731	.5004	−.0014	.0712	.0002	−.0006
4.4	3.5954	.9999	.0002	2.2231	.5003	−.0010	.0712	.0002	−.0004
4.5	3.6954	1.0000	.0001	2.2731	.5002	−.0007		.0000	−.0002
4.6			.0001	2.3231	.5001	−.0005			−.0002
4.7			.0000	2.3732	.5001	−.0004			−.0001
4.8					.5000	−.0002			−.0001
4.9						−.0001			−.0000
5.0						−.0001			
5.1						−.0001			
5.2						−.0000			

Table 16.1 continued

η	g_5	g_5'	g_5''	h_5	h_5'	h_5''	k_5	k_5'	k_5''	j_5	j_5'	j_5''	q_5	q_5'	q_5''
0.0	0	0	0.9054	0	0	0.0506	0	0	0.1768	0	0	0.0291	0	0	−0.0244
.2	.0168	.1612	.7075	.0010	.0101	.0500	.0029	.0255	.0790	.0006	.0058	.0278	−.0005	−.0049	−.0242
.4	.0619	.2838	.5210	.0040	.0198	.0467	.0090	.0324	−.0068	.0022	.0107	.0210	−.0019	−.0096	−.0230
.6	.1279	.3709	.3541	.0089	.0285	.0396	.0148	.0241	−.0724	.0047	.0137	.0074	−.0043	−.0140	−.0203
.8	.2082	.4270	.2123	.0153	.0354	.0289	.0179	.0051	−.1132	.0075	.0134	−.0104	−.0075	−.0176	−.0164
1.0	.2971	.4576	.0984	.0229	.0399	.0159	.0165	−.0195	−.1284	.0099	.0096	−.0280	−.0113	−.0204	−.0115
1.2	.3899	.4683	.0128	.0311	.0417	.0024	.0101	−.0447	−.1204	.0111	.0025	−.0412	−.0156	−.0222	−.0062
1.4	.4834	.4645	−.0459	.0394	.0409	−.0099	−.0011	−.0665	−.0948	.0108	−.0064	−.0467	−.0201	−.0229	−.0008
1.6	.5751	.4515	−.0808	.0473	.0379	−.0195	−.0161	−.0819	−.0585	.0085	−.0156	−.0437	−.0247	−.0226	.0043
1.8	.6637	.4335	−.0964	.0544	.0334	−.0256	−.0334	−.0897	−.0194	.0046	−.0234	−.0336	−.0291	−.0212	.0085
2.0	.7484	.4139	−.0974	.0606	.0279	−.0282	−.0515	−.0899	.0161	−.0006	−.0287	−.0191	−.0331	−.0192	.0117
2.2	.8293	.3952	−.0888	.0656	.0223	−.0277	−.0689	−.0838	.0432	−.0067	−.0310	−.0037	−.0367	−.0167	.0136
2.4	.9066	.3787	−.0750	.0695	.0170	−.0250	−.0847	−.0733	.0599	−.0129	−.0304	.0095	−.0398	−.0139	.0142
2.6	.9810	.3652	−.0594	.0724	.0124	−.0209	−.0981	−.0605	.0662	−.0187	−.0275	.0187	−.0423	−.0111	.0136
2.8	1.0530	.3549	−.0445	.0745	.0086	−.0165	−.1089	−.0474	.0642	−.0238	−.0232	.0234	−.0442	−.0085	.0122
3.0	1.1231	.3473	−.0317	.0760	.0058	−.0122	−.1171	−.0352	.0565	−.0279	−.0184	.0241	−.0457	−.0062	.0103
3.2	1.1920	.3420	−.0215	.0769	.0037	−.0086	−.1231	−.0249	.0460	−.0311	−.0137	.0219	−.0467	−.0044	.0082
3.4	1.2601	.3385	−.0139	.0775	.0023	−.0058	−.1272	−.0168	.0350	−.0334	−.0097	.0181	−.0474	−.0029	.0061
3.6	1.3275	.3363	−.0086	.0778	.0013	−.0037	−.1300	−.0109	.0251	−.0351	−.0065	.0138	−.0479	−.0019	.0044
3.8	1.3947	.3350	−.0051	.0780	.0008	−.0022	−.1317	−.0067	.0170	−.0361	−.0042	.0099	−.0482	−.0012	.0030
4.0	1.4616	.3342	−.0029	.0781	.0004	−.0013	−.1328	−.0040	.0109	−.0368	−.0025	.0067	−.0484	−.0007	.0019
4.2	1.5284	.3338	−.0016	.0782	.0002	−.0007	−.1334	−.0022	.0066	−.0372	−.0015	.0042	−.0485	−.0004	.0012
4.4	1.5951	.3336	−.0008	.0782	.0001	−.0004	−.1337	−.0012	.0039	−.0374	−.0008	.0026	−.0486	−.0002	.0007
4.6	1.6618	.3334	−.0004	.0782	.0000	−.0002	−.1339	−.0006	.0022	−.0375	−.0004	.0014	−.0486	−.0001	.0004
4.8	1.7285	.3334	−.0002	.0783	.0000	−.0001	−.1340	−.0003	.0012	−.0376	−.0002	.0008	−.0486	−.0000	.0002
5.0	1.7952	.3334	−.0001	.0783	.0000	−.0000	−.1340	−.0001	.0006	−.0376	−.0000	.0004	−.0486	−.0000	.0001
5.2	1.8618	.3334	−.0000	.0783	.0000	−.0000	−.1340	−.0000	.0003	−.0376	−.0000	.0002	−.0486	−.0000	.0001
5.4	1.9285	.3333	−.0000	.0783	.0000	−.0000	−.1340	−.0000	.0001	−.0376	−.0000	.0001	−.0486	−.0000	.0000
5.6									.0000			.0000			

to obtain

$$\frac{\tau_w}{\varrho u_\infty^2}\sqrt{\text{Re}_D} = 3.06710' - 0.98786(\theta')^3 - 0.23143(\theta')^5 \qquad (16.57)$$

With this latter expression, $\tau_w = 0$ at $\theta' = 1.444$ rad (82.75°). Hence

$$\bar{C}_f\sqrt{\text{Re}_D} = 2\int_0^{1.444} \frac{\tau_w}{\varrho u_\infty^2}\sqrt{\text{Re}_D}\sin^2\{\theta'\}d\theta' = 1.8065 \qquad (16.58)$$

The significant difference between Equations 16.55 and 16.58 is due primarily to the difference in Equations 16.30 and 16.56, hence in 16.54 and 16.57 in the region of $\theta \cong \pi/2$ for which $\sin^2\{\theta\} \cong 1$. Equation 16.56 is based on the pressure measurements of Fage [19] at $\text{Re}_D = 1.57 \times 10^5$ and Bernoulli's law (Equation 9.77).

Tomotika and Imai [20] used

$$\frac{U_x}{u_\infty} = 1.500' - 0.43707(\theta')^3 + 0.148097(\theta')^5 - 0.042329(\theta')^7 \qquad (16.59)$$

which is based on these same experimental measurements, and again Bernoulli's law, to derive a solution for $\bar{C}_f$ by integral boundary-layer theory.

NEWTON'S EQUATION FOR LARGE Re$_D$

Newton [21], p. 336, summarized his theoretical analysis of the terminal velocity of a falling sphere as follows:

> I have exhibited in this Proposition the resistance and retardation of spherical projectiles in mediums that are not continued, and shown that this resistance is to the force by which the whole motion of the globe may be destroyed or produced in the time in which the globe can describe two-thirds of its diameter, with a velocity uniformly continued, as the density of the medium is to the density of the globe, provided the globe and the particles of the medium be perfectly elastic, and are endued with the utmost force of reflection; and that this force, where the globe and particles of the medium are infinitely hard and void of any reflecting force, is diminished one-half. But in continued mediums, as water, hot oil, and quicksilver, the globe as it passes through them does not immediately strike against all the particles of the fluid that generate the resistance made to it, but presses only the particles that lie next to it, which press the particles beyond, which press other particles, and so on; and in these mediums the resistance is diminished one other half. A globe in these extremely fluid mediums meets with a resistance that is to the force by which its whole motion may be destroyed or generated in the time wherein it can describe, with that motion uniformly continued, eight third parts of its diameter, as the density of the medium is to the density of the globe. This I shall endeavor to show in what follows.

From this he concluded, p. 352, that "The resistances of globes in infinite compressed mediums are in a ratio compounded of the squared ratio of the velocity, and the squared ratio of the diameter, and the ratio of the density of the mediums." It follows quantitatively from these statements (see problem 13) that

$$\bar{C}_t = \frac{1}{4}$$ (16.60)

He confirmed the equivalent of Equation 16.60, which is called *Newton's law of drag*, experimentally by timing the fall of spheres of different density in St. Paul's Cathedral, London, in 1710 (see problem 18).

The reasoning that led to the proportionality of the drag force to the square of the diameter, the inverse of the density of the fluid, and the inverse of the square of the velocity is sound, but that which led to the numerical value of 1/4 is certainly specious. It may have been a rationalization for the observed behavior.

NUMERICAL SOLUTIONS

Just as for cylinders, numerical solutions provide a valuable theoretical supplement to the analytical solutions for creeping, slightly inertial, and thin-boundary-layer flow over a sphere, since the solutions for slightly inertial flow are not valid for $Re_D > 1$, and the thin-boundary-layer solutions are of limited utility with respect to the drag and other characteristics that depend upon the behavior in the wake.

Numerical solutions of increasing accuracy and range have followed the pioneering work of Kawaguti [22] in 1948 with the Galerkin (weighted residuals) method, and in 1950 with the finite-difference method. The first extensive results by the finite-difference method are those of Jenson [23] in 1959, which were carried out on a mechanical desk calculator! Solutions of confirmed accuracy have now been obtained for Re_D up to 1000, and extension in the near future up to the point of transition appears probable.

A comparison of some of the results for $Re_D \leq 40$ is reproduced in Table 16.2. Additional computed results are given in Tables 16.3–16.8. Some of the computed results are also illustrated in Figures 16–8, 16–10 to 16–14, and, subsequently, in 16–25. Figure 16–10 shows the Stokes stream functions

Table 16.2
Comparison by Dennis and Walker [24] of Computed Total Mean Drag Coefficients for Flow about a Sphere

Re_D	*Jenson* [23]	*Dennis & Walker* [24]	*LeClair, Hamielec, & Pruppacher*	*Rimon & Cheng* [28]	*Dennis & Walker* [24]
0.1	—	—	122.04	—	122.10
1	—	13.5	13.66	—	13.72
5	3.98	—	3.515	—	3.605
10	2.42	2.06	2.144	2.205	2.212
20	1.473	1.32	1.356	—	1.365
40	0.930	—	0.930	0.930	0.904

Table 16.3
Comparison by Dennis and Walker [24] of Analytical and Computed Total Mean Drag
Coefficients for Flow about a Sphere at Small Re_D

Re_D	Analytical			Computed
	Goldstein	*Proudman-Pearson*	*Chester-Breach*	*Dennis & Walker*
0.1	122.23	122.05	122.09	122.10
0.2	62.22	61.94	62.01	62.02
0.3	42.20	41.87	41.95	42.02
0.4	32.19	31.82	31.91	31.91
0.5	26.17	25.78	25.88	25.85
0.6	22.16	21.76	21.88	21.85
0.7	19.29	18.90	19.03	18.96
0.8	17.13	16.76	16.91	16.76
0.9	15.46	15.10	15.28	15.10
1.0	14.11	13.78	14.00	13.72

Table 16.4
Mean Values of Friction, Pressure, and Total Coefficients for Flow about a Sphere as
Computed by Dennis and Walker [24]

Re_D	$\bar{C}_f$	$\bar{C}_p$	$\bar{C}_t$	$k(0)$	$k(\pi)$
0.1	81.40	40.70	122.10	-60.07	62.03
0.2	41.34	20.67	62.02	-30.05	31.97
0.5	17.23	8.622	25.85	-12.02	13.86
0.8	11.17	5.598	16.76	-7.516	9.289
1	9.134	4.585	13.72	-6.017	7.753
5	2.369	1.236	3.605	-1.203	2.599
10	1.427	0.785	2.212	-0.654	1.878
20	0.854	0.512	1.365	-0.322	1.471
40	0.536	0.368	0.904	-0.192	1.261

Here $k\{0\} = 2(\mathscr{P}_0 - \mathscr{P}_\infty)/\varrho u_\infty^2$ and $k\{\pi\} = 2(\mathscr{P}_\pi - \mathscr{P}_\infty)/\varrho u_\infty^2$

Table 16.5
Mean Coefficients for Flow about a Sphere as
Computed Numerically by Ihme et al. [25]

Re_D	$\bar{C}_t$	$\dfrac{\bar{C}_f}{\bar{C}_p}$
0.1	126.15	(2.21)
1	13.95	2.01
10	2.190	1.88
20	1.365	1.70
40	0.865	1.76
60	0.660	1.80
80	0.550	1.97

Table 16.6
Mean Coefficients and Dimensionless Incident Pressure for Flow about a Sphere as Computed by Jenson [23]

Re_D	5	10	20	40
$\bar{C}_f$	2.65	1.55	0.9285	0.556
$\bar{C}_p$	1.325	0.87	0.5445	0.3745
$\bar{C}_t$	3.975	2.42	1.473	0.9305
$2(\mathscr{P}_0 - \mathscr{P}_\infty)/\varrho u_\infty^2$	2.742	1.962	1.508	1.264

Table 16.7
Total, Mean Drag Coefficients for Flow about a Sphere as Computed by Hamielec et al. [26]

Re_D	0.1	1.0	20	40	100
$\bar{C}_t$	131	13.7	1.39	0.93	0.56

Table 16.8
Mean Drag Coefficients for Flow about a Sphere as Computed by Lin and Lee [27]

Re_D	5	20	40	100
$\bar{C}_f$	2.4375	0.8924	0.5651	0.3090
$\bar{C}_p$	1.2586	0.5097	0.3372	0.2484
$\bar{C}_t$	3.6961	1.4021	0.9023	0.5574

computed by Jenson for Re_D = 5, 10, 20, and 40, and Figure 16–8 those of Brauer and Sucker [4] for Re_D = 100. Figures 16–11 and 16–12 show the computed variation of the dimensionless pressure and shear stress with angle computed by Ihme et al. [25] for several values of Re_D. Figures 16–13 and 16–14 show the local pressure and vorticity at the surface (proportional to the shear stress) as computed by Rimon and Cheng [28] for higher Re_D. These and other quantities are compared with experimental data in the next section.

COMPARISON OF SOLUTIONS WITH EXPERIMENTAL MEASUREMENTS

Experimental measurements are more difficult for a sphere than for a cylinder because of the need to use a support that interferes with the flow, or of letting a sphere fall in a stagnant fluid.

Pressure

The pressure distribution determined experimentally by Fage [19] at Re_D in the upper critical range is shown in Figure 16–15. The pressure distribution of

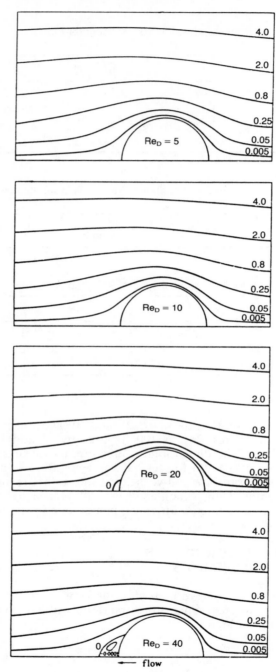

FIGURE 16–10   *Numerically computed Stokes stream functions for flow about a sphere. (From Jenson [23].)*

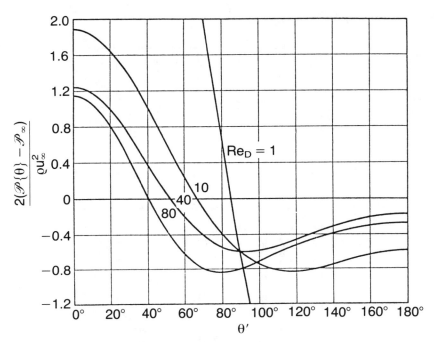

FIGURE 16–11 *Numerically computed pressure distributions on the surface of a sphere at low Re_D. (From Ihme et al. [25].)*

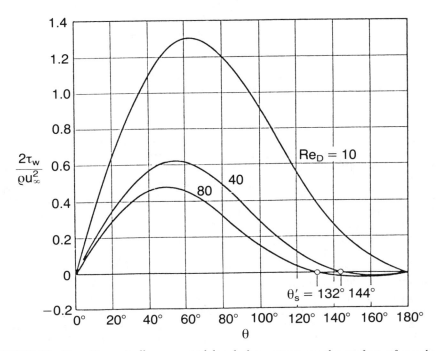

FIGURE 16–12 *Numerically computed local shear stress on the surface of a sphere. (From Ihme et al. [25].)*

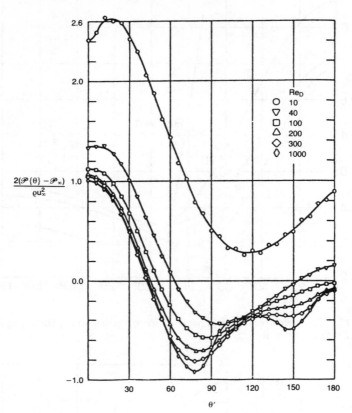

FIGURE 16–13   *Numerically computed pressure distributions on the surface of a sphere at moderate Re$_D$. (From Rimon and Cheng [28].)*

Equation 16.59 and of 16.6 [inviscid flow] are also plotted in Figure 16–15. Equation 16.59, which was actually based on these very measurements for Re$_D$ = 1.57 × 10^5, is seen to be an upper bound and Equation 16.6 a lower bound for these measurements. The point of separation is indicated on the upper set of data for subcritical flow (Re$_D$ = 1.57 × 10^5) and the point of transition from a laminar to a turbulent boundary layer on the three lower sets of data for supercritical flow.

The variation of the pressure at the point of incidence on the sphere is examined in Figures 16–16 and 16–17. In Figure 16–16, for very small Re$_D$, the numerically computed values of Dennis and Walker [24] are seen to approach Equation 16.11 of Stokes for Re$_D$ ≤ 0.2, as expected, and to approach Equation 16.48 of Homann as Re$_D$ increases. Equation 16.19 of Oseen is seen to be less accurate than Equation 16.11 for all Re$_D$. In Figure 16–17, for somewhat larger Re$_D$, Equation 16.48 of Homann is seen to provide a good representation for his own data. Stokes law (Equation 16.11) grossly underpredicts the pressure in this regime, whereas Equation 16.47 becomes a good approximation for Re$_D$ > 40. The numerically computed values of Jenson [23], Dennis and Walker [24], Ihme

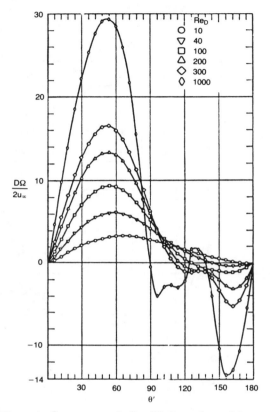

FIGURE 16–14 *Numerically computed distribution of vorticity on the surface of a sphere at moderate* Re_D. *(From Rimon and Cheng [28].)*

et al. [25], and Rimon and Cheng [28], which are not plotted because of overlap, agree closely with the experimental data in this range.

Wake

Experimental and computed values for the length of the stationary wake are seen in Figure 16–18 to be in good agreement. The minimum Reynolds number for formation of a wake appears to be 24. Other experimental estimates range from 20 to 25. The dotted curve, representing Equation 16.28 of Van Dyke [14], appears to provide a good representation only for $40 < Re_D < 120$, but the solid line, representing the empirical equation

$$\frac{l_w}{D} = 0.631 \ln\left\{\frac{Re_D}{24}\right\} \tag{16.61}$$

provides a good approximation from the onset of the wake up to $Re_D = 120$.

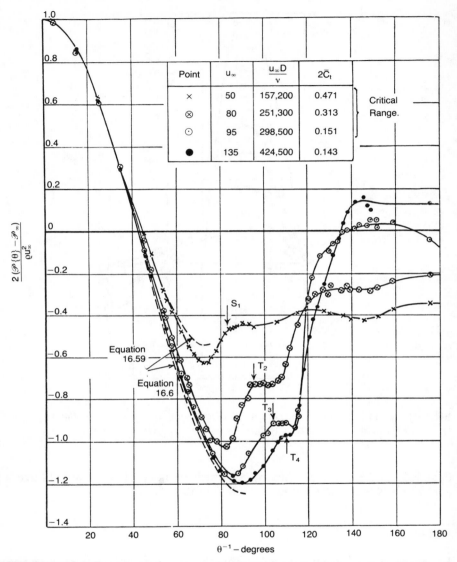

FIGURE 16–15 *Experimental pressure distributions on the surface of a sphere in the upper critical regime. S_1 designates point of separation, and T_2, T_3, and T_4 the points of transition. (From Fage [19].)*

For a cylinder the length of the wake was observed to vary linearly with Re_D (Equation 15.44), whereas for a sphere the dependence is logarithmic. Equation 16.25 appears in Figure 16–19 to provide a good representation for the experimental data of Taneda [30] for the entire boundary of the wake at $Re_D = 73.2$. However, Figure 16–20 indicates that such agreement may be fortuitous, since the corresponding prediction for the angle of separation (Equation 16.29)

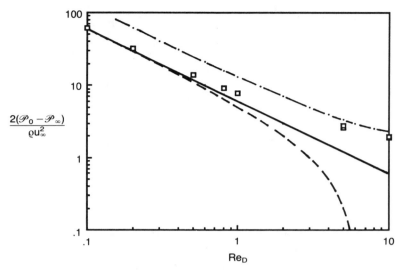

FIGURE 16–16   *Comparison of numerically computed values of Dennis and Walker [24] and theoretical expressions for the pressure at the point of incidence on a sphere at very low Re_D. ——— Equation 16.11 (Stokes); – – – – Equation 16.19 (Oseen); —·— Equation 16.48 (Homann).*

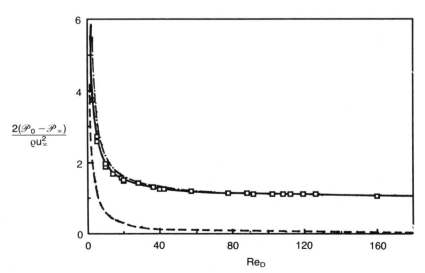

FIGURE 16–17   *Comparison of theoretical expressions with the experimentally measured values of Homann for the pressure at the point of incidence on a sphere for low Re_D. – – – – Equation 16.11 (Stokes); ——— Equation 16.48 (Homann); —·— Equation 16.47 (first-order boundary-layer theory).*

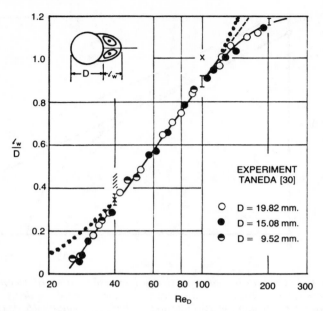

FIGURE 16–18   *Comparison of experimental data and computed values for the length of a stationary wake on a sphere with theoretical and empirical expressions.* ——— *correlating curve;* · · · · · *Equation 16.28 (theoretical);* – – – – *Equation 16.61 (empirical).* Computed values: *slant lines = Jenson [23]; crosses = Hamielec et al. [26]; vertical lines = Rimon and Cheng [28]. (Adapted from Cheng [29].)*

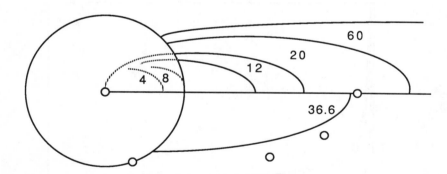

FIGURE 16–19   *Comparison of experimental data of Tanada [30] for* $Re_D = 73.2$ *with the predictions of Equation 16.25 for the boundary of the wake behind a sphere. Numbers designate* $Re_D/2$. *(After Van Dyke [14], p. 150.)*

is not satisfactory. The agreement of the experimental and *numerically* computed values for the angle of separation is, however, seen to be very good.

The frequency at which eddies are shed from a sphere has not been correlated as successfully as for a cylinder (see Figure 15–24). Müller [31] observed a vortex chain (Wirbelkette) with a constant Strouhal number, $Sr = \omega D/u_\infty = 0.4$, for $Re_D > 450$, as indicated by line B in Figure 16.21, plus

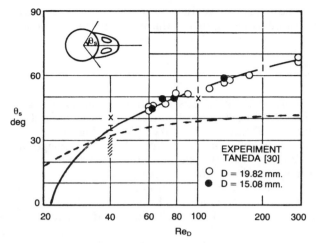

FIGURE 16–20 *Comparison of experimental data and computed values for the angle of separation on a sphere with Equation 16.29.* ——— *correlating curve;* ----- *Equation 16.29. Computed values: slant lines = Jenson [23]; crosses = Hamielec et al. [26]; vertical lines = Rimon and Cheng [28]. (Adapted from Cheng [29].)*

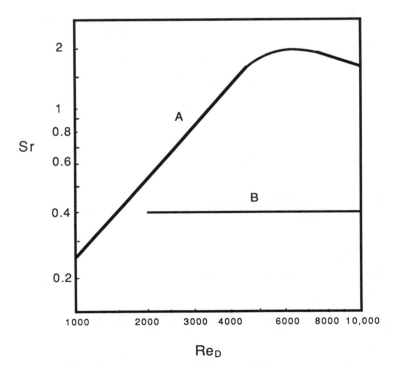

FIGURE 16–21 *Strouhal number for a sphere according to Müller [31]: curve A = Wirbelkolken (periodic balls of vorticity); curve B = Wirbelkette (vortex chain). (From Torobin and Gauvin [6].)*

discrete, periodic balls of vorticity (Wirbelkolken) at increasing Strouhal number as indicated by curve A, but this behavior has not been confirmed by all observers.

Drag Coefficients

The various theoretical expressions for the total mean drag are compared with computed and experimental results in Figure 16–22 and with the precise computed values of Dennis and Walker [24] in Figure 16–23. This choice of ordinate displays the deviations from Stokes' law and exaggerates the individual discrepancies. It is apparent that none of the approximations for slightly inertial flow is valid for $Re_D > 1$. The four-term Equation 16.22 of Chester and Breach appears the to be the best approximation up to $Re_D = 0.5$, but the three-term expansion of Proudman and Pearson's Equation 16.21 actually is better for $0.7 < Re_D < 1.0$. Figure 16–24 confirms the validity of the calculated values of Rimon and Cheng [28] up to $Re_D = 10^3$. A more extended range of experimental results is shown in Figure 16–25. The values for $Re_D < 10^5$ are well represented by the following empirical expression of Brauer [32]:

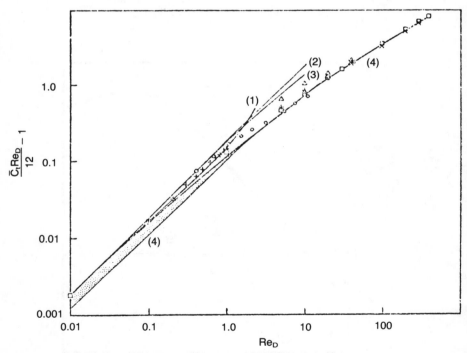

FIGURE 16–22 *Comparison of deviations of theoretical expressions, computed values, and experimental data for the total drag coefficient of a sphere at low Re_D from Stokes' law: (1) Proudman and Pearson; (2) Oseen; (3) Goldstein; (4) Pruppacher-Steinberger (theoretical), Pruppacher, Beard-Pruppacher (experimental); ■, experimental scatter; ○, Maxworthy (experimental); □, Le Clair, Hamielec and Pruppacher; △, Jenson; ×, Rimon and Cheng; +, present results (computed). (From Dennis and Walker [24].)*

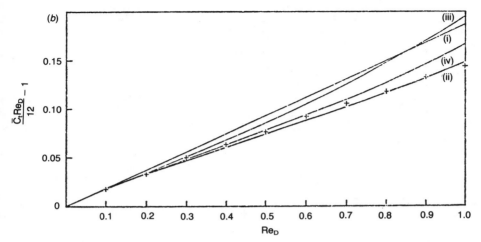

FIGURE 16–23 *Comparison of deviations of various theoretical expressions and their own computed values for the total drag coefficient of a sphere at low Re_D from Stokes' law: (i) Oseen (Equation 16.20); (ii) Proudman and Pearson (Equation 16.21); (iii) Chester and Breach (first three terms of Equation 16.22); (iv) Chester and Breach (Equation 16.22); +, computed values of Dennis and Walker. (From Dennis and Walker [24].)*

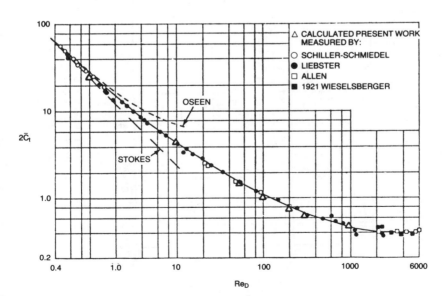

FIGURE 16–24 *Comparison of experimental and computed total drag coefficients for a sphere at intermediate Re_D. (From Rimon and Cheng [28].)*

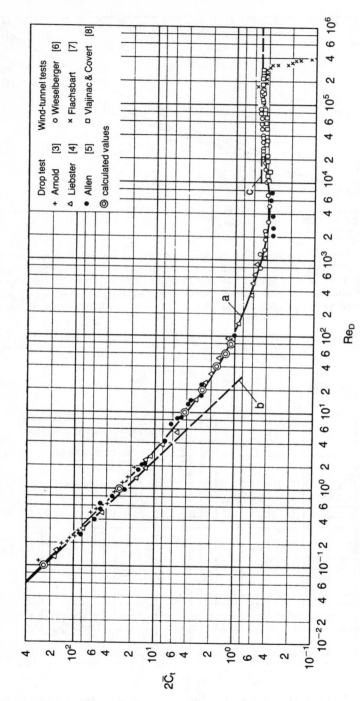

FIGURE 16–25 *Comparison of theoretical and empirical expressions for the total drag coefficient of a sphere with experimental and computed values for a wide range of Re_D: (a) Equation 16.62; (b) Equation 16.17 (Stokes' law); (c) Equation 16.60 (Newton's law). (From Brauer [32].)*

$$\bar{C}_t = \frac{12}{\mathrm{Re}_D} + \frac{1.865}{\mathrm{Re}_D^{1/2}} - \frac{2.415 \times 10^{-3}\,\mathrm{Re}_D^{1/2}}{1 + 3 \times 10^{-6}\,\mathrm{Re}_D^{3/2}} + 0.245 \qquad (16.62)$$

Equation 16.60 of Newton is an upper bound and a fair approximation for $10^3 < \mathrm{Re}_D < 3 \times 10^5$. The much simpler expression

$$\bar{C}_t^{1/2} = \left(\frac{12}{\mathrm{Re}_D}\right)^{1/2} + 0.371 \qquad (16.63)$$

derived by Weiner and Churchill [33] using the Churchill–Usagi model with Stokes' law and a pseudoasymptote of $\bar{C}_t = 0.138$, also provides a satisfactory representation of these values for $\mathrm{Re}_D < 3000$.

Theoretical results for $\bar{C}_f$ and $\bar{C}_p$ are plotted in Figure 16–26. Again, boundary-layer theory gives the correct trend for $\bar{C}_p$ for large Re_D, but unreliable absolute values because of the varying angle of separation and the effect of the wake on the velocity field outside the boundary layer.

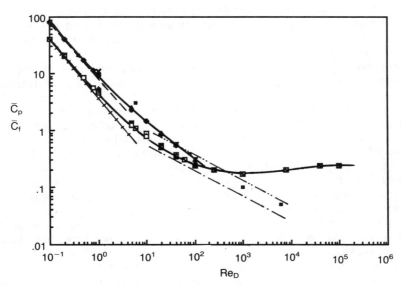

FIGURE 16–26 *Comparison of numerically computed values with theoretical expressions for the mean coefficients of shear stress and pressure on a sphere.* ++++++ *Equation 16.16 for* $\bar{C}_p$; ----- *Equation 16.15 for* $\bar{C}_f$; —··— *Equation 16.55 for* $\bar{C}_f$; —·—·— *Equation 16.58 for* $\bar{C}_p$.

Computed values:	$\bar{C}_p$	$\bar{C}_f$
Dennis and Walker	■	◆
Hamielec et al.	+	×
Ihme et al.	□	■

Velocity outside the Boundary Layer

Equations 16.30, 16.56, and 16.59 are compared in Figure 16–27 with one another and with the free-stream velocity distribution calculated from the experimentally measured pressure distribution of Fage [19] for $Re_D = 1.57 \times 10^5$. The small discrepancies here appear to be magnified in the boundary-layer solutions.

Critical Behavior

A sharp reduction in the drag at $Re_D = 3 \times 10^5$ was apparently first observed in 1911 by Eiffel [34] for spheres in air and by Costanzi [35] for spheres in water. Three years later Prandtl [36] explained this behavior in terms of the transition of the boundary layer and wake; that is, the transition to a turbulent boundary layer causes the point of separation to move backward, thereby reducing the size of the wake. Golf balls are dimpled in order to promote this transition at a lower Reynolds number and thereby increase their travel (see Figure 16–28). A swing that drives a smooth ball 50 yards is said to drive a dimpled ball 230 yards [37].

Brauer and Sucker [4] suggest that the correlation of Figure 15–25 and Equation 15.58 for prediction of the effect of free-stream turbulence on the critical Reynolds number for transition is valid for spheres as well as cylinders. However Dryden et al. [38] have shown, as indicated in Figure 16–29, that the scale as well as the intensity of free-stream turbulence influences the critical Reynolds number.

Achenbach [39] carried out very precise measurements in the higher transition regime. His measured drag coefficients for negligible free-stream turbulence are shown in Figure 16–30, from which he deduces and labels the several regimes as indicated in Figure 16–31. He compares his measurements with those of earlier investigators in Figure 16–32. Achenbach attributes the earlier transition observed by Maxworthy [40] to the insertion of a wire in the boundary layer at $\theta' = 55°$, and the higher minimum value of $\bar{C}_f$ observed by Wieselberger [41] to the effect of a support rod. Figure 16–32 also includes values of $\bar{C}_f$ determined by Achenbach from the measurements of the local skin friction and pressure shown in Figure 16–33. Remarkably good agreement on the forward half of the sphere may be noted with the theoretical pressure distribution for potential flow (Equation 16.6). Fair agreement can be noted in the same region with the shear stress calculated from Frössling's solution, Equation 16.54.

The point of transition from a laminar to a turbulent boundary layer (which begins at $Re_D \cong 2 \times 10^5$) is plotted in Figure 16–34, and the point of separation in Figure 16–35. The fraction of the total drag due to friction is plotted in Figure 16–36.

EJECTION AND SUCTION

Chuchottaworn et al. [42] investigated theoretically the effect of ejection and suction on the flow around a sphere. Flow from a sphere (ejection) may occur in

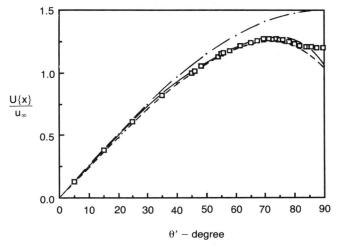

FIGURE 16–27 *Comparison of expressions for the velocity distribution outside the boundary layer on a sphere with values calculated from the experimental pressure distribution of Fage [19] for $Re_D = 1.57 \times 10^5$. —·— Equation 16.30, ——— Equation 16.56, ----- Equation 16.59.*

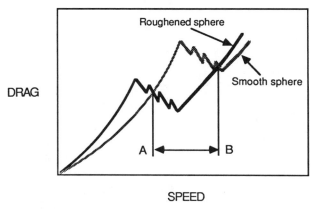

FIGURE 16–28 *Effect of roughness on the total drag of a sphere. (After Shapiro [37].)*

applications such as drying, evaporation, and combustion, although not at the uniform velocity postulated in this work. The vorticity-form of the equations of motion were solved by a finite-difference method. The validity of the results is demonstrated by agreement with the calculations of Jenson [23], Dennis and Walker [24], and others in the absence of flow normal to the sphere. The effect of ejection (labelled "injection") and suction at a uniform velocity u_o on the streamlines and the pressure at the surface is illustrated for $Re_D = 50$ in Figures 16–37 and 16–38. The effect on the skin friction and pressure drag is illustrated

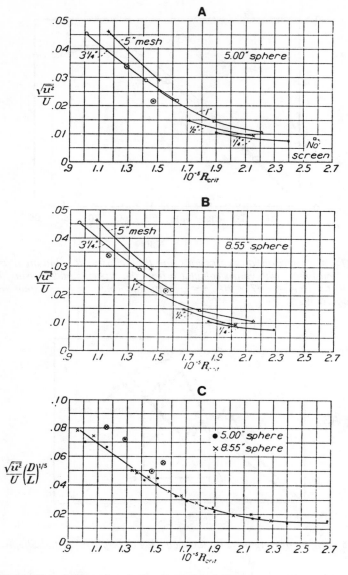

FIGURE 16–29   *Effect of scale (as determined by the mesh of screens) and intensity of turbulence on the upper critical Reynolds number of a sphere: (A) 5-inch sphere; (B) 8.55 inch sphere; (C) generalized correlation. L = mesh size; U = fluctuating component of velocity; u = time-mean value of velocity; R_{crit} = upper critical Reynolds number. (From Dryden et al. [38].)*

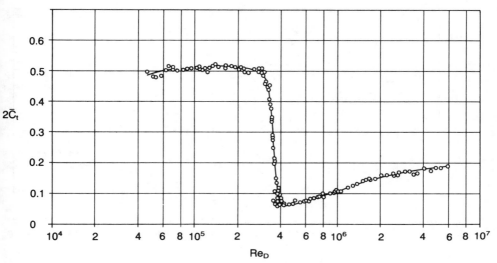

FIGURE 16–30 *Experimental measurements of the total drag coefficient of a sphere in the upper critical regime with negligible free-stream turbulence. (From Achenbach [39].)*

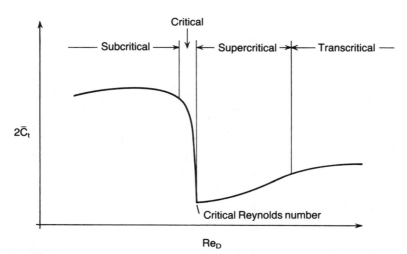

FIGURE 16–31 *Definition of regimes of flow about a sphere at high Re_D. (From Achenbach [39].)*

in Figure 16–39 for several Reynolds numbers, and the effect on the total drag in Figure 16–40. These latter data are successfully correlated by the expression

$$\frac{\bar{C}_t}{\bar{C}_{t0}} = 1 - \mathrm{Re}_D^{-0.43} \bigg/ \left(1 + \frac{0.328 u_\infty \bar{C}_{t0}}{v_a} \right) \tag{16.64}$$

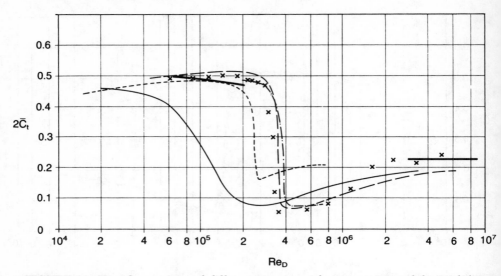

FIGURE 16–32   *Comparison of different experimental measurements of the total drag coefficient of a sphere at high* Re_D. ----- *Wieselsberger (1922);* ——— *Bacon and Reid (1924);* —·— *Millikan and Klein, free-flight (1933);* ——— *Maxworthy (1969). Present results:* ——— *from strain gauges;* ×, *from integration. (From Achenbach [39].)*

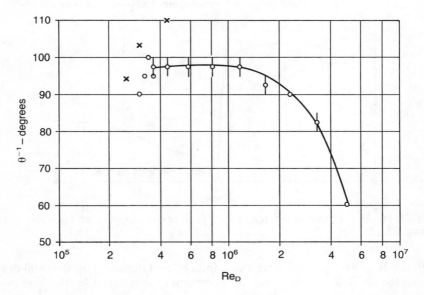

FIGURE 16–34   *Observed angle of transition from a laminar to a turbulent boundary layer on a sphere at low free-stream turbulence.* ×, *Fage (1936);* ○, *present results. (From Achenbach [39].)*

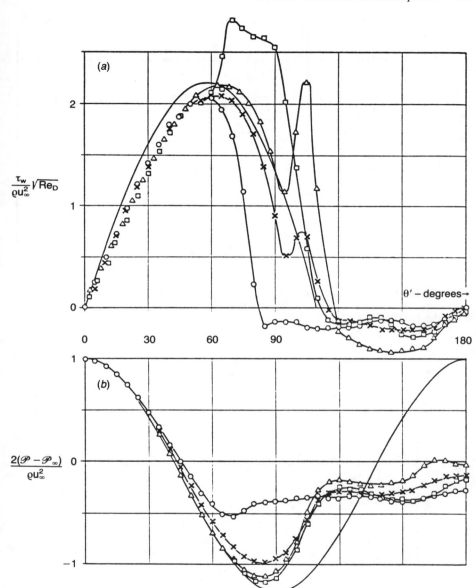

FIGURE 16–33 *Experimental measurements of (a) the local shear stress (skin friction) and (b) the pressure on the surface of a sphere at high Re_D and low free-stream turbulence. —————— represents Equation 16.54 in (a) and Equation 16.6 in (b). Experiment 9: —○—, $Re_D = 1.62 \times 10^5$; —×—, $Re_D = 3.18 \times 10^5$; —△—, $Re_D = 5.00 \times 10^6$. (From Achenbach [39].)*

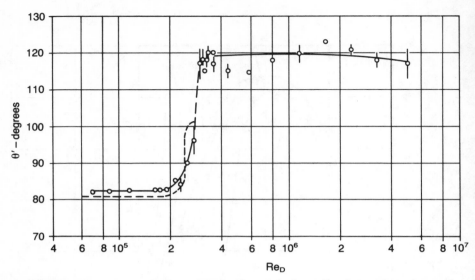

FIGURE 16–35   *Observed point of separation of the boundary layer on a sphere at low free-stream turbulence. ----- Raithby and Eckert (1968); ○, present results. (From Achenbach [39].)*

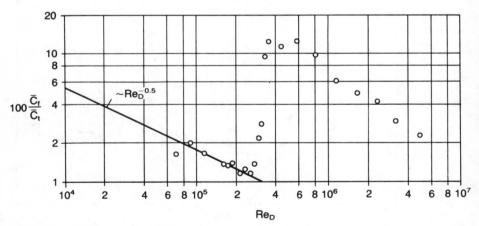

FIGURE 16–36   *Percentage of drag due to skin friction on a sphere at low free-stream turbulence. (From Achenbach [39].)*

as illustrated in Figure 16–41. Here

$\bar{C}_{t0}$ = drag for no ejection or suction
v_a = uniform velocity over surface of sphere due to ejection (positive) or suction (negative), m/s

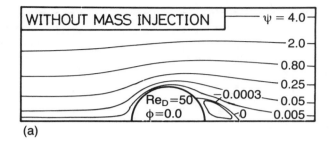

(a)

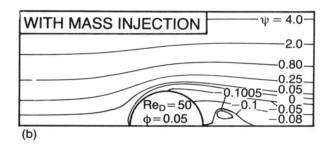

(b)

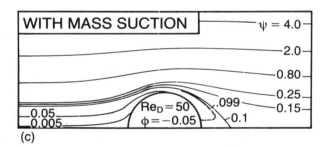

(c)

FIGURE 16–37 *Numerically computed effect of uniform ejection and suction on the streamlines about a sphere at $Re_D = 50$. $\psi = 4\bar{\psi}/u_\infty D^2$, $\phi = v_a/u_\infty$. (From Chuchottaworn, et al. [42].)*

MOVING SPHERES

The foregoing results all imply forced fluid motion pact a fixed sphere. Actually, as noted in the introductory paragraph, most applications involve motion of the sphere through a stagnant fluid. The velocity of the sphere relative to the fluid is then usually a dependent variable. For example, if the sphere is falling or rising steadily in a gravitational field, the force per unit area is

$$\frac{F}{A_p} = \frac{gV\Delta\varrho}{A_p} = \frac{2gD\Delta\varrho}{3} \tag{16.65}$$

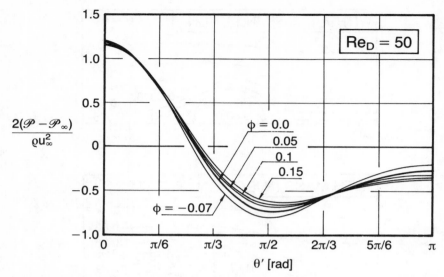

FIGURE 16–38   *Numerically computed effect of uniform ejection and suction on the pressure distribution on the surface of a sphere at $Re_D = 50$. $\phi = v_a/u_\infty$. (From Chuchottaworn, et al. [42].)*

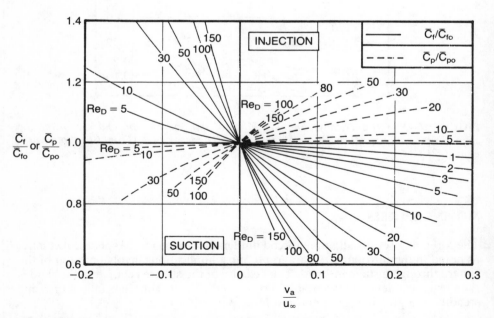

FIGURE 16–39   *Numerically computed effect of uniform ejection and suction on the mean drag coefficients for skin friction and pressure on a sphere. $Re_p = Re_D$, $C_{DF} = \bar{C}_f$, $C_{DP} = \bar{C}_p$, $\phi = v_a/u_\infty$, 0: $v_a = 0$. (From Chuchottaworn, et al. [42].)*

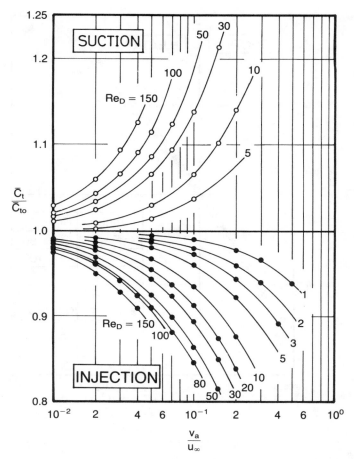

FIGURE 16–40 *Numerically computed effect of uniform ejection and suction on the total drag coefficient of a sphere. $\phi = v_a/u_\infty$, $Re_p = Re_D$, 0: $v_a = 0$. (From Chuchottaworn, et al. [42].)*

where F = total drag force, N
A_p = projected area of sphere = $\pi D^2/4$, m^2
g = acceleration due to gravity, m/s^2
V = volume of sphere = $\pi D^3/6$, m^3
$\Delta\varrho = |\varrho_d - \varrho|$ = absolute value of density difference between continuous and dispersed phase, kg/m^3
ϱ_d = density of dispersed phase, kg/m^3

Hence

$$\bar{C}_t \equiv \frac{F}{A_p \varrho u_T^2} = \frac{2gD\Delta\varrho}{3\varrho u_T^2} \tag{16.66}$$

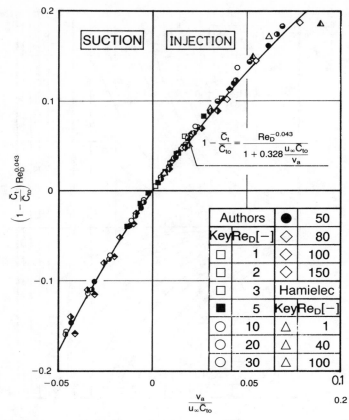

FIGURE 16–41   *Correlation for effect of uniform ejection and suction on the total drag coefficient of a sphere.* $\phi = v_a/u_\infty$, $Re_p = Re_D$, $C_D = \bar{C}_t$, $0: v_a = 0$. *(From Chuchotta-worn, et al. [42].)*

where u_T is the terminal (steady) velocity in meters per second of the sphere relative to fluid. Hence, relationships of the form

$$\bar{C}_t = f\{Re_D\} \tag{16.67}$$

can be rewritten in terms of the basic variables for a sphere moving at terminal velocity as

$$\frac{2gD\Delta\varrho}{3\varrho u_T^2} = f\left\{\frac{Du_T\varrho}{\mu}\right\} \tag{16.68}$$

These groups can then be recombined to yield the following explicit relationship for u_T:

$$\frac{Du_T\varrho}{\mu} = f\left\{\frac{g\varrho D^3 \Delta\varrho}{\mu^2}\right\} \tag{16.69}$$

which can be expressed as

$$\text{Re}_T = f\{\text{Ar}\} \tag{16.70}$$

where the *Archimedes number*, here based on diameter, is

$$\text{Ar} \equiv \frac{g\varrho D^3 \Delta\varrho}{\mu^2} = \frac{3}{2}\text{Re}_T^2 \bar{C}_t \tag{16.71}$$

The *Galileo number*, based on diameter,

$$\text{Ga} = \frac{gD^3}{\nu^2} \tag{16.72}$$

is sometimes used instead of the Archimedes number to characterize the terminal motion of spheres. Comparison of Equations 16.71 and 16.72 indicates that

$$\text{Ar} = \frac{\Delta\varrho}{\varrho}\text{Ga} \tag{16.73}$$

For $\varrho \gg \varrho_d$,

$$\text{Ar} \to \text{Ga} \tag{16.74}$$

For $\varrho \ll \varrho_d$,

$$\text{Ar} \to \frac{\varrho_d}{\varrho}\text{Ga} \tag{16.75}$$

Thus for $\varrho \gg \varrho_d$ these dimensionless groups are equivalent. Otherwise, the Archimedes number is to be preferred since ϱ_d/ϱ is thus avoided as a parameter.

Equation 16.70 can be further rearranged to give the following expression, explicit in D as well as in u_T:

$$u_T\left(\frac{\varrho^2}{g\mu\Delta\varrho}\right)^{1/3} = f\left\{D\left(\frac{g\varrho\Delta\varrho}{\mu^2}\right)^{1/3}\right\} \tag{16.76}$$

Equation 16.76 can be written symbolically as

$$u_T^* = f\{D^*\} \tag{16.77}$$

where

$$u_T^* \equiv u_T\left(\frac{\varrho^2}{g\mu\Delta\varrho}\right)^{1/3} = \left(\frac{2}{3}\frac{\text{Re}_T}{\bar{C}_t}\right)^{1/3} \tag{16.78}$$

and

$$D^* \equiv D\left(\frac{g\varrho\Delta\varrho}{\mu^2}\right)^{1/3} = \mathrm{Ar}^{1/3} = \left(\frac{3}{2}\mathrm{Re}_T^2\bar{C}_t\right)^{1/3} \tag{16.79}$$

Equations 16.71, 16.78, and 16.79 can be used to convert data or correlations in the form of Equation 16.67 to the explicit forms of 16.69 and 16.76. Thus Equation 16.63 becomes

$$\left(\frac{2}{3}\mathrm{Ar}\right)^{1/2} = 0.371\,\mathrm{Re}_T + (12\,\mathrm{Re}_T)^{1/2} \tag{16.80}$$

or

$$\left(\frac{g\varrho D^3\Delta\varrho}{\mu^2}\right)^{1/2} = 0.4544\left(\frac{Du_T\varrho}{\mu}\right) + 4.243\left(\frac{Du_T\varrho}{\mu}\right)^{1/2} \tag{16.81}$$

or

$$(D^*)^{3/2} = 0.4544D^*u_T^* + 4.243(D^*u_T^*)^{1/2} \tag{16.82}$$

Equations 16.76 and 16.77 are advantageous relative to say 16.67–16.70 in that the dependent variable u_T or u_T^* is expressed as an explicit function of the independent variable D or D^*. However, Equations 16.67 and 16.70 have another advantage in that they can be expressed explicitly in Re_T, which characterizes the transitions in the mode of flow.

Equations 16.76–16.82 imply that $\bar{C}_t$ for a free-falling sphere in a stagnant fluid is the same as for uniform flow over a fixed sphere. This is not necesarily so, as discussed at length by Torobin and Gauvin [6], Part III, p. 224. For example, they show that during acceleration the drag coefficient differs significantly from the stationary value. This behavior, however, has not yet been generalized.

Also, Lunnon [43] predicted that the periodic shedding of annular vortices

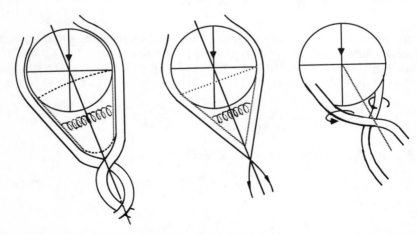

FIGURE 16–42   *Configuration of wake of freely moving spheres as sketched by Foch and Chartier [41]. (From Torobin and Gauvin [6].)*

for $Re_D > 450$ should produce fluctuations in the velocity of falling spheres. Such behavior has indeed been observed by Schmidt [44] and Liebster [45]. Lunnon also predicted [46] that helical vortices would induce a falling sphere to move in a helical path. Such vortices have been documented by Foch and Chartier [47], as sketched in Figure 16.42, and many observers, beginning with Newton [21], p. 355f, have observed helical motion.

SUMMARY

The general structure of the experimental and theoretical results for solid spheres follows that for cylinders except that a true regime of creeping flow exists. The value of boundary-layer theory is again somewhat limited owing to the presence and influence of the wake. The principal applications of fluid motion about a sphere arise from the movement of the latter through a stagnant fluid under the force of gravity. The velocity field and drag apparently differ only slightly from that for fluid motion over a fixed sphere, but the optimal formats for the data differ significantly.

PROBLEMS

1. Derive an expression for the Stokes stream function in cylindrical coordinates for a sphere moving uniformly in an infinite body of stagnant fluid. Determine the velocity components and the drag coefficient for skin friction. Compare with those for a moving fluid and fixed sphere. Explain.
2. Derive Equation 16.20 from 16.18 and 16.19.
3. Derive an expression for the local shear stress from Equation 16.24 and then expressions for $\bar{C}_f$ and $\bar{C}_t$.
4. Derive Equation 16.47 using spherical coordinates.
5. Derive Equation 15.26 using the procedure illustrated for a sphere.
6. Derive Equation 16.47 using the procedure illustrated for a cylinder.
7. Derive an expression for the local shear stress from Equation 16.18, and for the local pressure from 16.19.
8. Derive an expression for the pressure distribution on a sphere corresponding to the velocity distributions given by Equations 16.56 and 16.59. Compare with Equations 16.6, 16.9, and 16.19, and also with the experimental values plotted in Figure 16–15. Interpret the results.
9. Derive an expression with numerical coefficients for the stream function and velocity distribution corresponding to the values in Table 16.1.
10. Check the derivation of Equations 16.54 and 16.55 from Equation 16.53, and also the point of separation.
11. Derive the given expressions for $\tau_w/\varrho u_\infty^2$ and $\bar{C}_f$ and check the point of separation based on Equation 16.56.
12. Derive expressions for $\tau_w/\varrho u_\infty^2$ and $\bar{C}_f$ and the point of separation using Equation 16.59. Compare with Equations 16.57 and 16.58.
13. Show that Equation 16.60 follows from the quotation from Newton on p. 376.
14. Construct a correlation for the values in Figure 16–8 using Equations 16.11 and 16.48 as asymptotes.
15. Derive an expression for the terminal velocity of a falling spherical particle in the Stokes-law regime.

16. Derive an expression for the terminal velocity of a falling spherical particle in the Newton-law regime.

17. A sphere 0.5 m in diameter with a mass of 45 kg is placed in a nearly circular orbit around the earth with a period of 7.8 ks. The period is observed to decrease 1.2 s per day. What is the effective viscosity, assuming Stokes' law is applicable?

18. Newton [21] gives the following data for one of his experiments in St. Paul's Cathedral in June 1710. A globe full of air, 5.0 in. in diameter and weighing 483 grains, fell 220 ft. in 8.5 s. Calculate $\bar{C}_t$ and compare with Equation 16.60.

19. Derive Equation 16.5 from 16.1.

20. Derive the equivalent of Equation 16.37 for a power-law fluid.

21. Develop correlating equations for $\bar{C}_f$ and $\bar{C}_p$.

22. Rewrite Equation 16.17 in terms of the basic variables for a sphere moving under the force of gravity. Rewrite in terms of the Reynolds number and the Archimedes number and in terms of the two groups in Equation 16.76.

23. Rewrite Equation 16.20 in terms of the Reynolds number and the Archimedes number. Solve for the Reynolds number. Rewrite in terms of the groups in Equation 16.76.

24. Derive expressions from which the shear stress on the surface of a solid sphere can be calculated, presuming that

    a. the distribution of the Stokes stream function is known throughout the fluid

    b. the distribution of the components of the vorticity vector is known throughout the fluid

25. Repeat problem 23 for Equations 16.55 and 16.60.

26. Repeat problem 23 for Equation 16.62.

27. Replot Figure 16–24 in terms of the Archimedes and Reynolds numbers.

28. Replot Figure 16–24 in terms of the two dimensionless groups in Equation 16.76.

29. On a dry, 70°F day with a barometer of 740 mm Hg, a pitcher is able to throw a baseball with sufficient force so that it arrives at the plate 60 ft away with a velocity of 90 mi/h. Assuming he throws with the same force, what would be the maximum and minimum velocities at the plate if all combinations of weather from 40°F to 100°F, zero to saturated humidity, and 730 to 760 mm Hg were encountered during the season. Neglect the roughness provided by the seams and the effect of ambient conditions on the ball itself.

30. Develop a correlating equation for $\bar{C}_t$ for all Re_D by repeated application of the Churchill–Usagi expression [48].

31. Derive an expression for u_T^* as a function of D^* using the Churchill–Usagi model and the same asymptotes as used to construct Equation 16.63. Compare the result with that obtained by solving Equation 16.82 to obtain $u_T^* = f\{D^*\}$.

32. Repeat problem 31 for D^* as a function of u_T^*.

33. Repeat problem 31 for Ar as a function of Re_T, but compare with Equation 16.80 rather than 16.82.

34. Repeat problem 33 for Re_T as a function of Ar.

35. Slot [49] recently proposed a correlating equation for the terminal velocity of spheres, which can be rewritten in the form

$$\bar{C}_t = 0.15 + \frac{12}{\mathrm{Re}_T} + \frac{0.269}{(\bar{C}_t\,\mathrm{Re}_T^2)^{1/3}} \tag{16.83}$$

Compare with Equation 16.63 and with the experimental data. (Note: The coefficient 0.15 is erroneously given in [49] as 1.5.)

36. Reexpress Equation 16.83 in terms of u_T^* and D^*. Solve for $u_T^* = f\{D^*\}$ if possible. Compare with Equation 16.82.

37. The density of a sphere can be determined from its rate of fall through a viscous fluid. What is the density of a 4.27-mm sphere that falls at 17.8 mm/s through an oil with $\varrho = 0.835$ Mg/m^3 and $\mu = 0.92$ Pa·s?

38. A falling-ball viscometer operates by timing the fall of a sphere of known dimensions through the fluid of interest. If a 6.35-mm steel ball with $\varrho = 7.9$ Mg/m^3 falls 254 mm in 6.35 s through an oil with a density of 0.88 Mg/m^3, what is the viscosity?

39. Develop a correlation for the dimensionless pressure at the point of incidence of a sphere for all Re_D using the Churchill–Usagi method [48] with Equation 16.11 and 1.0 as asymptotes.

40. Look up the values of Hamielec et al. [26] for the effect of mass efflux on the drag coefficient of a sphere and compare with the results of Chuchottaworn et al. [42] as given by Equation 16.64.

41. Prepare a plot of the streamlines for flow past a fixed sphere at $\mathrm{Re}_D = 5$ using Equation 16.18 and compare the results with Figure 16–1.

42. Prepare a plot of the streamlines for a moving sphere in the regime of creeping flow and compare the results with Figure 16–4.

REFERENCES

1. M. Contanceau, *J. Méc.*, 7 (1968) 49 (according to Van Dyke [2]).
2. M. Van Dyke, *An Album of Fluid Motion*, Parabolic Press, Stanford, CA (1982).
3. S. Taneda, "Experimental Investigation of the Wakes behind Cylinders and Plates at Low Reynolds Numbers," *J. Phys. Soc. Japan*, 11 (1956) 1104.
4. H. Brauer and D. Sucker, "Umströmung von Platten, Zylindern und Kugeln," *Chem.-Ing.-Tech.*, 48 (1976) 665; English transl. "Flow about Plates, Cylinders and Spheres," *Int. Chem. Eng.*, 18 (1978) 367.
5. H. Werlé, "Le tunnel hydrodynamique au service de la recherche aérospatiale," *ONERA* Publ. No. 156, France (1974) (according to Van Dyke [2]).
6. L. B. Torobin and W. H. Gauvin, "Fundamental Aspects of Solids-Gas Flow," *Can. J. Chem. Eng.*, 37 (1959) 129, 167, 224.
7. G. C. Stokes, "On the Effect of the Internal Friction of Fluids on the Motion of Pendulums," *Trans. Camb. Phil. Soc.* 9 (1851) 8 (*Math. Phys. Papers*, Vol. III, Cambridge University Press (1901), p. 55).
8. H. Lamb, *Hydrodynamics*, Dover, New York (1945).
9. V. L. Streeter, *Fluid Dynamics*, McGraw-Hill, New York (1948).
10. C. W. Oseen, "Über die Stokes'sche Formel und über die verwandte Aufgabe in der Hydrodynamik," *Arkiv Math., Astronom. Fys.*, 6, (1910) 75.
11. Ian Proudman and J. R. A. Pearson, "Expansions at Small Reynolds Numbers for the Flow past a Sphere and Circular Cylinder," *J. Fluid Mech.*, 2, (1957) 237.

12. W. Chester and D. R. Breach, "On the Flow past a Sphere at Low Reynolds Number," *J. Fluid Mech.*, *37* (1969) 751.
13. S. Tomotika and T. Aoi, "The Steady Flow of Viscous Fluid past a Sphere and Circular Cylinder at Small Reynolds Numbers," *Quart. J. Mech. Appl. Math.*, *3* (1950) 140.
14. M. Van Dyke, *Perturbation Methods in Fluid Mechanics*, annotated ed., Parabolic Press, Stanford, CA (1978).
15. F. Homann, "Der Einfluss grosser Zähigkeit bei der Strömung um den Zylinder und um die Kugel," *Z. Angew Math. Mech.*, *16*, (1936) 153 and *Forsch. Gebiete Ingenieurw.* 7 (1936) 1; English transl., "The Effect of High Viscosity on the Flow around a Cylinder and around a Sphere," *NACA* TM 1334, Washington, D.C. (1952).
16. N. Frössling, "Verdunstung, Wärmeübergang und Geschwindigkeitsverteilung bei zweidimensionaler und rotationsymmetrischer laminarer Grenzschichtströmung," *Lunds Univ. Ärsskrift, N. F.*, Avd. 2, *36*, No. 4 (1940); English transl. "Evaporation, Heat Transfer and Velocity Distribution in Two-Dimensional and Rotationally Symmetrical Laminar Boundary-Layer Flow," *NACA* TM 1432, Washington, D.C. (1958).
17. H. Blasius, "Grenzschichten in Flüssigkeiten mit kleiner Reibung," *Z. Math. Phys.*, *56* (1908) 1; English transl., "The Boundary Layers in Fluids with Little Friction," NACA TM 1256, Washington, D.C. (1950).
18. T. Yuge, "Theory of Distributions of the Coefficients of Heat Transfer of Spheres," *Repts. Inst. High Speed Mech.*, *Tohoku Univ.*, *6* (1956) 115.
19. A. Fage, "Experiments on a Sphere at Critical Reynolds Numbers," *Aero. Res. Council Gt. Brit.*, R. & M. 1766 (1936).
20. S. Tomotika and I. Imai, "The Distribution of Laminar Skin Friction on a Sphere Placed in a Uniform Stream," *Proc. Japan Soc. Phys.-Math.*, *20* (1938) 288.
21. I. Newton, *Principia*, Vol. I. *The Motion of Bodies*, S. Pepys, London (1686); English transl. of 2nd ed. (1713) by A. Motte (1729); revised transl. by F. Cajori, University of California Press, Berkeley (1966).
22. M. Kawaguti, "Numerical Solution for the Viscous Flow past a Sphere," *Repts. Inst. Sci., Tokyo*, *2* (1948) 66; *4* (1950) 154.
23. V. G. Jenson, "Viscous Flow round a Sphere at Low Reynolds Numbers (< 40)," *Proc. Roy. Soc. (London)*, *A249* (1959) 346.
24. S. R. C. Dennis and J. D. A. Walker, "Calculation of the Steady Flow Past a Sphere at Low and Moderate Reynolds Numbers," *J. Fluid Mech.*, *48* (1971) 771.
25. F. Ihme, H. Schmidt-Traub, and H. Brauer, "Theoretische Untersuchung über die Umströmung und den Stoffübergang an Kugeln," *Chem.-Ing.-Tech.*, *44* (1972) 306.
26. A. E. Hamielec, T. W. Hoffman, and L. L. Ross, "Numerical Solution of the Navier–Stokes Equations for Flow past Spheres. Part I. Spheres with and without Radial Mass Efflux," *AIChE J.*, *13* (1967) 212.
27. C. L. Lin and S. C. Lee, "Transient State Analysis of Separated Flow around a Sphere," *Computers and Fluids*, *1* (1973) 235.
28. Y. Rimon and S.-I. Cheng, "Numerical Solution of a Uniform Flow over a Sphere at Intermediate Reynolds Numbers," *Phys. Fluids*, *12* (1969) 949.
29. S.-I. Cheng, "Accuracy of Difference Formulation of Navier–Stokes Equations," *Phys. Fluids*, *12*, Suppl. II (1969), II-34.
30. S. Taneda, "Studies on Wake Vortices. III. Experimental Investigation of the Wake behind a Sphere at Low Reynolds Numbers," *Rept. Res. Inst. Appl. Mech.*, Kyushu Univ., *4* (1956) 99.
31. W. Müller, "Experimentelle Untersuchung zur Hydrodynamik der Kugel," *Physik Z.*, *39* (1938) 57.
32. H. Brauer, "Impuls- Stoff- und Wärmetransport durch die Grenzfläche kugelförmiger Partikeln," *Chem.-Ing.-Tech.*, *45* (1973) 1099.
33. A. Weiner and S. W. Churchill, "Mass Transfer from Rising Bubbles of Carbon

Dioxide," p. 525 in *Physicochemical Hydrodynamics—V. G. Levich Festschrift*, D. B. Spaulding, Ed., Advance Publications, London (1977).

34. G. Eiffel, "Sur la résistance des sphères dans l'air en mouvement," *Compt. Rend, Acad. Sci., Paris*, *155* (1912) 1597.

35. S. Costanzi, *Alcune esperienze di idrodynamica*, 2, No. 4 (1912) (according to Torobin and Gauvin [6]).

36. L. Prandtl, "Über den Luftwiderstand von Kugeln," *Göttinger Nachr.* 177 (1914).

37. A. H. Shapiro, *Shape and Flow. The Fluid Dynamics of Drag*, Doubleday Anchor, New York (1961).

38. H. L. Dryden, G. B. Schubauer, W. C. Mock, and H. K. Shramstad, "Measurements of the Intensity and Scale of Wind-Tunnel Turbulence and their Relation to the Critical Reynolds Number of Spheres," *NACA* Rept. 581, Washington, D.C. (1937).

39. E. Achenbach, "Experiments on the Flow Past Spheres at very High Reynolds Numbers," *J. Fluid Mech.*, *54* (1972) 565.

40. T. Maxworthy, "Experiments on Flow around a Sphere at High Reynolds Numbers," *J. Appl. Mech.*, *36E* (1960) 598.

41. C. Wieselberger, "Weitere Festellungen über die Gesetze des Flüssigkeits- und Luftwiderstandes," *Phys. Z.*, *23* (1922) 219.

42. P. Chuchottaworn, A. Fujinami, and K. Asano, "Numerical Analyses of the Effect of Mass Injection or Suction on Drag Coefficients of a Sphere," *J. Chem. Eng. Japan*, *16* (1983) 18.

43. R. G. Lunnon, "Fluid Resistance to Moving Spheres," *Proc. Roy Soc. (London)*, *A110* (1926) 302.

44. F. S. Schmidt, "Zur beschleunigten Bewegung kugelförmiger Körper in widerstehenden Mitteln," *Ann. Phys.*, *61* (1920) 633.

45. H. Liebster, "Über den Widerstand von Kugeln," *Ann. Physik*, *82* (1927) 541.

46. R. G. Lunnon, "Fluid Resistance to Moving Spheres," *Proc. Roy. Soc., (London)*, *A118* (1928) 680.

47. A. Foch and C. Chartier, "Sur l'écoulement d'un fluide à l'aval d'un sphère," *Compt. Rend. Acad. Sci., Paris*, *200* (1935) 1178.

48. S. W. Churchill and R. Usagi, "A General Expression for the Correlation Rates of Transfer and Other Phenomena," *AIChE J.*, *18* (1972) 1121.

49. R. E. Slot, "Terminal Velocity Formula for Objects in a Viscous Fluid," *J. Hydraulic Res.*, *22* (1984) 235.

Chapter 17

The Motion of Bubbles and Droplets

The motion of bubbles through liquids and that of droplets through gases and immiscible liquids have many important applications. Their behavior differs from that of solid spheres owing to the finite viscosity of the dispersed gas or liquid, but even more so to the finite interfacial tension. Also, a solid sphere may be supported mechanically and thereby its motion relative to the fluid rendered independent of gravity; bubbles and liquids are invariably unsupported and thereby are subject to gravity.

The finite viscosity of a bubble or droplet permits a tangential velocity at the surface, leading to internal circulation and thereby decreased drag. The finite interfacial tension permits steady and oscillatory deviations from sphericity, controls the magnitude and frequency of these deviations, and limits the maximum stable size of a droplet or bubble.

The behaviors of moving bubbles and droplets have much in common, but they also differ significantly. Both the common and differing features will be noted.

QUALITATIVE CHARACTERISTICS OF BUBBLE MOTION

The viscosity and density of bubbles are ordinarily negligible compared to that of the surrounding liquid. However, owing to the presence of impurities in the liquid, bubbles, particularly small ones, rarely demonstrate the corresponding idealized behavior.

Surface-active contaminants in a liquid concentrate and accumulate at the surface of a bubble and, as a consequence, even a trace can have a significant effect. Such an accumulation readily stops the circulation of a *small* bubble, causing it to behave as a solid sphere. Aybers and Tapucu [1] and Weiner and Churchill [2] reported quantitative observations of a decrease in the velocity of small bubbles as they rose through water containing almost immeasurable traces of dissolved surface-active compounds. With large bubbles, which move at higher velocities, the surface-active contaminants are apparently forced to the rear and perhaps even discarded, thereby having less influence.

Haberman and Morton [3] observed a significant difference in the behavior of *large* bubbles of air in filtered and unfiltered tap water and attributed this discrepancy to particulate matter, which also accumulates at the interface.

Sufficient impurities to cause one or both of these effects probably exist in all waters and aqueous solutions except those especially prepared for their avoidance. Hence, a distinction must be made between the motion of a bubble in most practical applications and that measured in the laboratory or predicted theoretically.

Rosenberg [4] reported the following observations on the behavior of bubbles of air in water:

1.	$Re < 70$	spheres rising rectilinearly with a drag coefficient corresponding to solid spheres
2.	$70 < Re < 400$	spheres rising rectilinearly with a drag coefficient less than that of solid spheres
3.	$400 < Re < 500$	oblate spheroids[1] with a vertical axis of symmetry, rising rectilinearly
4.	$500 < Re < 1100$	oblate spheroids rising in a helical path
5.	$1100 < Re < 1600$	irregular oblate spheroids, rising almost rectilinearly
6.	$1600 < Re < 5000$	a transition from state 5 to state 7
7.	$Re > 5000$	irregular, horizontally oriented, mushroomlike shapes, called spherical or spheroidal caps, rising more or less rectilinearly and followed by a turbulent wake

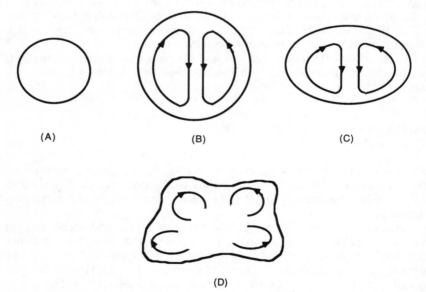

(A)   (B)   (C)

(D)

FIGURE 17–1 *Form of typical bubbles: (A) noncirculating sphere; (B) internally circulating sphere; (C) oblate spheroid with internal circulation; (D) irregular mushroom-like bubbles with multiple, internal circulation. (After Haas et al. [5].)*

---

[1] *Oblate (or planetary) ellipsoids* are geometrically defined bodies generated by rotation of an ellipse about its minor axis. Bubbles in these regimes have approximately such a form but are ordinarily not perfectly symmetrical, hence the designation "spheroid."

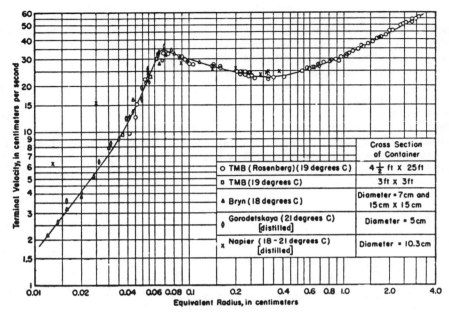

FIGURE 17-2   *Experimental measurements of terminal velocity of bubbles of air in filtered or distilled water. TMB = David Taylor Model Basin. (From Haberman and Morton [3].)*

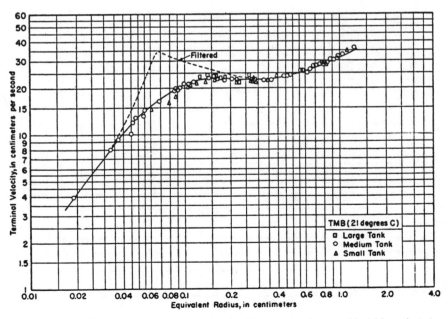

FIGURE 17-3   *Experimental measurements of terminal velocity of bubbles of air in tap water. TMB = David Taylor Model Basin. (From Haberman and Morton [3].)*

Stages 1–3 and 7 are sketched in Figure 17–1 (after Haas et al. [5]). Stage 1 implies the surface-active impurities already mentioned. The same general behavior has been observed by others, Aybers and Tapucu [6] for example, who, however, reported slightly different ranges of Re for the several stages and noted that the mushroom-shaped bubbles rocked back and forth during stage 6.

The velocity of solid spheres increases monotonically with diameter, but a more complicated dependence is observed for fluid bubbles because of the aforementioned effects of viscosity, interfacial tension, and contaminants. Figure 17–2 from Haberman and Morton [3] illustrates the dependence on diameter of the rate of rise of bubbles of air in filtered or distilled water as observed by several investigators. Here the *terminal velocity* u_T is the average vertical component in the case of helical or unsteady motion, and the *volume-equivalent diameter* D_V is that of a sphere of equal volume V; that is,

$$D_V = \left(\frac{6V}{\pi}\right)^{1/3} \tag{17.1}$$

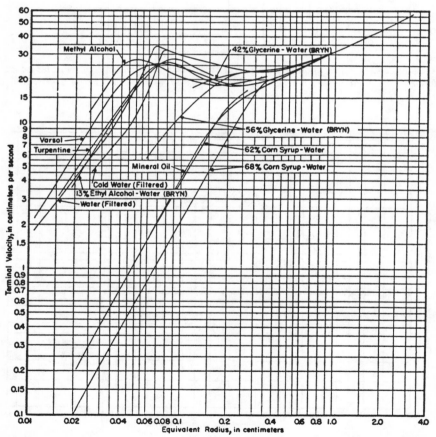

FIGURE 17–4   *Curves representing experimental data for terminal velocity of bubbles of air in various liquids. (From Haberman and Morton [3].)*

The velocity first increases rapidly with diameter, goes through a maximum $[(u_T)_{max} \cong 0.35$ m/s at $(D_V)_{max} \cong 1.4$ mm], a minimum $[(u_T)_{min} \cong 0.23$ m/s at $(D_V)_{min} \cong 7$ mm], and then increases at a slower rate.

The behavior of bubbles of air in tap water is illustrated in Figure 17–3. The velocity increases monotonically with diameter directly to the preceding "minimum" value of 0.23 m/s at $D_V = 7$ mm and then increases at the same rate as in purified water. Curves representing data for bubbles of air in a number of liquids are shown in Figure 17–4. A wide range of behavior is observed for small diameters, but for $D_V > 7$ mm the velocity is the same for all of the liquids and, hence, is independent of the physical properties.

QUALITATIVE CHARACTERISTICS OF DROPLET MOTION IN GASES AND LIQUIDS

Droplets of liquid falling in a gas constitute almost the converse limit of negligible density and viscosity for the continuous phase. Small droplets behave similarly to solid spheres, but larger droplets are deformed and demonstrate some of the behavioral characteristics of bubbles. However, in contrast to bubbles, droplets with a diameter above some limiting value disintegrate.

The behavior of small droplets rising or falling in another liquid is intermediate to that of droplets falling in gas and bubbles rising in liquids. Large droplets are distorted when moving in a liquid, and as in a gas, disintegrate above some maximum size.

COMPARATIVE BEHAVIOR OF BUBBLES, DROPLETS, AND SOLIDS

Figure 17–5 from Mersmann [7] compares the dependence of u_T on D_V for four different phase pairs:

1. A monotonic increase for solid spheres ($\varrho = 1$ Mg/m^3) in air
2. A slight maximum and a slight minimum for droplets of water in air, the curve being terminated by disintegration of the droplets
3. The same behavior as in Figure 17–1 for bubbles of air in pure water (an upturn would presumably have occurred if the experiments had been continued to larger diameters)
4. A monotonic increase followed by a constant value for droplets of nonyl alcohol in water, the curve again terminated by disintegration of the droplets

The maximum in velocity in Figures 17–2 to 17–5 is known to be associated with the onset of deformation, and the final increase in Figures 17–2 to 17–4 with the onset of the mushroom stage.

The quantitative results presented in this chapter as well as the general descriptions already given are limited to single bubbles and droplets in an unconfined fluid. The behavior of swarms and the effects of confinement are excluded on the basis of complexity and the limited theoretical structure which has been developed.

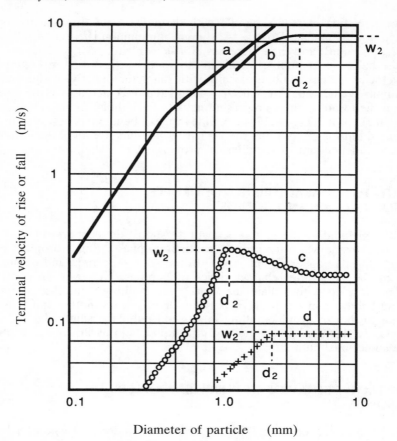

FIGURE 17-5 *Comparison of terminal velocities of bubbles and droplets: a = solid sphere ($\varrho = 1,000$ kg/m³) in air at 20°C; b = droplets of water in air at 20°C; c = bubbles of air in water at 20°C; d = droplets of nonyl alcohol in water at 20°C; w_2 = maximum attainable velocity; d_2 = diameter for maximum velocity. (From Mersmann [7].)*

DIMENSIONAL ANALYSIS

The theoretical solutions to date for the motion of droplets and bubbles almost all postulate sphericity. Hence guidance for correlation at high velocities, for which nonsphericity is expected to be significant, must come primarily from dimensional analysis. The added variables of interfacial tension and dispersed-phase viscosity increase the number of dimensionless groups from the two required to describe the terminal velocity or drag of a solid sphere to five. Straightforward dimensional analysis (see problem 1) then leads to a result such as

$$\frac{u_T^2}{gD_V} = \phi \left\{ \frac{D_V^3 \varrho_c^2 g}{\mu_c^2}, \frac{\mu_d}{\mu_c}, \frac{\varrho_d}{\varrho_c}, \frac{g\mu^4}{\varrho_c \sigma^3} \right\} \tag{17.2}$$

where g = acceleration due to gravity, m/s²
μ_d = viscosity of dispersed phase (bubble or droplet), Pa·s or kg/m·s
μ_c = viscosity of continuous phase, Pa·s or kg/m·s
ϱ_d = density of dispersed phase (bubble or droplet), kg/m³
ϱ_c = density of continuous phase, kg/m³
σ = interfacial tension, N/m or kg/s²

D_V and u_T were defined earlier.

The recognition that the gravitational force on a bubble or droplet is $g|\varrho_d - \varrho_c|\pi D_V^3/6$ suggests that g must always be multiplied by $|\varrho_d - \varrho_c|$, leading to the improved form

$$\frac{\varrho_c u_T^2}{g|\varrho_d - \varrho_c|D_V} = \phi\left\{\frac{D_V^3 \varrho_c g|\varrho_d - \varrho_c|}{\mu_c^2}, \frac{\mu_d}{\mu_c}, \frac{\varrho_d}{\varrho_c}, \frac{g|\varrho_d - \varrho_c|\mu_c^4}{\varrho_c^2 \sigma^3}\right\} \quad (17.3)$$

For bubbles, $|\varrho_d - \varrho_c| \to \varrho_c$, while both ϱ_d/ϱ_c and μ_d/μ_c approach zero. Therefore, Equation 17.3 might be expected to reduce to

$$\frac{u_T^2}{gD_V} = \phi\left\{\frac{D_V^3 \varrho_c^2 g}{\mu_c^2}, \frac{g\mu_c^4}{\varrho_c \sigma^3}\right\} \quad (17.4)$$

For droplets falling in a gas, $|\varrho_d - \varrho_c| \to \varrho_d$, while both ϱ_d/ϱ_c and μ_d/μ_c are very large. Therefore Equation 17.3 might be expected to reduce to

$$\frac{\varrho_c u_T^2}{\varrho_d g D_V} = \phi\left\{\frac{D_V^3 \varrho_c \varrho_d g}{\mu_c^2}, \frac{g\varrho_d \mu_c^4}{\varrho_c^2 \sigma^3}\right\} \quad (17.5)$$

Thus the terms μ_d/μ_c and ϱ_d/ϱ_c would be expected to be possible significant parameters in Equation 17.3 only for liquid droplets rising or falling in an immiscible liquid.

The foregoing dimensionless groups and combinations thereof will be given the following symbols and names:

$\varrho_c u_T^2/D_V g\Delta\varrho$ = Fr = *Froude number (of a droplet or bubble)*

$D_V u_T \varrho_c/\mu_c$ = Re = *Reynolds number (of a droplet or bubble)*

$\mu_c^4 g\Delta\varrho/\varrho_c^2 \sigma^3$ = M = *Morton number*

$\zeta = \mu_d/\mu_c$ = *viscosity ratio*

$\Delta\varrho/\varrho_c = |\varrho_d - \varrho_c|/\varrho_c = |\varrho_d/\varrho_c - 1|$ = *fractional density difference*

$2D_V g\Delta\varrho/3\varrho_c u_T^2 = \bar{C}_t$ = *drag coefficient (of a droplet or bubble)*

$D_V u_T^2 \varrho_c/\sigma$ = We = *Weber number (of a droplet or bubble)*

$D_V(\varrho_c g\Delta\varrho/\mu_c^2)^{1/3} = D_V^*$ = *dimensionless equivalent diameter of a droplet or bubble*

$u_T(\varrho_c^2/\mu_c g\Delta\varrho)^{1/3} = u_T^*$ = *dimensionless terminal velocity*

$D_V^3 \varrho_c g \Delta \varrho / \mu_c^2 = \mathrm{Ar} = $ *Archimedes number (of a droplet or bubble)*

$D_V^2 g \Delta \varrho / \sigma = \mathrm{E\ddot{o}} = $ *Eötvös number (of a droplet or bubble)*; also called the *Bond number* (Bo)

Care is advised in reading and interpreting the literature on bubbles and droplets since these named groups have often been given slightly different definitions. They are necessarily interrelated, since according to Equation 17.3 only five can be independent. For example,

$$\mathrm{Fr} = \frac{2}{3\bar{C}_t} \tag{17.6}$$

$$\mathrm{We} = \left(\frac{2M\mathrm{Re}^4}{3\bar{C}_t}\right)^{1/3} = M^{1/3} D_V^*(u_T^*)^2 \tag{17.7}$$

$$\mathrm{E\ddot{o}} = \frac{3}{2}\mathrm{We}\,\bar{C}_t = \left(\frac{9}{4}M\mathrm{Re}^4\bar{C}_t^2\right)^{1/3} = M^{1/3}(D_V^*)^2 \tag{17.8}$$

$$\mathrm{Ar} = \frac{3}{2}\mathrm{Re}^2\,\bar{C}_t = (D_V^*)^3 \tag{17.9}$$

$$\mathrm{Re} = u_T^* D_V^* \tag{17.10}$$

and

$$\bar{C}_t = \frac{2D_V^*}{3(u_T^*)^2} \tag{17.11}$$

The shape of a droplet or bubble is also a dependent variable and hence dependent on four of the groups in Equation 17.3 or their combinations. For purposes of analysis, bubbles and droplets are usually postulated to have the form of regular ellipsoids. Their shape can then be represented in terms of D_{max}/D_V or

$$D_{max}/D_{min} = E = \textit{eccentricity ratio}$$

where D_{max} = maximum diameter, m
$\;\;\; D_{min}$ = minimum diameter, m

For oblate (planetary) ellipsoids,

$$\frac{\pi D_V^3}{6} = \frac{\pi D_{max}^2 D_{min}}{6} \tag{17.12}$$

or

$$D_V = D_{max}^{2/3} D_{min}^{1/3} \tag{17.12A}$$

It follows that

$$\frac{D_{\max}}{D_V} = \left(\frac{D_{\max}}{D_{\min}}\right)^{1/3} = E^{1/3} \tag{17.13}$$

For prolate (ovary) ellipsoids,[1]

$$\frac{\pi D_V^3}{6} = \frac{\pi D_{\max} D_{\min}^2}{6} \tag{17.14}$$

or

$$D_V = D_{\max}^{1/3} D_{\min}^{2/3} \tag{17.14A}$$

Hence

$$\frac{D_{\max}}{D_V} = \left(\frac{D_{\max}}{D_{\min}}\right)^{2/3} = E^{2/3} \tag{17.15}$$

Bubbles and droplets ordinarily form oblate spheroids with a vertical minor axis, although prolate spheroids with a major vertical axis have been predicted and occasionally observed. The mushroom shapes observed for large droplets and bubbles are ordinarily approximated quantitatively as oblate spheroids.

Three possible fluid-mechanical regimes are to be anticipated for all fluid pairs:

1.   *Creeping motion.*   For very low velocities inertial effects become negligible. Hence the densities should not be significant variables except in the density difference. Also, the surface tension would not be expected to be a significant variable, and droplet distortion would not be expected to be appreciable. Eliminating these variables permits Equation 17.3 to be reduced to

$$\frac{D_V^2 g \Delta \varrho}{\mu_c u_T} = \phi \left\{ \frac{\mu_d}{\mu_c} \right\} \tag{17.16}$$

For a liquid droplet falling in a gas and for a bubble, Equation 17.16 can be approximated as

$$\frac{D_V^2 g \Delta \varrho}{\mu_c u_T} = \text{a constant} \tag{17.17}$$

which is equivalent to

$$\bar{C}_t = \frac{A}{Re} \tag{17.18}$$

---

[1] *Prolate* (or *ovary*) ellipsoids are geometric bodies generated by rotation of an ellipse about its major axis.

or

$$u_T^* = B(D_V^*)^2 \tag{17.19}$$

2. *A malleable shape.* The size of a completely malleable bubble, and hence D_V, might not be expected to be a significant factor in determining the velocity. This postulate reduces Equation 7.3 to

$$\frac{\mu_c g \Delta \varrho}{\varrho_c^2 u_T^3} = \phi \left\{ \frac{\mu_d}{\mu_c}, \frac{\varrho_d}{\varrho_c}, \frac{\mu_c^4 g \Delta \varrho}{\varrho_c^2 \sigma^3} \right\} \tag{17.20}$$

which further reduces for a droplet in a gas or for a bubble to

$$\frac{\mu_c g \Delta \varrho}{\varrho_c^2 u_T^3} = \phi \left\{ \frac{\mu_c^4 g \Delta \varrho}{\varrho_c^2 \sigma^3} \right\} \tag{17.21}$$

which is equivalent to

$$\bar{C}_t = \text{Re} \, \phi \{M\} \tag{17.22}$$

or

$$u_T^* = \phi \{M\} \tag{17.23}$$

3. *Viscosity-free motion.* As the velocity increases, inertial effects increase with respect to viscous effects. As they become dominant, Equation 17.3 reduces to

$$\frac{\varrho_c u_T^2}{D_V g \Delta \varrho} = \phi \left\{ \frac{D_V^2 g \Delta \varrho}{\sigma}, \frac{\varrho_d}{\varrho_c} \right\} \tag{17.24}$$

Then for a bubble or for a droplet in a gas,

$$\frac{\varrho_c u_T^2}{D_V g \Delta \varrho} = \phi \left\{ \frac{D_V^2 g \Delta \varrho}{\sigma} \right\} \tag{17.25}$$

which is equivalent to

$$\bar{C}_t = \phi \{E\ddot{o}\} = \phi \{MRe^4\} \tag{17.26}$$

or

$$u_T^* = (D_V^*)^{1/2} \phi \{E\ddot{o}\} \tag{17.27}$$

a. *Inertial force dominant.* If the inertial force is dominant over surface forces, Equation 17.25 reduces to

$$\frac{\varrho_c u_T^2}{D_V g \Delta \varrho} = \text{constant} \tag{17.28}$$

which is equivalent to

$$\bar{C}_t = \text{constant} \tag{17.29}$$

or

$$u_T^* = A\,(D_V^*)^{1/2} \tag{17.30}$$

b. *Surface effects dominant.* If surface forces are wholly balanced by inertial forces, thereby eliminating the effect of the gravitational force, Equation 17.25 reduces to

$$\frac{\varrho_c D_V u_T^2}{\sigma} = \text{constant} \tag{17.31}$$

which is equivalent to

$$\bar{C}_t = A\,\text{Eö} = B\text{MRe}^4 \tag{17.32}$$

or

$$u_T^* = \frac{A'}{(D_V^*)^{1/2}\text{M}^{1/6}} \tag{17.33}$$

Equations 17.17, 17.20, 17.28, and 17.31 imply that u_T should be proportional to D_V^2, independent of D_V, proportional to $D_V^{1/2}$ and inversely proportional to $D_V^{1/2}$, respectively, in these four limiting conditions. Figures 17–2 to 17–5 indicate that limiting condition 1 is attained for all fluid pairs for sufficiently small diameters, that condition 3b is attained for bubbles in pure liquids and for droplets under some circumstances at intermediate diameters, that condition 2 is attained for most fluid pairs at still larger diameters, and that condition 3a may be approached for sufficiently large diameters.

Equations 17.18, 17.22, 17.29, and 17.32 similarly imply that $\bar{C}_t$ should be inversely proportional to Re, proportional to Re, independent of Re, and proportional to Re^4, respectively. However, the behavior is more difficult to interpret in terms of these two dimensionless groups, since both incorporate the dependent variable u_T and the principal independent variable D_V, and neither includes σ. Two speculations in terms of Re and $\bar{C}_t$ are, however, suggested. For small Re, the effect of the finite viscosity of the dispersed fluid might be expected to reduce the drag coefficient below that for a solid sphere, which has an effectively infinite viscosity. For very large Re, the effect of deformation might be expected to increase the drag coefficient above that for a solid sphere because of an increased surface area.

The foregoing dimensional and speculative analyses, despite their naivete and limitations, provide a most useful framework for the detailed quantitative examination to follow, particularly with respect to empirical correlations for the range of nonsphericity. More attention than in previous chapters has been afforded to such techniques because of the lesser theoretical structure.

THEORETICAL EXPRESSIONS FOR SPHERES

Completely theoretical expressions have been derived for some of the limiting behavior described thus far. Some such results follow. Most of these theoretical solutions postulate a spherical shape, which is a valid approximation for sufficiently small bubbles and droplets. Such models are examined before models for nonspherical shapes.

Spherical Models

Inviscid Flow

The solution for inviscid flow *outside* a spherical bubble or droplet is identical to that for a solid sphere (Equations 16.1–16.6). The corresponding solution for inviscid motion *inside* the fluid sphere was posed as problems 104 and 105 in Chapter 10.

Creeping Flow

In 1911 Hadamard [8] and Rybczynski [9] independently derived a solution for flow past a fixed *spherical droplet or bubble*, neglecting the inertial terms and assuming continuity of velocity and viscous shear at the interface (see, for example, Lamb [10], p. 600f, and Levich [11], p. 395f). The results presented in Section 10EC1 for fluid spheres are repeated here for convenience. In spherical coordinates, the *Stokes stream function* in the *continuous* fluid is

$$\tilde{\psi} = -\frac{u_\infty R^2}{2}\left[1 - \frac{1}{2}\left(\frac{2 + 3\zeta}{1 + \zeta}\right)\frac{a}{R} + \frac{1}{2}\left(\frac{\zeta}{1 + \zeta}\right)\left(\frac{a}{R}\right)^3\right]\sin^2\{\theta\} \qquad (17.34)$$

where, as usual, $\zeta = \mu_d/\mu_c$.

The corresponding velocity components of the continuous fluid are

$$u_R = \frac{-1}{R^2\sin\{\theta\}}\frac{\partial\tilde{\psi}}{\partial\theta}$$

$$= u_\infty\left[1 - \frac{1}{2}\left(\frac{2 + 3\zeta}{1 + \zeta}\right)\frac{a}{R} + \frac{1}{2}\left(\frac{\zeta}{1 + \zeta}\right)\left(\frac{a}{R}\right)^3\right]\cos\{\theta\} \qquad (17.35)$$

and

$$u_\theta = \frac{1}{R\sin\{\theta\}}\frac{\partial\tilde{\psi}}{\partial R}$$

$$= -u_\infty\left[1 - \frac{1}{4}\left(\frac{2 + 3\zeta}{1 + \zeta}\right)\frac{a}{R} - \frac{1}{4}\left(\frac{\zeta}{1 + \zeta}\right)\left(\frac{a}{R}\right)^3\right]\sin\{\theta\} \qquad (17.36)$$

The pressure distribution outside the bubble or droplet is

$$\mathscr{P} - \mathscr{P}_\infty = -\left(\frac{2 + 3\zeta}{1 + \zeta}\right)\frac{\mu_c u_\infty a\cos\{\theta\}}{2R^2} \qquad (17.37)$$

At the surface it is

$$\mathcal{P} - \mathcal{P}_\infty = -\left(\frac{2 + 3\zeta}{1 + \zeta}\right)\frac{\mu_c u_\infty \cos\{\theta\}}{2a} \tag{17.38}$$

For the *dispersed* fluid (inside the droplet or bubble),

$$\tilde{\psi} = \frac{u_\infty R^2}{4(1 + \zeta)}\left[1 - \left(\frac{R}{a}\right)^2\right]\sin^2\{\theta\} \tag{17.39}$$

$$u_R = -\frac{u_\infty}{2(1 + \zeta)}\left[1 - \left(\frac{R}{a}\right)^2\right]\cos\{\theta\} \tag{17.40}$$

$$u_\theta = \frac{u_\infty}{2(1 + \zeta)}\left[1 - 2\left(\frac{R}{a}\right)^2\right]\sin\{\theta\} \tag{17.41}$$

and

$$\mathcal{P} - \mathcal{P}_\infty = \frac{2\sigma}{a} - \frac{5\mu_d u_\infty R \cos\{\theta\}}{a^2(1 + \zeta)} \tag{17.42}$$

It follows that the total drag coefficient is

$$\bar{C}_t = \frac{4}{\mathrm{Re}}\left(\frac{2 + 3\zeta}{1 + \zeta}\right) \tag{17.43}$$

The Hadamard–Rybczynski solution, which postulates sphericity, satisfies all of the boundary conditions for both phases, implying that a deformation from sphericity can occur only when the inertial terms become significant.

Surface Effects.   In 1913 Boussinesq [12] extended the Hadamard–Rybcyznski solution to take into account the effect of interfacial tension and obtained

$$\bar{C}_t = \frac{12}{\mathrm{Re}}\left(\frac{\varkappa + 2 + 3\zeta}{\varkappa + 3 + 3\zeta}\right) \tag{17.44}$$

where $\varkappa = \mu_s/\mu_c$ and μ_s is the "surface viscosity" in pascal seconds (see Levich [11], p. 413).

As $\varkappa$ increases, Equation 17.35 approaches Stokes' law (Equation 16.17) and the bubble or droplet behaves as a solid sphere.

Bubbles.   The viscosity of gases is far less than that of liquids; hence a good approximation for a *completely mobile but spherical bubble* in creeping flow is obtained by setting $\zeta = 0$ in Equations 17.34–17.43. Thus, for the continuous phase,

$$\tilde{\psi} = -\frac{\theta_\infty R^2}{2}\left(1 - \frac{a}{R}\right)\sin^2\{\theta\} \tag{17.45}$$

$$u_R = u_\infty \left(1 - \frac{a}{R}\right)\cos\{\theta\} \tag{17.46}$$

and

$$u_\theta = -u_\infty \left(1 - \frac{a}{2R}\right)\sin\{\theta\} \tag{17.47}$$

Also

$$\bar{C}_t = \bar{C}_p = \frac{8}{\text{Re}} \tag{17.48}$$

As implied by Equation 17.48, $\bar{C}_f = 0$, and the drag is entirely due to the pressure distribution.

Liquid Droplets in Gases.   Conversely, letting $\zeta \to \infty$ in Equations 17.34–17.43 gives good approximations for a small liquid droplet falling in a gas in the regime of creeping flow and for a small bubble in a contaminated liquid. This limit is nothing more than the *Stokes solution* for a solid sphere (Equations 16.8, 16.12, 16.13, and 16.17). Thus the Stokes solution is a special case of the Hadamard–Rybczynski solution.

Power-Law Fluids.   Hirose and Moo-Young [13] derived a solution for creeping flow of a power-law fluid over mobile bubbles, obtaining

$$\tilde{\psi} = -\frac{u_\infty R^2}{2}\sin^2\{\theta\}\left\{\left[1 - \frac{a}{R}\right] + \frac{6\alpha(\alpha - 1)}{2\alpha + 1}\left[\frac{R}{a}\ln\left\{\frac{R}{a}\right\}\right.\right.$$
$$\left.\left. + \frac{1}{6}\left(\frac{a}{R} - \frac{R}{a}\right)\right]\right\} \tag{17.49}$$

and

$$\bar{C}_t \, \text{Re}_{\alpha B} = 8(12)^{(\alpha-1)/2}\left(\frac{13 + 4\alpha - 8\alpha^2}{(2\alpha + 1)(\alpha + 2)}\right) \tag{17.50}$$

where $\text{Re}_{\alpha B} = D_V^\alpha u_\infty^{2\alpha} \varrho / M$.

Boundary-Layer Regime

Levich [11] and Ackeret [14] independently utilized thin-boundary-layer theory for *spherical bubbles* with $\zeta = 0$ to obtain

$$\bar{C}_t = \frac{24}{\text{Re}} \tag{17.51}$$

The dependence of $\bar{C}_t$ on Re^{-1} is in contrast to the dependence of solid spheres on $\text{Re}^{-1/2}$ in this same regime (see problem 16).

Harper and Moore [15] derived the following improved thin-boundary-layer solution for *spheres* with density ratios of the order of unity and small viscosity ratios:

$$\bar{C}_t = \frac{24}{\text{Re}}\left[1 + \frac{3}{2}\zeta + \frac{C_1\zeta^2\lambda^2(1+\xi)}{\xi}\frac{\ln\{\text{Re}\}}{\text{Re}^{1/2}}\right.$$

$$\left. + \frac{\lambda\zeta(1+\xi)}{\xi\,\text{Re}^{1/2}}\left(\frac{C_2\lambda\zeta(1+\xi)}{\xi} - C_3\left(1 + \frac{3\zeta}{2}\right)\right)\right] \qquad (17.52)$$

where

$$\xi = \sqrt{\frac{\rho_d\mu_d}{\rho_c\mu_c}} \qquad \text{and} \qquad \lambda = \frac{3 + 2/\zeta}{3 + 2/\xi}$$

The coefficients C_1, C_2, and C_3 are given as functions of ξ in Table 17.1. Equation 17.52 is obviously limited to values of ζ and ξ such that $\bar{C}_t$ does not exceed the value for a solid sphere (see problem 15).

Table 17.1
Coefficients in Equation 17.52 of Harper and Moore [15]

ξ	5	2	1	0.5	0.2	0
C_1	0.0275	0.0935	0.177	0.262	0.325	0.390
C_2	8.89	7.56	6.15	5.15	4.22	3.56
C_3	9.72	9.00	8.22	7.41	6.59	5.77

For bubbles $\zeta \rightarrow 0$, $\xi \rightarrow 0$, and $\lambda \rightarrow \xi/\zeta$. In this limit, Equation 17.52 reduces to

$$\bar{C}_t = \frac{24}{\text{Re}}\left(1 - \frac{2.21}{\sqrt{\text{Re}}}\right) \qquad (17.53)$$

which was previously derived by Moore [16].

NUMERICAL SOLUTIONS FOR SPHERES

Several numerical solutions of the equations of motion have also been carried out for *spherical bubbles and droplets* for conditions ranging from creeping to boundary-layer flow. The first was apparently by Hamielec and Johnson [17], who used the Galerkin method. They compared their results with experimental data but did not provide explicit values for the computed functions. Their computed streamlines for $\zeta = 0$ and $\zeta = 10$ at Re = 80 are compared in Figure 17–6. The latter high viscosity ratio (corresponding to a liquid droplet in another, less viscous liquid) is seen to produce a stationary vortex in the wake and a secondary circulation inside the droplet.

Hamielec et al. [18] subsequently concluded from calculations for flow about a solid sphere that finite-difference calculations were more accurate than those by the Galerkin method over a wide range of Re. Hamielec et al. [19] then used

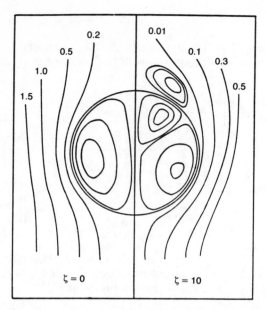

FIGURE 17-6 *Stream functions for a spherical bubble and a spherical droplet at Re = 80 as computed by the Galerkin method. ζ = 0 corresponds to a bubble; ζ = 10 corresponds to a droplet in another liquid. (From Hamielec and Johnson [17].)*

such calculations to obtain the values listed in Table 17.2 for a bubble ($\zeta = 0$). Note that $\bar{C}_f/\bar{C}_p$ decreases from ~2.0 to ~1.0 as Re increases, and that even for Re = 0.1 the values of $\bar{C}_t$ significantly exceed those given by the Hadamard–Rybczynski relationship for $\zeta = 0$ (Equation 17.48).

Yamaguchi et al. [20] used essentially the same Galerkin method as Hamielec et al. [17] to obtain the total mean drag coefficients in Table 17.3 for

Table 17.2
Coefficients Computed by Finite Differences for Spherical Bubbles ($\zeta = 0$) (from Hamielec et al. [19])

Re	0.1	1.0	50	100	200
$\bar{C}_f$	57.73	6.115	0.2134	0.11205	0.06590
$\bar{C}_p$	30.43	3.071	0.1438	0.09025	0.06705
$\bar{C}_t$	88.16	9.186	0.3572	0.2023	0.13295

Table 17.3
Total Mean Drag Coefficients for Fluid Spheres as Computed by the Galerkin Method (from Yamaguchi et al. [20])

ζ	Re				
	0.5	1.0	5.0	10.0	20.0
0	16.5	8.10	1.873	1.036	0.546
0.1	16.77	8.46	1.942	1.072	0.565
1.0	20.04	10.08	2.233	1.217	0.646
10.0	23.30	11.69	2.501	1.346	0.729

Re = 0.5, 1.0, 5.0, 10.0, and 20.0 and ζ = 0, 0.1, 1.0, and 10. The values in Table 17.3 shift from those given by the Hadamard–Rybczynski equation to those given by Stokes' law (Equation 16.17) as ζ increases. The only directly comparable value (for ζ = 0 and Re = 1.0) is considerably below that computed by Hamielec et al. [19].

Haas et al. [5] carried out finite-difference calculations for almost the same

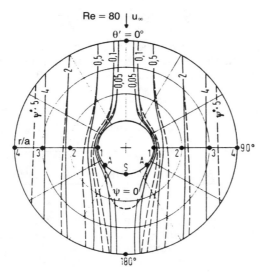

FIGURE 17–7   *Comparison of computed Stokes' stream functions for a spherical bubble (solid curves) and a solid sphere (dashed curves) at Re = 80. $\psi^* = 4\tilde{\psi}/u_\infty D^2$. (From Haas et al. [5].)*

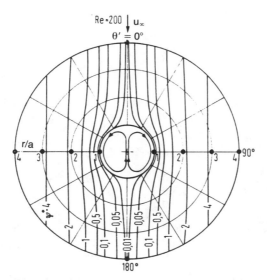

FIGURE 17–8   *Computed internal and external Stokes' stream functions for a spherical bubble at Re = 200. $\psi^* = 4\tilde{\psi}/u_\infty D^2$. (From Haas et al. [5].)*

conditions as those of Hamielec et al. [19] ($\zeta = 0$ and $1 \leq Re \leq 200$). Their computed streamlines for $Re = 80$ are compared with those for a solid sphere ($\zeta \rightarrow \infty$) in Figure 17–7, and those for $Re = 200$ are shown in Figure 17–8. Values of $\bar{C}_t$ were not tabulated but were shown graphically to be well represented by the empirical expression

$$\bar{C}_t = \frac{8}{Re} + \frac{0.745}{Re^{0.18}(1 + 0.1\,Re^{0.6})} \tag{17.54}$$

Brabston and Keller [21] used a series of Legendre polynomials and a finite-difference method with the resulting ordinary differential equations to obtain the total mean drag coefficients in Table 17.4 for $\zeta = 0$.

Table 17.4
Total Mean Drag Coefficients Computed for Spherical bubbles ($\zeta = 0$) (from Brabston and Keller [21])

Re	0.1	0.5	1.0	5.0	10	20	40	60	120	200
$\bar{C}_t$	80.83	16.85	8.795	2.184	1.175	0.6810	0.4156	0.3001	0.1647	0.0985

Abdel-Alim and Hamielec [22] used a finite-difference method to calculate the drag coefficients in Table 17.5 for spherical droplets falling in another liquid. The values of ζ of Part I correspond to droplets of water in cyclohexanol, with small quantities of carboxymethylcellulose added to the water to vary the

Table 17.5
Mean Drag Coefficients Computed for Spherical Droplets Falling in Liquids (from Abdel-Alim and Hamielec [22])

	ζ	Re	$\bar{C}_f$	$\bar{C}_p$	$\bar{C}_t$
Part I	0.0995	1.0	5.80	2.95	8.75
	0.301	1.0	6.42	3.16	9.60
	0.554	1.0	6.64	3.36	10.00
	0.0995	5.0	1.39	0.805	2.19
	0.301	5.0	1.59	0.815	2.40
	0.554	5.0	1.65	0.910	2.56
	0.0995	10.0	0.825	0.425	1.25
	0.301	10.0	0.950	0.495	1.44
	0.554	10.0	1.02	0.535	1.55
Part II	0.266	5.0	1.52	0.745	2.27
	0.708	5.0	1.62	0.830	2.45
	1.40	5.0	1.63	0.880	2.51
	0.266	25.0	0.440	0.215	0.655
	0.708	25.0	0.575	0.275	0.800
	1.40	25.0	0.560	0.285	0.845
	0.266	50.0	0.285	0.135	0.420
	0.708	50.0	0.365	0.190	0.555
	1.40	50.0	0.390	0.225	0.615

viscosity without varying the density or interfacial tension significantly. Part II corresponds to droplets of similarly modified water in *n*-butyl lactate. Figures 17–9A and 17–9B show the computed dependence of the velocity and pressure, respectively, at the surface of the droplets on ζ for Re = 50.

These several solutions are subsequently compared with one another and with experimental data. First, theoretical solutions for deformed bubbles and droplets will be described.

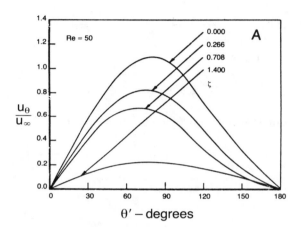

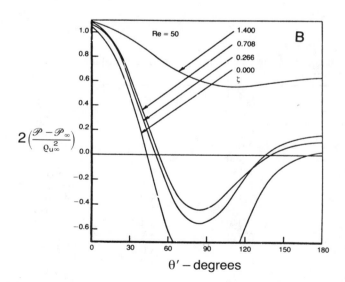

FIGURE 17–9 *Computed tangential velocity and normal pressure at the surface of a spherical droplet for Re = 50 and several viscosity ratios: (A) velocity; (B) pressure. (From Abdel-Alim and Hamielec [22].)*

THEORETICAL SOLUTIONS FOR DEFORMED BUBBLES AND DROPLETS

Slightly Deformed Bubbles

Apparently, the only complete solution taking into account deformation is that of Moore [23] who derived the following implicit relationship for the eccentricity of an oblate ellipsoid with $\zeta = 0$ as a function of the Weber number for the *thin-boundary-layer regime and* $We = \mathcal{O}\{1\}$:

$$We\{E\} = \frac{4(E^3 + E - 2)[E^2 \sec^{-1}\{E\} - (E^2 - 1)^{1/2}]^2}{E^{4/3}(E^2 - 1]^3} \qquad (17.55)$$

which for $E \to 1$ reduces to

$$E = 1 + \frac{9}{64}We + \mathcal{O}\{We^2\} \qquad (17.56)$$

For the drag coefficient in terms of E, he then obtained

$$\bar{C}_t = \frac{24}{Re} G\{E\}\left[1 + \frac{H\{E\}}{Re^{1/2}} + \mathcal{O}\left\{\frac{1}{Re^{1/2}}\right\}\right] \qquad (17.57)$$

where

$$G\{E\} = \frac{E^{4/3}(E^2 - 1)^{3/2}[(E^2 - 1)^{1/2} - (2 - E^2)\sec^{-1}\{E\}]}{3[E^2 \sec^{-1}\{E\} - (E^2 - 1)^{1/2}]^2} \qquad (17.58)$$

$H\{E\}$, which was obtained by numerical integration, is given in Table 17.6. For $E \to 1$, $G\{E\} \to 1$ and

Table 17.6
Function $H\{E\}$ in Equation 17.57 of Moore [23] for Deformed Bubbles

E	$H\{E\}$	E	$H\{E\}$
1.0	−2.211	2.6	+1.499
1.1	−2.129	2.7	+1.884
1.2	−2.025	2.8	+2.286
1.3	−1.899	2.9	+2.684
1.4	−1.751	3.0	+3.112
1.5	−1.583	3.1	+3.555
1.6	−1.394	3.2	+4.013
1.7	−1.186	3.3	+4.484
1.8	−0.959	3.4	+4.971
1.9	−0.714	3.5	+5.472
2.0	−0.450	3.6	+5.987
2.1	−0.168	3.7	+6.517
2.2	+0.131	3.8	+7.061
2.3	+0.448	3.9	+7.618
2.4	+0.781	4.0	+8.189
2.5	+1.131		

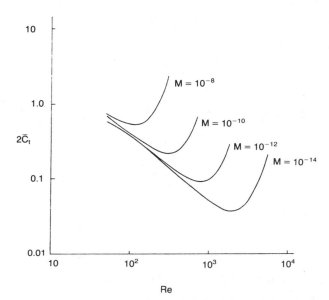

FIGURE 17–10   *The drag coefficient for a slightly deformed bubble in the boundary-layer regime as predicted by Equations 17.55 and 17.57. The curves are arbitrarily terminated at E = 4. (From Moore [23].)*

$$H\{E\} \rightarrow \frac{-4\sqrt{2}\,(6\sqrt{3} + 5\sqrt{2} - 14)}{5\sqrt{\pi}} = -2.2107$$

which is in agreement with Equation 17.53.

Equation 17.57 is plotted in Figure 17–10 as $2\bar{C}_t$ versus Re with M rather than E or W as a parameter per Equation 17.7. The upturn in these curves represents the effect of deformation whose onset is governed by M.

Large, Significantly Deformed Bubbles and Droplets

Davies and Taylor [24] derived an expression for the terminal velocity of a large bubble or droplet by approximating the observed "spheroidal-cap" or "mushroom" shape as the *segment* of a sphere (see Figure 17–11). Postulating that the pressure distribution on the *outside* of the curved surface follows that for inviscid flow gives

$$p - p_o = \pm ga\varrho_c(1 - \cos\{\theta'\}) - \frac{9}{8}\varrho_c u_T^2 \sin^2\{\theta'\} \qquad (17.59)$$

where p_o = pressure at the forward point of incidence, Pa
a = radius of curvature, m
θ' = polar (cone) angle measured from the point of incidence, rad

and the + and − signs indicate upward (bubble) and downward (droplet) movement, respectively. The pressure distribution inside the bubble or droplet is postulated to be that due to the hydrostatic pressure only; that is,

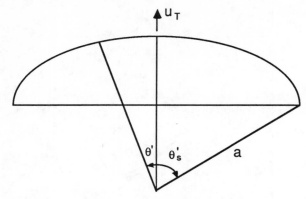

FIGURE 17–11   *Sketch of variables for analysis of the drag of a "spheroidal-cap."*

$$p - p_o = \pm g a \varrho_d (1 - \cos\{\theta'\})$$ (17.60)

Equating these two expressions and solving for u_T^2 gives

$$u_T^2 = \frac{8ga\Delta\varrho}{9\varrho_c}\left(\frac{1 - \cos\{\theta'\}}{\sin^2\{\theta'\}}\right)$$ (17.61)

Equation 17.61 cannot be satisfied exactly for all θ' (see problem 36), implying that the shape corresponding to these pressure distributions is not exactly a segment of a sphere (see problem 37). If Equation 17.61 is arbitrarily satisfied for $\theta' \to 0$, then

$$u_T = \frac{2}{3}\sqrt{\frac{ga\Delta\varrho}{\varrho_c}}$$ (17.62)

which is known as the *Davies–Taylor equation.*

The terminal velocity predicted by Equation 17.62 is seen to be independent of the wake angle, θ_s', defining the aspect ratio of the segment. However, this angle must be known to determine D_V and hence $\bar{C}_t = \bar{C}_p$. As shown subsequently with experimental data, $\theta_s' \cong 50\pi/180$ rad for Re > 100. From geometrical considerations (see problem 38), a $50\pi/180$-rad spherical segment gives

$$D_V = 0.877a$$ (17.63)

Hence

$$u_T = 0.712\sqrt{\frac{gD_V\Delta\varrho}{\varrho_c}}$$ (17.64)

and

$$\bar{C}_t = \bar{C}_p = 1.315$$ (17.65)

For ellipsoidal segments of eccentricity

$$\varepsilon = \sqrt{1 - \frac{1}{E^2}} = \sqrt{1 - \left(\frac{b}{a}\right)^2} \tag{17.66}$$

where b is the length of the vertical semiaxis in meters.

An analogous derivation gives

$$u_T = f\{\varepsilon\} \sqrt{\frac{gb\Delta\varrho}{\varrho_c}} \tag{17.67}$$

For oblate ellipsoidal caps, Wairegi and Grace [25] derived

$$f\{\varepsilon\} = \frac{1}{\varepsilon^3}(\sin^{-1}\{\varepsilon\} - \varepsilon\sqrt{1 - \varepsilon^2}) \tag{17.68}$$

and for prolate ellipsoidal caps, Grace and Harrison [26] obtained

$$f\{\varepsilon\} = \frac{\sqrt{1 - \varepsilon^2}}{\varepsilon^3}(\varepsilon - (1 - \varepsilon^2)\tanh^{-1}\{\varepsilon\}) \tag{17.69}$$

Determination of the corresponding drag coefficients is examined in problems 39–42.

Mendelson [27] suggested that the movement of large malleable bubbles through a liquid might be considered as a disturbance analogous to a wave on a free surface. The velocity would then be that derived by Lamb [10] p. 459, for a capillary-gravity wave:

$$u_T = \left(\frac{2\pi\sigma}{\varrho_c\lambda} + \frac{g\lambda}{2\pi}\right)^{1/2} \tag{17.70}$$

where λ is the wavelength. Arbitrarily taking as the wavelength the circumference of the volume-equivalent bubble; that is,

$$\lambda = \pi D_V \tag{17.71}$$

then gives

$$u_T = \left(\frac{2\sigma}{\varrho_c D_V} + \frac{g D_V}{2}\right)^{1/2} \tag{17.72}$$

Marrucci et al. [28] extended Equation 17.72 for droplets as

$$u_T = \left(\frac{2\sigma}{\varrho_c D_V} + \frac{D_V g \Delta\varrho}{2\varrho_c}\right)^{1/2} \tag{17.73}$$

Since neither μ_c or μ_d is present, they suggested that Equation 17.72 should be applicable to non-Newtonian liquids as well.

Equation 17.73 reduces to Equation 17.64, except for a coefficient of $1/\sqrt{2}$ = 0.707 instead of 0.712 as D_V increases and/or σ decreases; and to Equation 17.31 with a constant of 2, or to the equivalent of Equation 17.32 with $A = 1/3$, or to Equation 17.33 with $A' = \sqrt{2}$ as σ increases and/or D_V decreases.

Equation 17.73 can be rewritten in dimensionless form as

$$u_T^* = \left(\frac{2}{D_V^* M^{1/3}} + \frac{D_V^*}{2} \right)^{1/2} \tag{17.74}$$

or in terms of the total mean drag coefficient as

$$\bar{C}_t = \frac{4}{3}\left(1 - \frac{2}{\mathrm{We}} \right) \tag{17.75}$$

or

$$\bar{C}_t = \frac{4}{3} \bigg/ \left(1 + \frac{4}{\mathrm{E\ddot{o}}} \right) \tag{17.76}$$

or

$$\bar{C}_t = \frac{4}{3}\left[1 - \left(\frac{12\bar{C}_t}{\mathrm{M\,Re}^4} \right)^{1/3} \right] \tag{17.77}$$

Setting the derivative of u_T with respect to D_V in Equation 17.73 to zero provides a prediction of the value and location of a *minimum* in velocity such as that observed in Figure 17–2. The resulting expression for the diameter at the minimum in velocity is

$$D_V = \left(\frac{4\sigma}{g\,\Delta\varrho} \right)^{1/2} \tag{17.78}$$

which is equivalent to

$$D_V^* = \frac{2}{M^{1/6}} \tag{17.79}$$

or

$$\mathrm{E\ddot{o}} = 4 \tag{17.80}$$

The corresponding minimum velocity is

$$u_T = \left(\frac{4\sigma g\,\Delta\varrho}{\varrho_c^2} \right)^{1/4} \tag{17.81}$$

which is equivalent to

$$u^*_T = \frac{2^{1/2}}{M^{1/2}} \tag{17.82}$$

Lehrer [29] used an argument, based on the conversion of the potential energy of the displaced continuous phase to kinetic energy, followed by dissipation of that kinetic energy in the wake, to derive for the same regime

$$u_T = \left(\frac{3\sigma}{\varrho_c D_V} + \frac{D_V g \Delta \varrho}{2 \varrho_c} \right)^{1/2} \tag{17.83}$$

Equation 17.83 differs from 17.72 only in the coefficient of 3 instead of 2 in the first term on the right side.

Lehrer [30] subsequently derived an expression for prediction of a maximum in the velocity, such as that observed in Figure 17–2. He postulated that the force required to move the interface and thereby deform a bubble or droplet is proportional to $\mu_c + \mu_d$. Accordingly, he modified Equation 17.51 to obtain the following expression for the drag coefficient at the point of maximum velocity as a function of the Reynolds number:

$$\bar{C}_t = \frac{24}{Re} \left(\frac{\mu_c + \mu_d}{\mu_c} \right) \tag{17.84}$$

or

$$u_T = \frac{D_V^2 g \Delta \varrho}{36 (\mu_c + \mu_d)} \tag{17.85}$$

which can be rewritten as

$$u^*_T = \frac{(D_V^*)^2}{36 (1 + \zeta)} \tag{17.86}$$

The maximum in the velocity and its location are then obtained by the intersection of the curves represented by Equations 17.86 and 17.83, that is, by their simultaneous solution. Eliminating u_T gives

$$D_V^5 = \frac{648 \sigma}{\varrho_c} \left(\frac{\mu_c + \mu_d}{g \Delta \varrho} \right)^2 \left(6 + \frac{D_V^2 g \Delta \varrho}{\sigma} \right) \tag{17.87}$$

which can be rewritten in canonical form as

$$(D_V^*)^5 = 3888 (1 + \zeta)^2 \left(\frac{1}{M^{1/3}} + \frac{(D_V^*)^2}{6} \right) \tag{17.88}$$

Equation 17.88 can be solved numerically for D_V^* as a function of M and ζ, and the result substituted in Equation 17.86 to give the corresponding maximum in the dimensionless velocity itself. In order for a maximum in velocity to occur, the value of D_V^* given by Equation 17.88 must be less than the value of D_V^* for the minimum in velocity as given by Equation 17.79, namely $D_V^* = 2M^{-1/6}$. The corresponding maximum allowable value of M according to Equation 17.88 is $1.29 \times 10^{-4}/(1 + \zeta)^4$. For larger values of M, behavior such as that illustrated by the rightmost curves of Figure 17–4, in which there is no maximum in U_T, would be expected.

If Equation 17.73 rather than 17.83 were used to derive an expression for the maximum in the velocity, the coefficient of 6 in the rightmost term of Equation 17.88 would be replaced by 4 and the coefficient of 3888 would be replaced by 2592. The maximum value of M for the occurrence of a maximum in velocity would then be predicted to be $3.81 \times 10^{-5}/(1 + \zeta)^4$.

For $\sigma \to 0$ or $D_V \to \infty$ the solutions of both Mendelson [27] and Lehrer [29] predict the identical asymptotic behavior, which can be expressed as

$$u_T^* = \left(\frac{D_V^*}{2}\right)^{1/2} \tag{17.89}$$

or

$$\bar{C}_t = \frac{4}{3} \tag{17.90}$$

The coefficients of Equations 17.89 and 17.90 differ negligibly from those of Equations 17.64 and 17.65. Equations 17.89 and 17.90 are independent of the viscosity as well as of the surface tension of the liquid, and constitute quantitative expressions for the purely inertial and gravitational motion predicted qualitatively by Equations 17.29 and 17.30.

COMPARISON OF THEORETICAL AND NUMERICAL SOLUTIONS FOR BUBBLES

Before comparing the various theoretical solutions mentioned above with experimental data, comparison of the solutions with one another is worthwhile to indicate the general behavior to be expected and to identify the expected range of applicability of particular solutions without the distraction of experimental scatter. This comparison is arbitrarily limited to bubbles ($\zeta = 0$) for which more extensive solutions are available, but the equivalent can be constructed to some extent for droplets (see problems 54 and 55).

Spherical Bubbles

The four above-mentioned sets of numerical solutions for the drag coefficients of spherical bubbles are compared in Figure 17–12A with the solution of Stokes for a *solid sphere* in creeping motion, the solution of Hadamard-Rybczynski for a *completely mobile sphere* in creeping motion, and the solutions of Levich-Ackeret and of Moore for a *completely mobile sphere* in the boundary-layer regime.

The numerically computed values of Haas et al. [5], as represented by

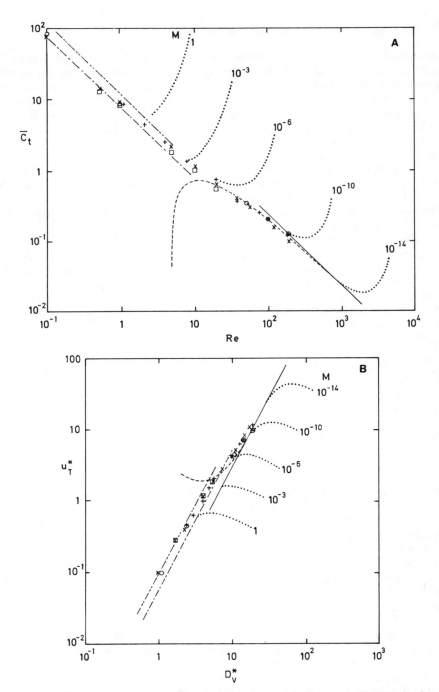

FIGURE 17–12 *Comparison of theoretical solutions for the drag coefficient of spherical bubbles: (A) drag coefficient; (B) dimensionless terminal velocity.* —··— *Stokes;* —·— *Hadamard-Rybczynski;* ——— *Levich-Ackeret;* – – – – *Moore (spheres);* ······ *Moore (oblate ellipsoids);* ○, *Hamielec et al. [19];* □, *Yamaguchi et al. [20], ζ = 0;* ×, *Brabston and Keller [21];* +, *Haas et al. [5] per Equation 17.54.*

Equation 17.54, and those of Hamielec et al. [19] and Brabston and Keller [21] from Tables 17.2 and 17.4, respectively, suggest a gradual transition from Equation 17.48 of Hadamard-Rybczynski to Equation 17.51 of Levich-Ackeret as Re increases. The values of Yamaguchi et al. [20] from Table 17.3 show the same trend, but appear to be in slight error on the low side for all Re.

The improved boundary-layer solution of Moore [16], Equation 17.53, describes the latter portion of this transition, but becomes singular as Re decreases and approaches $2.212^2 = 4.88$.

The dotted curves in Figure 17–12A which represent Equation 17.57, the solution of Moore [23] for distorted bubbles, are included to indicate the value Re for which the postulate of sphericity is no longer valid for a liquid with the indicated property ratio. Thus the assumption of sphericity would be expected to be valid and Equation 17.53 to be applicable up to Re = 10^3 for a liquid for which M = 10^{-14}. On the other hand, for a liquid for which M = 1, the postulate of sphericity is implied to be valid only for Re < 1, and the boundary-layer solutions for spheres (Equations 17.51 and 17.53) then have no applicability.

The same computed values and solutions are plotted in Figure 17–12B in terms of the dimensionless diameter D_V^* and the dimensionless velocity u_T^*. A transition from creeping to boundary-layer motion is observed as D_V^* increases from 1 to 10. Bubbles are predicted to remain spherical up to D_V^* of 30 in a liquid for which M = 10^{-14}, but only up to D_V^* of 3 in a liquid for which M = 1.

In summary, Equation 17.54 of Haas et al. [5] appears to provide an approximate expression for the drag of a *completely mobile, spherical bubble* for Re < 300, and Equation 17.53 of Moore [16] for Re > 30. These equations can readily be re-expressed in terms of u_T^* and D_V^* and are then applicable for D_V^* < 19 and D_V^* > 9, respectively. Possible, alternative expressions for all Re are posed in problems 83 and 84.

Distorted Bubbles

The solutions of Moore, Mendelson, and Lehrer for the drag coefficient of distorted bubbles are compared in Figure 17–13A. Equation 17.48 of Hadamard-Rybczynski and Equation 17.51 of Levich-Ackeret for *spherical bubbles* are included as a frame of reference.

On the one hand, the solutions of Mendelson [27] and Lehrer [29], which are based on wave motion, provide a rationalization for the abrupt increase in the drag coefficient of bubbles, which is observed in some liquids as well as the subsequent gradual transition to a universal constant value. Even more, they predict the quantitative dependence of this overall transition on the dimensionless grouping M.

On the other hand, the solution of Moore [23], which postulates an oblate ellipsoidal shape for the bubbles and boundary-layer-type motion, predicts quantitatively the transition from the boundary-layer regime to the wave regime. For the illustrative parametric values of M = 10^{-10} and 10^{-14} this solution of Moore is seen to become nearly contiguous with the solution of Lehrer [29]. For larger M, as illustrated in Figure 17–13A for 10^{-6}, the solution of Moore fails to approach that of Lehrer. Moore explained this failure as due to the complete distortion of the bubble (which cannot therefore be represented as an oblate ellipsoid) at values of Re below that required for the formation of a boundary

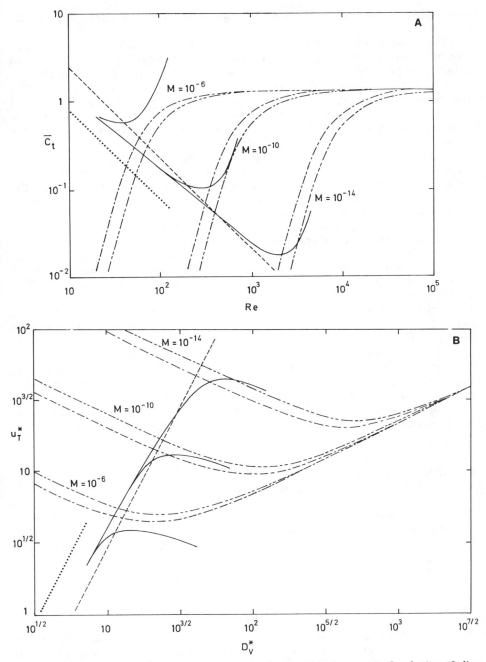

FIGURE 17–13 *Comparison of theoretical solutions for the terminal velocity of distorted bubbles: (A) drag coefficient; (B) dimensionless terminal velocity.* ——— *Moore;* —·— *Mendelson;* —··— *Lehrer;* ----- *Levich-Ackeret (solid spheres);* ······ *Hadamard-Rybczynski (spheres, $\zeta = 0$).*

layer. He suggests $M = 10^{-8}$ as the minimum value for application of his solution.

The solution of Mendelson [27] predicts the same qualitative behavior as that of Lehrer [29], as would be expected since they only differ in the numerical value of one coefficient, but agrees less well with the solution of Moore [23].

Comparison of Figures 17–12A and 17–13A suggests that for $M > 10^{-3}$ a transition may occur directly from the regime of creeping flow, as represented by Equation 17.48, to the regime of purely inertial-gravitational flow, as represented by Equation 17.90, without a pronounced minimum or an abrupt increase.

Just as for spherical bubbles, each of the theoretical solutions has only a narrow range of validity, but in combination they closely predict the entire detailed behavior. An overall correlating equation for the regime in which the bubbles are distorted significantly is suggested in problem 49 and one for all regimes in problem 86.

The same solutions are plotted in Figure 17–13B in terms of u_T^* and D_V^*. The above-mentioned relationships between the several solutions are also apparent in this form. The maximum in velocity for small M is seen to coincide with the minimum in the drag coefficient. The intersection of Equation 17.83 of Lehrer with Equation 17.48 of Levich-Ackeret was suggested by Lehrer [30] as an approximation for the maximum in the velocity. However, Figure 17–13B suggests that the maximum revealed in the solution of Moore [23] might be more accurate. It may be inferred from Figure 17–13B that a maximum will not occur for $M \gg 10^{-6}$.

The minimum in u_T^*, which is predicted by the solutions of Lehrer [29] and Mendelson [27], is evident in Figure 17–13B. This minimum, which does not apparently have a characteristic counterpart in $\bar{C}_t$, must also vanish with the maximum in u_T^* as M increases.

Correlating equations for $u_T^* \{D_V^*\}$ for the regime of distorted bubbles and for all regimes are posed in problems 48 and 85, respectively.

COMPARISON OF PREDICTED AND MEASURED VALUES FOR BUBBLES AND DROPLETS

The various theoretical solutions above are compared with experimental values in this section. These comparisons include the shape of the bubbles and droplets as well as the terminal velocity and drag coefficient. Several generalized empirical correlating equations are also examined.

Shape

The shape of bubbles and droplets has generally been correlated in terms of $E \equiv D_{max}/D_{min}$ or in terms of D_V/D_{max}. These two ratios can be expressed in terms of one another for a particular geometric shape (see, for example, Equations 17.13 and 17.15).

Some guidance as to the dependence of E on the various independent variables and parameters can be obtained from dimensional analysis just as was done for u_T and $\bar{C}_t$. For example, eliminating u_T from the right side of Equation 17.3 and then replacing u_T by E as the dependent variable yield

$$E = \phi\left\{\frac{D_V^3 \varrho_c g \Delta \varrho}{\mu_c^2}, \frac{\mu_d}{\mu_c}, \frac{\varrho_d}{\varrho_c}, \frac{\mu_c^4 g \Delta \varrho}{\varrho_c^2 \sigma^3}\right\} \qquad (17.91)$$

The Hadamard–Rybczynski derivation indicates that deformation is negligible for creeping flow. It follows that deformation must be a consequence of the inertial forces. Eliminating μ_c from Equation 17.91 therefore indicates the limiting behavior for dominance of inertial over viscous forces. This result is

$$E = \phi\left\{\frac{D_V^2 g \Delta \varrho}{\sigma}, \frac{\varrho_d}{\varrho_c}\right\} \qquad (17.92)$$

For both bubbles in liquid and droplets in gas, the term ϱ_d/ϱ_c becomes negligible. For these systems

$$E = \phi\left\{\frac{D_V^2 g \Delta \varrho}{\sigma}\right\} \qquad (17.93)$$

Thus, in this limit the eccentricity ratio would be expected to be a function of Eö only. On the other hand, Moore [23], in his derivation for small deformations in the thin-boundary-layer regime, found E to be a function of We only, given by Equation 17.55.

Aybers and Tapucu [1] correlated their measured values of E for bubbles of air in water graphically as a function of We as shown in Figure 17–14, whereas Reinhart [31] correlated his own data for droplets of various liquids in air in terms of

$$E = 1 + 0.130 \text{ Eö} \qquad (17.94)$$

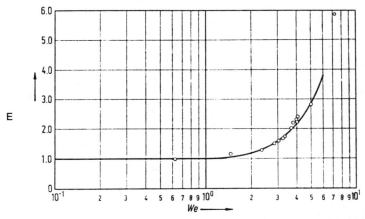

FIGURE 17–14 *Experimental measurements of the eccentricity ratio of bubbles. (From Aybers and Tapucu [1].)*

as shown in Figure 17–15. The relevant physical properties of his fluids are listed in Table 17.7. Wellek et al. [32] correlated data of their own and others for 45 liquid–liquid pairs in terms of

$$E = 1 + 0.091 \; \text{We}^{0.95}, \qquad 0.194 \leq \text{We} \leq 12.6 \tag{17.95}$$

and almost as well by

$$E = 1 + 0.129 \; \text{Eö}, \qquad 0.144 \leq \text{Eö} \leq 9.59 \tag{17.96}$$

The latter result agrees closely with Equation 17.94 of Reinhart.

Takahashi et al. [33] developed the graphical representation of Figure 17–16 for E as a function of Re $\text{M}^{0.23}$ for bubbles of air in aqueous solutions of glycerine and in glycerine–ethanol solutions with the properties given in Table 17.8. They represented this behavior segmentally with the empirical equations listed in Table 17.9. From Equations 17.7 and 17.8 for bubbles ($\Delta\varrho \cong \varrho_c$):

$$\text{Re M}^{1/4} = \left(\frac{3\,\bar{C}_t}{2}\right)^{1/4} \text{We}^{3/4} = \left(\frac{2}{3\,\bar{C}_t}\right)^{1/2} \text{Eö}''^{3/4} \tag{17.97}$$

The slight difference in the exponent of M, 0.25 as opposed to 0.23, is probably an artifact of the statistical evaluation and the scatter in the data. Hence this

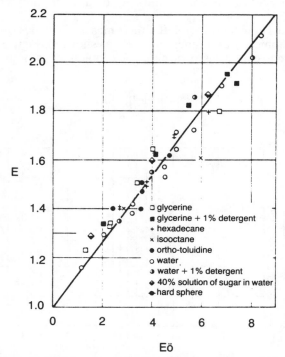

FIGURE 17–15   *Experimental measurements of the eccentricity ratio of droplets of various liquids falling in air.* ——— *Equation 17.94. (From Reinhart [31].)*

Table 17.7
Physical Properties of Fluids Investigated by Reinhart [31]

	Density ϱ (g/cm^3)	Interfacial Tension σ $(dyne/cm)$	Viscosity $\mu \times 10^2$ $(g/cm \cdot s)$	$\left(\dfrac{\varrho_c \sigma^3}{\mu_c^4 g}\right) \times 10^{-12}$ for 760 torr	$\dfrac{\Delta\varrho}{\varrho_c}$
Dry air, 760 torr	1.205×10^{-3}	—	1.819×10^{-2}	—	—
Isoamylalcohol	0.812	25.3	4.25	18.41	668
Chlorobenzene	1.112	35.3	0.800	51.0	892
Diethylene glycol	1.118	50.2	21.0	142.0	928
Glycerine[a] (5% H_2O)	1.232	65.8	1195	319.8	1023
Glycerine 0.80	1.203	67.0	42.7	337.5	998
Glycerine 0.55	1.144	68.3	7.78	356.0	953
Glycerine 0.25	1.059	71.3	1.94	403.5	886
Glycerine[a] + detergent	1.232	35.0	1195	48.1	1023
Heptane	0.684	21.6	0.404	12.51	512
Hexadecane[a]	0.775	29.0	3.67	27.38	643
Hexanol	0.820	27.7	0.552	24.11	673
Isooctane[a]	0.693	20.1	0.497	9.50	551
Paraffin oil	0.879	33.0	177.6	40.35	729
ortho-toluidine[a]	0.998	42.3	4.40	84.9	828
Distilled water[a]	0.998	72.7	1.00	427.8	834
Water[a] + deterg. 2000:1	0.998	35.2	1.00	48.5	834
Water[a] + deterg. 40:1	0.998	33.3	1.00	41.1	834
21% Sugar solution	0.091	74.0	2.14	451.0	912
44% Sugar solution	1.202	75.7	8.40	482.5	1006
40% Sugar solution	1.196	67.8[b]		346.8	1001

[a] These droplets were photographed to determine their deformation.
[b] This solution was prepared with ordinary tap water.

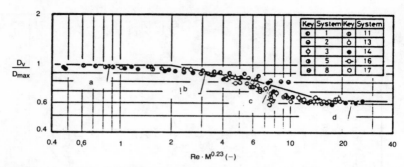

FIGURE 17–16   *Correlation for eccentricity ratio of bubbles of air in various liquids. The system designations refer to Table 17.8. (From Takahashi et al. [33].)*

Table 17.8
Physical Properties of Liquids Investigated by Takakashi et al. [33]

System	ϱ_c (g/cm^3)	μ_c $(g/cm \cdot s)$	σ $(dyne/cm)$	M $[--]$	Liquid
1	1.26	13.3	61.5	1.04×10^2	
2	1.25	6.39	61.5	5.62×10^0	
3	1.22	1.10	59.0	5.78×10^{-3}	
4	1.22	0.811	62.0	1.46×10^{-3}	
5	1.22	0.805	60.0	1.56×11^{-3}	
6	1.21	0.780	62.0	1.24×10^{-3}	
7	1.21	0.762	58.7	1.34×10^{-3}	Aqueous
8	1.21	0.599	58.1	5.33×10^{-4}	solutions
9	1.21	0.564	62.0	3.45×10^{-4}	of
10	1.21	0.532	60.0	3.04×10^{-4}	glycerine
11	1.20	0.472	62.0	1.69×10^{-4}	
12	1.19	0.418	55.5	1.46×10^{-4}	
13	1.16	0.244	53.0	2.02×10^{-5}	
14	1.16	0.0906	59.0	2.79×10^{-7}	
15	1.14	0.0703	56.0	1.19×10^{-7}	
16	1.14	0.0574	59.0	5.94×10^{-8}	
17	1.16	1.46	39.0	6.47×10^{-2}	Glycerine-ethanol mixture

Table 17.9
Empirical Equations Proposed by Takakashi et al. [33] to
Represent Eccentricity Ratio of Bubbles

Designation in Figure 17–16	Re $M^{0.23}$	D_V/D_{max}
a	< 2	1
b	2–6	$1.14/(Re\ M^{0.23})^{0.176}$
c	6–16.5	$1.36/(Re\ M^{0.23})^{0.28}$
d	> 16.5	0.62

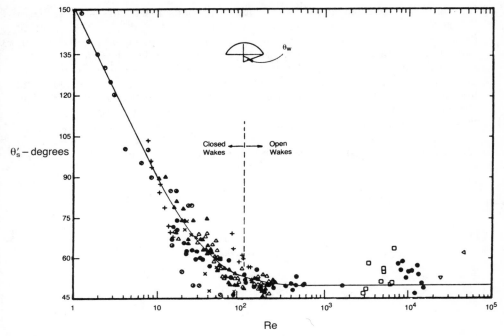

FIGURE 17–17   *Graphical correlation for wake angle of spherical-cap bubbles. (From Clift et al. [34], p. 204.)*

correlation is essentially equivalent to one in We or Eö insofar as the variation in $\bar{C}_t$ is negligible or a function of Re $M^{1/4}$.

Experimental data for the wake angle θ'_s of bubbles (see Figure 17–11) are plotted in Figure 17–17 as a function of Re by Clift et al. [34], p. 204. E can be calculated from these data if the shape is assumed to be a true segment of a sphere (see problem 61). As mentioned in connection with the derivation of Equation 17.64, θ'_s approaches $50\pi/180$ rad as Re increases. The corresponding value of E is 0.572. The deviation of this value from the observed asymptotic values of about 0.62 in Figure 17–16 implies that the bubbles are not pure segments of a sphere.

Drag Coefficients

A bewildering array of data has been collected for the drag coefficient of bubbles and droplets, much of it of uncertain accuracy and/or for fluids of undefined interfacial tension. Several representative sets will be examined.

Bubbles

The precise determinations of Redfield and Houghton [35] for the drag coefficient of bubbles of CO_2 in aqueous solutions of dextrose are shown in Figure 17–18. For Re < 0.3, the data follow the Hadamard–Rybczynski relationship for completely mobile spheres. A transition to Stokes' law occurs between Re = 0.3 and Re = 2, followed by an upward transition signifying the onset of deformation at a different value of Re for each solution. These transitions cannot be characterized in terms of M since the interfacial tension was not specified. The maximum measured value of $\bar{C}_t$ thereafter appears to be about 1.5, which is slightly above the prediction of Equations 17.65 and 17.90. (The line in Figure 17–18 labeled "Boussinesq" corresponds to Equation 17.51. The curves marked "Chao" and Hamielec-Johnson" are not discussed here since they do not appear to be particularly useful.)

The experimentally determined drag coefficients of Hacker and Hussein [36] for bubbles of nitrogen in ethylene glycol and silicone oil in Figure 17–19 show a

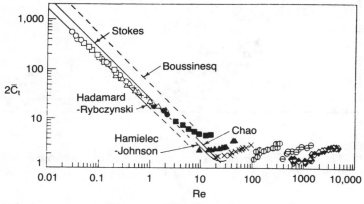

FIGURE 17–18   *Experimental drag coefficients for bubbles of CO_2 in aqueous solutions of dextrose. (From Redfield and Houghton [35].)*

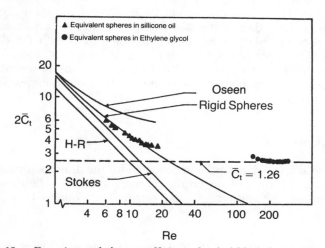

FIGURE 17–19   *Experimental drag coefficients for bubbles of nitrogen in silicone oil and ethylene glycol. H-R designates Hadamard-Rybczynski. (From Hacker and Hussein [36].)*

smooth and monotonic variation with Re from Stokes' law to an asymptotic value of $\bar{C}_t = 1.26$.

An extensive and general correlation for bubbles was developed by Miyahara and Takahashi [37] in terms of a modified drag coefficient.

$$\bar{C}_t' = \left(\frac{D_V}{D_{max}}\right)^2 \bar{C}_t = \frac{2}{3}\frac{D_V^3 g \Delta \varrho}{D_{max}^2 \varrho_c u_T^2} = \frac{2}{3}\frac{D_V^3 g}{D_{max}^2 u_T^2} \qquad (17.98)$$

and a modified Reynolds number

$$\text{Re}' = \left(\frac{D_{max}}{D_V}\right) \qquad \text{Re} = \frac{D_{max} u_T \varrho_c}{\mu_c} \qquad (17.99)$$

as suggested by Takahashi et al. [33]. Here, as before, D_{max} is the major axis of the bubble or droplet in meters. The correlation, which is shown graphically in Figure 17–20, is represented by the following set of empirical but coherent equations:

1. For $M > 10^{-7}$ and all Re', and for $\text{Re}' < 10$ and all M,

$$\bar{C}_t' = \frac{8}{\text{Re}'} + 0.5 \qquad (17.100)$$

2. For $M < 10^{-7}$ and $10 < \text{Re}' < 15.8\,M^{-0.136}$,

$$\bar{C}_t' = \frac{6.5}{(\text{Re}')^{0.7}} \qquad (17.101)$$

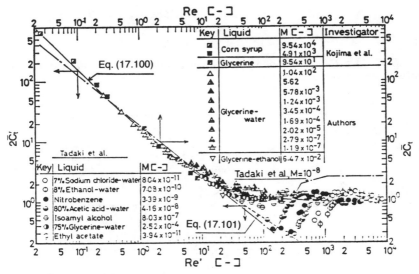

FIGURE 17–20 *Correlation of Miyahara and Takahashi [37] for the drag coefficient of bubbles.*

3.   For $M < 10^{-7}$ and $Re' > 15.8\,M^{-0.136}$,

$$\bar{C}_t' = 0.015\,(Re')^{3/2}M^{0.3} \tag{17.102}$$

or

$$\bar{C}_t' = 0.5 \tag{17.103}$$

whichever is greater.

Equation 17.101 represents the lower bounding behavior for $Re' > 10$ and sufficiently large surface tensions (small M) such that the bubbles are relatively undeformed, Equation 17.102 the transition due to the onset of deformation, and Equation 17.103 the purely inertial regime (independent from viscosity). The representation of the transition by Equation 17.102 is confirmed in Figure 17–21. The structure of the correlation is sketched in Figure 17–22.

The method of Churchill and Usagi [38] can be used to develop the following single expression for the correlation of Miyahara and Takahashi:

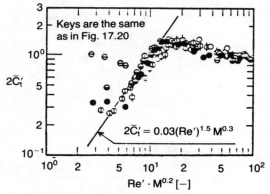

FIGURE 17–21   *Correlation of Miyahara and Takahashi [37] for the drag coefficient of bubbles in the regime of transition.*

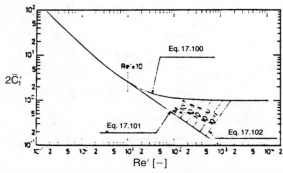

FIGURE 17–22   *Structure of correlation of Miyahara and Takahashi [37] for the drag coefficient of bubbles.*

$$\bar{C}_t' = \frac{1}{2}\left[\left(\frac{16}{\text{Re}'}\right)^4 + \left(\frac{13}{(\text{Re}')^{0.7}}\right)^4 + \left(\left[1 + \left(\frac{16}{\text{Re}'}\right)^{3/4}\right]^{-40/3}\right.\right.$$
$$\left.\left. + \left[0.03\,(\text{Re}')^{1.5}\,\text{M}^{0.3}\right]^{-10}\right)^{-2/5}\right]^{1/4} \tag{17.104}$$

The exponents in Equation 17.104 are arbitrary, but $\bar{C}_t'$ is relatively in sensitive to their choice.

Figure 17–16 or the expressions in Table 17.9 must be used with Figure 17–20, Equations 17.100–17.103, or Equation 17.104 to determine $\bar{C}_t$ and Re from $\bar{C}_t'$ and Re'. As Re'$\text{M}^{0.23} \rightarrow \infty$, $D_V/D_{\max} \rightarrow 0.62$. Hence from Equations 17.98 and 17.102,

$$\bar{C}_t \rightarrow \frac{0.5}{(0.62)^2} = 1.30 \tag{17.105}$$

which is in close agreement with Equations 17.65 and 17.90.

The data of Figures 17–18 and 17–19 are readily rationalized in terms of Equations 17.100–17.103. An alternative generalized representation for the drag coefficient for bubbles is examined below and another is posed in problem 49.

Droplets in Air

The precise measurements by Reinhart [31] of the drag coefficient for droplets of water falling in air are seen in Figure 17–23 to follow his own data for hard spheres up to Re $\cong$ 800 and then to turn up, indicating deformation. His data for a variety of liquid droplets falling in air, as plotted in Figure 17–24, show similar behavior but with different points of transition corresponding to the values of M in Table 17.7. He tabulated empirical equations for $\bar{C}_t$ as a function of $D_V^*(\varrho_c/\Delta\varrho)^{1/2}$, $\text{M}\varrho_c/\Delta\varrho$, and $\Delta\varrho/\varrho_c$.

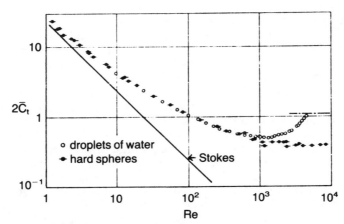

FIGURE 17–23   *Experimental drag coefficients for droplets of water in air. (From Reinhart [31].)*

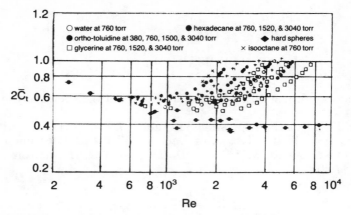

FIGURE 17–24   *Experimental drag coefficients for droplets of various liquids in air. (From Reinhart [31].)*

Droplets in Liquid

Droplets rising or falling in another immiscible liquid would be expected to demonstrate more complex and less readily generalized behavior because of the wide range of possible combinations of finite values of μ_d/μ_c and ϱ_d/ϱ_c above and below unity. This expectation is generally confirmed.

The drag coefficients measured by Satapathy and Smith [39] for a variety of liquid droplets in aqueous solutions of glycerine are reproduced in Figure 17–25. Because of the high viscosity of pure glycerine, very low values of Re were attained. They did not document the physical properties, but they did describe the shape and internal circulation of the droplets. For Re < 4 the drag followed the Hadamard–Rybczynski relationship for a completely mobile sphere ($\zeta = 0$), and the droplets moved in a straight line. For 4 < Re < 10 the drag underwent a gradual transition to that for solid spheres. A single ring vortex was formed at the rear and the fluid within the droplet continued to circulate. Some flattening of the droplet occurred at Re = 8. For 10 < Re < 30 the drag coefficient followed Stokes' law; then from Re = 30 to 45 it remained constant and thereby exceeded the value for solid spheres. Above Re = 40 the droplet moved unsteadily; the vortex remained attached but moved from side to side. At Re = 45 the vortex began to detach on alternate sides; the droplet was somewhat deformed but moved steadily up to Re = 100; internal circulation was minimal. Above Re = 100 the motion was unsteady and the surface oscillated. Above Re = 500 the droplet followed a helical path induced by the detachment of vortices on alternating sides. The drag coefficient turned up at different Re depending on the dispersed liquid. Note that all of the measured drag coefficients, except for aniline, fall above the curve for solid spheres for all Re > 10. The lower values for aniline suggest that these droplets were still circulating internally even at Re = 600. All of the droplets were observed to break up below Re = 1500.

The drag coefficients determined by Hu and Kintner [40] for droplets of *more dense liquids falling in water* are plotted in Figure 17–26. These measurements, which are confined to Re > 10, are consistent with those of Satapathy and Smith [39], including the singular deviation of the values for

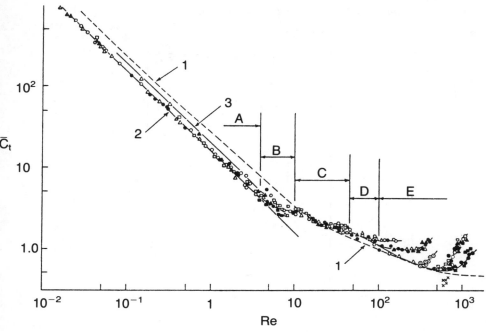

FIGURE 17–25 *Experimental drag coefficients for droplets of various liquids in aqueous solutions of glycerine. 1, solid spheres; 2, Hadamard–Rybczynski ($\zeta = 0$); 3, $\zeta = 1$. Experimental:* $\bigcirc$, *tetrachloroethylene;* $\bullet$, *carbon tetrachloride;* $\triangle$, *chloroform;* $\blacktriangle$, *bromobenzene;* $\square$, *ethyl bromide;* $\blacksquare$, *carbon disulphide;* $\otimes$, *chlorobenzene;* $\odot$, *benzyl alcohol;* $\times$, *aniline. (From Satapathy and Smith [39].)*

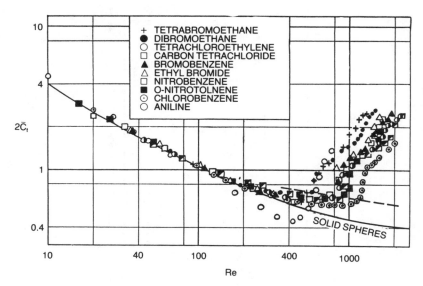

FIGURE 17–26 *Experimental drag coefficients for droplets of various liquids falling in water. The dashed line represents the observed maximum in terminal velocity. (From Hu and Kintner [40].)*

Table 17.10
Physical Properties of Liquids Investigated by Hu and Kintner [40]

System	ϱ_o	ϱ	$\dfrac{\Delta \varrho}{\varrho}$	$\mu_o \times 10^2$	$\mu \times 10^2$	σ_i	P	T
1 Aniline	1.0176	0.9979	0.0197	3.5416	0.9471	2.8[a]	1.41×10^8	25.0
2 Tetrabromoethane	2.9474	0.9973	1.9553	9.2888	0.8968	35.9	3.73×10^9	24.9
3 Dibromoethane	2.1541	0.966	1.1614	1.5852	0.8968	31.9	4.39×10^9	25.0
4 Ethyl bromide	1.4478	0.9977	0.4511	0.4908	0.8814	30.0	1.01×10^{10}	26.2
5 Nitrobenzene	1.1947	0.9972	0.1981	1.7379	0.8835	24.1	1.18×10^{10}	25.6
6 Bromobenzene	1.4881	0.9971	0.4924	1.0719	0.8958	37.9	1.75×10^{10}	25.0
7 o-Nitrotoluene	1.1576	0.9970	0.1609	2.0360	0.8996	26.5	1.80×10^{10}	24.7
8 Tetrachloroethylene	1.6143	0.9970	0.6192	0.8903	0.8946	44.4	2.25×10^{10}	25.1
9 Carbon tetrachloride	1.5770	0.9957	0.5838	0.8702	0.7797	40.6	3.14×10^{10}	30.4
10 Chlorobenzene	1.0995	0.9969	0.1029	0.7606	0.9036	36.7	7.30×10^{10}	24.8

ϱ = density; μ = viscosity; σ_i = interfacial tension; (cal-g-s units) sub o = organic phase; others = water phase; $P = \sigma_i^3 \varrho^2 / g\mu^4 \Delta\varrho$; T = temperature, °C.
[a] System unstable

droplets of aniline below the curve for a solid sphere. Examination of the physical properties of these liquids in Table 17.10 suggests that the low interfacial tension is responsible. The droplets disintegrated at approximately Re = 1500 and approached or appeared to be approaching an asymptotic value of $\bar{C}_t \cong 1.3$. (The dashed line in Figure 17–26 represents the observed maximum in the terminal velocity.)

On the other hand, as shown in Figure 17–27, the drag coefficients determined by Klee and Treybal [41] for *droplets of less dense liquids rising in water* fall below the curve for solid spheres for $10 < \text{Re} \le 100$. An explanation in terms of the physical properties listed in Table 17.11 is not obvious.

The experimental values of Hamielec [42] for droplets of water (with additions of carboxymethyl cellulose to vary the viscosity without varying the density or interfacial tension significantly) falling in *m*-butyl lactate are compared in Figure 17–28 with curves representing the numerically computed values of Abdel-Alim and Hamielec [22] from Table 17.5. A similar comparison for droplets of the enhanced water in cyclohexanol is shown in Figure 17–29. In both instances the agreement is remarkably good. The correction for finite ζ provided by Equation 17.43 is in good agreement with the experimental values for Re = 1.0 but is insufficient for higher Re, whereas the correction provided by Equation 17.52 is too great. Accordingly Abdel-Alim and Hamielec proposed a correction equivalent to

$$\frac{\bar{C}_t\{\zeta\}}{\bar{C}_t\{0\}} = \frac{2.60}{1.19}\left(\frac{(1.3 + \zeta)^2 - 0.5}{(1.3 + \zeta)(2 + \zeta)}\right) \tag{17.106}$$

Hu and Kintner [40] showed that all of their own data fall along a single curve with an intermediate break in slope when plotted as $2\bar{C}_t \text{We M}^{-0.15}$ versus $\text{Re M}^{0.15}$, as in Figure 17–30. They represented this relationship by the pair of empirical equations

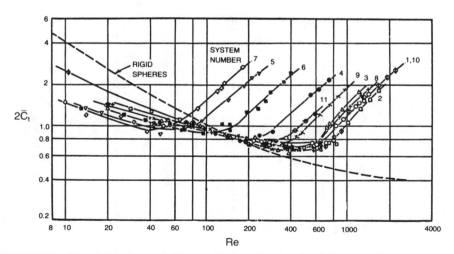

FIGURE 17–27 *Experimental drag coefficients for droplets of various liquids rising in water. System numbers refer to Table 17.11. (From Klee and Treybal [41].)*

Table 17.11
Physical Properties of Liquids Investigated by Klee and Treybal [41]

System	Continuous liquid	Dispersed liquid	Densities		Viscosities		Interfacial tension (σ)	Temp. (°C)
			ϱ_c	ϱ_d	μ_c	μ_d		
1	Water	Benzene	0.9975	0.8870	0.0114	0.0068	30.0	23
2	Water	Kerosene	0.9986	0.8071	0.0108	0.0147	40.4	18
3	Water	S.A.E. 10W oil	0.9975	0.8650	0.0106	0.721	18.5	23
4	Water	Nonyl alcohol	0.9982	0.8242	0.0100	0.162	4.9	20
5	Water	sec-Butyl alcohol	0.9705	0.8660	0.0156	0.0278	0.6	28
6	Furfural	Water	1.1450	1.0110	0.0134	0.0096	1.5	26.5
7	Water	Methyl ethyl ketone	0.960	0.8370	0.0145	0.0060	0.3	24
8	20% Aqueous sucrose	Benzene	1.0600	0.8720	0.0139	0.0059	30.1	26
9	Water	Methyl isobutyl ketone	0.9947	0.8155	0.0093	0.0060	9.8	20
10	Water	Pentachloroethane	0.9978	1.6740	0.0095	0.0203	42.4	25
11	Water	n-Heptylic acid	0.9980	0.9200	0.0095	0.0427	6.9	24.5

cal-g-s units.

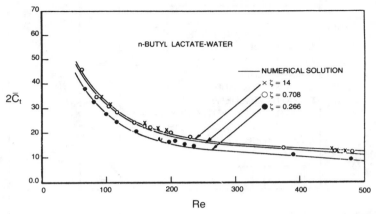

FIGURE 17–28   *Comparison of computed and experimental drag coefficients for droplets of water (with additions of carboxymethyl cellulose) falling in n-butyl lactate. (From Abdel-Alim and Hamielec [22].)*

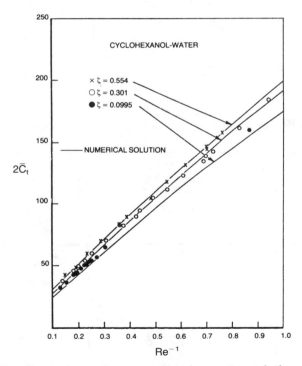

FIGURE 17–29   *Comparison of computed and experimental drag coefficients for droplets of water (with additions of carboxymethyl cellulose) falling in cyclohexanol. (From Abdel-Alim and Hamielec [22].)*

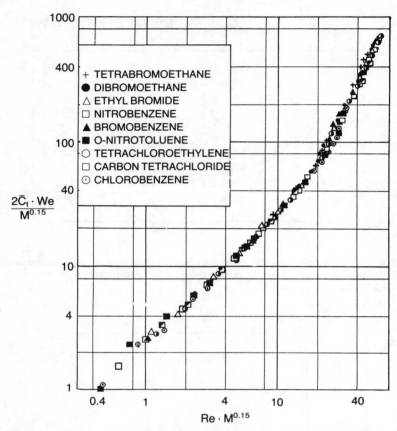

FIGURE 17–30 *Correlation of Hu and Kintner [40] for droplets falling in water.*

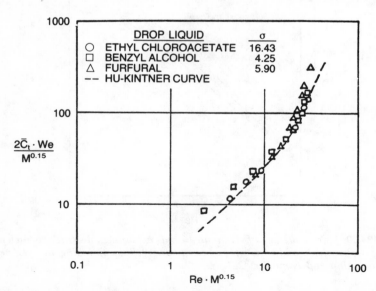

FIGURE 17–31 *Comparison by Warshay et al. [43] of their data for the drag coefficient of droplets of organic liquids falling in water with the correlating equations of Hu and Kintner. (σ in dynes/cm)*

$$Y = \frac{4}{3}X^{1.275}, \qquad 2 < Y \le 70 \tag{17.107}$$

and

$$Y = 0.045X^{2.37}, \qquad Y \ge 70 \tag{17.108}$$

where

$$Y \equiv 2\,\bar{C}_t\,\text{We}\,\text{M}^{-0.15} \tag{17.109}$$

and

$$X \equiv 0.75 + \text{Re}\,\text{M}^{0.15} \tag{17.110}$$

Warshay et al. [43] tested this correlation for other fluid pairs (see Figures 17–31 to 17–34) and found it satisfactory except for continuous fluids of high viscosity and pairs of liquids with very low interfacial tension, as shown in Figures 17–35

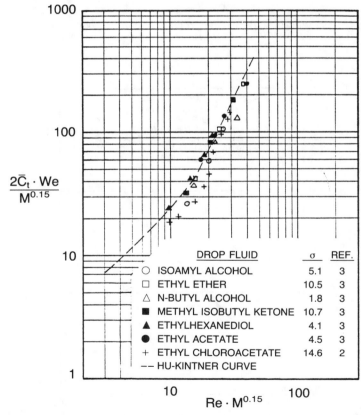

FIGURE 17–32 *Comparison by Warshay et al. [43] of the data of Licht and Narasimhamurty (labeled REF 2) and of Keith and Hixson (labeled REF 3) for the drag coefficient of droplets with a low interfacial tension with the correlating equations of Hu and Kintner. (σ in dynes/cm)*

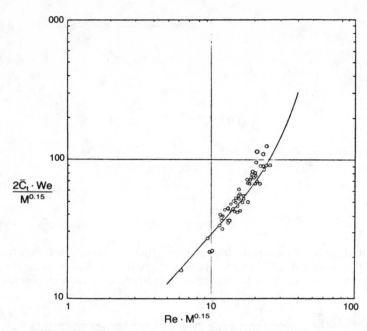

FIGURE 17-33 *Comparison by Warshay et al. [43] of the drag coefficients of Smirnov and Rubin, after correction for wall effects, with the correlating equations of Hu and Kintner.*

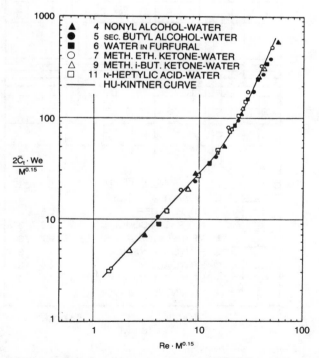

FIGURE 17-34 *Comparison by Warshay et al. [43] of the drag coefficients of Klee and Treybal with the correlating equations of Hu and Kintner. Reference numbers refer to Table 17.11.*

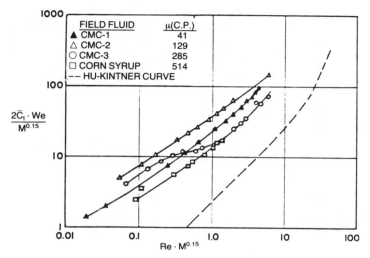

FIGURE 17–35 *Comparison of Warshay et al. [43] of their own data for the drag coefficient of droplets of C_2Cl_4 in various liquids with the correlating equations of Hu and Kintner.*

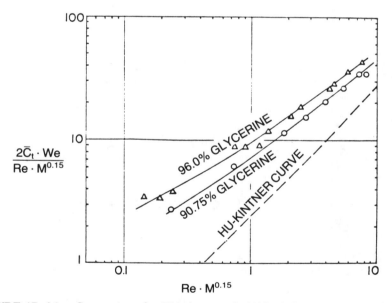

FIGURE 17–36 *Comparison by Warshay et al. [43] of their own data for the drag coefficient of droplets of C_2Cl_4 in glycerine with the correlating equations of Hu and Kintner.*

and 17–36. They suggested that the drag coefficient for a solid sphere provides an adequate representation for these exceptions to the correlation of Hu and Kintner. The properties of the special fluid pairs that they tested are listed in Table 17.12.

Table 17.12
Physical Properties of Liquids Investigated by Warshay et al. [43]

System (Drop Field)	Drop Density ϱ_d (g/cm³)	Field Density ϱ (g/cm³)	Density Difference $\Delta\varrho$ (g/cm³)	Drop Viscosity μ_d (centipoises)	Field Viscosity μ (centipoises)	Interfacial tension σ (dyne/cm)	1/M (Dimensionless)	Temperature (°C)
Furfural—water	1.154	0.9966	0.1574	1.50	0.9075	5.9	1.96×10^8	26.50
Benzyl alcohol—water	1.040	0.9966	0.0434	5.15	0.8631	4.25	3.225×10^8	26.50
Ethylchloroacetate—water	1.141	0.9966	0.1444	1.38	0.8585	16.43	57.5×10^8	26.75
Tetrachloroethylene—CMC 1	1.612	1.000	0.612	0.926	41.3	19.15	67×10^8	26.50
Tetrachloroethylene—CMC 2	1.612	1.00	0.612	0.926	128.7	11.99	162.3	26.00
Tetrachloroethylene—CMC 3	1.612	1.00	0.612	0.926	285.0	30.15	1.355	26.25
Tetrachloroethylene—corn syrup	1.614	1.395	0.219	0.926	514	35.2	5.26	26.50
Tetrachloroethylene—96% glycerine	1.615	1.243	0.372	0.926	287 to 386	9.71		28.2 to 30.2
Tetrachloroethylene—90.75% glycerine	1.615	1.231	0.384	0.926	141	13.35	4.30	27.0

Substitution for We from Equation 17.7 reveals that the ordinate of Figures 17–30 to 17–36 is equivalent to

$$\left(\frac{16}{3}\right)^{1/3}\left(\frac{C_t^{2/3}}{M^{1/60}}\right)(\text{Re } M^{0.15})^{4/3}$$

The appearance of the predominant independent variable to the 4/3 power in the ordinate and the first power in the abscissa suggests that the apparent success of the correlation may be deceptive (see problem 68).

Similar graphical and empirical correlations have been developed by Klee and Treybal [41], Johnson and Braida [44], Grace et al. [45], and others, but they are subject to the same criticism.

Terminal Velocity

Plots of experimental data for u_T versus D_V have already been illustrated in Figures 17–2 to 17–5. Additional data for droplets falling and rising in water are shown in Figures 17–37 and 17–38, respectively. The various equations and plots of $\overline{C}_t$ as a function of Re can readily be converted to plots of u_T^* versus D_V^* by combining Equations 17.10 and 17.11 to obtain

$$u_T^* = \left(\frac{2}{3}\frac{\text{Re}}{\overline{C}_t}\right)^{1/3} \tag{17.111}$$

and

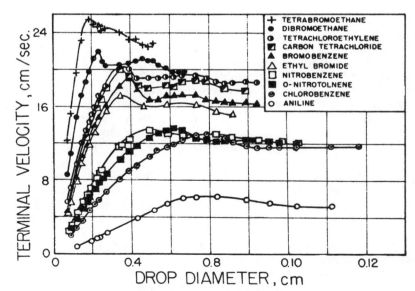

FIGURE 17–37   *Terminal velocities of droplets of various liquids falling in water as measured by Hu and Kintner [40].*

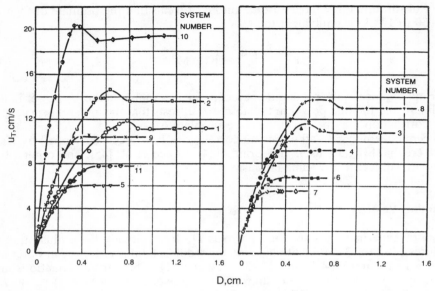

FIGURE 17–38   *Terminal velocities of droplets of various liquids rising in water as measured by Klee and Treybal [41]. System numbers refer to Table 17.11.*

$$D_V^* = \left(\frac{3\,\mathrm{Re}^2\,\bar{C}_t}{2}\right)^{1/3} \tag{17.112}$$

(see problem 4).

Direct comparisons of the boundary-layer solution for slightly deformed bubbles (Equations 17.57 with 17.55), and the wave equations for greatly distorted bubbles (Equations 17.73 and 17.83), with the experimental data of Haberman and Morton [3] for bubbles of air in various liquids are shown in Figures 17–39 to 17–42. The properties of these liquids are listed in Table 17.13.

For distilled or filtered water, Equations 17.73 and 17.83 are seen in Figure 17–39 to bound the experimental data for all diameters above that for the maximum in the velocity. Equation 17.57, which is not plotted in Figure 17–39, provides a reasonably good approximation for all diameters up to that for the minimum in velocity.

For Varsol, as shown in Figure 17–40, Equation 17.83 represents the data near the maximum somewhat better than does 17.73, but near the minimum Equation 17.73 provides the better representation. Equation 17.57 provides a good overall prediction up to the minimum. For methyl alcohol, in Figure 17–41, Equation 17.73 is slightly better than 17.83, and Equation 17.57 does not provide as close a prediction as Varsol. For turpentine, in Figure 17.42, Equations 17.57 and 17.73 fall below the data at the maximum in velocity, whereas Equation 17.83 is too high near the minimum.

Moore [23] concludes that the reasonably good predictions for diameters beyond the maximum in velocity where the shape is not ellipsoidal as postulated indicates that the drag in this regime is insensitive to shape.

Table 17.13
Physical Properties of Liquids Investigated by Haberman and Morton [3]

Liquid	Temperature, (°C)	Viscosity, μ (poises)	Density, ϱ (g/cm³)	Surface tension, σ (dyne/cm)	M
Water	19	0.0102	0.998	72.9	0.26×10^{-10}
Water	21	0.0098	0.998	72.6	0.24×10^{-10}
Cold water	6	0.0147	0.999	74.8	1.08×10^{-10}
Hot water	49	0.0056	0.989	68.1	0.307×10^{-11}
Glim solution	19	0.0103	1.000	32.8	2.78×10^{-10}
Mineral oil	27.5	0.580	0.866	20.7	1.45×10^{-2}
Varsol	28	0.0085	0.782	24.5	4.3×10^{-10}
Turpentine	23	0.0146	0.864	27.8	24.1×10^{-10}
Methyl alcohol	30	0.0052	0.782	21.8	$0.89 + 10^{-10}$
62% corn syrup and water	22	0.550	1.262	79.2	0.155×10^{-3}
68% corn syrup and water	21	1.090	1.288	79.9	0.212×10^{-2}
56% glycerine and water (T. Bryn)	18	0.0915	1.143	69.9	1.75×10^{-7}
42% glycerine and water (T. Bryn)	18	0.043	1.105	71.1	4.18×10^{-8}
13% ethyl alcohol and water (T. Bryn)	22	0.0176	0.977	43.5	1.17×10^{-8}
Olive oil (H. D. Arnold)	22	0.73	0.925	34.7	0.716×10^{-2}
Syrup (W. N. Bond and D. A. Newton)	17	180	1.48	91	0.92×10^{6}

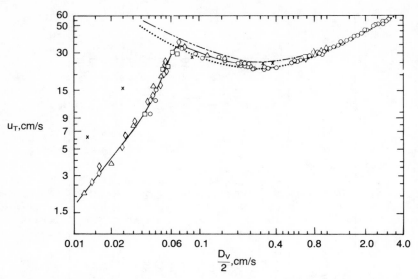

FIGURE 17–39   *Comparison of wave-type solutions with experimental terminal veloci-ties of bubbles of air in distilled or filtered water. Data are the same as in Figure 17–2.* ———— *graphical representation (Haberman and Morton);* ······ *Equation 17.72 (Mendelson);* —··— *Equation 17.83 (Lehrer).*

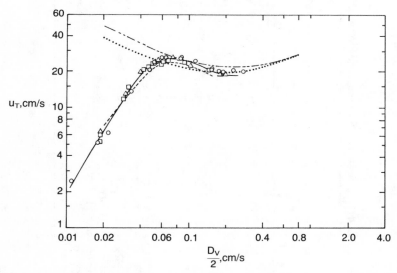

FIGURE 17–40   *Comparison of wave and boundary-layer solutions with experimental terminal velocities of Haberman and Morton for bubbles of air in Varsol.* ———— *graphical representation (Haberman and Morton);* ----- *Equations 17.55 and 17.57 (Moore);* ······ *Equation 17.72 (Mendelson);* —··— *Equation 17.83 (Lehrer). (Adapted from Moore [23].)*

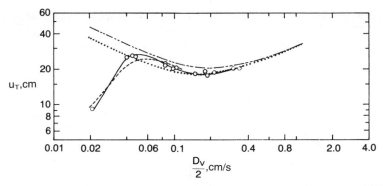

FIGURE 17-41 *Comparison of wave and boundary-layer solutions with experimental data of Haberman and Morton for bubbles of air in methyl alcohol.* ——— *graphical representation (Haberman and Morton);* ———— *Equations 17.55 and 17.57 (Moore);* ······ *Equation 17.72 (Mendelson);* —··— *Equation 17.83 (Lehrer). (Adapted from Moore [23].)*

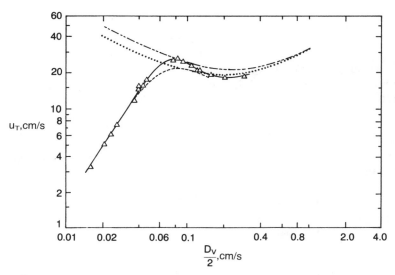

FIGURE 17-42 *Comparison of wave and boundary-layer solutions with experimental data of Haberman and Morton for bubbles of air in turpentine.* ——— *graphical representation (Haberman and Morton);* ———— *Equation 17.55 and 17.57 (Moore);* ······ *Equation 17.72 (Mendelson);* —··— *Equation 17.83 (Lehrer). (Adapted from Moore [23].)*

The Maximum and Minimum in the Terminal Velocity

The maximum and minimum observed in the terminal velocity are critical characteristics of the behavior of bubbles and droplets, and are of direct practical interest.

Peebles and Garber [46] concluded from their graphical correlation for the drag coefficient for bubbles in various liquids that the Reynolds number corresponding to the maximum in the velocity could be represented by

$$Re_{max} = 4.02 \, M^{-0.215} \tag{17.113}$$

and that corresponding to the minimum by

$$Re_{min} = 3.1 \, M^{-0.25} \tag{17.14}$$

Hu and Kintner [40] concluded that the maximum in the terminal velocity of liquid droplets in other liquids corresponded to the break in their graphical correlation, thus to the intersection of the lines represented by Equations 17.107 and 17.108 or $Y = 68.96$ and $X = 22.08$. These values are equivalent to

$$Re_{max} = 21.33 \, M^{-0.15} \tag{17.115}$$

$$(D_V^*)_{max} = 7.192 \, M^{-0.0917} \tag{17.116}$$

and

$$(u_T^*)_{max} = 2.966 \, M^{-0.0583} \tag{17.117}$$

The wave equation of Mendelson [27] led to Equation 17.79 for the dimensionless diameter at the minimum in the velocity and to Equation 17.82 for the dimensionless velocity itself. The product of these two expressions yields

$$Re_{min} = M^{-1/4} = 2.83 \, M^{-1/4} \tag{17.118}$$

Combining Equations 17.74 and 17.51 yields the corresponding prediction for the maximum in the velocity:

$$(D_V^*)_{max} = 2 \cdot 3^{4/5} \, M^{-1/15} = 4.82 \, M^{-1/15} \tag{17.119}$$

$$(u_T)_{max} = 3^{-2/5} \, M^{-2/15} = 0.644 \, M^{-2/15} \tag{17.120}$$

and

$$Re_{max} = 2 \cdot 3^{2/5} \, M^{-1/5} = 3.10 \, M^{-1/5} \tag{17.121}$$

The modified wave equation of Lehrer [29, 30] similarly yields

$$(D_V^*)_{min} = 6^{1/2} \, M^{-1/6} = 2.449 \, M^{-1/6} \tag{17.122}$$

$$(u_T^*)_{min} = 6^{1/4} \, M^{-1/12} = 1.565 \, M^{-1/12} \tag{17.123}$$

and

$$Re_{min} = 6^{3/4} \, M^{-1/4} = 3.833 \, M^{-1/4} \tag{17.124}$$

Combining Equations 17.88 for $M^{1/3}(D_V^*)^2 \ll 6$ with 17.86 produces

$$
\begin{aligned}
(D_V^*)_{max} &= 3 \cdot 2^{4/5}(1 + \zeta)^{2/5} M^{-1/15} \\
&= 5.223(1 + \zeta)^{2/5} M^{-1/15}
\end{aligned}
\tag{17.125}
$$

$$
\begin{aligned}
(u_T^*)_{max} &= 2^{-2/5}(1 + \zeta)^{-1/5} M^{-2/15} \\
&= 0.7579(1 + \zeta)^{-1/5} M^{-2/15}
\end{aligned}
\tag{17.126}
$$

and

$$
Re_{max} = 3 \cdot 2^{2/5}(1 + \zeta)^{1/5} M^{-1/5} = 3.958(1 + \zeta)^{1/5} M^{-1/5}
\tag{17.127}
$$

which are proposed to represent droplets as well as bubbles.

The empirical expressions of Peebles and Garber for the Reynolds number at the maximum and minimum in velocity are in good accord *functionally* with the approximate theoretical expressions of Mendelson and of Lehrer, but those of Hu and Kintner differ radically.

These several expressions are compared quantitatively in Table 17.14 with the experimental determinations of Haberman and Morton for bubbles of air in clean water [$M = 2.5 \times 10^{-11}$]. As expected from the prior comparison in Figure 17–39, the predictions of Equations 17.122–17.127 are in the best accord. The discrepancies for Equations 17.115–17.117 were also to be expected because of the development of these expressions for droplets in liquids rather than for bubbles.

Table 17.14
Comparison of Experimental Data, Correlations, and Solutions for the Maximum and Minimum in the Velocity of Bubbles of Air in Pure Water ($M = 2.5 \times 10^{-11}$)

	At Maximum in Velocity			At Minimum in Velocity		
	u_T^*	D_V^*	Re	u_T^*	D_V^*	Re
Haberman & Morton [3] (experimental)	16.3	29.9	489	10.5	141	1,482
Peebles & Garber [46] Equations 17.113–17.114 (empirical)			765			1,386
Hu and Kintner [40] Equations 17.115–17.117 (empirical)	12	67	830			
Mendelson [27] Equations 17.79 and 17.82, 17.118–17.121 (theoretical)	17	25	409	11	117	1,265
Lehrer [29, 30] Equations 17.122–17.127 (theoretical)	20	27	522	12	143	1,714

Table 17.15
Comparison of Predictions of $(D_V)_{min}$ and $(D_V)_{max}$ from Equations 17.83 and 17.88 with Experimental Values for Bubbles and Droplets (from Lehrer [30])

Dispersed Phase	*Continuous Phase*	*Observed Range of D_V for $u_T > 0.9\ u_{max}$ (mm)*	*Predicted*	
			Eq. 17.88 (mm)	*Eq. 17.83 (mm)*
Drops in water (Thorsen et al.)				
Carbon tetrachloride	Water	1.90– 2.60	1.74	1.99
Ethylene bromide	Water	1.35– 1.93	1.43	1.64
Methylene bromide	Water	1.34– 1.71	1.19	1.36
Bromoform	Water	1.31– 1.71	1.29	1.48
Tetrabromoethane	Water	1.65– 2.29	2.12	2.44
Ethyl bromide	Water	1.87– 2.78	1.64	1.88
o-Dichlorobenzene	Water	2.3 – 3.51	2.43	2.79
Drops in water (Krishna et al.)				
n—Amyl phthalate	Water	14.0 –15.3	15.7	18.0
Aniline	Water	7.5	6.44	7.40
m—Creosol	Water	6.2 – 7.2	6.72	7.72
Bubbles in liquids (Haberman & Morton)				
Air	Water	1.34– 1.90	1.25	1.44
Air	Methanol	0.80– 1.40	0.88	1.01
Air	Mineral oil	4.0 – 6.0	5.34	6.13

Table 17.15 is a comparison prepared by Lehrer [30] of values of $(D_V)_{max} = (6\sigma/g\Delta\varrho)^{1/2}$ and $(D_V)_{min}$ computed from Equations 17.87 and 17.86 with experimental observations for both bubbles and droplets in the range of values of D_V for which $u_T > 0.9\ (u_T)_{max}$. The agreement is generally satisfactory, including the prediction of the absence of a minimum for bubbles of air in mineral oil.

Equation 17.57 would be expected to provide a more accurate prediction of the maximum in the velocity of bubbles than would the other theoretical expressions because of the better representation of conditions just before the maximum. However, it is implicit in Re and $\bar{C}_t$ and hence in u_T^* and D_V^*.

GENERALIZED CORRELATIONS FOR BUBBLES AND DROPLETS

Mersmann [7] developed a generalized correlation for bubbles, droplets in gas, and droplets in immiscible liquids in terms of u_T and D_V as follows. First, he expanded and rearranged Stokes' law (Equation 16.17 or Equation 17.43 for $\zeta \to \infty$) in the form

$$u_T^* = \frac{(D_V^*)^2}{18} \tag{17.128}$$

He then rewrote Equation 17.128 as

$$u_T^* f_1 = \frac{f_1}{18 f_2^2} (D_V^* f_2)^2 \tag{17.129}$$

and attempted to find functions f_1 and f_2, depending only on the physical properties, such that the experimental data for bubbles in liquids, droplets in gas, and droplets in liquid fall on a single curve at large D_V when plotted as $u_T^{**} = f_1 u_T^*$ versus $D_V^{**} = f_2 D_V^*$. For this he chose $f_1 = M^{1/15}$ and $f_2 = (\varrho_c/\varrho_d)^{2/15} f_1 = (M\varrho_c^2/\varrho_d^2)^{1/15}$. The effective parameter of the correlation (the multiplier of the numerical constant 18) is thus

$$P = \frac{f_2^2}{f_1} = \left(\frac{\varrho_c}{\varrho_d}\right)^{2/15} f_2 = \left(\frac{\varrho_c}{\varrho_d}\right)^{4/15} f_1 = \left(M\frac{\varrho_c^4}{\varrho_d^4}\right)^{1/15} \tag{17.130}$$

Experimental data for liquid droplets in gases and in more viscous liquids, such that $\zeta \to 0$, when plotted as $\log\{u_T^{**}\}$ versus $\log\{D_V^{**}\}$, would be expected to fall along a single curve for large D_V^{**} but approach a set of lines with a common slope of 2.0 and a location governed by the parameter P in the limit as $D_V^{**} \to 0$ (Re $\to 0$). Data for bubbles and for liquid droplets in less viscous liquids would, however, be expected to deviate up to 50% on the high side of the predicted limiting behavior, owing to internal circulation.

Mersmann's test of this correlation with the data of Reinhart [31] for liquid droplets falling in gases is reproduced in Figure 17–43. The correlation appears to be very successful, the scatter being no more than might be expected from uncertainties in the data. However, the parameter P for these several fluid pairs ranges only from 0.025 to 0.04. Also, because of the difficulty in generating

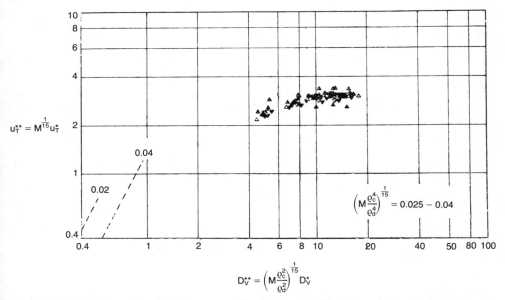

FIGURE 17–43 *Test by Mersmann [7] of his generalized correlation with the experimental data of Reinhart for the terminal velocity of droplets in air.*

droplets with $D_V < 0.5$ mm, none of the data extend to the lower range of $D_V^{**} < 4$, for which a dependence on P might be expected. No experimental data for droplets in liquids were discovered for $D_T^{**} > 20$, suggesting this as an upper limit imposed by the disintegration of droplets.

The corresponding test by Mersmann of the data of Hu and Kintner [40], Klee and Treybal [41], and Licht and Narasimhamurty [47], all for liquid droplets in more viscous liquids, is seen in Figure 17–44 to be almost equally successful. The parameter P ranges from only 0.17 to 0.60 for these sets of data, and very few data extend into the range of small D_V^{**} in which a parametric dependence would be expected, again due to the aforementioned difficulty in generating droplets with $D_V < 0.5$ mm. Only one experimental point was found for $D_V^{**} > 40$, suggesting this as the upper limit imposed by disintegration of droplets in liquids.

Finally, Figure 17–45 is the analogous plot of the data of Peebles and Garber [46] and Haberman and Morton [3] for air bubbles in various liquids. As contrasted with the data in the previous two plots, D_V^{**} extends from below 2 to above 600. The parametric dependence, corresponding to P from 1 to 2.5 is seen to be quite significant, the former value representing Stokes' law. A maximum occurs at $D_V^{**} = 12–15$ (just as in Figure 17–41) and a minimum at $D_V^{**} = 70–100$. For still larger D_V^{**} a proportionality to $(D_V^{**})^{1/2}$ is apparent, as expected.

The resulting generalized correlation itself is shown in Figure 17–46. The limiting value of D_V^{**} for the breakup of liquid droplets is indicated by the long-dashed line representing

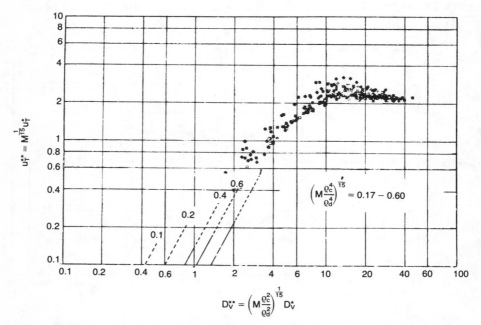

FIGURE 17–44   *Test by Mersmann [7] of his generalized correlation with the experimental data of Hu and Kintner, Klee and Treybal, and Licht and Narasimhamurty for the terminal velocity of liquid droplets in more viscous liquids.*

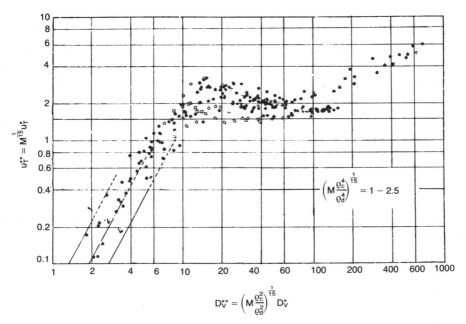

FIGURE 17–45 *Test by Mersmann [7] of his generalized correlation with the experimental data of Peebles and Garner and Haberman and Morton for the terminal velocity of bubbles of air in various liquids.*

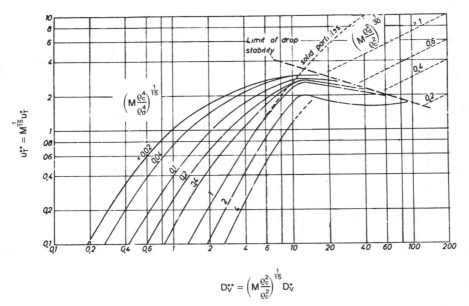

FIGURE 17–46 *Generalized correlation of Mersmann [7] for the terminal velocity of bubbles and droplets.*

$$u_T^{**} = 8(D_V^{**})^{-1/3} \qquad (17.131)$$

A different parameter, $Q = M^{1/30}(f_d/\varrho_c)^{1/15}$, is required for D_V^{**} above this limit in order to attain the required independence from μ_c. The behavior represented by the series of short-dashed lines can be expressed much more simply by

$$u_T = 0.714\left(\frac{D_V g \Delta\varrho}{\varrho_c}\right)^{1/2} \qquad (17.132)$$

or

$$\bar{C}_t = 1.308 \qquad (17.133)$$

Maxima and Minima. Figure 17–46 indicates that a maximum in terminal velocity for all particles occurs at

$$D_V^{**} \cong 12 \qquad (17.134)$$

or

$$D_V = 12\left(\frac{\mu_c}{g\Delta\varrho}\right)^{2/5}\left(\frac{\sigma}{\varrho_c}\right)^{1/5}\left(\frac{\varrho_d}{\varrho_c}\right)^{2/15} \qquad (17.135)$$

The corresponding terminal velocity for droplets in gas is

$$(u_T^{**})_{\max} \cong 3 \qquad (17.136)$$

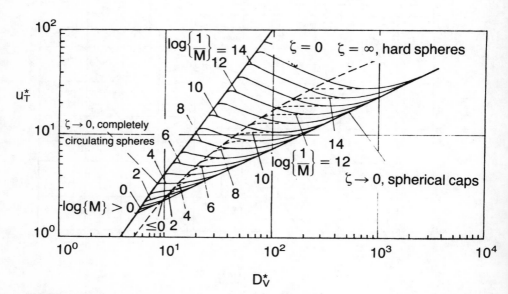

FIGURE 17–47 *Generalized correlation of Mersmann et al. [48] for the terminal velocity of bubbles and droplets.*

or

$$(u_T)_{max} = 3\left(\frac{\mu(g\Delta\varrho)^4\sigma^3}{\varrho_c^8}\right)^{1/15} \tag{17.137}$$

For droplets in liquids the coefficient in Equations 17.136 and 17.137 varies from 3 to 2.5, and for bubbles from 2.5 to 1.8.

Droplets may disintegrate before a minimum in the velocity can occur, but for bubbles a minimum occurs at $D^*_{min} = 30-100$, and the corresponding velocity is $(u_T^{**})_{min} = 1.5-2.0$, by Equation 17.131.

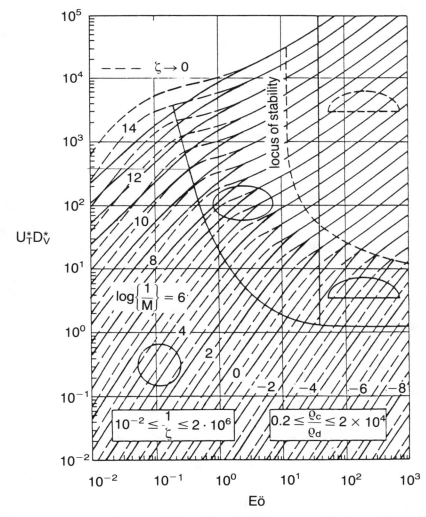

FIGURE 17-48   *Regimes of shape for a bubble at terminal velocity. (From Mersmann et al. [48].)*

Mersmann et al. [48] subsequently proposed the somewhat simpler graphical correlation shown in Figure 17–47 in which u_T^* is plotted versus D_T^* with $(\varrho_d/\varrho_c)^{1/2}/M$ as a parameter. The bounding curves correspond to Equation 17.33 for completely mobile spheres, Equation 16.62 for solid spheres, and Equation 17.132 for spherical caps, respectively. The dashed parametric curves represent droplets; the solid ones, bubbles. They also prepared the diagrams reproduced as Figures 17–48 and 17–49, which show the regimes of particle shape as a function of Re, Eö, and 1/M, for bubbles and droplets, respectively. Although the correlation of Figure 17–47 as considered by the authors to have superceded that of Figure 17–46 the process of construction of the latter remains of interest.

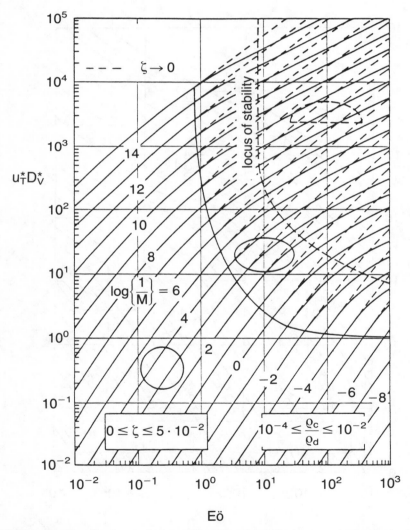

FIGURE 17–49   *Regimes of shape for a droplet falling at terminal velocity in a gas. (From Mersmann et al. [48].)*

SUMMARY

The behavior of bubbles and droplets is much more complex than that for solid spheres because of internal circulation and deformation. Nevertheless, the theoretical solutions for particular conditions appear accumulatively to approximate closely the entire gamut of behavior and to provide a viable alternative to the various empirical and semitheoretical correlations.

The terminal velocity as a function of the volume-equivalent diameter provides a more explicit relationship for bubbles and droplets than $\bar{C}_t$ as a function of Re, since u_T is generally a dependent variable. The terminal velocity of bubbles and droplets, as contrasted with solid spheres, generally goes through a maximum with diameter owing to deformation and may also go through a subsequent minimum. Moving droplets disintegrate above some maximum size. These particular characteristics are of great importance in applications.

Many discrepancies still exist in the data for bubbles and droplets. The deviations for bubbles have generally been attributed to particulate matter and surface-active contaminants in the liquid. Those for droplets are more often associated with their generation.

In retrospect the fluid-mechanical behavior of bubbles and droplets resembles that of solid spheres only for small Reynolds numbers (small velocities *and* diameters). The drag of large bubbles and droplets is dominated by their deformation, whereas that of solid spheres is controlled by the formation and behavior of their wake.

PROBLEMS

1. Derive Equation 17.2 from a list of the variables.
2. Restate the regimes identified by Rosenberg in terms of D_V^* and Ar.
3. To what extent does the terminal velocity of a rising bubbles depend on its density and its viscosity? Explain.
4. Reexpress the following equations in terms of u_T^* and D_V^* plus parameters:

a.	17.43	**f.**	17.54	**k.**	17.101
b.	17.48	**g.**	17.56	**l.**	17.102
c.	17.51	**h.**	17.57	**m.**	17.103
d.	17.52	**i.**	17.98	**n.**	17.104
e.	17.53	**j.**	17.100	**o.**	17.105

5. Repeat problem 4 in terms of Ar and Re.
6. Repeat problem 4 in terms of u_T^* and Re.
7. Reexpress the values in Tables 17.2–17.5 in terms of u_V^* and D_T^* plus parameters.
8. Explain the reduction of the Hadamard–Rybcyznski solution to that for inviscid flow when ζ is set equal to $-2/3$.
9. Derive an expression for $\bar{C}_p$ from Equations 17.38 and 17.42.
10. Derive Equation 17.37 from Equation 17.34.
11. Derive Equation 17.42 from Equation 17.39.
12. Derive an expression for $\tau_w/\varrho u_\infty^2$ from Equations 17.36 and 17.41.
13. Derive an expression for $\bar{C}_f$ from the result of problem 12.

14. Why is the drag of a bubble greater than of a solid sphere in the thin-boundary-layer regime?
15. For what values of Re, if any, would Equation 17.52 predict a greater drag coefficient than that for a solid sphere?

    a.   for bubbles ($\zeta \to 0, \xi = 0$)
    b.   for $\xi = 5$ and $\zeta = 5.5$

16. Explain the dependence of $\bar{C}_t$ on Re^{-1} for bubbles in the thin-boundary-layer regime as contrasted with the dependence on $Re^{-1/2}$ for solid spheres.
17. Correlate the ratio $\bar{C}_p/\bar{C}_t$ for bubbles as a function of Re based on the computed values in Table 17.2.
18. Compare the computed dependence of $\bar{C}_t$ on ζ in Tables 17.3 and 17.5 with the predictions of Equations 17.43, 17.52, and 17.104.
19. Derive an alternative to Equation 17.104 using the method of Churchill and Usagi [38], with Equation 17.43 and the first-order correction of Equation 17.52 as asymptotes.
20. Prepare a plot indicating the dependence of $\bar{C}_t$ on $\Delta\varrho/\varrho_c$ for $\zeta = 1$ as predicted by Equation 17.52.
21. Develop an empirical representation for $H\{E\}$ as given in Table 17.6, using the method of Churchill and Usagi [38].
22. Plot $\bar{C}_t$ versus Re with E as a parameter as given by Equation 17.57.
23. Plot $\bar{C}_t$ versus Re with We as a parameter as given by Equations 17.57 and 17.55.
24. Plot $\bar{C}_t$ versus Re with $M = 10^{-4}, 10^{-8}$, and 10^{-12} as a parameter as given by Equation 17.57.
25. Develop a correlating equation for $W\{E\}$ to represent Equation 17.55, using the method of Churchill and Usagi [38].
26. Repeat problem 22 in terms of u_T^* versus D_V^*.
27. Repeat problem 23 in terms of u_T^* versus D_V^*.
28. Repeat problem 24 in terms of u_T^* versus D_V^*.
29. Derive asymptotic expressions for $G\{E\}$ as given by Equation 17.58 and construct a simple correlating equation, using the method of Churchill and Usagi [38].
30. Use asymptotic expressions to construct an expression for the limiting behavior of Equation 17.57 for E approaching unity in terms of W. Recast in terms of M. Recast in terms of u_T^*, D_V^*, and M.
31. Use the result of problem 30 to determine Re and D_V^* for the onset of deformation, as defined by $E = 1.1$, as a function of M.
32. Repeat problem 31 using the full expressions for $W\{E\}, G\{E\}$, and $H\{E\}$ and compare.
33. Explain how $\bar{C}_t$ can be a function of either We or Eö as implied by Equations 17.75 and 17.76.
34. Devise an efficient algorithm for the calculations of problem 24.
35. Devise an efficient algorithm for the calculations of problem 28.
36. Determine the net coefficient of Equation 17.61 as a function of θ' and the average value.
37. Determine the radius of curvature as a function of θ', which satisfies Equations 17.59 and 17.60.
38. Derive Equation 17.63.

39. Determine and plot $\bar{C}_t$ for oblate ellipsoidal caps as a function of E, using Equation 17.68.

40. Determine and plot $\bar{C}_t$ for prolate ellipsoidal caps as a function of E, using Equation 17.69.

41. Derive an asymptotic expression for $f\{\varepsilon\}$ as $\varepsilon \to 0$ for oblate ellipsoidal caps from Equation 17.68 and the corresponding expressions for $\bar{C}_t\{\varepsilon\}$ and $\bar{C}_t\{E\}$. Compare with the plot of Figure 17–39.

42. Repeat problem 41 for prolate ellipsoidal caps.

43. Develop, as an alternative to Equation 17.77, a correlating equation explicitly in $\bar{C}_t$, using the method of Churchill and Usagi [38].

44. Solve Equation 17.88 for $D_V^*\{M, \zeta\}$ and plot the results for fixed values of ζ.

45. Prepare a plot of u_T^* as a function of M for fixed values of ζ based on the results of problem 44.

46. Indicate the range of applicability of the approximation obtained by dropping the $(D_V^*)^2$ term in Equation 17.88 by comparison with the plots of problems 44 and 45.

47. Determine the analogs of Equations 17.75–17.76 according to Equation 17.83.

48. Develop a continuous correlating equation for u_T^* as a function of D_V^* for distorted bubbles using the Churchill–Usagi method [38] with Equations 17.83 and 17.51 as asymptotes. (Use Equation 17.57 to evaluate n. Extend for droplets. Compare your expression with experimental data.

49. Develop a correlating equation corresponding to that of problem 48 for $\bar{C}_t\{Re\}$. Compare this impression with experimental data.

50. Reexpress Equation 17.88 in terms of Eö rather than D_V^*. Solve and prepare a plot of Eö as a function of M and $1 + \zeta$. Indicate the error due to dropping the term corresponding to that in $(D_V^*)^2$.

51. Determine the relationship between Eö and We corresponding to Equations 17.73 and 17.83.

52. Develop a correlating Equation for $D_V^*\{M, \zeta\}$ as determined in problem 44, using the method of Churchill and Usagi [38].

53. Identify the behavior of $\bar{C}_t$ in terms of Re corresponding to the minimum in u_T.

54. Prepare plots corresponding to Figures 17–12 and 17–13 for droplets in air ($\zeta \to \infty$) and compare.

55. Prepare plots corresponding to Figures 17–12 and 17–13 for droplets in liquids with $\zeta = 0.5$, 1.0, and 2.0, and compare.

56. Explain the successful correlation of the same data by Equations 17.95 and 17.96.

57. Compare Equation 17.95 with Figure 17–14.

58. Compare Equation 17.95 and Figure 17–14 with Equation 17.55 and explain.

59. Derive a continuous correlating equation for the data of Figure 17–16, using the method of Churchill and Usagi [38].

60. Convert the correlation of Figure 17–16 into a function of We only using Equation 17.97 and an approximate expression for $\bar{C}_t$.

61. Calculate E as a function of θ_s' according to Figure 17–17, assuming the bubbles are segments of a sphere. Correlate the results.

62. Develop a correlating equation for the transition in Figure 17–12 from the Hadamard–Rybczynski relationship to Equation 17–53.

63. Rationalize the data of Hacker and Hussein in Figure 17–19 in terms of M.

64. Compare Equation 17.104 (together with E from Figure 17–16) with Equations 17.57 and 17.83.

65. Construct a plot of u_T^* versus D_V^* with M as a parameter based on Equations 17.100–17.103 or 17.104.

66. Reduce Equation 17.106 for $\zeta \to 0$ and compare with Equations 17.43 and 17.52.

67. Construct a continuous correlating equation to represent the data of Hu and Kintner in Figure 17–30, using the method of Churchill and Usagi [38]. (*Suggestion*: Avoid the linear dependence of Equation 17.110.)

68. Explain the criticism on page 461 of the Hu–Kintner plot.

69. Explain the radically different functional dependence of Equations 17.115–7.117.

70. Derive expressions for $\mathrm{Re}_{\max}$, $(D_V^*)_{\max}$, and $(u_T^*)_{\max}$ analogous to Equations 17.125–17.127 for $\zeta = 0$, but using Equation 17.53 instead of 17.51.

71. Calculate $\mathrm{Re}_{\max}$, $(D_V^*)_{\max}$, and $(u_T^*)_{\max}$ for $\zeta = 0$ and M $= 0.25 \times 10^{-10}$, using Equation 17.83 with Equation 17.53 instead of 17.51, and compare with the values in Table 17.12.

72. Derive expression for $\mathrm{Re}_{\max}$, $(D_V^*)_{\max}$, and $(u_T^*)_{\max}$ analogous to Equations 17.125–17.127 but using the first terms on the right side of Equation 17.52 instead of 17.86.

73. Use the result of problem 72 to calculate and plot $\mathrm{Re}_{\max}$, $(D_V^*)_{\max}$, and $(u_T^*)_{\max}$ as a function of M with ζ as a parameter. Compare with Equations 17.125–17.127.

74. Use Equation 17.54 with 17.83 to predict $(\mathrm{Re})_{\max}$, $(D_V^*)_{\max}$, and $(u_T^*)_{\max}$ for M $= 2.5 \times 10^{-11}$. Compare with the results in Table 17.12.

75. Compare the criterion of Mersmann [7] of $D_T^{**} = 20$ for the breakup of droplets in other liquids with that of Hu and Kintner [40]: Eö $= 14.2$.

76. Compare the criterion of Mersmann [7] of $D_T^{**} = 40$ for the breakup of droplets in air with that of Reinhart [31]: $D_V^* = 2.24\mathrm{M}^{-1/6}$.

77. Compare the prediction of Figure 17–42 (and Equations 17.131, 17.132, 17.134, and 17.136) with

   a. Equation 17.43 for different values of ζ and small D_V^{**}
   b. Equation 17.51 for $\zeta = 0$ and intermediate D_V^{**}
   c. Equation 17.51 for different values of ξ and ζ and intermediate D_V^{**}
   d. the values in Table 17.2
   e. the values of $\bar{C}_t$ in Table 17.3
   f. the values of Table 17.4
   g. the values of $\bar{C}_t$ in Table 17.5
   h. Equation 17.54 for M $= 0$
   i. Equation 17.59 for $\zeta = 0$ and different values of M
   j. Equation 17.83 for different values of M
   k. Equations 17.107–17.110
   l. Equations 17.122–17.123
   m. Equations 17.125–17.126

78. Calculate and plot the eccentricity ratio and velocity of a raindrop falling in air at 20°C and 1 atm as a function of diameter. What is its maximum possible size (D_V)?

79. Calculate the rate at which a 1.5-mm droplet of water will fall through air at 15°C.

80. Calculate the velocity at which a 5-mm bubble of air will rise through water at 15°C. Note any assumptions.

81. A Jeep traveling at 15 km/h along an open beach area into a wind that averages 2.2 m/s discharges an insecticidal spray into 20°C air 2 m above the ground. What would be the diameter of the particles striking the ground 30 m from the point of discharge? The particles may be assumed to be spherical, and evaporation can be neglected. The density of the insecticide is 0.83 Mg/m^3.

82. Interpret Equation 17.74 in terms of the model of Churchill and Usagi [38].

83. Develop a correlating equation for the drag coefficient of spherical bubbles by the method of Churchill and Usagi [38] using as asymptotes Equation 17.48 and the following modified form of Equation 17.53 to eliminate the singularity:

$$\bar{C}_t = \frac{24}{\text{Re}} e^{-2.21/\sqrt{\text{Re}}} \tag{17.138}$$

84. Develop a correlating equation for u_T^* as a function of D_T^* equivalent to that of problem 83.

85. Combine the expressions of problems 48 and 84 to obtain an expression for all Re and M for completely mobile bubbles. Compare with experimental data.

86. Combine the expressions of problems 49 and 83 to obtain an expression for all Re and M for completely mobile bubbles. Compare with experimental data.

REFERENCES

1. N. M. Aybers and A. Tapucu, "Studies on the Drag and Shape of Gas Bubbles Rising through a Stagnant Liquid," *Wärme- u. Stoffübertragung*, 2 (1969) 171.

2. Aaron Weiner and S. W. Churchill, "Mass Transfer from Rising Bubbles of Carbon Dioxide," p. 525 in *Physicochemical Hydrodynamics—V. G. Levich Festschrift*, D. B. Spalding, Ed., Advance Publications, London (1977).

3. W. L. Haberman and R. K. Morton, "An Experimental Study of Bubbles Moving in Liquids," Paper No. 2799, *Trans. ASCE*, 121 (1956) 227; David Taylor Model Basin Report No. 802, U.S. Dept. of the Navy, Washington, D.C. (1953).

4. B. Rosenberg, *The Drag and Shape of Air Bubbles Moving in Liquids*, David Taylor Model Basin Report No. 727, U.S. Dept. of the Navy, Washington, D.C. (1950).

5. V. Haas, H. Schmidt-Traub, and H. Brauer, "Umströmung kugelförmiger Blasen mit innerer Zirkulation," *Chem.-Ing.-Tech.*, 44 (1972) 1060.

6. N. M. Aybers and A. Tapucu, "The Motion of Gas Bubbles Rising through Stagnant Liquid," *Wärme- u. Stoffübertragung*, 2 (1969) 118.

7. A. Mersmann, "Rate of Rise or Fall of Fluid Particles," Preprint, Section G6-3, p. 1, *Joint GVC/AIChE Meeting*, München (1974); "Zur stationären bewegung Fluider Partikel," *Chem.-Ing.-Tech.*, 46 (1974) 251.

8. J. Hadamard, "Mouvement permanent lent d'une sphere liquide visqueuse dans un liquid visqueux," *Compt. Rend. Acad. Sci.*, Paris, 152 (1911) 1735.

9. W. Rybczynski, "Über die fortschreitende Bewegung einer flüssigen Kugel in einem zähen Medium," *Bull. Acad. Sci. Cracovie*, Ser. A, 1 (1911) 40.

10. H. Lamb, *Hydrodynamics*, Dover, New York (1945).
11. V. G. Levich, "The Motion of Bubbles at High Reynolds Numbers," *Zh. Eksperim. i Teor. Fiz.* 19 (1949) 18; also see *Physicochemical Hydrodynamics*, English transl. by Scripta Technica, Prentice-Hall, Englewood Cliffs, NJ (1962), p. 436f.
12. J. Boussinesq, "Vitesse de la chute lente, devenue uniforme d'une goutte liquide sphérique dans une fluid visqueux de poids specifique moindre," *Compt. Rend. Acad. Sci., Paris*, 156 (1913) 983, 1035, 1040, 1124.
13. T. Hirose and M. Moo-Young, "Bubble Drag and Mass Transfer in Non-Newtonian Fluids; Creeping Flow with Power-Law Fluids," *Can. J. Chem. Eng.*, 47 (1969) 265.
14. J. Ackeret, "Über exakte Lösungen der Stokes-Navier-Gleichungen inkompressibler Flüssigkeiten bei veränderten Grenzbedingungen," *Z. Angew. Math. Phy.*, 3 (1952) 259.
15. J. F. Harper and D. W. Moore, "The Motion of a Spherical Liquid Drop at High Reynolds Number," *J. Fluid Mech.*, 32 (1968) 367.
16. D. W. Moore, "The Boundary Layer on a Spherical Gas Bubble," *J. Fluid Mech.*, 16 (1963) 161.
17. A. E. Hamielec and A. I. Johnson, "Viscous Flow around Fluid Spheres at Intermediate Reynolds Numbers," *Can. J. Chem. Eng.*, 40 (1962) 41.
18. A. E. Hamielec, T. W. Hoffman, and L. L. Ross, "Numerical Solution of the Navier–Stokes Equation for Flow Past Spheres. Part I. Viscous Flow Around Spheres with and without Radial Mass Efflux," *AIChE J.*, 13 (1967) 212.
19. A. E. Hamielec, A. I. Johnson, and W. T. Houghton. "Part II. Viscous Flow Around Circulating Spheres of Low Viscosity," *AIChE J.*, 13 (1967) 220.
20. M. Yamaguchi, T. Katayama, and K. Ueyama, "Drag Coefficients and Mass Transfer in the Continuous Phase for Single Drops at Low Reynolds Numbers," *J. Chem. Eng., Japan*, 7 (1974) 334; 9 (1976) 79.
21. D. C. Brabston and H. B. Keller, "Viscous Flow past Spherical Gas Bubbles," *J. Fluid Mech.*, 69 (1975) 179.
22. A. W. Abdel-Alim and A. E. Hamielec, "Theoretical and Experimental Investigation of the Effect of Internal Circulation on the Drag of Spherical Droplets Falling at Terminal Velocity in Liquid Media," *Ind. Eng. Chem. Fundam.*, 14 (1975) 308.
23. D. W. Moore, "The Velocity Rise of Distorted Bubbles in a Liquid of Small Viscosity," *J. Fluid Mech.*, 23 (1965) 749.
24. R. M. Davies and G. I. Taylor, "The Mechanics of Large Bubbles Rising through Extended Liquids and through Liquids in Tubes," *Proc. Roy. Soc. (London)*, A200 (1950) 375.
25. T. Wairegi and J. R. Grace, "The Behavior of Large Drops in Immiscible Liquids," *Int. J. Multiphase Flow*, 3 (1976) 67.
26. J. R. Grace and D. Harrison, "The Influence of Bubble Shape on the Rising Velocities of Large Bubbles," *Chem. Eng. Sci.*, 22 (1967) 1337.
27. H. D. Mendelson, "The Prediction of Bubble Terminal Velocities from Wave Theory," *AIChE J.*, 13 (1967) 250.
28. G. Marrucci, G. Apuzzo, and G. Astarita, "Motion of Liquid Drops in Non-Newtonian Systems," *AIChE J.*, 16 (1970) 538.
29. I. H. Lehrer, "A Rational Terminal Velocity Equation for Bubbles and Drops at Intermediate and High Reynolds Numbers," *J. Chem. Eng., Japan*, 9 (1976) 237.
30. I. H. Lehrer, "A Theoretical Criterion of Transition in the Free Motion of Single Bubbles and Drops," *AIChE J.*, 26 (1980) 170.
31. A. Reinhart, "Das Verhalten fallender Tropfen," *Chem.-Ing.-Tech.*, 36 (1964) 740.
32. R. M. Wellek, G. S. K. Agarwal, and A. H. P. Skelland, "Shape of Liquid Drops Moving in Liquid Media," *AIChE J.*, 12 (1966) 854.
33. T. Takahashi, T. Miyahara, and H. Izawa, "Drag Coefficient and Wake Volume of Single Bubbles Rising through Quiescent Liquids," *Kagaku Kogaku Ronbunshu*, 2 (1976) 480 (in Japanese).

34. R. Clift, J. R. Grace, and M. E. Weber, *Bubbles, Drops and Particles*, Academic Press, New York (1978).
35. J. A. Redfield and G. Houghton, "Mass Transfer and Drag Coefficients for Single Bubbles at Reynolds Numbers of 0.02–5000," *Chem. Eng. Sci.*, 20 (1965) 131.
36. D. S. Hacker and F. D. Hussein, "The Application of a Laser-Schlieren Technique to the Study of Single Bubble Dynamics," *Ind. Eng. Chem. Fundam.*, 17 (1978) 277.
37. T. Miyahara and T. Takahashi, "Drag Coefficients of a Single Bubble Rising through a Quiescent Liquid," *Kagaku Kogaku Ronbunshu* 9 (1983) 592; English transl., *Int. Chem. Eng.*, 25 (1985) 146.
38. S. W. Churchill and R. Usagi, "A General Expression for the Correlation of Rates of Transfer and Other Phenomena," *AIChE J.*, 18 (1972) 1121.
39. R. Satapathy and W. Smith, "The Motion of Single Immiscible Drops through a Liquid," *J. Fluid Mech.*, 10 (1961) 561.
40. S. Hu and R. C. Kintner, "The Fall of a Single Liquid Drop through Water," *AIChE J.*, 1 (1955) 42.
41. A. S. Klee and R. E. Treybal, "The Rate of Rise of Fall of Liquid Drops," *AIChE J.* 2 (1956) 444.
42. A. E. Hamielec, *Studies of Fluid Flow and Mass Transfer in Droplets*, Ph.D. Thesis, University of Toronto (1961).
43. M. Warshay, E. Bogusz, M. Johnson, and R. C. Kintner, "Ultimate Velocity of Drops in Stationary Liquid Media," *Can. J. Chem. Eng.*, 37 (1959) 29.
44. A. I. Johnson and L. Braida, "The Velocity of Fall of Circulating and Oscillating Drops through Quiescent Liquid Phases," *Can. J. Ch. E.*, 35 (1957) 165.
45. J. R. Grace, T. Wairegi, and T. N. Nguyen, "Shapes and Velocities of Single Drops and Bubbles Moving Freely through Immiscible Liquids," *Trans. Inst. Chem. Engr. (London)*, 54 (1976) 167.
46. F. N. Peebles and H. J. Garber, "Studies on the Motion of Gas Bubbles in Liquids," *Chem. Eng. Progr.*, 49 (1953) 88.
47. W. Licht and G. S. R. Narasimhamurty, "Rate of Fall of Single Liquid Droplets," *AIChE J.*, 1 (1955) 366.
48. A. Mersmann, I. B. von Morgenstern, and A. Diexler, "Deformation, Stabilität und Geschwindigkeit," *Chem.-Ing.-Tech.*, 55 (1983) 865.

Chapter 18

Generalized Methods and Other Geometries

Attention in the previous seven chapters has been confined to steady flow over surfaces with only four basic shapes: (1) wedges, including the limiting cases of flat plates parallel and perpendicular to the direction of flow, (2) circular cylinders perpendicular to the flow, (3) spheres, and (4) mobile ellipsoids. Although these are the shapes of the greatest simplicity and practical importance, theoretical results have also been obtained for a few other situations. Some of these are examined in this chapter.

Just as in the previous seven chapters, theoretical attention is confined to the simplest and most successful methods. A few generalized methods that are applicable to several geometries, including those just mentioned, are described first.

OTHER GEOMETRIES AND CONDITIONS

Theoretical solutions have been developed for a few unsteady flows, including flow over accelerating, rotating, and oscillating bodies, but such cases are arbitrarily omitted here. Theoretical solutions have also been developed for steady flow over cones, noncircular cylinders, yawed cylinders, spheroids, and airfoils. Some of these solutions are discussed in the references found in the introduction to the third section. The results of Gluckman et al. [1] for creeping flow over arbitrary convex bodies of revolution should also be mentioned.

Experimental data are shown in Figure 18–1 (after Rouse and Howe [2], p. 181) for the variation of the total drag coefficient with Re_D for fixed objects of several common shapes. (The drag of *freely falling* objects of both regular and irregular shapes is examined in Chapter 20.) The pattern is roughly the same for all of the shapes, but the qualitative and quantitative differences are significant. The limiting values of the total drag coefficient for large Re_D in both the laminar boundary layer and turbulent regimes from Gupta and Gupta [3] and others are listed in Table 18.1 for a number of objects. Since the drag at high Re_D is primarily due to pressure rather than skin friction, $\bar{C}_t$ is then essentially a measure of the cross section of the wake as compared to the cross section of the object. The coefficients in Figure 18–1 and Table 18.1 are based on the projected area in the direction of flow, and the Reynolds number is based on the characteristic dimension.

The pioneering numerical solution of Fromm and Harlow [4] for develop-

483

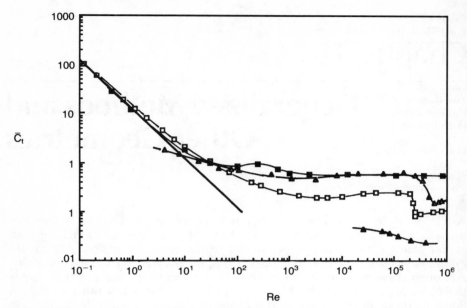

FIGURE 18-1   *Total drag coefficients for fixed objects of regular geometrical shape.* □ *sphere;* △ *Infinite cylinder;* ■ *Disk;* ▲ *Airship hull;* ———— *Stokes' law. (After Rouse and Howe [2], p. 181.)*

ing flow perpendicular to a finite strip can be noted as an illustration of both unsteady flow and another geometry. It represents perhaps the first detailed results for a vortex street. Their computed streaklines for $Re_D = 200$, where D designates the width of the strip, are shown in Figure 18–2. The flow and the bounding walls were impulsively accelerated at time zero from no motion to the final motion.

GENERALIZED METHODS

Goldstein [5] and Mangler [6] showed that similarity transformations are possible only for problems which can be represented by the Falkner–Skan equations, i.e., for wedge flows such that the inviscid velocity is proportional to x^m. Lee and Ames [7] examined the applicability of these transformations to non-Newtonian fluids. Methods and solutions of problems subject to a similarity transformation have been reviewed by Dewey and Gross [8].

The series solution developed by Blasius [9] (see Chapters 15 and 16) is applicable for problems that are not subject to a similarity transformation, but this procedure is tedious and requires extensive computing. Finite-difference solution of the partial differential equations has also proven to require extensive computing even by modern standards, at least for Reynolds numbers such that wake becomes nonperiodic. Hence, there remains a strong incentive to develop better approximate methods. Dewey and Gross also reviewed some of the proposed techniques for nonsimilar problems. A few of the more important generalized methods are described here.

Table 18.1
Total Mean Drag Coefficients for Several Shapes at Large Re_D[a]

		$\bar{C}_t$	
		Laminar	*Turbulent*
Two-dimensional shapes			
Circular cylinder		0.6	0.165
Half-circular cylinder		0.58	
		0.85	
Half-circular tube		0.60	
		1.15	
Flat strips			
Height to width			
0		1.0	
0.05		0.75	
0.10		0.65	
0.20		0.60	
1.00		0.59	
Elliptical cylinders			
Height to depth			
4/3		0.30	0.105
1		0.60	0.165
3/4			0.065
1/2		0.30	0.10
1/4		0.175	0.10
1/8		0.14	0.05
Square cylinders			
		1.05	
		0.8	
Equilateral triangles		0.8	
		1.0	
Circular disk		0.55	0.55
Three-dimensional shapes			
Sphere		0.22	0.10
Ellipsoid			
Length/diameter			
2/1		0.135	0.03
4/1		0.10	0.03
8/1		0.125	0.065
Hemisphere		0.19	
		0.585	
Half-cup		0.19	
		0.71	
60° cone		0.245	
Cube		0.525	
		0.400	

[a] After Gupta and Gupta [3] and others.

Mangler Transformation

In 1948 Mangler [10] showed that all axisymmetrical, thin-boundary-layer flows could be transformed to those of two-dimensional planar flow by the change of variables

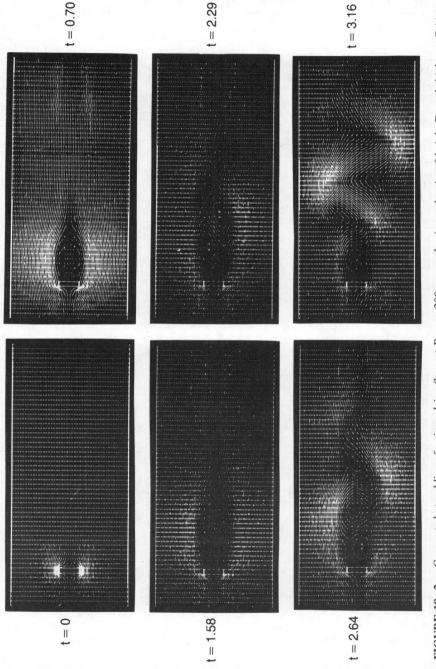

FIGURE 18–2   *Computed streaklines for impulsive flow at* $Re_D = 200$ *over a horizontal strip of height D and thickness D/4 in a channel of width 6D at relative times t. (From Fromm and Harlow [4].)*

$$\bar{x} = \frac{1}{L^2} \int_0^x r^2\{x\}\, dx \tag{18.1}$$

$$\bar{y} = \frac{r\{x\}\, y}{L} \tag{18.2}$$

$$\bar{u}_x = u_x \tag{18.3}$$

$$\bar{u}_y = \frac{L}{r}\left(u_y + \frac{y u_x}{r\{x\}} \cdot \frac{dr\{x\}}{dx}\right) \tag{18.4}$$

and

$$\bar{U}\{x\} = U\{x\} \tag{18.5}$$

where x = distance along surface from forward point, m
y = distance normal to surface, m
r = radius of surface from axis of symmetry, m

L is an arbitrary constant with the dimension of length. For example, this process can be used to transform Equations 14.24 and 14.25 to 11.1 and 14.2 (see problems 1 and 2). The transformed problem may remain difficult to solve, but it is at least subject to all the techniques developed for two-dimensional planar flow.

Von Mises Transformation

Von Mises [11], in 1927, proposed a change of variables that converts the differential force-momentum balance of the thin-boundary-layer model for any two-dimensional flow into the equation for transient thermal conduction with a variable thermal diffusivity. The starting point is

$$u_x \frac{\partial u_x}{\partial x} + u_y \frac{\partial u_x}{\partial y} = -\frac{1}{\varrho} \frac{d\mathscr{P}}{dx} + \nu \frac{\partial^2 u_x}{\partial y^2} \tag{18.6}$$

with x measured along the surface and y normal to it. $d\mathscr{P}/dx$ is presumed to be a known function of x from inviscid flow theory. Retaining x but replacing y by the Lagrange stream function, $\psi\{x, y\}$, requires that

$$\frac{\partial}{\partial x} = \frac{\partial}{\partial x} + \frac{\partial}{\partial \psi} \frac{\partial \psi}{\partial x} = \frac{\partial}{\partial x} + u_y \frac{\partial}{\partial \psi} \tag{18.7}$$

and

$$\frac{\partial}{\partial y} = \frac{\partial}{\partial \psi} \cdot \frac{\partial \psi}{\partial y} = -u_x \frac{\partial}{\partial \psi} \tag{18.8}$$

Hence, Equation 18.1 becomes

$$u_x \frac{\partial u_x}{\partial x} = -\frac{1}{\varrho} \frac{d\mathscr{P}}{dx} + \nu u_x \frac{\partial}{\partial \psi} \left(u_x \frac{\partial u_x}{\partial \psi} \right) \tag{18.9}$$

A *total dynamic pressure* can be defined as

$$\mathscr{P}_t = \mathscr{P} + \frac{\varrho v^2}{2} = \mathscr{P} + \frac{\varrho u_x^2}{2} + \frac{\varrho u_y^2}{2} \cong \mathscr{P} + \frac{\varrho u_x^2}{2} \tag{18.10}$$

It follows that

$$\frac{\partial \mathscr{P}_t}{\partial x} = \frac{d\mathscr{P}}{dx} + \varrho u_x \frac{\partial u_x}{\partial x} \tag{18.11}$$

and

$$\frac{\partial \mathscr{P}_t}{\partial \psi} = \varrho u_x \frac{\partial u_x}{\partial \psi} \tag{18.12}$$

Substituting from Equations 18.11 and 18.12 in 18.9 permits the derivation of

$$\frac{\partial \mathscr{P}_t}{\partial x} = \nu u_x \frac{\partial^2 \mathscr{P}_t}{\partial \psi^2} \tag{18.13}$$

Finally eliminating u_x through Equation 18.10 gives

$$\frac{\partial \mathscr{P}_t}{\partial x} = \nu \left(\frac{\mathscr{P}_t - \mathscr{P}}{\varrho/2} \right)^{1/2} \frac{\partial^2 \mathscr{P}_t}{\partial \psi^2} \tag{18.14}$$

which is analogous to the equation for transient conduction with x as time and ψ as distance, and with a variable thermal diffusivity $\nu[2(\mathscr{P}_t - \mathscr{P})/\varrho]^{1/2}$. The boundary conditions are

$$\mathscr{P}_t = \mathscr{P}\{x\} \qquad \text{at } \psi = 0 \tag{18.15}$$

and

$$\mathscr{P}_t = \mathscr{P}\{x\} + \frac{\varrho U^2\{\xi\}}{2} \qquad \text{as } \psi \to \infty \tag{18.16}$$

The large body of closed-form solutions for conduction are not generally applicable because of the particular variability of $\nu[2(\mathscr{P}_t - \mathscr{P})/\varrho]^{1/2}$, but the finite-difference methods of solution developed for variable conductivity are adaptable. Once the total pressure is calculated from Equation 18.14, u_x can be calculated from Equation 18.10 and u_y from the equation of continuity. The potential of this transformation does not appear to have been fully explored or exploited.

Görtler Method

A general change of variables for curved surfaces was proposed by Görtler [12] in the hope of obtaining a series solution which would converge more rapidly than the Blasius-type series illustrated in the previous chapters for boundary-layer flow over flat plates, cylinders, and spheres. He defined the new variables

$$\xi = \frac{1}{\nu} \int_0^x U_x\{x\}\, dx \tag{18.17}$$

and

$$\eta = \frac{y U_x\{x\}}{\nu\sqrt{2\xi}} \tag{18.18}$$

Then the assumption that

$$\psi = -\nu\sqrt{2\xi}\, F\{\xi, \eta\} \tag{18.19}$$

gives for the force-momentum balance

$$\frac{\partial^3 F}{\partial \eta^3} + F\frac{\partial^2 F}{\partial \eta^2} + \beta\{\xi\}\left(1 - \frac{\partial F^2}{\partial \xi}\right) = 2\xi\left(\frac{\partial F}{\partial \eta}\frac{\partial^2 F}{\partial \eta \partial \xi} - \frac{\partial F}{\partial \xi}\frac{\partial^2 F}{\partial \eta^2}\right) \tag{18.20}$$

where

$$\beta\{\xi\} = \frac{2\xi\nu}{U_x^2\{x\}}\frac{dU\{x\}}{dx} \tag{18.21}$$

The boundary conditions on $F\{\xi, \eta\}$ are $F = \partial F/\partial \eta = 0$ at $\eta = 0$ and $\partial F/\partial \eta \to 1$ for $\eta \to \infty$. In this form the specified information, $U_x\{x\}$, appears only in the function $\beta\{\xi\}$. A series solution can be developed for all geometries with only the coefficients in a series expansion of $\beta\{\xi\}$ required for each new case. Unfortunately, this scheme has not proven to be as efficient as expected.

Meksyn Transformation and Merk Expansion

In 1947–1948, Meksyn [13] developed a transformation of variables that leads to exact solutions for wedge flows and good approximations for arbitrary shapes. Merk [14] extended its usefulness for nonwedge flows by the inversion and expansion of the nonsimilar terms as described next.

Planar and rotationally symmetrical, thin-boundary-layer flows can both be represented approximately by

$$u_x\frac{\partial u_x}{\partial x} + u_y\frac{\partial u_y}{\partial y} = -\frac{1}{\varrho}\frac{\partial \mathscr{P}}{\partial x} + \nu\frac{\partial_2 u_x}{\partial y^2} \tag{18.22}$$

and

$$\frac{\partial (ru_x)}{\partial x} + \frac{\partial (ru_y)}{\partial y} = 0 \tag{18.23}$$

where x and y are measured along the surface and perpendicular to it, respectively. Also

$$\frac{\partial \mathscr{P}}{\partial x} \simeq \frac{d\mathscr{P}_\infty}{dx} = -\varrho U_x \frac{dU_x}{dx} \tag{18.24}$$

where $\mathscr{P}_\infty$ and U_x are the values outside the boundary layer. For two-dimensional, planar flows, r is a constant and hence drops out of Equation 18.23. The boundary conditions are then those previously described for a flat plate ($u_x = u_y = 0$ at $y = 0$; $u \rightarrow U$, as $y \rightarrow \infty$), and the following generalized stream function can be introduced to satisfy Equation 18.23:

$$u_x = -\frac{L}{r} \frac{\partial \psi}{\partial y} \tag{18.25}$$

$$u_y = \frac{L}{r} \frac{\partial \psi}{\partial x} \tag{18.26}$$

where L is an arbitrary reference length. For two-dimensional planar flows L is set equal to r in Equations 18.25 and 18.26.

The general transformation for all flows is

$$\xi = \int_0^x \frac{U_x\{x\}}{u_\infty} \left(\frac{r}{L}\right)^2 d\left(\frac{x}{L}\right) \tag{18.27}$$

and

$$\eta = \left(\frac{\mathrm{Re}_L}{2\xi}\right)^{1/2} \frac{U_x\{x\}}{u_\infty} \left(\frac{r}{L}\right) \frac{y}{L} \tag{18.28}$$

with

$$\psi = -u_\infty L \left(\frac{2\xi}{\mathrm{Re}_L}\right)^{1/2} f\{\xi, \eta\} \tag{18.29}$$

Here

$$\mathrm{Re}_L = \frac{u_\infty L}{\nu} \tag{18.30}$$

The *Mangler transformation* can be recognized as a special case of this transformation for rotationally symmetric boundary layers.

From Equations 18.25–18.28 it follows that

$$u_x = U\{x\} \frac{\partial f}{\partial \eta}$$ (18.31)

and

$$u_y = -\frac{r}{L} \frac{U_x\{x\}}{(2\xi \mathrm{Re}_L)^{1/2}} \left(f + 2\xi f + (\Lambda - 1)\eta \frac{\partial f}{\partial \eta} \right)$$ (18.32)

where the "wedge variable" is

$$\Lambda\{\xi\} \equiv \frac{2\xi}{U_x} \frac{dU_x}{d\xi}$$

$$= 2 \left(\frac{L}{r} \frac{u_\infty}{U_x} \right)^2 \frac{d(U_x/u_\infty)}{d(x/L)} \int_0^x \left(\frac{r}{L} \right)^2 \frac{U_x}{u_\infty} d\left(\frac{x}{L} \right)$$ (18.33)

For a pure wedge flow, $r = L$, $U_x = cx^m$, as in Chapter 14, and

$$\Lambda = \frac{2m}{m + 1}$$ (18.34)

is a constant. From Equations 18.31 and 18.32,

$$u_x \frac{\partial u_x}{\partial x} + u_y \frac{\partial u_x}{\partial y} = \frac{r^2 U_x^2}{2L^3 \xi u_\infty^2} \left[2\xi \left(\frac{\partial f}{\partial \eta} \frac{\partial u_x}{\partial \xi} - \frac{\partial f}{\partial \xi} \frac{\partial u_x}{\partial \eta} \right) - f \frac{\partial u_x}{\partial \eta} \right]$$ (18.35)

and it follows that

$$f''' + ff'' + \Lambda(1 - f'^2) = 2\xi \left(f' \frac{\partial f'}{\partial \xi} - f'' \frac{\partial f}{\partial \xi} \right)$$ (18.36)

The primes here refer to differentiation with respect to η. The boundary conditions become

$$f + 2\xi \frac{\partial f}{\partial \xi} = 0, \qquad \eta = 0$$ (18.37)

$$f' = 0, \qquad \eta = 0$$ (18.38)

and

$$f' = 1 \qquad \text{for } \eta \to \infty$$ (18.39)

Since r within the boundary layer and U_x depend only on x, ξ depends only on x. For a given body, x and U_x can be considered to be known functions of ξ. Hence, Λ is a known function of ξ. Merk suggested inverting this relationship conceptually and expanding in series as

$$f\{\xi, \eta\} = f_0\{\Lambda, \eta\} + 2\xi\frac{d\Lambda}{d\xi}f_1\{\Lambda, \eta\} + \cdots \tag{18.40}$$

Substituting 18.40 in 18.36–18.39 gives

$$f_0''' + f_0 f_0'' + \Lambda[1 - (f_0^1)^2] = 0 \tag{18.41}$$

with

$$f_0 = f_0' = 0 \qquad \text{at } \eta = 0 \tag{18.42}$$

$$f_0' = 1 \qquad \text{for } \eta \to \infty \tag{18.43}$$

and

$$f_1''' + f_0 f_1'' - 2\Lambda f_0' f_1' + f_0'' f_1 = f_0' \frac{\partial f_0'}{\partial \Lambda} - f_0'' \frac{\partial f_0}{\partial \Lambda} \tag{18.44}$$

with

$$f_1 = -\frac{\partial f_0}{\partial \Lambda} \qquad \text{at } \eta = 0 \tag{18.45}$$

$$f_1' = 0 \qquad \text{at } \eta = 0 \tag{18.46}$$

and

$$f_1' = 0 \qquad \text{for } \eta \to \infty \tag{18.47}$$

It follows that

$$C_f\sqrt{\mathrm{Re}_L} = \frac{1}{\sqrt{2\xi}}\frac{rU_x}{Lu_\infty}\left(A_0 + 2\xi\frac{d\Lambda}{d\xi}A_1 + \cdots\right) \tag{18.48}$$

where

$$A_n = f_n''\{\Lambda, 0\}$$

For pure wedge flows, $A_1 = A_2 = \cdots = A_k = 0$. For other bodies the A_1 term eventually becomes appreciable at some distance from the point of forward stagnation. The principal value of the method lies in its rapid rate of convergence. Merk demonstrated that the first-term (A_0) approximation was reasonably accurate for all elliptical cylinders, even up to the point of separation.

The *Mangler transformations* (Equations 18.24–18.29) and the *Merk expansion* (Equation 18.40) are closely related to the *Görtler transformation*. However, Görtler effectively expands Λ in terms of ξ. If this is done, Λ remains as an independent variable replacing ξ.

Sparrow–Quack–Boerner Method

Sparrow et al. [15] devised a method called *local nonsimilarity* for two-dimensional geometries. This method can be extended to axisymmetrical flows through the Mangler transformation. It begins with Equation 18.36 of the *Meksyn–Merk development* with the wedge variable Equation 18.33 simplified to

$$\Lambda = \frac{2}{U_x} \frac{dU_x}{dx} \int_0^x U_x \, dx \tag{18.49}$$

and the boundary conditions 18.37–18.39 simplified to

$$f = f' = 0 \qquad \text{at } \eta = 0 \tag{18.50}$$

and

$$f' = 1 \qquad \text{as } \eta \to \infty \tag{18.51}$$

Local similarity solutions are obtained by setting the right side of Equation 18.36 to zero. Equation 18.36 can then be solved *locally* (i.e., for any value of ξ). This procedure produces the first-order approximation of Merk. The approximation of local similarity can obviously be rationalized for small ξ and possibly for large ξ on the basis that the derivatives with respect to ξ are negligible in magnitude.

Sparrow et al. developed their improved approximation by introducing

$$g\{\xi, \eta\} = \frac{\partial f}{\partial \xi} \tag{18.52}$$

which transforms Equation 18.36 to

$$f''' + ff'' + \Lambda(1 - f'^2) = 2\xi(f'g' - f''g) \tag{18.53}$$

Differentiating Equations 18.50, 18.51, and 18.53 with respect to ξ and substituting from 18.52 then give

$$g''' + fg'' - 2\Lambda f'g' + f''g + \frac{d\Lambda}{d\xi}(1 - f'^2)$$

$$= 2(f'g' - f''g) + 2\xi \frac{\partial}{\partial \xi}(f'g' - f''g) \tag{18.54}$$

with

$$g = g' = 0 \qquad \text{at } \eta = 0 \tag{18.55}$$

and

$$g' = 1 \qquad \text{as } \eta \to \infty \tag{18.56}$$

The term $\xi(\partial/\partial\xi)(f'g' - f''g)$ is now postulated to be negligible, reducing Equation 18.54 to an ordinary differential equation that can be solved locally (i.e., for any value of ξ), just as in the method of local similarity. The neglect of $\xi(\partial/\partial\xi)(f'g' - f''g)$ can readily be rationalized for small ξ. For large ξ the approximation seems less severe than that of local similarity, since the original equation of momentum 18.53 remains intact. The two-equation model represented by Equations 18.53 and 18.54, with the indicated simplification, constitutes a pair of simultaneous ordinary differential equations that can be solved numerically for any ξ.

A three-equation model can be developed by letting

$$h = \frac{\partial g}{\partial \xi} = \frac{\partial^2 f}{\partial \xi^2} \tag{18.57}$$

The same process as before produces

$$h''' + fh'' - 2\Lambda(g'^2 + f'h') + f''h + 2gg'' - 4\frac{d\Lambda}{d\xi}f'g' + \frac{d^2\Lambda}{d\xi^2}(1 - f'^2)$$

$$= 4(g'^2 - gg'' + f'h' - f''h) + 2\xi\frac{\partial}{\partial\xi^2}(f'g' - f''g) \tag{18.58}$$

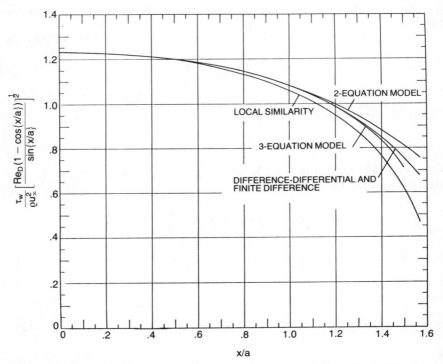

FIGURE 18–3   *Comparison of the local shear stress on a circular cylinder in cross-flow as obtained by different methods of solution. (From Sparrow et al. [15].)*

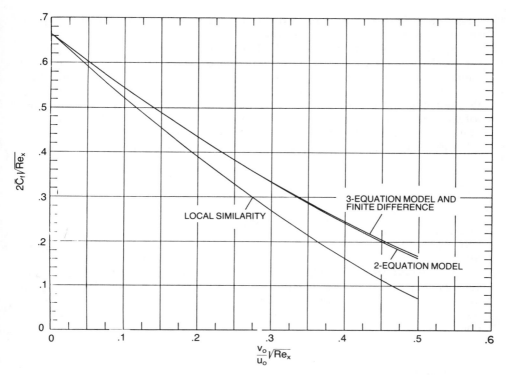

FIGURE 18–4. *Comparison of the local drag coefficient for flow along flat plate with uniform blowing as obtained by different methods of solution. (From Sparrow et al. [15].)*

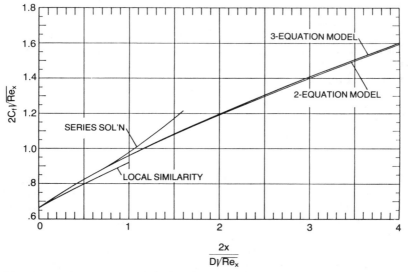

FIGURE 18–5 *Comparison of the local drag coefficient for longitudinal flow along a circular cylinder as obtained by different methods of solution. (From Sparrow et al. [15].)*

This time it is only necessary to drop $\xi(\partial^2/\partial\xi^2)(f'g' - f''g)$ to obtain a set of three ordinary differential equations.

Sparrow et al. applied this method to a number of problems. The results for the local shear stress on a circular cylinder as calculated by the method of local similarity and by the two- and three-equation models are compared in Figure 18–3 with the "exact" solution obtained by Schönauer [16], using finite-difference methods. The agreement is seen to be good for all methods near the forward stagnation point and to improve noticeably with the order of approximation at $\theta = \pi/2$. Similar comparisons are shown in Figures 18–4 and 18–5 for a flat plate with uniform blowing and for flow along a cylinder, respectively.

SUMMARY

The drag coefficients for different shapes have a structural similarity but differ owing to the relative contributions of skin friction and the wake. Those geometries with axial or planar symmetry are now within the capability of numerical solution, at least for conditions such that the wake is stationary.

The generalized methods proposed for determination of the flow around various shapes, by and large reduce the mathematical model from partial to ordinary differential equations that must be solved numerically or by series. Such methods are ordinarily restricted to the regime and region of the thin boundary layer.

PROBLEMS

1. Transform Equations 14.24 and 14.25 to 11.1 and 14.2 using the Mangler transformation.
2. Transform the boundary-layer equations for a sphere using the Mangler transformation.
3. Develop a solution for flow along a flat plate using the von Mises transformation and assuming a mean value of $2(\mathscr{P}_t - \mathscr{P})/\varrho$. Compare the results with the exact solution.
4. Derive expressions for ξ, η, and $\beta\{\eta\}$ in the Görtler transformation for

    a. flow along a flat plate
    b. flow over a wedge
    c. flow over a cylinder
    d. flow over a sphere

5. Derive expressions for ξ, η, and Λ in the Meksyn transformation for the cases in problem 4. Compare the results with those of problems 3 and 4.
6. Use Equation 18.48 with the A_0 coefficient only to derive solutions for the the cases in problem 4. Compare the solutions for a cylinder and a sphere with the solutions in Chapters 15 and 16.
7. Replot Figure 18–1 in terms of Re_D and $\mathrm{Re}_D^2\bar{C}_t$. Interpret.
8. Replot Figure 18–1 in terms of $\mathrm{Re}_D/\bar{C}_t$ and $\mathrm{Re}_D^2\bar{C}_t$. Interpret.
9. Compare the values of $\bar{C}_t$ in Table 18.1 for a circular cylinder with those of Tables 15.4–15.9 and Figures 15–20, 15–26, and 18–1. Interpret.

10. Compare the values of $\bar{C}_f$ in Table 18.1 for a sphere with Equation 16.60 and those of Tables 16.2–16.8 and Figures 16–24, 16–25, 16–30, and 18–1. Interpret.
11. Compare the values in Table 18.1 for a disk and a long strip with those of Chapter 10. Also compare the values for a disk with Figure 18–1. Interpret.
12. Correlate the values of $\bar{C}_f$ in Table 18.1 for

   a. flat strips
   b. elliptical cylinders in the laminar regime
   c. elliptical cylinders in the turbulent regime
   d. ellipsoids in the laminar regime
   e. ellipsoids in the turbulent regime

REFERENCES

1. M. J. Gluckman, S. Weinbaum, and R. Pfeffer, "Axisymmetric Slow Viscous Flow past an Arbitrary Convex Body of Revolution," *J. Fluid Mech.*, 55 (1972) 667.
2. H. Rouse and J. W. Howe, *Basic Mechanics of Fluids*, John Wiley, New York (1953).
3. V. Gupta and S. K. Gupta, *Fluid Mechanics and Its Applications*, Wiley Eastern, New Delhi (1984).
4. J. E. Fromm and F. H. Harlow, "Numerical Solution of the Problem of Vortex Street Developments," *Phys. Fluids*, 6 (1963) 975.
5. S. Goldstein, "A Note on the Boundary Layer Equations," *Proc. Camb. Phil. Soc.*, 35 (1939) 388.
6. W. Mangler, "Die "ähnlichen" Lösungen der Prandtlschen Grenzschichtglei-chungen," *Z. Angew. Math. Mech.*, 23 (1943) 243.
7. S. Y. Lee and W. F. Ames, "Similarity Solutions for Non-Newtonian Fluids," *AIChE J.*, 12 (1966) 700.
8. C. F. Dewey, Jr., and J. F. Gross, "Exact Similar Solutions of the Laminar Boundary Layer Equations," p. 317 in *Advances in Heat Transfer*, Vol. 4, Academic Press, New York (1967).
9. H. Blasius, "Grenzschichten in Flüssigkeiten mit kleiner Reibung," *Z. Math. Phys.*, 56 (1908) 1; English transl., "The Boundary Layers in Fluids with Little Friction," *NACA* TM 1256, Washington, D.C. (1950).
10. W. Mangler, "Zusammenhang zwischen ebenen und rotationssymetrischen Grenz-schichten in kompressibeln Flüssigkeiten," *Z. Angew. Math. Mech.*, 28 (1948) 97.
11. R. von Mises, "Bemerkungen zur Hydrodynamik," *Z. Angew. Math. Mech.*, 7 (1927) 425.
12. H. Görtler, "A New Series for the Calculation of Steady Laminar Boundary-Layer Flow," *J. Math. Mech.*, 6 (1957) 1.
13. D. Meksyn, *New Methods in Laminar Boundary Layer Theory*, Pergamon, Oxford (1961).
14. H. J. Merk, "Rapid Calculations for Boundary-Layer Transfer Using Wedge Solutions and Asymptotic Expansions," *J. Fluid Mech.*, 5 (1959) 460.
15. E. M. Sparrow, H. Quack, and C. J. Boerner, "Local Non-Similarity Boundary Layer Solutions," *AIAA J.*, 8 (1970) 1936.
16. W. Schönauer, "Ein Differenzenverfahren zur Lösung der Grenzschichtgleichung für stationäre, laminare, inkompressible Strömung," *Ing.-Arch.*, 33 (1964) 173.

PART IV

Flow Relative to Dispersed Solids

Here the term *dispersed solids* implies multiple, interacting particles, in contact or suspension, and includes porous, consolidated media. Flow past dispersed solids might have been considered as a special case of flow through channels (but hardly one-dimensional) *or* as a limiting case of flow over single, unconfined objects (although in general not geometrically regular), and hence treated as an extension of parts I or III, respectively. However, the characteristic behavior of multiple-particle systems differs sufficiently from both of these limiting cases to justify the separate consideration of this part. On the other hand, concepts borrowed from one-dimensional flow in channels and two- or three-dimensional flow over cylinders, spheres, and so on, are both found to be surprisingly useful for interpretation, correlation, and even prediction of flow relative to dispersed solids.

The geometric complexity and randomness that characterize most dispersed systems preclude the exact formulations that typify Parts I, II, and III. Multidimensional modeling is necessarily confined to geometrically regular arrangements or statistical distributions of the solid particles and to date has been very limited in scope. The majority of modeling of dispersed systems has instead been one-dimensional or quasi-one-dimensional through the use of effective, mean dimensions and velocities. Such semitheoretical methods and results are given primary attention.

Fixed solids (packed beds and consolidated media) are considered in Chapter 19, and free solids (suspensions and fluidized beds) in Chapter 20.

Chapter 19

Flow through
Porous Media

The flow of fluids through porous media is of great practical importance in many diverse applications, including the production of oil and gas from geological structures, the gasification of coal, the retorting of shale oil, filtration, groundwater movement, regenerative heat exchange, surface catalysis of chemical reactions, adsorption, coalescence, drying, ion exchange, and chromatography.

Some of the applications mentioned above involve two or even three fluids, and multidimensional and unsteady flows. Attention here will be confined to steady one-dimensional flow of a single fluid relative to a fixed solid phase. In some of the applications, the details of the local velocity field are of concern. Here, in the interest of simplicity and in keeping with the objectives of this book, attention will be focused primarily on the use of theory in predicting the pressure drop.

TYPES OF POROUS MEDIA

A porous medium is defined as a solid structure or an array of solid particles with continuous channels in the direction of flow. Such media can arbitrarily be classified as follows:

1. Channels of varying cross section but regular design, such as the checkerwork and matrices used in thermal regenerators
2. Packed beds of spheres
3. Packed beds of nonspherical but regular geometric shapes, such as "saddles" and annular rings, which are often used to promote mixing between two fluid phases
4. Packed beds of catalyst pellets, usually cylinders
5. Packed beds of irregular granular materials, such as those encountered in filtration, percolation, retorting, and drying
6. Regular consolidated structures such as reticulants
7. Irregular consolidated granular materials, such as porous metals, metal sponges, and natural limestones, sandstones, and shales

QUANTITATIVE CHARACTERIZATION OF POROUS MEDIA

The pressure gradient through a bed of porous media would be expected to be a function of the viscosity, density, and rate of flow of the fluid, and some characteristic dimension of the voids occupied by the fluid, just as for flow through a pipe, but additionally of the extent, shape, arrangement, and size distribution of the voids, and of the dimension(s) of the container. Several of these variables and parameters, including the characteristic dimensions, shape, arrangement and size-distribution of the voids, are difficult to define unambiguously.

Volumetric Fraction of Fluid

The volumetric fraction of the media occupied by the fluid is often called the *void fraction* or *porosity* and is usually symbolized by ε. The micropores within a porous particle, which may be important in catalysis, adsorption, ion exchange, and so on, and any totally enclosed voids are excluded insofar as possible from the values of ε utilized for the description of flow through porous media. The void fraction can be determined by measuring the mass or volume of the liquid required to fill the open voids. Some difficulty may be encountered in such measurements in terms of getting the liquid to enter and leave the voids.

Characteristic Dimension

The flow through porous media depends primarily on the dimensions of the void fraction. Since the size of the pores is difficult to characterize, a dimension of the solid phase in the case of discrete particles and the surface area of the solid phase in the case of consolidated media are usually used for characterization.

For a *packed bed* of uniformly sized spheres, the dimensions of the voids would be expected to be proportional to the diameter of a sphere, which thereby might be utilized simplistically as the characteristic dimension. For a bed of spheres with a distribution of diameters some average value may be utilized. Very small spheres will fit completely within the void space between large spheres and hence lead to a lower void fraction. The *combination* of void fraction and average diameter of the spheres might be expected to provide at least a first-order characterization of the packed bed.

The dimension (or dimensions) of other regular particles, such as cylinders, may similarly be used as a first-order characterization of the size of the voids in packed beds of such shapes. The average dimension of granular particles as determined by screening is often similarly utilized. The symbol D_p is used for all of these *particulate dimensions*.

An alternative approach for either discrete particles or consolidated media is to utilize the *specific surface*—that is, surface area per unit volume of solid, a_S, or the surface area per unit volume of packing, a_V—as the characteristic dimension. As an example, for a sphere

$$a_S = \frac{\pi D_p^2}{\pi D_p^3 / 6} = \frac{6}{D_p} \qquad (19.1)$$

and

$$a_V = \left(\frac{\pi D_p^2}{\pi D_p^3/6}\right)(1 - \varepsilon) = \frac{6(1 - \varepsilon)}{D_p} \qquad (19.2)$$

The former quantity would appear to preferable owing to its independence from the porosity but the latter quantity arises more directly in some of the derivations that follow. For a known ε they are obviously interchangeable.

An equivalent diameter for any porous medium can be defined by Equations 19.1 or 19.2. This quantity is the diameter of a sphere with the same surface area per unit volume as the medium. For a granular particle the *hydrodynamic surface*—that is, the area of an elastic film stretched over the granule—is used, at least conceptually, for a_S and a_V. The increased external surface of granular particles due to pebbling, and so on, and the internal surface of porous materials or regular geometrical shapes, which may be more important in surface reactions or interphase transfer than in flow itself, are thereby neglected.

A further alternative is to determine the *effective diameter* for which a correlation based on packings of uniformly sized spheres will predict the observed pressure drop for the porous medium in question.

Characteristic Dimension of Voids

The concept of a hydraulic diameter D_h equal to four times the cross-sectional area for flow divided by the wetted perimeter has often been suggested for the channels within porous media of all varieties. Thus

$$D_h = \frac{4A_x}{P_w} = \frac{4A_x L}{P_w L} = \frac{4(\text{volume of channel})}{\text{surface area of channel}}$$

$$= \frac{4(\text{volume of channel/volume of solid})}{\text{surface area/volume of solid}}$$

$$= \frac{4\varepsilon/(1 - \varepsilon)}{a_S} = \frac{4\varepsilon}{a_S(1 - \varepsilon)} \qquad (19.3)$$

or

$$D_h = \frac{4(\text{volume of channel/volume of packing})}{\text{surface area/volume of packing}}$$

$$= \frac{4\varepsilon}{a_V} \qquad (19.4)$$

For uniformly sized spheres, the combination of 19.1 with 19.3, or 19.2 with 19.4 gives

$$D_h = \frac{2}{3}\frac{\varepsilon D_p}{1 - \varepsilon} \qquad (19.5)$$

The intuitive use of the particle diameter as a characteristic dimension of the void space is thus confirmed qualitatively by Equation 19.5. On the other hand, Equation 19.5 predicts the quantitative dependence of the characteristic dimension of the void space on the void fraction as well.

Sphericity

The *shape* and *arrangement* of the void space is difficult to describe and generalize (see, for example, Dullien [1] and Greenkorn [2, 3]). One widely used measure of shape is the *sphericity*, ψ, defined as the surface area of a sphere divided by the surface area of the solid portion of the packing that occupies the same volume. The hydrodynamic surface, as defined here, is used to determine the sphericity of granular particles. Sphericity is ordinarily independent of the characteristic dimension of the particles.

Relationship between Porosity and Sphericity

The porosity is closely related to the sphericity, as indicated in Figure 19-1 from Brown and Associates [4], in which are plotted experimentally observed values for random packings of uniformly sized particles. The data used to prepare Figure 19-1 are listed in Table 19.1. Granular materials usually have sphericities from 0.70 to 0.80 and porosities from 0.32 to 0.40. The range of values of ε for a given ψ provides some indication of the unpredictability of random packings.

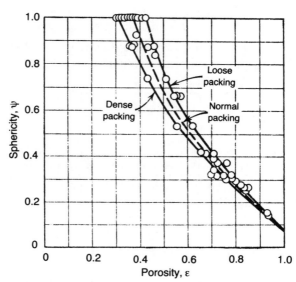

FIGURE 19-1   *Void fractions of typical packed beds. (From Brown and Associates [4], p. 214.)*

Table 19.1
Typical Void Fractions and Sphericities for Random Packings (from Brown and Associates [4], p. 214).

Material	Diameter (in.)	Void Fraction (ε)	Sphericity (ψ)
Spheres	0.217	0.3781–0.468	1
Glass spheres	0.209	0.412	1
Lead shot, uniform size	0.25	0.375–0.421	1
	0.058	0.363–0.375	1
	0.121	0.370–0.390	1
Lead shot mixtures	0.066	0.303	1
	0.078	0.325	1
	0.076	0.320	1
Celite cylinders	0.267	0.361–0.461	0.877
Berl saddles	1.97	0.780	0.314
	1.38	0.785	0.297
	0.985	0.750	0.317
	0.590	0.758	0.296
	0.472	0.710	0.342
	0.390	0.694	0.329
	1.00	0.725	0.370
	0.5	0.7125–0.761	0.370
Nickel saddles	0.132	0.931	0.140
	0.1295	0.935	0.140
Raschig rings	1.97	0.853	0.260
	1.38	0.835	0.262
	0.985	0.826	0.272
	0.390	0.655	0.420
	1.00	0.707	0.391
	0.385	0.554–0.620	0.531
Glass rings	0.228	0.67	0.411
	0.273	0.72	0.370
	0.3875	0.80	0.294
	0.4715	0.845	0.254

Orientation

Orientation is a possible characteristic of packed beds of *regular* particles in addition to their sphericity, void fraction, and characteristic dimension. For example, uniformly sized spheres may be packed in the several *ordered arrangements* (some with identical porosities) illustrated in Figure 19–2 from Martin et al. [5], as well as randomly. The values of ε in Figure 19–2 extend above and below the maximum and minimum values in Table 19.1 for randomly packed beds.

Tortuosity

Shape is also sometimes characterized by the *tortuosity*, τ, usually defined as the ratio of the average distance traversed by a particle of fluid, L_e, to the direct

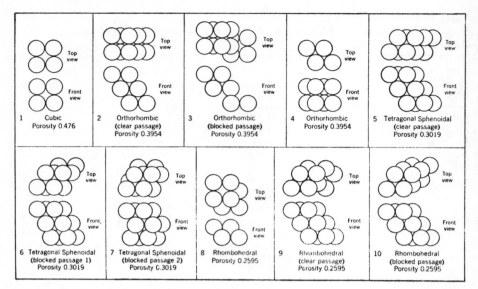

FIGURE 19–2.   *Ordered arrangements of spheres after Martin et al. [5]. (From Brown and Associates [4], p. 215).*

distance through the bed, L. However, Carman [6] defined it as the square of that ratio. The mean tortuosity could presumably be determined experimentally, for example with a tracer such as a dye, or theoretically from a specified structure for the voids. Dullien [1] notes that this quantity has generally been used merely as a "fudge factor" in correlations.

Connectivity

In consolidated porous media such as limestones, some of the void space may be completely closed and hence inaccessible to the moving fluid, or partially closed, thereby constituting a dead end for the flowing fluid. This characteristic is expressed quantitatively in terms of the *connectivity*, defined as the average number of pores at a junction.

Roughness

Roughness, ē, in meters is found to be a parameter affecting the pressure drop through packed beds of regular geometrical shapes, but is obviously difficult to quantify for granular or consolidated media.

Anisotropy

Anisotropy—that is, a variation in the properties of a porous medium with direction–is an important characteristic of geological materials such as shale, which are usually stratified.

Characteristic Velocity

The velocity through porous media is usually expressed in terms of the *superficial velocity*, u_0, i.e., the velocity that would exist for the same volumetric rate of flow in the absence of the porous media. The corresponding *mean normal velocity* through the void space is then

$$u_n = \frac{u_0}{\varepsilon} \tag{19.6}$$

Carman [6] noted that the *mean tortuous velocity* through the void space is actually

$$u_t = \frac{u_0}{\varepsilon}\left(\frac{L_e}{L}\right) \tag{19.7}$$

u_t is a more *characteristic velocity* than u_0. It is used in the subsequent derivations even though the final results are expressed in terms of u_0.

Wall Effects

Completely regular (ordered) arrangements with a uniform distribution of the mean void fraction even adjacent to the wall of a container are possible only if the cross section of the container conforms to that arrangement: for example, square cross sections for ordered arrangements 1, 2, 3, and 8 of Figure 19–2, and hexagonal ones for arrangements 4, 5, 6, 7, 9, and 10.

   Except for completely ordered arrangements of spheres or geometrical matrices, the void fraction (on-the-mean) is highest *at* the wall of the container and then oscillates with decreasing amplitude toward a uniform value. Figure 19–3 is a plot prepared by Cohen and Metzner [7] to represent the data of a

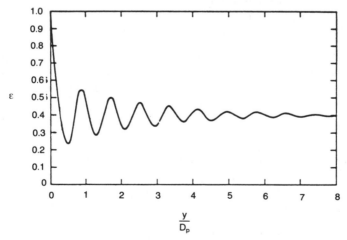

FIGURE 19–3   *Variation of the average void fraction near the wall of a cylindrical packed bed of uniformly sized spheres. (From Cohen and Metzner [7].)*

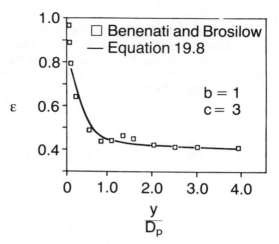

FIGURE 19-4   *Comparison of Equation 19.8 with the experimental data of Benenati and Brosilow [9] for the variation of the void fraction with distance from the wall in a random packing of uniformly sized spheres with $D_t/D_p = 20.3$. (From Chandrasekhara and Vortmeyer [8].)*

number of investigators for the radial distribution of the void fraction of uniformly sized spheres in cylindrical enclosures. Such a highly organized structure occurs near the wall because almost all of the outer particles actually touch the wall.

Chandrasekhara and Vortmeyer [8] proposed representing this behavior, minus the oscillations, by an empirical equation of the form

$$\varepsilon = \varepsilon_0(1 + be^{-cy/D_p}) \tag{19.8}$$

where ε_0 = void fraction far from a wall
y = distance from the wall, m
b, c = arbitrary, dimensionless constants

In Figure 19-4, Equation 19.8 with $b = 1$ and $c = 3$ is compared with the experimental data of Benenati and Brosilow [9] for a random packing of spheres with $D_t/D_p = 20.3$, where D_t is the diameter of the enclosure.

Functional Relationships

From the foregoing considerations and dimensional analysis, it follows that

$$\left(\frac{-\Delta\mathscr{P}}{L}\right)\frac{D_p^2}{\varrho u_0^2} = \left\{\frac{D_p u_0 \varrho}{\mu}, \frac{D_p}{D_t}, \varepsilon_0, \frac{\tilde{e}}{D_p}, \text{shape}\right\} \tag{19.9}$$

where, for uniformly sized particles, D_p may be chosen according to Equations 19.1 or 19.2. The choice of an average dimension for particles with a size distribution or for a consolidated medium is clearly arbitrary, although Equations

19.1 or 19.2 could again be utilized. For granular particles of distributed size, some arbitrary average of the distribution determined by screening is often used. An alternative, and perhaps the best, choice is to determine the effective dimension for which some correlation predicts the observed pressure drop. The effective roughness $\bar{e}$ can also only be evaluated by comparison of measured pressure drops with a correlation. All things considered, it is improbable that the shape of porous media, apart from that represented by the effective diameter and porosity, can be characterized simply and generally.

In view of all these factors, the prediction of flow through porous media from first principles or even the construction of general correlations would not be expected to be successful.

MODELS AND SOLUTIONS FOR CREEPING (LAMINAR) FLOW

Stokes' Equation

For very low rates of flow, such that the inertial terms can be neglected, the equations of steady motion for the fluid within the void space of a porous medium can be approximated as

$$\nabla \cdot \varrho v_0 = 0 \tag{19.10}$$

and

$$\nabla \mathscr{P} = \frac{\mu}{\varepsilon} \nabla^2 v_0 \tag{19.11}$$

where v_0 is the *superficial velocity vector*, that is, the velocity vector in the absence of a solid phase in meters per second, and, as before,

$$\mathscr{P} = p + g\varrho h - p_0 \tag{19.12}$$

In the literature of flow through porous media, Equation 19.11 is often called *Stokes' equation*.

It follows from Equation 19.11 that for an incompressible fluid

$$\nabla^2 \mathscr{P} = \frac{\mu}{\varepsilon} \nabla^2 (\nabla \cdot v_0) = 0 \tag{19.13}$$

The dynamic pressure $\mathscr{P}$ can thus in principle be determined by using the various techniques of potential theory (see problems 28–31). However, the velocity field must still be determined from Equation 19.11.

Darcy's Law

The gradient of the dynamic pressure due to the drag of the surface of the porous medium is usually expressed as

$$-\nabla \mathscr{P} = \frac{\mu \mathbf{v}_0}{K} \qquad (19.14)$$

Equation 19.14 is the vectorial form of *Darcy's law*, which he formulated in one-dimensional form in 1856 [10]. The coefficient K with dimensions of square meters is known as the *permeability*. The square root of the permeability may be considered to be a characteristic dimension of the pore space. Different related coefficients are also called the permeability by some authors, so caution is advised relative to units and numerical values. This form was chosen here because it depends only on the properties of the porous medium and not on those of the fluid.

Equation 19.14 was originally defined heuristically. It can also be interpreted as an analog of Poiseuille's law (Equation 4.8). Slattery [11] and others (see, for example, Greenkorn [2, 3]) have rationalized Darcy's law on the basis of volume averaging, but Larson [12] has recently demonstrated, without invoking the heuristics of volume averaging, that Equation 19.14 satisfies the general equations of motion. He further showed that in one dimension, Darcy's law can be generalized for power-law fluids as

$$u_0^\alpha = \frac{-K'}{M} \frac{d\mathscr{P}}{dx} \qquad (19.15)$$

Bird et al. [13], p. 150, discuss the extension of Darcy's law to compressible and unsteady flows.

Equation 19.14 rather than the more general Equation 19.11 is usually used as the starting point for the description of flow in porous media, because the boundaries of the channels must be well defined to apply the latter.

Stokes–Darcy Equation

Brinkman [14] added the pressure gradients of Equations 19.11 and 19.14 to obtain empirically the following expression for conditions under which the confining walls are important:

$$-\nabla \mathscr{P} = \frac{\mu \mathbf{v}_0}{K} - \frac{\mu}{\varepsilon} \nabla^2 \mathbf{v}_0 \qquad (19.16)$$

He suggested that the first term on the right side of Equation 19.16 represents the drag of the porous media and that the second term represents the distortion of the flow by the walls of the container. Equation 19.16, known as the *Stokes–Darcy equation*, was subsequently derived rigorously by Tam [15] and Lundgren [16].

Azzam and Dullien [17] proposed a model for the estimation of the permeability based on the bivariant distribution of the pore volume, but this quantity is ordinarily determined experimentally according to Equations 19.14 or 19.16.

Extensions of Darcy's Law Based on Flow through a Pipe

Dupuit [18] in 1863 suggested that the mean velocity in the pores $u_n = u_0/\varepsilon$ rather than the superficial velocity be used in Darcy's law. Blake [19] further suggested that ε/a_V be used as the characteristic dimension of the channels through the porous media. He then used dimensional analysis to derive for a finite packed bed of length L,

$$\frac{\varepsilon^3(-\Delta\mathscr{P})}{a_V\varrho u_0^2 L} = f\left\{\frac{u_0\varrho}{a_V\mu}\right\} \tag{19.17}$$

For the regime of creeping flow the density is not a variable and the *Blake equation* 19.17 reduces to

$$\frac{-\Delta\mathscr{P}}{L} = \frac{k\mu u_0 a_V^2}{\varepsilon^3} \tag{19.18}$$

Ergun [20] recommended the use of the hydrodynamic surface for a_V in Equation 19.18. The dimensionless coefficient k is called the *Kozeny constant* after Kozeny [21], who independently derived the equivalent of Equation 19.18 and improved it by replacing L with the tortuous length, L_e, actually traversed by a particle of fluid.

Introducing u_t from Equation 19.7 as the characteristic velocity, according to Carman [6], and L_e for L, according to Kozeny, converts Equation 19.18 to

$$\frac{-\Delta\mathscr{P}}{L} = \frac{k\mu u_0 a_V^2}{\varepsilon^3}\left(\frac{L_e}{L}\right)^2 \tag{19.19}$$

Introducing a_V from Equation 19.2 then gives, for uniformly sized spheres,

$$\frac{-\Delta\mathscr{P}}{L} = \frac{36k\mu u_0(1-\varepsilon)^2}{\varepsilon^3 D_p^2}\left(\frac{L_e}{L}\right)^2 \tag{19.20}$$

The simple adaption of Poiseuille's law for a round tube (Equation 4.18) to flow through porous media by postulating a mean velocity $(u/\varepsilon)(L_e/L)$, according to Equation 19.7, through a channel of length L_e and hydraulic diameter $2\varepsilon D_p/3(1-\varepsilon)$, according to Equation 19.5, gives

$$\frac{-\Delta\mathscr{P}}{L} = \frac{72\mu(1-\varepsilon)^2 u_0}{\varepsilon^3 D_p^2}\left(\frac{L_e}{L}\right)^2 \tag{19.21}$$

Comparison of Equations 19.21 and 19.20 implies that the Kozeny constant for a packed bed of spheres equals 2.

Carman chose $(L_e/L)^2 = 5/2$ on the basis of experimental data and utilized $k = 2$ to derive

$$\frac{-\Delta\mathcal{P}}{L} = \frac{5\mu u_0 a_V^2}{\varepsilon^3} \tag{19.22}$$

For spheres Equation 19.22 then becomes

$$\frac{-\Delta\mathcal{P}}{L} = \frac{180\mu(1 - \varepsilon)^2 u_0}{\varepsilon^3 D_p^2} \tag{19.23}$$

Equation 19.23 is usually called the *Carman–Kozeny equation*.

The derivation of Equations 19.22 and 19.23 can be faulted on several grounds, particularly for the use of the hydraulic diameter concept for laminar flow, for which it is known to be unreliable even for straight channels (see Chapters 3 and 4). Even so, as shown here, these expressions have been found to provide good first-order predictions for the effects of all the variables for a wide range of conditions.

For example, Ergun [20] demonstrated excellent agreement for the dependence on void fraction predicted by Equation 19.23 with experimental data for coke. He first plotted, as illustrated in Figure 19–5, measurements of the pressure drop across packings of a single size of particle at series of void

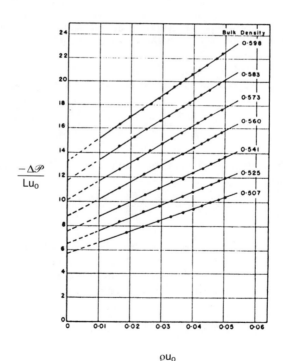

FIGURE 19–5   *Demonstration of the linear dependence of the pressure drop through 16–20 mesh coke (packed at a series of void fractions) on the mass rate of flow of nitrogen. Density of particles = 1.06 gm/cm³. No units were indicated by the author. (From Ergun [20].)*

fractions in the form of $-\Delta\mathscr{P}/Lu_0$ versus $u_0\varrho$. He then plotted, as illustrated in Figure 19–6, the intercepts of straight lines through these data, as determined by least squares, versus $(1 - \varepsilon)^2/\varepsilon^3$. Similar results were obtained for other packings.

Despite the apparent success indicated in Figure 19–6 for the dependence on ε according to Equation 19.23, contradictory results have been obtained, and alternative functional dependences have been proposed by others. For example, Foscolo et al. [22] noted that Equation 19.23 is based on a fixed tortuosity, whereas this quantity might itself be expected to be a function of the void fraction. In order to derive such a relationship, they postulated that at any cross section of the packed bed a fraction of fluid equal to the void fraction ε moves a small distance l in the forward direction, and that the remaining fraction, $1 - \varepsilon$, moves an equal distance sidewise. Of this latter quantity, a fraction of the fluid equal to ε moves forward in the next void and a fraction equal to $1 - \varepsilon$ moves sidewise. The total distance $l\varepsilon$ traversed by the fluid in moving forward a distance l is then $l\varepsilon + 2l\varepsilon(1 - \varepsilon) + 3l\varepsilon(1 - \varepsilon)^2 + \cdots$, and the corresponding tortuosity is

$$\tau = \frac{l_\varepsilon}{l} = \sum_{n=1}^{\infty} n\varepsilon(1 - \varepsilon)^{n-1} = \frac{1}{\varepsilon} \tag{19.24}$$

This very idealized model for the pattern of flow correctly yields $\tau = 1$ for $\varepsilon = 1$. Substituting τ from Equation 19.24 for $(L_e/L)^2$ in Equation 19.22 gives, for spheres,

$$-\frac{\Delta\mathscr{P}}{L} = \frac{72\mu u_0(1 - \varepsilon)^2}{D_p^2 \varepsilon^4} \tag{19.25}$$

(The distance traveled rather than its square is used in this model of Foscolo et al. for the pressure drop, since the corresponding change in local velocity is

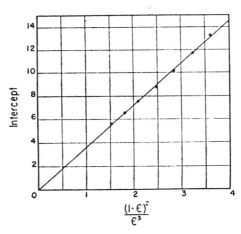

FIGURE 19–6 *Test of the dependence on void faction predicted by Equation 19.23 for the pressure drop across a packed bed; the ordinate represents the intercepts of Figure 19–5. (From Ergun [20].)*

not taken into account.) Equation 19.25 coincides with Equation 19.23 for $\varepsilon = 0.4$, since Equation 19.24 then gives $\tau = 5/2$, which is the same value as used by Carman [6]. Equation 19.25 might be expected to provide a better prediction than Equation 19.23 for $\varepsilon < 0.4$, owing to its correct behavior in the limit.

The precision of the data of Rumpf and Gupte [23] for both the flow of air and the flow of oil through packings of uniformly sized glass spheres is demonstrated in Figure 19–7, in which $-(\Delta\mathcal{P}/L)(D_p/\varrho u_0^2)$ is plotted versus $D_p u_0 \varrho / \mu$ in log-log coordinates with void fraction as a parameter. The data for air for a series of values of $D_p u_0 \varrho / \mu > 1$ are replotted versus ε in Figure 19–8, again in log-log coordinates. The latter plot indicates a proportionality of $-\Delta\mathcal{P}$ to $\varepsilon^{-5.5}$ for all $D_p u_0 \varrho / \mu$. Figure 19–7 indicates that this proportionality is also applicable for $D_p u_0 \varrho / \mu < 1$. The implied relationship for all $D_p u_0 \varrho / \mu$ is thus

$$\left(-\frac{\Delta\mathcal{P}}{L}\right)\frac{D_p \varepsilon^{5.5}}{\varrho u_0^2} = \phi\left\{\frac{D_p u_0 \varrho}{\mu}\right\}$$ (19.26)

For $D_p u_0 \varrho / \mu \leq 1$ the data for Figures 19–7 and 19–8 indicate that

$$-\frac{\Delta\mathcal{P}}{L} = \frac{5.6\mu u_0}{D_p^2 \varepsilon^{5.5}}$$ (19.27)

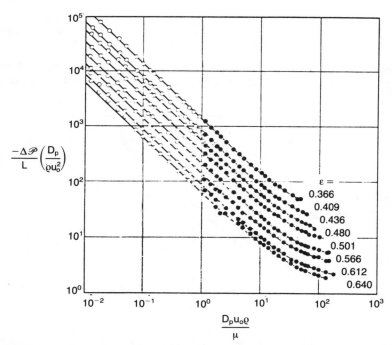

FIGURE 19–7   *Experimental data for the pressure drop for the flow of oil ($\bigcirc$) and air ($\bullet$) through packed beds of uniformly-sized glass spheres. (From Rumpf and Gupte [23].)*

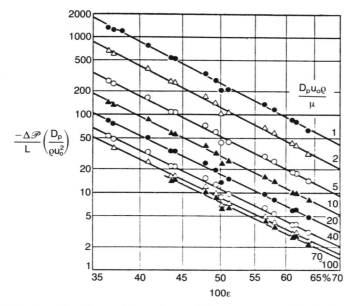

FIGURE 19–8 *Experimental dependence of the pressure drop for the flow of air through packed uniformly sized spheres on the void fraction for fixed values of* $D_p u_o \varrho / \mu$. *(From Rumpf and Gupte [23].)*

where 5.6 is a purely empirical coefficient. Figure 19–9 indicates that Equation 19.27 provides a better representation than either Equation 19.23 or Equation 19.25 for the data of Rumpf and Gupte for $D_p u_0 \varrho / \mu < 1$.

Kyan et al. [24] have shown (see Figure 19–10) that in randomly packed beds of fibers the coefficient k of Equation 19.19 depends critically upon ε, particularly for $\varepsilon > 0.9$. It may be inferred (see problem 7) that Equations 19.23, 19.25, and 19.27 all fail to predict the indicated dependence of the data on ε. This failure for $\varepsilon \to 1$ can presumably be rationalized on the basis of the wide spacing of the fibers and hence on a shift from flow in "channels" to flow over isolated cylinders.

Comparison of Equations 19.22 and 19.23 with Equation 19.14 indicates that the predicted *permeability* is

$$K = \frac{\varepsilon^3}{5a_V^2} \tag{19.28}$$

or, for uniformly sized spheres,

$$K = \frac{\varepsilon^3 D_p^2}{180(1 - \varepsilon)^2} \tag{19.29}$$

The permeabilities corresponding to Equation 19.25 and 19.27 are left to problem 8.

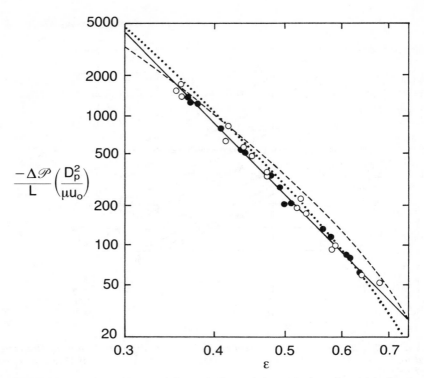

FIGURE 19-9   *Comparisons of theoretical, semi-theoretical, and empirical expressions for the dependence of the pressure drop through packed beds on the void fraction with the experimental data of Rumpf and Gupte [23] for $D_p u_0 \varrho / \mu < 1$. ———— Equation 19.27 (Rumpf and Gupte), empirical; · · · · · Equation 19.25 (Foscolo et al.), semi-theoretical; ---- Equation 19.23 (Carman–Kozeny), theoretical.*

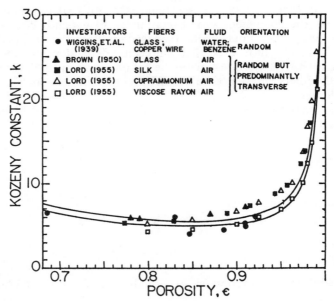

FIGURE 19-10   *Dependence of the Kozeny constant on the void fraction of beds of fibers. (From Kyan et al. [24].)*

Numerical Solutions

Snyder and Stewart [25] derived numerical solutions for creeping flow through two regular arrays of uniformly sized spheres using the Galerkin method of approximation by weighted residuals. For *dense cubic packing* ($\varepsilon = 0.2595$) and *simple cubic packing* ($\varepsilon = 0.476$) they computed the equivalent of 175 and 150, respectively, for the coefficient of Equation 19.23. Illustrative computed velocity and pressure fields about a sphere are shown in Figures 19–11 and 19–12 for their calculations for the dense cubic arrangement.

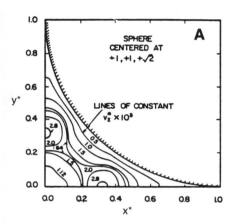
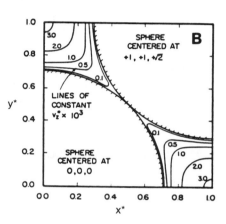
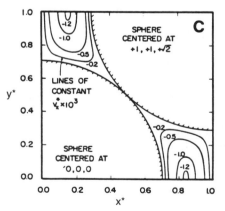

FIGURE 19–11   *Computed velocity profiles for a dense cubic packing of uniformly sized spheres. (From Snyder and Stewart [25].)*

$$x_2^* = \frac{2x_i}{D_p}, \qquad v_i^* = \frac{2u_0\mu}{(-\Delta\mathscr{P})D_p}$$

(A) v_z^* at $z^* = \sqrt{2}$; (B) v_z^* at $z^* = \sqrt{2}/2$; (C) v_x^* at $z^* = \sqrt{2}/2$. (From Snyder and Stewart [25].)

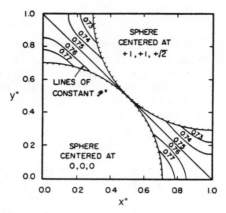

FIGURE 19–12   *Computed pressure profiles for a dense cubic packing of uniformly sized spheres with* $z^* = \sqrt{2}/2$. *(From Snyder and Stewart [25].)*

$$\mathscr{P}^* = \frac{\mathscr{P} - \mathscr{P}_0}{\mathscr{P}_L - \mathscr{P}_0}, \qquad x_i^* = \frac{2x_i}{D_p}$$

COMPLETELY INERTIAL FLOW

For very large rates of flow through porous media the pressure drop has been observed to approach linearity with the square of the velocity and independence from the viscosity, implying that the pressure loss is due to inertial rather than viscous effects. Such behavior is often attributed to "turbulence" by analogy to flow in channels. However, Bakhmateff and Feodoroff [26] noted, by analogy to flow through an orifice, that the increased losses might be attributed wholly to inertia rather than turbulence.

Burke and Plummer [27] utilized the analogy to flow in a channel at constant friction factor, with a velocity u_0/ε and the effective diameter given by Equation 19.5, to derive the equivalent of

$$\left(\frac{-\Delta\mathscr{P}}{L}\right)\left(\frac{\varepsilon^3}{1-\varepsilon}\right)\frac{D_p}{6\varrho u_0^2} = f_{\text{pipe}} \tag{19.30}$$

They recommended a value of 1.75 for $6f$ in Equation 19.30 based on experimental data for packed beds. The result

$$\left(\frac{-\Delta\mathscr{P}}{L}\right)\left(\frac{\varepsilon^3}{1-\varepsilon}\right)\frac{D_p}{\varrho u_0^2} = 1.75 \tag{19.31}$$

is known as the *Burke–Plummer equation*. Equation 19.31, as reexpressed in terms of a_V corresponding to Equation 19.2, is

$$\left(\frac{-\Delta\mathscr{P}}{L}\right)\frac{\varepsilon^3}{\varrho u_0^2 a_V} = \frac{1.75}{6} = 0.292 \tag{19.32}$$

The form of Equation 19.32 follows directly from the Blake equation, Equation 19.17, for the limiting condition of independence from the viscosity.

Subsequent observations of $6f$ have ranged from 1.2 up to 4.0, with the latter corresponding to rough particles. De Nevers [28] rationalized the order of magnitude of the observed values of f by suggesting that the behavior was analogous to that for completely developed turbulent flow in a tube with an effective roughness $\bar{e}$ equal approximately to the diameter of a particle, and a channel width equal to approximately half the diameter of a particle, thereby producing a roughness ratio $\bar{e}/a \cong 4$. The friction factor for fully developed turbulent flow in a rough pipe can be represented (see, e.g., Churchill [29]) by

$$\frac{1}{\sqrt{f}} = 4.75 + 2.5 \ln\left\{\frac{a}{e}\right\} \tag{19.33}$$

Hence, for the packing, according to the concept of de Nevers,

$$\frac{1}{\sqrt{f}} = 4.75 + 2.5 \ln\left\{\frac{1}{4}\right\} = 1.28 \tag{19.34}$$

and

$$6f = 3.64 \tag{19.34A}$$

A value of 1.75 rather than 3.64 could have attained merely by postulating $\bar{e}/a = 3.18$ instead of 4.0.

Equation 19.31 is subsequently compared with experimental data in conjunction with expressions for a complete range of flows.

Foscolo et al. [22] conclude that since the "friction factor" for completely inertial flow through a packed bed represents expansion/contraction losses rather than shear stress on a surface, it should depend additionally on the void fraction. They postulated that this dependence should be proportional to $1 - \varepsilon$. Introducing the factor of $1 - \varepsilon$ as well as the tortuosity of Equation 19.24 into Equation 19.31 and adjusting the coefficient to give the same pressure drop for $\varepsilon = 0.4$, give

$$-\frac{\Delta \mathcal{P}}{L} = \frac{1.17\varrho u_0^2 (1 - \varepsilon)^2}{D_p \varepsilon^4} \tag{19.35}$$

Equation 19.35 has the same dependence on ε as Equation 19.25, which suggests a possible universal dependence for all rates of flow, just as does Equation 19.26.

MODELS AND SOLUTIONS FOR THE COMPLETE REGIME OF FLOW

Dimensional Correlations

The earliest attempts to correlate data for flow through porous media did not include the properties of the media (ε, D_p, and shape). Hence, they included *dimensional* constants and were specific to the particular media. The first such expression was apparently that of Forchheimer [30] in 1901:

$$\frac{-\Delta \mathscr{P}}{L} = \alpha \mu u_0 + \beta \varrho u_0^n \tag{19.36}$$

He and most subsequent investigators have concluded that 2 is the best value for n, giving

$$\frac{-\Delta \mathscr{P}}{L} = \alpha \mu u_0 + \beta \varrho u_0^2 \tag{19.37}$$

Equation 19.37, known as the *Forchheimer equation*, was based on experimental observations. However Ahmed and Sunada [31] have rationalized this form on the basis of heuristic arguments concerning turbulent effects, and Dullien and Azzam [32] on the basis of the volume-averaging theorem of Slattery [11].

Green and Duwez [33] successfully used the Forchheimer equation to correlate data for flow through porous metals with different values of α and β for each material. Ahmed and Sunada [31] similarly correlated the data of different investigators by first determining α and β for each packing (material, void fraction, and particle diameter or the equivalent) and then plotting $-\Delta\mathscr{P}/\beta\varrho u_0^2$ versus $\beta\varrho u_0/\alpha\mu$, as shown in Figure 19–13.

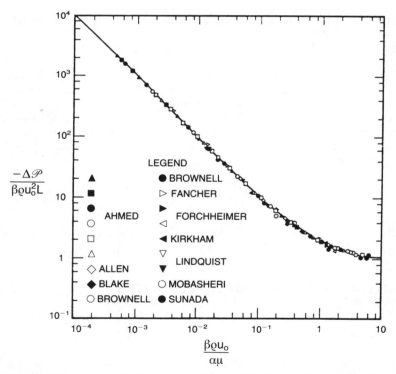

FIGURE 19–13   *Comparison of experimental pressure drops through packed beds with Equation 19.37 using different experimental values of α and β for each packing. (From Ahmed and Sunada [31].)*

Macdonald et al. [34] used the Forchheimer equation (which they called the Ahmed–Sunada equation) to correlate data separately for different types of porous media. They determined α and β for each packing by least squares, and presented a tabulation of these values as well as the effective values of D_p. The results for the randomly packed, uniformly sized spheres of Rumpf and Gupte [23] are shown in Figure 19–14; for the uniformly sized fibers of Kyan et al. [24] in Figure 19–15; for the unconsolidated granular particles of Dudgeon [35] in Figure 19–16; and for the consolidated media of Fancher and Lewis [36] in Figure 19–17. The correlations are reasonably good. However, the data of Fancher and Lewis, which are the only ones extending to asymptotically high rates of flow, show the greatest scatter, particularly at intermediate rates of flow.

Figures 19–13 to 19–17 demonstrate the successful representation of the dependence on velocity by the Forchheimer equation. It is apparent that the discrete transition from laminar to turbulent flow in a round pipe at Re $\cong$ 2100 does not occur in flow through porous media. Instead, there is a smooth transition from the completely viscous to the completely inertial regime.

Comparison of the Forchheimer equation 19.37 with Darcy's law (Equation 19.14) indicates that the coefficient α, with the dimensions of (square meters)$^{-1}$,

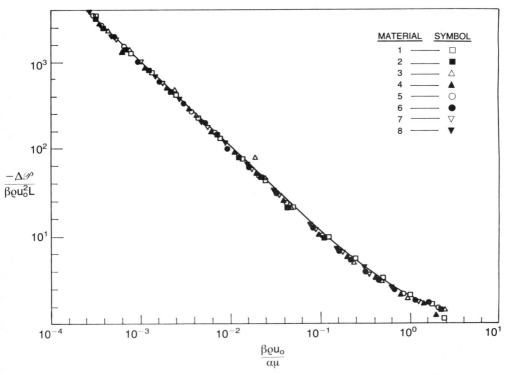

FIGURE 19–14   *Comparison of experimental data of Rumpf and Gupte [23] for the pressure drop through packed beds of uniformly sized spheres with Equation 19.37 using values of α and β determined for each packing by least squares. (From Macdonald et al. [34].)*

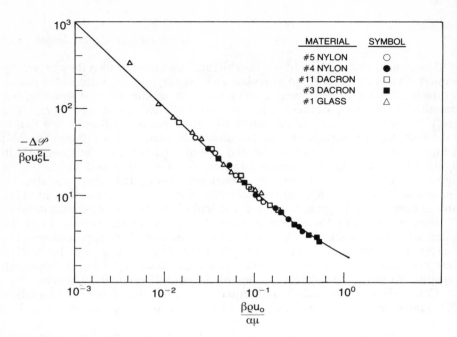

FIGURE 19–15 *Comparison of experimental data of Kyan et al. [24] for the pressure drop through beds of uniformly sized fibers with Equation 19.37 using values of α and β determined for each packing by least squares. (From Macdonald et al. [37].)*

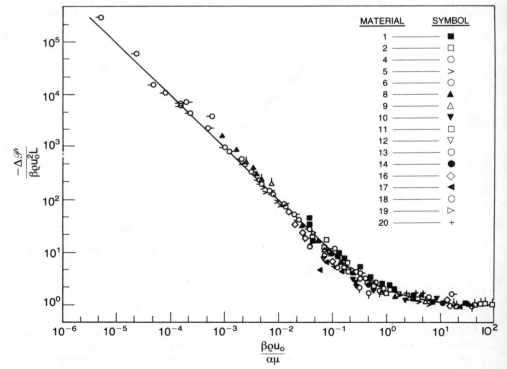

FIGURE 19–16 *Comparison of experimental data of Dudgeon et al. [35] for the pressure drop through beds of unconsolidated granular particles with Equation 19.37 using values of α and β determined for each packing by least squares (From Macdonald et al. [34].)*

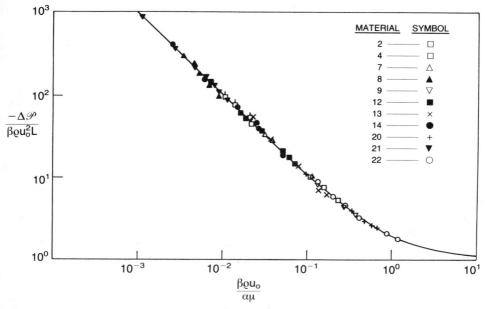

FIGURE 19–17 *Comparison of experimental data of Fancher and Lewis [36] for the pressure drop through consolidated media with Equation 19.37 using values of α and β determined for each medium by least squares. (From Macdonald et al. [34].)*

is equivalent to the reciprocal of the permeability. Similarly, comparison of the Forchheimer equation with the generalized expressions of the following section indicates that β, with the dimensions of reciprocal meters, is proportional to the friction factor for porous media divided by the equivalent diameter. The principal limitation of the Forchheimer equation is its failure to provide any direct indication of the dependence of the pressure drop on ε, D_p, shape, and arrangement.

Dimensional and Generalized Models Based on Flow through a Pipe

Expression of a modified friction factor as a function of a modified Reynolds number by direct or indirect *analogy* with the corresponding relationship for flow through a round pipe has been widely used to correlate and even predict the pressure drop in flow through porous media. This procedure has two aspects: (1) the determination of the appropriate form for the friction factor and Reynolds number, preferably incorporating the parametric dependence on the fraction and shape (including roughness), as well as the size, of the void spaces; and (2) determination of the functional relationship between this friction factor and the Reynolds number.

The generally accepted forms of the modified friction factor and Reynolds number, except for a numerical factor of 6, were actually identified by Blake [19] in 1922, and follow from Equation 19.17:

$$f_p = \frac{\varepsilon^3(-\Delta\mathscr{P})}{6a_V\varrho u_0^2 L} = \frac{\varepsilon^3(-\Delta\mathscr{P})D_p}{(1-\varepsilon)\varrho u_0^2 L} \tag{19.38}$$

and

$$\mathrm{Re}_p = \frac{u_0\varrho}{6a_V\mu} = \frac{u_0\varrho D_p}{(1-\varepsilon)\mu} \tag{19.39}$$

Note the correspondence of 19.38 with 19.30. The factors $\varepsilon^3/(1-\varepsilon)$ and $1-\varepsilon$ are frequently omitted from the definitions of f_p and Re_p, respectively, and then included explicitly in the resulting correlations. Hence, caution is recommended in interpreting numerical values of f_p and Re_p or the equivalent in various articles and texts. Also, as noted, the determination of D_p is arbitrary for nonspherical particles, and the determination of either a_V or a_S is difficult, and perhaps arbitrary, for granular particles or consolidated media.

Early investigators, including Chilton and Colburn [37], Carman [6], and Rose [38], developed correlations of the general form

$$f_p = \frac{A}{\mathrm{Re}_p} + \frac{B}{\mathrm{Re}_p^n} + C \tag{19.40}$$

with $n > 0$, and Brownell and Katz [39] actually forced the relationship between f_p and Re_p to conform to the accepted graphical relationship between f and Re for smooth pipe by setting $B = 0$ and correlating A and C with ε and ψ. Recent correlations, as exemplified by the following one of Ergun, have followed the dependence of $-\Delta\mathscr{P}$ on u_0 in the Forchheimer equation, thereby setting B equal to zero. An exception is that of Talmadge [40], who proposed $n = 1/6$ and $C = 0$.

Ergun [20] combined the Carman–Kozeny equation, Equation 19.23, but with a coefficient of 150 instead of 180, and the Blake–Plummer equation, Equation 19.31 in the form of the Forchheimer equation as

$$\frac{-\Delta\mathscr{P}}{L}\left(\frac{\varepsilon^3}{1-\varepsilon}\right)\frac{D_p}{\varrho u_0^2} = 150\frac{\mu(1-\varepsilon)}{D_p u_0\varrho} + 1.75 \tag{19.41}$$

The excellent representation obtained by the *Ergun equation* for the data of a number of investigators is shown in Figure 19–18.

From Figure 19–18 it may be inferred that Equation 19.23 with a coefficient of 150 instead of 180 is a good approximation (accurate within 10%) for $D_p u_0\varrho/\mu(1-\varepsilon) < 8.6$, and that Equation 19.31 is an equally good approximation for $D_p u_0\varrho/\mu(1-\varepsilon) > 860$.

Equation 19.41 can be rewritten in terms of a_V by substituting for D_p from Equation 19.2, thereby obtaining

$$\left(\frac{-\Delta\mathscr{P}}{L}\right)\frac{\varepsilon^3}{a_V\varrho u_0^2} = 150\frac{\mu a_V}{u_0\varrho} + 1.75 \tag{19.42}$$

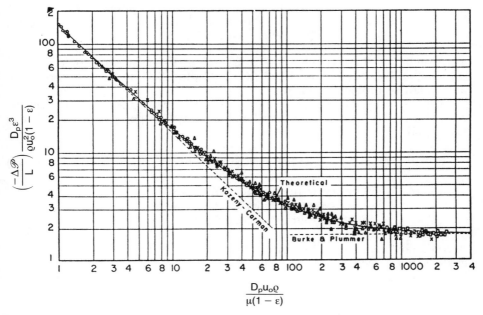

FIGURE 19-18   *Comparison of experimental data for the pressure drop through packed beds of spheres, cylinders, sand, pulverized coke, and Raschig rings with Equation 19.41. (From Ergun [20].)*

Ergun [41, 42] utilized Equation 19.42 to estimate a_V and the apparent value of ϱ_s (which differs from the value of the solid itself due to the micropores) for crushed solids.

Churchill [43] presents plots of the Ergun equation and the data of Figure 19-18 in a number of different forms, and Churchill and Usagi [44] have critically tested the simple addition of the pressure drops from Equations 19.23 and 19.31 using these same data.

Macdonald et al. [34] examined the success of the Ergun Equation 19.41 in representing a large number of sets of data, including those in Figures 19-14 to 19-17. They redetermined the coefficients for Equation 19.41 that best represent each of these sets of data and recommended a mean value of 180, corresponding to Equation 19.23, rather than 150, and, depending on the roughness of the medium, values of the constant term from 1.8 to 4.0 instead of 1.75. They also proposed a correlation for the first coefficient in the Ergun equation (A in Equation 19.40) as a function of D_p and for the second coefficient (C in Equation 19.40) as a function of ε. The first of these can hardly be general because of its dimensionality.

Equations 19.25 and 19.35 can be combined in the form of Equation 19.41 as

$$-\frac{\Delta\mathscr{P}}{L}\left(\frac{\varepsilon^4 D_p}{(1-\varepsilon)^2\varrho u_0^2}\right) = 72\frac{\mu}{D_p u_0 \varrho} + 1.17 \qquad (19.43)$$

The analogous expression based on Equations 19.26 and 19.27 is

$$-\frac{\Delta\mathscr{P}}{L}\left(\frac{\varepsilon^{5.5}D_p}{\varrho u_0^2}\right) = 5.6\frac{\mu}{D_p u_0 \varrho} + A \tag{19.44}$$

The evaluation of the coefficient A in Equation 19.44 is left to problem 9, and the comparison of Equations 19.43 and 19.44 with 19.41 and with experimental data is left to problems 10 and 11.

Equations 19.36, 19.37, and 19.41–19.44 all have the form, for the special case of $p = 1$, of the general correlating equation proposed by Churchill and Usagi [44]:

$$y^p\{x\} = y_0^p\{x\} + y_\infty^p\{x\} \tag{19.45}$$

where $y_0\{x\}$ = asymptotic behavior for $x \to 0$
$y_\infty\{x\}$ = asymptotic behavior for $x \to \infty$
p = an arbitrary exponent

A crude physical explanation for a value of unity for p for flow through porous media is that viscous flow is occurring in a fraction of the void space and inertial flow in the balance, and further that the fraction in inertial flow is proportional to the superficial velocity.

A Semitheoretical Model for Drag Based on Flow over Immersed Objects

A completely different approach is to develop an expression for the pressure drop across a packed bed based on the flow over immersed objects rather than by analogy to flow through a pipe. The derivation of Ranz [45] for the pressure drop through a rhombohedral arrangement of uniformly sized spheres can be paraphrased as follows:

$$\frac{-\Delta\mathscr{P}}{L} = F_p n = \varrho u_\infty^2 A_p \bar{C}_t n \tag{19.46}$$

where n = particles per unit volume, m^{-3}
$\bar{C}_t$ = drag coefficient for a single sphere
F_p = drag force on a single particle, N or $\mathrm{kg \cdot m \cdot s^{-2}}$
u_∞ = "free-stream" velocity in the void space, m/s

The corresponding fraction factor for the packed bed can then arbitrarily be defined as

$$f_0 \equiv \left(\frac{-\Delta\mathscr{P}}{L}\right)\frac{D_p}{4\varrho u_0^2} = \left(\frac{u_\infty}{u_0}\right)^2\frac{D_p A_p n}{4}\bar{C}_t \tag{19.47}$$

where u_0 = superficial velocity, m/s
A_p = projected area of a sphere, m^2
D_p = diameter of a sphere, m

For a rhombohedral packing

$$n = \frac{\sqrt{2}}{D_p^3}, \qquad A_p = \frac{\pi D_p^2}{4}$$

and

$$\varepsilon = 1 - \frac{\pi\sqrt{2}}{6} = 1 - \frac{\pi}{3\sqrt{2}} = 0.2595$$

Based on the minimal cross section open to flow, Ranz proposed letting

$$\frac{u_0}{u_\infty} = 1 - \frac{\pi}{2\sqrt{3}} = 0.0931$$

Substituting from these several expressions in 19.47 gives

$$f_0 = 32.04\,\bar{C}_t \tag{19.48}$$

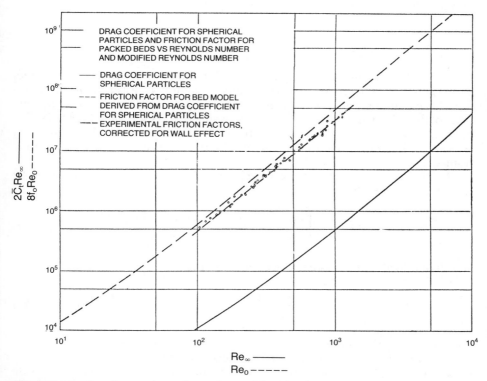

FIGURE 19–19 *Comparison of experimental data for the pressure drop across a packed bed of spheres with Equation 19.49. (From Ranz [45].)*

or

$$f_0 \text{Re}_0^2 = 0.278 \bar{C}_t \text{Re}_\infty^2 \qquad (19.49)$$

where

$$\text{Re}_0 = \frac{D_p u_0 \varrho}{\mu} \qquad (19.50)$$

and

$$\text{Re}_\infty = \frac{D_p u_\infty \varrho}{\mu} \qquad (19.51)$$

In Figure 19–19 Ranz compares experimental values of $2f_0 \text{Re}_0^2$ versus Re_0 with values computed from Equation 19.49, using the indicated curve for $2\bar{C}_t \text{Re}_\infty^2$ versus Re_∞ (see problem 35). The prediction gives the correct trend and is less than 10% too high. This model has perhaps less theoretical justification than Equation 19.41, and may be less accurate and general, but does demonstrate that relatively successful correlations can sometimes be developed by radically different and incompatible approaches.

Numerical Calculations

Lahbabi and Chang [46] used the Galerkin method to reduce the three-dimensional Navier–Stokes and continuity equations to a set of ordinary differential equations, which were then solved numerically for low and intermediate rates of flow through a simple cubic array of spheres. As shown in Figure 19–20, the results agree reasonably well with the Ergun equation up to $D_p u_0 \varrho / \mu (1 - \varepsilon)$ $\cong 300$; at higher values numerical instability apparently resulted in erratic results.

Chandrasekhara and Vortmeyer [8] generalized the Ergun equation to the following vectorial form:

$$-\nabla \mathscr{P} = \mathbf{v}_0 \left[\frac{150(1 - \varepsilon)^2 \mu}{\varepsilon^3 D_p^2} + \frac{1.75(1 - \varepsilon) \varrho |\mathbf{v}_0|}{\varepsilon^3 D_p} \right] \qquad (19.52)$$

and then added a viscous-resistance term to obtain

$$-\nabla \mathscr{P} = \mathbf{v}_0 \left[\frac{150(1 - \varepsilon)^2 \mu}{\varepsilon^3 D_p^2} + \frac{1.75(1 - \varepsilon) \varrho |\mathbf{v}_0|}{\varepsilon^3 D_p} \right] - \frac{\mu}{\varepsilon} \nabla^2 \mathbf{v}_0 \qquad (19.53)$$

The latter term is asserted to be necessary to account for the viscous stress near the containing walls. Equation 19.53 reduces to the Stokes–Darcy equation, Equation 19.16, for sufficiently low velocities such that the term on the right side within the brackets becomes negligible.

Chandrasekhara and Vortmeyer solved Equation 19.53 numerically for

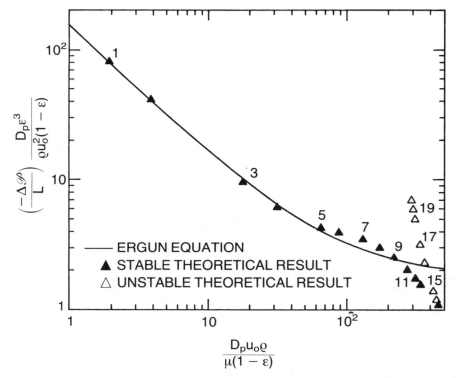

FIGURE 19–20 *Comparison of numerical solutions for the pressure drop through a cubic array of spheres with Equation 19.41. The numerals identify the parameters of the numerical procedure. (From Lahabi and Chang [46].)*

several representative conditions. Their computed results for the conditions in the legend of Figure 19–21, and using Equation 19.8 with $b = 1$, $c = 3$, and $\varepsilon_0 = 0.4$ for the distribution of the void fraction, were found to be in serious disagreement with the experimental results of Schwartz and Smith [47], which are represented by the continuous curved line, yielding a maximum value four times the observed value and located virtually at the wall. Therefore, based on the observation that the peak in velocity was actually at $y \cong D_p/4$, they modified Equation 19.8 as follows:

$$\varepsilon = \varepsilon_0[1 + be^{-c(1/4-y/D_p)}] \tag{19.54}$$

Using $b = 0.4$ and $c = 0.93$ then yielded the dashed curve in Figure 19–21. The predicted peak is nearer the wall than the observed value, but the magnitude is in satisfactory agreement. Better agreement is demonstrated for the experimental results of Schertz and Bischoff [48] in Figure 19–22. Chandrasekhara and Vortmeyer specifically tested the effect of the inertial term in Equation 19.53 and found it to be very significant at $\mathrm{Re}_p(1 - \varepsilon) = 100$ but barely significant at $\mathrm{Re}_p(1 - \varepsilon) = 20$. The effect of the inertial term is to lower the peak velocity.

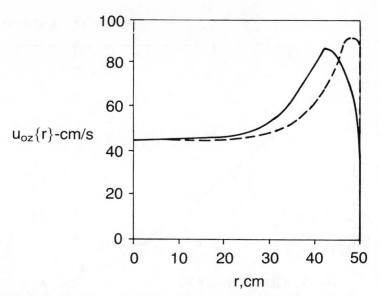

FIGURE 19–21 *Comparison of the modified numerical solution of Chandrasekhara and Vortmeyer for the averaged distribution of the velocity near the wall of a cylindrical bed of uniformly-sized spheres with the experimentally observed distribution of Schwartz and Smith for $Re_D = 190$, $D_t = 10$ cm, $d_p = 0.635$ cm, $\varepsilon_0 = 0.40$, and $u_0 = 65.23$ cm/s: ——— observed; ———— Equation 19.54 with $b = 0.4$ and $c = 0.93$. (From Chandrasekhara and Vortmeyer [8].)*

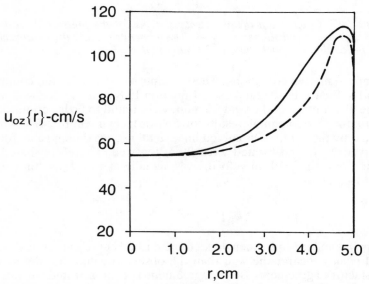

FIGURE 19–22 *Comparison of the modified numerical solution of Chandrasekhara and Vortmeyer for the averaged distribution of the velocity near the wall of a cylindrical bed of uniformly sized spheres with the experimental observations of Schertz and Bischoff for $Re_0 = 277$, $D_t = 10$ cm, $D_p = 0.761$ cm, and $\varepsilon_0 = 0.42$: ——— observed; ———— Equation 19.54 with $b = 0.4$ and $c = 0.93$. (From Chandrasekhara and Vortmeyer [8].)*

MISCELLANEOUS EFFECTS

Compressible Flow

The Ergun equation can be rewritten for compressible flow as

$$\frac{\varepsilon^3 \varrho D_p}{(1 - \varepsilon) G_0^2}\left(-\frac{dp}{dx}\right) + \frac{\varepsilon^3 \varrho^2 D_p g_x}{(1 - \varepsilon) G_0^2} + \frac{D_p}{(1 - \varepsilon)\varrho}\left(\frac{d\varrho}{dx}\right) \qquad (19.55)$$

$$= 150\frac{\mu(1 - \varepsilon)}{D_p G_0} + 1.75$$

where $G_0 = \varrho u_0$. Equation 19.45 can be integrated for horizontal isothermal flow of an ideal gas ($\varrho = pm/RT$) from p_1, ϱ_1 at $x = 0$ to p_2, ϱ_2 at $x = L$ to obtain

$$\frac{\varepsilon^3 \varrho_1 D_p}{(1 - \varepsilon) G_0^2 p_1}\left(\frac{p_1^2 - p_2^2}{2}\right) = \frac{D_p}{1 - \varepsilon}\ln\left\{\frac{p_1}{p_2}\right\}$$

$$+ \left[\frac{150\mu(1 - \varepsilon)}{DG_0} + 1.75\right]L \qquad (19.56)$$

Slip Flow

For the flow of a gas at high velocity the mean free path of the molecules may exceed the dimensions of the pores, resulting in slip flow and deviations from Equation 19.14.

Wall Effects

Mehta and Hawley [49] proposed to correct the Ergun equation to account for the drag of the confining wall simply by including the surface area of the wall in the definition of D_h. Then

$$D_h = 4\left(\frac{\text{volume of voids}}{\text{volume of packing}}\right)\left(\frac{\text{surface area of packing plus wall}}{\text{volume of packing}}\right)^{-1}$$

$$= \frac{4\varepsilon}{a_V + (4/D_t)} \qquad (19.57)$$

Substituting from Equation 19.2 for a_V gives

$$D_h = \frac{2\varepsilon D_p}{3(1 - \varepsilon) + 2(D_p/D_t)} = \frac{2\varepsilon D_p}{3(1 - \varepsilon)M} \qquad (19.58)$$

where

$$M = 1 + \frac{2}{3}\frac{D_p}{D_t(1 - \varepsilon)} \qquad (19.59)$$

Then D_p/M is simply substituted for D_p in Equation 19.41. Equation 19.59 suggests that in the viscous regime with $\varepsilon = 0.4$, $D_t/D_p < 23$ will result in a 10% decrease in flow over that predicted by Equation 19.23.

Cohen and Metzner [7] note that the equivalent correction, except for the factor of 2/3, was derived earlier by Coulson [50], Carman [6], and others, and, subsequently, by Dolejs [51]. They developed a more detailed but less explicit, three-zone model that gives somewhat better agreement with experimental data for different D_p/D_t than does Equation 19.59.

SUMMARY

Despite the geometric variety and complexity of porous media, theoretical considerations have proven to be more helpful in developing expressions for the dependence of the pressure drop on the rate of flow, the fluid properties, and even the properties of the porous media than might have been expected.

Equations 19.23 and 19.31, which have a complete theoretical structure, except for the coefficients themselves, provide reasonable predictions for the limiting behavior for small and high velocities, respectively, for a wide range of parametric conditions.

Equation 19.41, which results from the arbitrary addition of these two limiting expressions, provides an equally good prediction for intermediate rates of flow. The success of this additive form (i.e., the absence of a discrete transition) supports the hypothesis that the dependence of the pressure drop on the square of the velocity is a consequence of inertial effects rather than of turbulence.

Data for a particular packing can be correlated even more precisely by Equation 19.37, which eliminates the main sources of error in the predictions of Equation 19.35, namely, the postulated coefficients and dependence on ε and D_p. On the other hand, Equation 19.37 has limited value for a priori prediction.

The pressure drop in flow through porous media is very sensitive to the value of ε. Hence ε must be determined with high accuracy directly from mass or volume balances or indirectly from measurements of the pressure drop. D_p must also be determined indirectly from measurements of the pressure drop for irregular particles or consolidated media.

Wall effects may be important for values of $D_t/D_p < 23$. Slip flow may be important for flow through fine pores.

Although most of the expressions are for incompressible flow, compressibility is easily taken into account, as illustrated by Equations 19.55 and 19.56.

PROBLEMS

1. Can the void fraction ε be determined reliably by measuring the volume of liquid required to fill a section of porous media? Explain.
2. Determine a_S and a_V for cylinders with height equal to

   a. the diameter
   b. half the diameter
   c. twice the diameter

3. Determine D_p for cylinders with height equal to

   a. the diameter
   b. half the diameter
   c. twice the diameter

4. Determine the sphericity of cylinders with height equal to

   a. the diameter
   b. half the diameter
   c. twice the diameter

5. What other shapes of containers besides square and hexagonal could be used to give regular arrangements near the wall for the orientations in Figure 19–2?

6. Will the void fraction be different near the wall for square containers with arrangements 1, 2, 3, and 8 of Figure 19–2? If so, what could be done to minimize this variation?

7. Compare the dependence predicted by Equations 19.23, 19.25, and 19.27 with the data of Figure 19–10.

8. Derive expressions for the permeability defined by Equation 19.14 corresponding to Equations 19.25 and 19.27.

9. Evaluate the coefficient A in Equation 19.44

   a. to fit the data of Figure 19–7
   b. to yield the same pressure drop as Equation 19.41 for $\varepsilon = 0.4$ and large Re_0, compare the resulting expression with the data of Figure 19–7

10. Compare the dependence on ε predicted by Equations 19.44, with an appropriate value of A, 19.41, and 19.43 for varying Re_0.

11. Compare Equations 19.41 and 19.43 with the data of Figure 19–7.

12. Add Equation (10) of Gibilaro et al. [52] to the comparisons of problem 10.

13. Compare Equation (10) of Gibilaro et al. [52] with the data of Figure 19–7.

14. The pressure gradient for horizontal flow of water at 21°C through a bed of crushed rock is found to be 3.93 kPa/m for a superficial water velocity of 1.83 mm/s. The void fraction is found to be 0.42. What is the effective diameter of the crushed rock as defined by Equation 19.41?

15. Data obtained for the pressure drop of air ($\mu = 17.8 \ \mu Pa \cdot s$), $\varrho = 1.28$ kg/m^3 through a packed bed appear to yield a straight line of slope 322 m^2/kg when $(-\Delta \mathcal{P})/LG_0$ in s^{-1} is plotted versus G_0 (kg/m$^2 \cdot$ s) in Cartesian coordinates. The porosity is determined to be 0.43.

   a. What is the effective particle diameter as defined by Equation 19.41?
   b. Could the effective particle diameter also be determined from the intercept of the plot? Would you expect the same answer? Explain.

16. Data are obtained for the pressure drop as a function of the rate of flow of air through a bed of crushed rock of relatively uniform size. Explain how to prepare a working plot of pressure drop versus the rate of flow for beds of the same material but different particle size.

17. The measured pressure drop across a 0.9-m-high bed of packing is 42 kPa for an upward rate of flow of 1.36 kg water/m$^2 \cdot$ s at 21°C. The void fraction is 0.39. The packing is retained in place with screens. Calculate the pressure drop for a fourfold increase in flow.

18. The following measurements were obtained for flow of air through a bed of crushed coke with a void fraction of 0.493.

u_0 (mm/s)	$-\Delta \mathscr{P}$ (kPa)
17.7	2.07
177.3	59.4

Predict the pressure drop to be attained at a superficial velocity of 121 mm/s if the porosity decreases to 0.402 owing to settling and compaction.

19. Calculate the pressure drop for the flow of air at 20°C and 0.27 MPa at a velocity of 9 m/s through a 0.6-m length of a packed bed composed of 6.35 mm spheres, if

a. randomly packed ($\varepsilon = 0.420$)
b. in a cubic-stacked array ($\varepsilon = 0.476$)
c. in an orthorhombic-stacked array ($\varepsilon = 0.3954$)
d. in a rhombohedral-stacked array ($\varepsilon = 0.2595$)

Compare the calculated values with the experimental results of Martin et al. [5].

20. In order to reduce the pressure drop through a bed of randomly packed and randomly mixed spheres, the spheres are screened and then repacked systematically with respect to size (i.e., with the sphere size increasing regularly from top to bottom). The size distribution found by screening is given below. The original bed was 0.3 m in diameter and 3 m high. Estimate the increase in bed height and percentage decrease in pressure drop obtained by the rearrangement.

Screen Mesh	D_{ave} (mm)	Mass fraction
−0.441 + 0.525	2.27	0
−0.371 + 0.441	0.31	0.01
−0.312 + 0.371	8.69	0.02
−2½ + 3	7.32	0.05
−3 + 3½	6.15	0.22
−3½ + 4	5.16	0.32
−4 + 5	4.34	0.28
−5 + 6	3.66	0.07
−6 + 7	3.07	0.03
−7 + 8	2.59	0

21. Turbid water is clarified by pumping it through a bed of Ottawa sand 0.6 m deep. The sand has a void fraction of 0.40 and an effective diameter of 0.25 mm. Calculate the capacity in m^3/m$^2 \cdot$ s.

22. Calculate the capacity of the filter in problem 21 if the water flows under gravity with the liquid level maintained just above the sand.

23. Estimate the permeability of the packings in problems 14, 15, 17, and 18.

24. Oil flows radially through a very large porous structure to a spherical cavity from which it flows through a vertical pipe to the surface. Estimate the effect of doubling the diameter of the cavity on the rate of production of the well. (The pressure of the formation at a large distance from the cavity and the pressure at the cavity may be assumed to be unchanged.)

25. A rotary vacuum filter is used to remove the solids from an aqueous slurry. Calculate the rate of production of filtrate at a point where the filter cake is 25 mm thick. The permeability of the cake is 1.1×10^{-10} m^2 and the operating vacuum is 4 in. Hg.

26. An oil well in a large bed of sand terminates in spherical cavity. The pressure in the bed far from the well is 13.7 MPa. Calculate the rate of flow in kilograms per second if the pressure at the cavity is maintained at 3.45 MPa. Is the pressure drop determined primarily by viscous or inertial losses? Use $\mu = 1.0$ mPa·s, $\varrho = 0.93$ Mg/m³, $\varepsilon = 0.34$, $D_p = 0.25$ mm.

27. The void fraction of a packed bed decreases from 0.45 to 0.36 due to settling. Estimate the resulting fractional increase in pressure drop for the same mass rate of flow

    a. in the creeping regime
    b. in the purely inertial regime

28. Derive expressions for the superficial velocity field and the dynamic pressure distribution for radial flow to a point sink in an infinite bed of porous material. (*Hint*: See Chapter 10.) Adapt this solution to predict the mass rate of flow to a spherical well of radius r_0 and dynamic pressure $\mathscr{P}_0$ from an oil field at $\mathscr{P}_\infty$.

29. Repeat problem 30 for a line sink and a cylindrical well.

30. Derive expressions for the superficial velocity field and dynamic pressure distribution for flow through porous media from a small spherical source at a pressure of $\mathscr{P}_0$ to an infinite plane at a pressure of $\mathscr{P}_\infty$.

31. Repeat problem 30 for a cylindrical source with its axis parallel to the plane.

32. A tarry material equal in volume to 0.01% of the feed is deposited on the surface of a catalyst for the polymerization of propylene. The increased resistance to flow due to the deposit eventually requires replacement of the catalyst. The catalytic bed consists of 12 m of 6-mm by 6-mm cylinders and has an initial porosity of 38%. A reciprocating pump forces the feed stream down through the bed at a rate of 175 kg/m²·s. A mean viscosity of 10^{-5} kg/m·s and a mean density of 240 kg/m³ may be used for the fluid. The tar may be assumed to be deposited uniformly throughout the bed. Estimate the operating time before the pressure drop exceeds 0.7 MPa.

33. Determine the sphericity of Raschig rings with $\varrho = 62.37$ lb/ft³, $\varepsilon = 0.845$, $D_0 = 0.484$ in., $L = 0.449$ in., and $\delta = 0.030$ in.

34. Thirty minutes are required for the level of water to fall from 25 to 50 mm above a 0.60-mm-thick bed of soil supported by a screen. The test is initiated by opening a valve just below the screen. What is the permeability of the soil if $\varepsilon = 0.22$ and $\varrho_s = 1600$ kg/m³?

35. Prepare a plot of $2 \bar{C}_t \mathrm{Re}_\infty^2$ versus Re_∞ using Equation 16.62. Compare with the curve in Figure 19–19.

REFERENCES

1. F. A. L. Dullien, "Single Phase Flow through Porous Media and Pore Structure," *Chem. Eng. J.*, *10* (1975) 1.
2. R. A. Greenkorn, "Steady Flow through Porous Media," *AIChE J.*, *27* (1981) 529.
3. R. A. Greenkorn, "Single-Fluid Flow through Porous Media," Chap. 11 in *Handbook of Fluids in Motion*, N. R. Cheremisinoff and R. Gupta, Eds., Ann Arbor Science, Ann Arbor, MI (1983).
4. G. G. Brown and Associates, *Unit Operations*, John Wiley, New York (1950).
5. J. J. Martin, W. L. McCabe, and C. C. Monrad, "Pressure Drop Through Stacked Spheres—The Effect of Orientation," *Chem. Eng. Progr.*, *41*, No. 2 (1951) 91.
6. P. C. Carman, "Fluid Flow through Granular Beds," *Trans. Inst. Chem. Engr. (London)*, *15* (1937) 150.
7. Y. Cohen and A. B. Metzner, "Wall Effects in Laminar Flow of Fluids through Packed Beds," *AIChE J.*, *27* (1981) 705.
8. B. C. Chandrasekhara and D. Vortmeyer, "Flow Model for Velocity Distribution in Fixed Porous Beds under Isothermal Conditions," *Wärme- und Stoffübertragung*, *12* (1979) 105.
9. R. F. Benenati and C. B. Brosilow, "Void Fraction Distribution in Packed Beds," *AIChE J.*, *8* (1962) 359.
10. H. P. G. Darcy, *Les fontaines publiques de la ville de Dijon*, Victor Dalmont, Ed., Paris (1856).
11. J. C. Slattery, "Single-Phase Flow through Porous Media," *AIChE J.*, *15* (1969) 866.
12. R. G. Larson, "Derivation of Generalized Darcy Equations for Creeping Flow in Porous Media," *Ind. Eng. Chem. Fundam.*, *20* (1981) 132.
13. R. B. Bird, W. E. Stewart, and E. N. Lightfoot, *Transport Phenomena*, John Wiley, New York (1960).
14. H. C. Brinkman, "A Calculation of the Viscous Force Exerted by a Flowing Fluid on a Dense Swarm of Particles," *Appl. Sci. Res.*, *A1* (1947) 27.
15. C. K. W. Tam, "The Drag on a Cloud of Spherical Particles in Low Reynolds Number Flow," *J. Fluid. Mech.*, *38* (1969) 537.
16. T. S. Lundgren, "Slow Flow through Stationary Random Beds and Suspensions of Spheres," *J. Fluid Mech.*, *51* (1972) 273.
17. M. I. S. Azzam and F. A. L. Dullien, "Calculation of the Permeability of Porous Media from the Navier–Stokes Equation," *Ind. Eng. Chem. Fundam.*, *15* (1976) 281.
18. A. J. E. J. Dupuit, *Etudes théoretiques et pratiques sur le mouvement des eaux*, Paris (1863).
19. F. E. Blake, "The Resistance of Packing in Fluid Flow," *Trans. Amer. Inst. Chem. Engr.*, *14* (1921–1922) 415.
20. S. Ergun, "Fluid Flow through Packed Columns," *Chem. Eng. Progr.*, *48* (1952) 89.
21. J. Kozeny, "Über kapillare Leitung des Wassers im Boden," *S. Ber. Weiner Akad*, Abt. IIa, *136* (1927) 271.
22. P. U. Foscolo, L. B. Gibilaro, and S. P. Waldram, "A Unified Model for Particulate Expression of Fluidized Beds and Flow in Fixed Porous Media," *Chem. Eng. Sci.*, *38* (1983) 1251.
23. H. Rumpf and A. R. Gupte, "Einflüsse der Porosität und Korngrössenverteilung im Widerstandsgesetz der Porenströmung," *Chem.-Ing.-Tech.*, *43* (1971) 367.

24. C. P. Kyan, D. T. Wasan, and R. C. Kintner, "Flow of Single-Phase Fluids through Fibrous Beds," *Ind. Eng. Chem. Fundam.*, *9* (1970) 596.
25. L. J. Snyder and W. E. Stewart, "Velocity and Pressure Profiles for Newtonian Creeping Flow in Regular Packed Beds of Spheres," *AIChE J.*, *12* (1966) 167.
26. B. A. Bakhmateff and N. V. Feodoroff, "Flow through Porous Media," *J. Appl. Mech.*, *Trans. ASME*, *59* (1937) A97.
27. S. P. Burke and W. B. Plummer, "Gas Flow through Packed Columns," *Ind. Eng. Chem.*, *20* (1928) 1196.
28. N. de Nevers, *Fluid Mechanics*, Addison-Wesley, Reading, MA (1970), p. 384.
29. S. W. Churchill, *The Practical Use of Theory in Fluid Flow. Book IV. Turbulent Flows*, Notes, The University of Pennsylvania (1981).
30. P. Forchheimer, "Wasserbewegung durch Boden," *Z. Ver. Deut. Ing.*, *45* (1901) 1781.
31. N. Ahmed and D. K. Sunada, "Nonlinear Flow in Porous Media," *J. Hyd. Div.*, *ASCE*, *95* (1969) 1847.
32. F. A. L. Dullien and M. I. S. Azzam, "Flow Rate-Pressure Gradient Measurement in Periodically Non-Uniform Capillary Tubes," *AIChE J.*, *19* (1972) 222.
33. L. Green and P. Duwez, "Fluid Flow through Porous Metals," *Trans. ASME*, *73*, *J. Appl. Mech.*, *18* (1950) 39.
34. T. F. Macdonald, M. S. El-Sayer, K. Mow, and F. A. L. Dullien, "Flow through Porous Media—The Ergun Equation Revisited," *Ind. Eng. Chem. Fundam.*, *18* (1979) 199.
35. C. R. Dudgeon, "An Experimental Study of the Flow of Water through Coarse Granular Media," *Houille Blanche*, *21* (1966) 785.
36. G. H. Fancher and J. A. Lewis, "Flow of Simple Fluids through Porous Media," *Ind. Eng. Chem.*, *25* (1933) 1139.
37. T. H. Chilton and A. P. Colburn, "Pressure Drop in Packed Tubes," *Trans. Amer. Inst. Chem. Engr.*, *26* (1931) 128.
38. H. E. Rose, "Flow of Liquids through Beds of Granular Materials," *Engineering*, *148* (1939) 536.
39. L. E. Brownell and D. L. Katz, "Flow of Fluids through Porous Media—Part I," *Chem. Eng. Progr.*, *43* (1947) 537.
40. J. A. Talmadge, "Packed Bed Pressure Drop—An Extension to Higher Reynolds Numbers," *AIChE J.*, *16* (1970) 1092.
41. S. Ergun, "Determination of the Geometric Surface Area of Crushed Porous Solids—Gas Flow Method," *Anal. Chem.*, *24* (1952) 388.
42. S. Ergun, "Determination of Particle Density of Crushed Porous Solids—Gas Flow Method," *Anal. Chem.*, *23* (1951) 151.
43. S. W. Churchill, *The Interpretation and Use of Rate Data—The Rate Process Concept*, rev. printing, Hemisphere, Washington, D.C. (1979), pp. 269–271.
44. S. W. Churchill and R. Usagi, "A General Expression for the Correlation of Rates of Transfer and other Phenomena," *AIChE J.*, *18* (1972) 1121.
45. W. E. Ranz, "Friction and Transfer Coefficients for Single Particles and Packed Beds," *Chem. Eng. Progr.*, *48* (1952) 247.
46. A. Lahbabi and H.-C. Chang, "High Reynolds Number Flow through Cubic Arrays of Spheres—Steady-State Solution and Transition to Turbulence," *Chem. Eng. Sci.*, *40* (1985) 435.
47. C. E. Schwartz and J. M. Smith, "Flow Distribution in Packed Beds," *Ind. Eng. Chem.*, *45* (1953) 1209.
48. W. M. Schertz and K. B. Bischoff, "Thermal and Material Transport in Non-Isothermal Packed Beds," *AIChE J.*, *15* (1969), 597.
49. D. Mehta and M. C. Hawley, "Wall Effects in Packed Columns," *Ind. Eng. Chem. Proc. Des. Dev.*, *8* (1969) 280.
50. J. M. Coulson, "The Flow of Fluids through Granular Beds: Effects of Particle Shape and Voids," *Trans. Inst. Chem. Engr.*, *(London)*, *27* (1949) 237.

51. V. Dolejs, *Chemický Průmysl*, *27* (1977) 275; English transl. "Pressure Drop in Viscous Flow of Newtonian Fluid through a Fixed Random Bed of Spherical Particles," *Int. Chem. Eng.*, *18* (1978) 718.
52. L. G. Gibilaro, R. DiFelice, and S. P. Waldram, "Generalized Friction Factor and Drag Coefficient Correlations for Fluid-Particle Interactions," *Chem. Eng. Sci.*, *40* (1985) 1817.

Chapter 20

The Relative Motion of Fluids and Dispersed Solids

The relative motion of fluids and fully dispersed solids is of great importance in separations, including classification and sedimentation; in solid–fluid contacting for drying, regenerative heat exchange, and surface-catalyzed chemical conversions; and in the transport of solids. The role of theory in such fluid–solid motions is somewhat limited because of geometrical complexities and randomness, but even so is worth examining as a structure for correlation and interpolation.

SINGLE PARTICLES

The drag coefficient of a single *fixed*, solid sphere was examined in Chapter 16, that of fluid spheres and spheroids in Chapter 17, and that of fixed, regular solids in Chapter 18. As noted, the drag coefficient for the free motion of a sphere (falling, rising, or suspended) may not be the same as for a fixed sphere because of the asymmetry of the shedding of the eddies. This difference is much more significant for nonspherical particles since they may assume a preferred orientation or may rotate and/or oscillate as they move. Such rotation and/or oscillation may in turn produce an irregular motion. Most of the solid particles encountered in practical applications have not only a nonspherical but also an irregular shape.

Experimental drag coefficients for the *free settling* at terminal velocity, u_T, of a number of *solid* objects are plotted versus Re_T in Figure 20–1. These drag coefficients are defined per Equation 16.66. For the geometrically regular particles the volume-equivalent diameter

$$D_V = \left(\frac{6V}{\pi}\right)^{1/3} \tag{20.1}$$

where V is the volume of particle in cubic meters, was used for D in Equation 16.66, thereby implying a projected area of $\pi D_V^2/4$. For the irregular particles (crushed quartz) of Figure 20–1 the average diameter obtained by screening, D_s, was used with Equation 16.66 to calculate the values of $\bar{C}_t$ in Figure 20–1, which implies a volume of $\pi D_s^3/6$ and a projected area of $\pi D_s^2/4$. D_V and D_s were correspondingly used in the Reynolds number, $Re_T = Du_T\varrho/\mu$, as well.

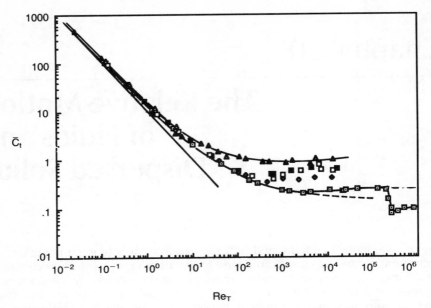

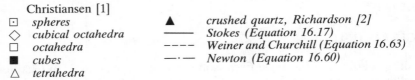

FIGURE 20–1   *Experimental drag coefficient of freely falling particles:*

Christiansen [1]
⊡	*spheres*	▲	*crushed quartz, Richardson [2]*
◇	*cubical octahedra*	——	*Stokes (Equation 16.17)*
□	*octahedra*	----	*Weiner and Churchill (Equation 16.63)*
■	*cubes*	—·—	*Newton (Equation 16.60)*
△	*tetrahedra*		

Experimental data for spheres are included in Figure 20–1, and Equations 16.17, 16.60, and 16.63 for fixed spheres are plotted for reference. Data are most commonly plotted in the coordinates of Figure 20–1 since Re_T is the criterion for transition between different regimes of flow and since $\bar{C}_t$ has a fairly limited range of variation. The advantages of the form of Figure 20–1 are, however, somewhat countered by inconvenience in that the primary dependent variable, the terminal velocity, u_T, appears in both coordinates. These same data are therefore replotted in Figure 20–2 in the form suggested in Chapter 16—as $Du_T\varrho/u = Re_T$ versus $Ar = g\varrho D^3\Delta\varrho/\mu^2 = (3/2)\bar{C}_t Re_T^2$, thus retaining Re_T while eliminating u_T from the other coordinate.

The data in Figures 20–1 and 20–2 are further replotted in Figure 20–3, again as suggested in Chapters 16 and 17, as

$$u_T^* = u_T\left(\frac{\varrho^2}{g\mu\Delta\varrho}\right)^{1/3} = \left(\frac{2}{3}\frac{Re_T^2}{\bar{C}_t}\right)^{1/3}$$

versus

$$D^* = D\left(\frac{g\varrho\Delta\varrho}{\mu^2}\right)^{1/3} = \left(\frac{3}{2}\bar{C}_t Re_T^2\right)^{1/3} = Ar^{1/3}$$

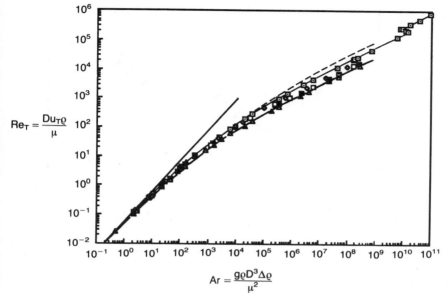

$$Re_T = \frac{Du_{T}\varrho}{\mu}$$

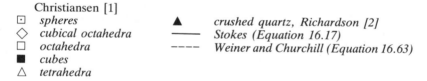

$$Ar = \frac{g\varrho D^3 \Delta\varrho}{\mu^2}$$

FIGURE 20–2   *Reynolds number of freely falling particles as a function of the Archimedes number:*

Christiansen [1]
⊡   *spheres*   ▲   *crushed quartz, Richardson [2]*
◇   *cubical octahedra*   ⸺   *Stokes (Equation 16.17)*
☐   *octahedra*   ----   *Weiner and Churchill (Equation 16.63)*
■   *cubes*
△   *tetrahedra*

in order to confine both the terminal velocity and the diameter to separate groups and thereby provide an explicit relationship between the primary dependent and independent variables. This plot implies the existence of three possible terminal velocities for $1090 < D^* < 2500$. The intermediate values that correspond to a decrease in velocity with increasing diameter would be expected to be unstable if the diameter changed dynamically, as is possible for a bubble. This prediction of instability is a particular advantage of these coordinates.

It is evident that different sets of coordinates may each have particular advantages and that in important applications several of them should be used for analysis and correlation.

The separation of solid particles from a liquid by *settling* is called *sedimentation*. The separation of solid particles of different sizes or densities based on their relative rates of terminal settling is called *classification*. Separation by size is called *sizing*, that by density is called *sorting*. *Elutriation* is a laboratory sizing of small solid particles. Insofar as suspensions are and remain so dilute that individual particles do not hinder one another, plots such as Figures 20–1 to 20–3 can be used to predict these separations (see problems 41–43).

Most of the dispersions encountered in industrial practice involve a sufficient concentration of particles such that the motion of one is affected by the

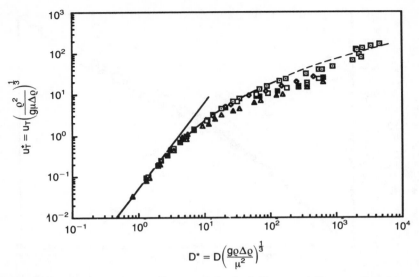

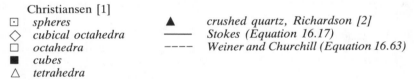

FIGURE 20–3   *Dimensionless velocity of freely falling particles as a function of the dimensionless diameter.*

Christiansen [1]

⊡	*spheres*	▲	*crushed quartz, Richardson [2]*
◇	*cubical octahedra*	——	*Stokes (Equation 16.17)*
□	*octahedra*	- - - -	*Weiner and Churchill (Equation 16.63)*
■	*cubes*		
△	*tetrahedra*		

surrounding ones. The general characteristics of such hindered motion are described in this section. The *quantitative* theoretical relationships derived here are generally based on the postulate of uniformly sized spheres and are thereby limited. However, just as with single particles and packed beds, these relationships are expected to be applicable qualitatively to nonspherical particles of reasonably uniform size, and perhaps semiquantitatively in terms of an efffective diameter such as D_V or D_s. (Care is taken to identify the postulate of uniformly sized spheres whenever it is made in a particular derivation.)

Experimental data for the *mean relative* motion of *liquids*, and dispersions of *coarse*, solid particles of all shapes have long been observed to yield an approximately linear relationship when the superficial velocity is plotted versus the volume fraction of the liquid phase in logarithmic coordinates. Since the relative velocity must approach the terminal velocity of a single solid particle as the volume fraction of liquid approaches unity, this relationship can be expressed as

$$\frac{u_\varepsilon}{u_T} = \varepsilon^n \tag{20.2}$$

where u_ε = mean, relative, superficial velocity between the particles and the fluid in a dispersion, m/s

u_T = terminal velocity of a single particle, m/s

ε = volumetric fraction of fluid, often called the *void fraction* or *porosity*

n = slope of a plot of $\log\{u_\varepsilon\}$ versus $\log\{\varepsilon\}$

The restriction of Equation 20.2 to coarse particles is to exclude the influence of factors such as flocculation and electrical charges.

The behavior represented by Equation 20.2 was apparently first noted in connection with the separation of minerals in a suspension in water by Hancock [3], and in terms of fluidization in the modern sense by Wilhelm and Kwauk [4].

The applicability of Equation 20.2 to both sedimentation and fluidization was noted by Lewis et al. [5] who obtained the data in Figure 20–4 for glass spheres in water. The results for fluidization are seen to overlap those for sedimentation and, for $\varepsilon < 0.8$, to demonstrate the linearity predicted by Equation 20.2. Significant deviations between the results for sedimentation and fluidization, and of both from Equation 20.2 are, however, to be observed for $\varepsilon > 0.8$.

Foscolo et al. [6] recently provided some theoretical rationalization for the form of Equation 20.2 in terms of fluidization. For a fluidized bed with a sufficiently large diameter D_t such that wall effects can be neglected, the total buoyant weight of the particles per unit cross-sectional area of the bed must just be balanced by the *dynamic pressure drop* $-\Delta\mathscr{P}$. Since this weight remains constant as u_ε and ε vary and the bed expands, the dynamic pressure drop must remain constant. That is,

$$-\Delta\mathscr{P}\{u_\varepsilon, \varepsilon\} = \text{constant} \qquad (20.3)$$

Therefore

$$d(-\Delta\mathscr{P}) = \frac{\partial(-\Delta\mathscr{P})}{\partial u_\varepsilon} du_\varepsilon + \frac{\partial(-\Delta\mathscr{P})}{\partial \varepsilon} d\varepsilon = 0 \qquad (20.4)$$

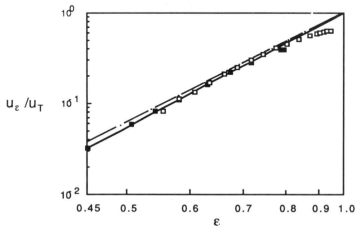

FIGURE 20–4 *Comparison of Equation 20.2 with the experimental data of Lewis et al. [5] for the sedimentation and fluidization of 150-μm glass spheres in water with $Re_T = 2.19$ and $D/D_t = 0.00257$; □, sedimentation; ■, fluidization; —·— Equation 20.2 with n = 4.11 from Table 20.1.*

Foscolo et al. next postulated that

$$-\Delta\mathcal{P}\{u_\varepsilon, \varepsilon\} \propto u_\varepsilon^a \varepsilon^b \tag{20.5}$$

The postulate of such a dependence of $-\Delta\mathcal{P}$ on u_ε is justified, at least in the limits of creeping and purely inertial flow, for which $b = 1$ and 2, respectively. The postulated power dependence of $-\Delta\mathcal{P}$ on ε in this situation is, however, purely conjectural.

From Equations 20.4 and 20.5,

$$\frac{du_\varepsilon}{d\varepsilon} = \frac{-bu_\varepsilon}{a\varepsilon} \tag{20.6}$$

Equation 20.6 can be integrated from $u_\varepsilon = u_T$ at $\varepsilon = 1$ to obtain

$$\frac{u_\varepsilon}{u_T} = e^{-b/a} \tag{20.7}$$

For creeping flow $a = 1$, and, as noted later, $n \cong 4.65$. Hence, $b = -4.65$ and

$$-\Delta\mathcal{P} \propto u_\varepsilon \varepsilon^{-4.65} \tag{20.8}$$

For purely inertial flow $a = 2$, and, as noted later, $n \cong 2.4$. Hence, $b = -4.8$ and

$$-\Delta\mathcal{P} \propto u_\varepsilon^2 \varepsilon^{-4.8} \tag{20.9}$$

A common power dependence of the dynamic pressure drop on the void fraction for all ranges of flow *in packed beds* has indeed been claimed by several investigators (see Chapter 19), thereby giving some credence to this derivation for dispersions.

Dimensional Considerations

Some guidance to the dependence of the exponent n in Equation 20.2 on the other variables governing the hindered motion of particles can be obtained by dimensional analysis.

For a single particle, from Chapter 16,

$$\frac{Dg\Delta\varrho}{\varrho u_T^2} = \phi\left\{\frac{Du_T\varrho}{\mu}\right\} \tag{20.10}$$

where for brevity $\Delta\varrho = |\varrho_s - \varrho|$. From now on, D with no subscript will be used as the characteristic dimension for dispersions of particles, since the ambiguity that exists for packed beds relative to channels and particles is here less of a problem, and for nonspheres D_V or D_s is implied.

For a dispersion of particles the analog of Equation 20.10 is

$$\frac{Dg\Delta\varrho}{\varrho u_\varepsilon^2} = \phi\left[\frac{Du_\varepsilon\varrho}{\mu}, \varepsilon, \frac{D}{D_t}\right] \tag{20.11}$$

Equations 20.10 and 20.11 can be combined functionally as

$$\frac{u_\varepsilon}{u_T} = \phi\left\{\frac{Du_T\varrho}{\mu}, \varepsilon, \frac{D}{D_t}\right\} \tag{20.12}$$

For the creeping regime of flow (that is, for $Re_T \equiv Du_T\varrho/u \to 0$), u_T and u_ε must become independent of ϱ, except of course in $\Delta\varrho$. Hence,

$$\frac{u_\varepsilon}{u_T} = \phi\left\{\varepsilon, \frac{D}{D_t}\right\} \tag{20.13}$$

Conversely, for purely inertial flow (that is, for $Re_T \to \infty$), u_T and u_ε must become independent of μ, and, for this condition as well, Equation 20.12 reduces to 20.13. Thus, u_ε/u_T must depend only on Re_T for intermediate values. It follows, insofar as Equation 20.2 is valid, that the exponent n is also a function only of Re_T for intermediate values.

Richardson and Zaki [7], using a more detailed process of dimensional analysis, were apparently the first to come to that conclusion. On the basis of their analysis and their own experimental data, they constructed a set of correlating equations for n as a function of Re_T and D/D_t *for spheres*. These expressions, as slightly modified by the subsequent work of Richardson and Meilke [8], are listed in Table 20.1. For fluidization they found that

$$u_t = u_{T\infty}e^{-D/D_t} \tag{20.19}$$

where $u_{T\infty}$ is the terminal velocity in meters per second of a particle in a vessel of asymptotically large diameter. For sedimentation, u_T was not found to differ significantly from $u_{T\infty}$.

Table 20.1
Exponent n of Equation 20.2 for the Hindered Motion of Spheres [7, 8]

Re_T	n	Equation Number
< 0.2	$4.65 + 20\dfrac{D}{D_t}$	20.14
0.2–1	$\left(4.4 + 18\dfrac{D}{D_t}\right)Re_T^{-0.3}$	20.15
1–200	$\left(4.4 + 18\dfrac{D}{D_t}\right)Re_T^{-0.1}$	20.16
200–500	$4.4\,Re_T^{-0.1}$	20.17
> 500	2.4	20.18

The validity of Equation 20.2 for any particular set of conditions and, insofar as this validity is established, the value of n is best determined experimentally. In the absence of such data, Equation 20.2 together with Equations 20.14–20.18 can be expected to provide a good first-order prediction for any Re_T and D/D_t. For the conditions of Figure 20–4 ($Re_T = 2.19$ and $D/D_t = 0.00257$) the predicted value of n, as represented by the dashed line, is 4.11 as compared with the least-squares value of 4.31, as represented by the solid line.

Alternative Expression for $u_\varepsilon\{\varepsilon\}$

Foscolo et al. [6] developed an alternative set of expressions for the expansion of a fluidized bed of *uniformly sized spheres*. They started by noting that the rate of dissipation of energy (that is, the work done) in the passage of a stream of fluid through a dispersion is equal to the dynamic pressure drop, $-\Delta\mathscr{P}$, times the cross-sectional area, A_x, times the superficial velocity, u_ε, and also equal to the number of particles, N_p, times the actual relative velocity, $u_\varepsilon/\varepsilon$, times the drag force, F_p, on each particle. Equating these two expressions and canceling out u_ε gives

$$A_x(-\Delta\mathscr{P}) = \frac{N_p F_p}{\varepsilon} \tag{20.20}$$

The number of particles in a dispersion of uniformly sized spheres is equal to the total volume of solids in the dispersion divided by the volume of a sphere; that is,

$$N_p = \frac{6A_x L(1 - \varepsilon)}{\pi D^3} \tag{20.21}$$

If D/D_t is small enough so that wall effects can be neglected, the dynamic pressure drop is equal to the total buoyant weight of the particles *in the liquid*; that is,

$$-\Delta\mathscr{P} = L(1 - \varepsilon)g\Delta\varrho \tag{20.22}$$

(Equation 20.22 is derived rigorously in the next section.) Using Equations 20.21 and 20.22 to eliminate N_p and $-\Delta\mathscr{P}$ from Equation 20.20 gives

$$F_p = \frac{\pi D^3 \varepsilon g \Delta\varrho}{6} \tag{20.23}$$

The right side of Equation 20.23 is *the effective buoyant weight of a single sphere in a dispersion*, which equals the weight of the sphere, $\pi D^3 g\varrho_s/6$, minus the weight of the *dispersion that it displaces*, $\pi D^3 g[\varrho_s(1 - \varepsilon) + \varrho\varepsilon]/6$.

The *creeping flow past an isolated sphere* (from Equation 16.17, which is known as *Stokes' law for the drag coefficient*) is

$$F_p = 3\pi\mu u_\varepsilon D \tag{20.24}$$

whereas for *creeping flow through a packed bed* the pressure drop given by Equation 19.25 can be substituted in Equation 20.20 and N_p eliminated through Equation 20.21 to obtain

$$F_p = \frac{12\pi\mu(1 - \varepsilon)u_\varepsilon D}{\varepsilon^3} \qquad (20.25)$$

For *creeping flow through a dispersion of spheres*, Foscolo et al. suggested interpolating between Equations 20.24 and 20.25 by adding the right sides to obtain

$$F_p = 3\pi\mu u_\varepsilon D \left[1 + \frac{4(1 - \varepsilon)}{\varepsilon^3} \right] \qquad (20.26)$$

This interpolation corresponds to the special case of $p = 1$ in Equation 19.45.
    Eliminating F_p between Equations 20.23 and 20.26 gives

$$u_\varepsilon = \frac{D^2 \varepsilon^4 g \Delta\varrho}{18\mu[4(1 - \varepsilon) + \varepsilon^3]} \qquad (20.27)$$

Letting $\varepsilon = 1$ in Equation 20.27 gives

$$u_T = \frac{D^2 g \Delta\varrho}{18\mu} \qquad (20.28)$$

which is known as *Stokes' law for the terminal velocity*. From the ratio of Equations 20.27 and 20.28,

$$\frac{u_\varepsilon}{u_T} = \frac{\varepsilon^4}{4(1 - \varepsilon) + \varepsilon^3} \qquad (20.29)$$

The effective value of n per Equation 20.29, that is, the slope of a plot of $\log\{u_\varepsilon\}$ versus $\log\{\varepsilon\}$, is

$$n_{\text{eff}} = \frac{\varepsilon}{u_\varepsilon} \frac{du_\varepsilon}{d\varepsilon} = 4 + \frac{\varepsilon(4 - 3\varepsilon^2)}{4(1 - \varepsilon) + \varepsilon^3} \qquad (20.30)$$

According to Equation 20.30, n varies from 4 at $\varepsilon = 0$ to 4.57 at $\varepsilon = 0.4$ to 5 at $\varepsilon = 1$, rather than being invariant, as implied by Equation 20.2. On the other hand, as demonstrated in Figure 20–5, Equation 20.29 agrees remarkably well with Equation 20.2 when using the value of $n = 4.65$ given for creeping flow by Equation 20.14.
    Foscolo et al. proposed an empirical coefficient of 10/3 in place of 4 in Equation 20.29 in order to obtain even better agreement. The dashed line in Figure 20.5 represents this modified version. However, the most significant accomplishment of the derivation of Equation 20.29 is the attainment of a theoretical relationship rather than an expression for quantitative predictions.

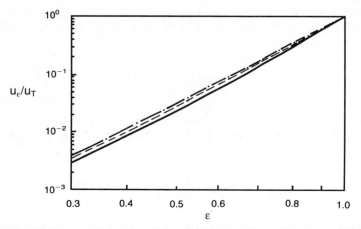

FIGURE 20–5   *Comparison of relationships of Foscolo et al. [6] and Richardson and Zaki [7] for the expansion of a dispersion of uniformly sized spheres in the creeping regime of flow: —·— Richardson (Equation 20.2 with* n = 4.65*); ———— Foscolo et al. (Equation 20.29 with 10/3 replacing 4); ——— Foscolo et al. (Equation 20.29).*

The only empiricism in Equation 20.29 arises from the form of interpolation between Equations 20.24 and 20.25.

The same procedure yields an analogous result for the regime of purely inertial flow. The limiting drag coefficient given by Equation 16.62 for $\text{Re}_T \to \infty$ yields for the drag of an isolated sphere

$$F_p = 0.06125\pi\varrho u_\varepsilon^2 D^2 \tag{20.31}$$

Also, Equation 19.35 can be combined with Equations 20.20 and 20.21, as for creeping flow, to obtain the following expression for the drag on a single sphere in a packed bed:

$$F_p = \frac{0.195\pi\varrho u_\varepsilon^2 D^2 (1 - \varepsilon)}{\varepsilon^3} \tag{20.32}$$

Equations 20.31 and 20.32 can arbitrarily be combined according to Equation 19.45 with $p = 1$ to obtain an expression for F_p for all values of ε; that result can be combined with Equation 20.23 to obtain an expression that can be rearranged as

$$\frac{u_\varepsilon}{u_T} = \frac{\varepsilon^2}{[3.18(1 - \varepsilon) + \varepsilon^3]^{1/2}} \tag{20.33}$$

Equation 20.33 is seen in Figure 20–6 to agree closely with Equation 20.2 for $n = 2.4$, given by Equation 20.18 for large Re_D. Again, the principal ac-

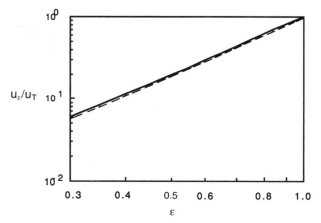

FIGURE 20–6 *Comparison of relationships of Foscolo et al. [6] and Richardson [7] for expansion of a bed of uniformly-sized spheres in the regime of purely inertial flow: ———— Richardson (Equation 20.2 with n = 2.4);* ——— *Foscolo et al. (Equation 20.33).*

complishment of this derivation is the attainment of a semitheoretical expression for the relationship between u_ε and ε in the regime of purely inertial flow. The only empiricism is in the coefficients in the expressions for the drag on an isolated particle and on a sphere in a packed bed, and in the arbitrary form of the interpolation between these expressions.

Finally, the same procedure can be carried out for a complete range of flow, using Equation 19.43 for a sphere in a packed bed and Equation 16.63, which is much simpler and almost as accurate as Equation 16.62 for all $Re_D < 3000$, for an isolated sphere. The result is

$$\frac{u_\varepsilon}{u_T} = \frac{(0.371 + [12/Re_T]^{1/2})\,\varepsilon^2}{[(1 - \varepsilon)(0.78 + 48/Re_T[u_T/u_\varepsilon]) + \varepsilon^3(0.371 + [(u_T/u_\varepsilon)(12/Re_T)]^{1/2})^2]^{1/2}}$$

$$(20.34)$$

Equation 20.34 provides a semitheoretical dependence for u_ε/u_T on Re_T as well as on ε for all $Re_T < 3000$. It converges exactly to Equation 20.29 for $Re_T \to 0$ and would converge exactly to Equation 20.33 for $Re_T \to \infty$ if the coefficient of 0.371 were changed in both places to 0.495. However, Equation 16.63 and therefore the coefficient of 0.371 is believed to provide sufficient accuracy for $Re_T < 3000$ (which exceeds the range of most fluidized beds).

Equation 20.34 must be solved by trial and error. Foscolo et al. derived an analogous but explicit relationship by using an empirical relationship for the dependence of the dynamic pressure drop on the void fraction, which allowed omission of an expression for the drag on an isolated sphere (see problem 9).

Equation 20.34 as plotted in Figure 20–7A for $Re_T = 0$, 1, 10, 100, and ∞ can be compared with the relationships given by Equation 20.2 and Table 20.1 as plotted in Figure 20–7B for the same values of Re_T. The agreement is only fair.

Several conclusions can be drawn concerning Equations 20.29, 20.33, and 20.34:

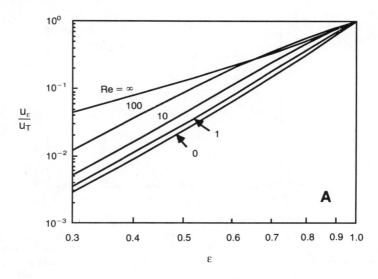

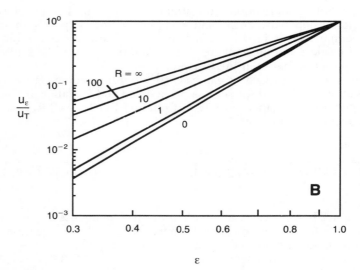

FIGURE 20–7   *Predictions of the expansion of a dispersion of uniformly sized spheres: (A) by Equation 20.34; (B) by Equation 20.2 with n from Table 20.1.*

1. Such complicated functional relationships could hardly be deduced from experimental data.
2. The power-type relationships of Equations 20.2 and 20.15–20.17 are probably artifacts of the use of graphical correlation and have no theoretical basis.
3. More precise data than those now available are needed to assess the merits, if any, of Equations 20.29, 20.33, and 20.34 over Equations 20.2 and 20.14–20.18.

FIGURE 20–8   *Batch sedimentation for a suspension with a narrow range of particle size.*

4. The continued use of Equation 20.2 with 20.14–20.18 is probably justified on grounds of simplicity, particularly when they are used as approximations for nonspherical particles.

BATCH SEDIMENTATION

The interface of a dilute suspension initially falls at the terminal velocity of the smallest particle. As the concentration increases, the rate u_ε decreases in accordance with Equation 20.2.

If the range of particle sizes is not greater than 6/1, a concentrated suspension settles with a sharp interface, as implied by Figure 20–8, and all the particles fall at the same rate; the larger particles are retarded, and the smaller ones are accelerated.

The rate of settling of suspensions of different initial heights is observed to be equal for equal fractions settled. This relationship provides a basis for scaling up laboratory-scale measurements.

FLUIDIZATION OF SOLID PARTICLES

If a fluid is passed upward through a dispersion of solids, the individual particles remain essentially motionless as long as the gravitational force exceeds the drag force for hindered motion at the existing void fraction. In this regime of flow the pressure drop is given by expressions for packed beds, such as Equation 19.41.

Incipient Fluidization

At the point of incipient fluidization the combined weight of the fluid and solids must be supported by the total pressure difference across the bed. That is,

$$(-\Delta p_i)A_x \equiv (-\Delta \mathscr{P}_i + g\varrho L_i)A_x = gL_iA_x[\varrho\varepsilon_i + \varrho_s(1 - \varepsilon_i)] \quad (20.35)$$

where here ϱ_s is the density of the particles of solid. Equation 20.35 can be reduced to

$$-\Delta \mathscr{P}_i = L_i(1 - \varepsilon_i)g(\varrho_s - \varrho) \tag{20.35A}$$

Wilhelm and Kwauk [4] apparently first derived Equation 20.35A. They also noted that the volume of solids $L_i A_x(1 - \varepsilon_i)$ must remain constant as the bed expands; that is,

$$L(1 - \varepsilon) = L_i(1 - \varepsilon_i) \tag{20.36}$$

Hence, Equation 20.35A can be generalized as Equation 20.22. Equations 20.35 and 20.35A invoke several idealizations not yet mentioned: (1) a negligible drag on the wall, corresponding to $D/D_t \to 0$; (2) incompressibility of the fluid; and (3) points of measurement for $-\Delta p$ or $-\Delta \mathscr{P}$ that are both within or both without the dispersion, in order to cancel out the inertial change on entering and exiting. The validity of idealization (1) will be examined subsequently.

The minimum superficial velocity for incipient fluidization u_i could be calculated as a function of the associated void fraction ε_i simply by eliminating $-\Delta \mathscr{P}$ between Equations 20.22 and 19.41 and solving the resulting quadratic to obtain

$$\mathrm{Re}_{pi} = 42.86\left[\left(1 + 3.11 \times 10^{-4}\frac{\varepsilon_i^3 \mathrm{Ar}}{(1 - \varepsilon_i)^2}\right)^{1/2} - 1\right] \tag{20.37}$$

where

$$\mathrm{Re}_{pi} \equiv \frac{Du_i\varrho}{\mu(1 - \varepsilon_i)} \tag{20.38}$$

which is a special case of the effective Reynolds number for a packed bed implied by Equation 19.41 and defined in general by Equation 19.39.

The choice of ε_i for Equations 20.37 and 20.38 poses a major difficulty. Packed beds often have a nonrandom arrangement owing to settling or tapping and a correspondingly lower void fraction than does an incipiently fluidized bed. Then, when the pressure drop slightly exceeds the value given by Equation 20.37 with ε_p substituted for ε_i, the particles rearrange themselves such that the resistance to flow is decreased and a higher velocity is possible with the same pressure drop. As the velocity is further increased the bed expands, the pressure drop remaining constant. If the velocity is then decreased, a decrease in pressure drop will occur at a higher velocity and higher void fraction than before. Such behavior is illustrated in the sketch of idealized behavior in Figure 20–9. In the lower portion (Part B) $-\Delta \mathscr{P}/L(1 - \varepsilon)$ is plotted versus u_ε for the packed-bed regime before and after rearrangement as indicated by the arrows, and $g\Delta\varrho$ is plotted as a constant ordinate. The velocity for incipient fluidization is nominally defined by the intersection of the horizontal line representing pressure drop given by Equation 20.35A, and the lower curve representing the pressure drop in the *rearranged* packed bed. The horizontal line might be inferred from Equation 20.35A to depend on ε_i, but as indicated by Equation 20.36 the

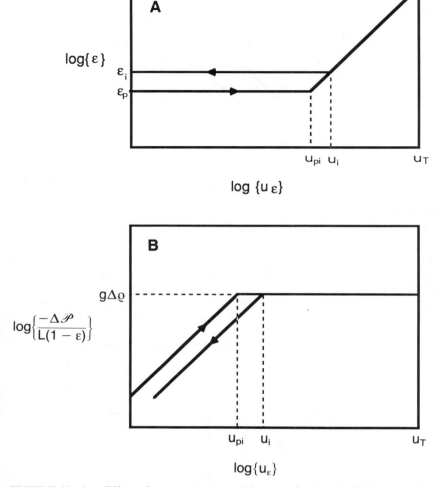

FIGURE 20–9   *Effect of rearrangement of the particles on the fluidization of a mono-dispersion:* u_{pi} = *velocity for rearrangement;* u_i = *velocity for incipient fluidization after rearrangement;* ε_p = *void fraction of packed bed;* ε_i = *void fraction for incipient fluidization after rearrangement; (A) void fraction; (B) pressure drop.*

product of L and $1 - \varepsilon$ is invariant. Likewise, the void fraction for incipient fluidization ε_i is nominally defined in the upper portion (Part A) by the intersection of the upper horizontal line representing the rearranged packed bed and the curve representing expansion of the fluidized bed.

Because of the tendency of particles to interlock, "bridging" may occur, with the consequence that the pressure drop must overcome the frictional forces thereby exerted by the walls of the container as well as weight of the solids. Such behavior results in the "hump" sketched in Figure 20–10.

With nonuniformly sized particles some will fluidize before others, resulting

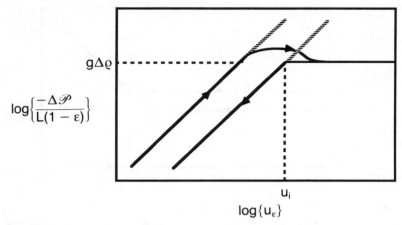

FIGURE 20–10   *Effect of interlocking of the particles on the fluidization of a mono-dispersion:* u_i = *nominal velocity for incipient fluidization.*

in the behavior sketched in Figure 20–11 in which u_{ii} represents the velocity at which fluidization begins and u_{if} is the velocity for complete fluidization. u_i itself is defined as before.

Figures 20–12 and 20–13 illustrate with experimental data of Chen and Keairns [9][1] the behavior sketched in Figures 20–10 and 20–11, respectively,

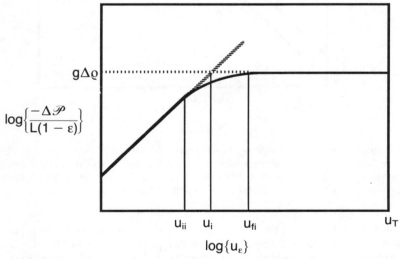

FIGURE 20–11   *Effect of a distribution in particle size on the fluidization of a packed bed:* u_{ii} = *velocity for beginning of fluidization;* u_{fi} = *velocity for complete support of all of the particles;* u_i = *nominal velocity for incipient fluidization.*

---

[1] The paper from which these plots were reproduced won an award for the excellence of the experimental work and its analysis.

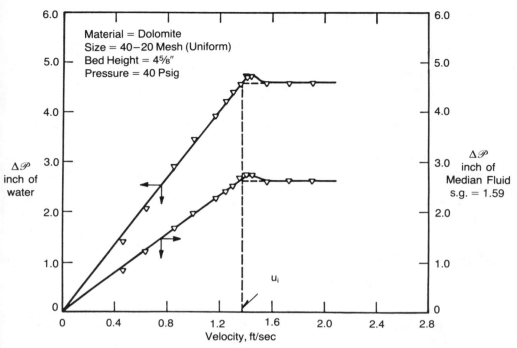

FIGURE 20–12 *Experimental data for fluidization of a monodispersion of interlocked particles with compressed air. (From Chen and Keairns [9].)*

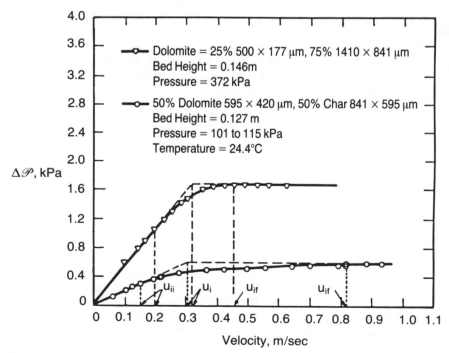

FIGURE 20–13 *Experimental data for the fluidization of two mixtures of two particle sizes with compressed air. (From Chen and Keairns [9].)*

and Figure 20–14 from Shirai [10] the behavior sketched in both Figures 20–9 and 20–10.

Van Heerden et al. [11] and Lewis et al. [5] recommend using a general value of 0.406 for ε_i. The adoption of this value reduces Equation 20.37 to

$$\text{Re}_i \equiv \frac{Du_i\varrho}{\mu} = 25.46[(1 + 5.90 \times 10^{-5}\,\text{Ar})^{1/2} - 1] \qquad (20.39)$$

Since by Equation 16.17, $\text{Ar} = (3/2)\bar{C}_t\,\text{Re}_T^2$, and since $\bar{C}_t = \phi\{\text{Re}_T\}$ (as for example, according to Equation 16.62), Equation 20.39 may be considered to give Re_i as a function of Re_T only.

Wen and Yu [12] arbitrarily adjusted the coefficients of Equation 20.39 in order to fit better a variety of experimental data for both spheres and nonspheres. This adjustment is equivalent to letting $\varepsilon_i = 0.414$ in the factor ε^3 in the viscous term and $\varepsilon_i = 0.382$ in the factor $\varepsilon^3/(1 - \varepsilon)$ in the inertial term of Equation 19.41. Their resulting expression

$$\text{Re}_i = 33.7[(1 + 3.59 \times 10^{-5}\,\text{Ar})^{1/2} - 1] \qquad (20.40)$$

is seen in Figure 20–15 to provide a reasonable representation for these data.

Using a coefficient of 180 corresponding to Equation 19.23 rather than 150 in Equation 19.41 has the effect of revising Equation 20.39 to

$$\text{Re}_i = 30.55[(1 + 4.10 \times 10^{-5}\,\text{Ar})^{1/2} - 1] \qquad (20.41)$$

which differs negligibly from Equation 20.40 for all Ar. This exercise illustrates how two fundamentally different rationalizations can be invoked to achieve similar improvements in the representation of experimental data by a semitheoretical equation.

The apparent success of Equation 20.40 in Figure 20–15, and by implication of Equation 20.41, is remarkable considering the widely differing values of ε_i observed by Leva [13] and compiled in Table 20.2.

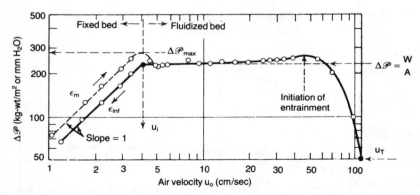

FIGURE 20–14   *Experimental data for fluidization of uniformly sized sand by air demonstrating entrainment due to bubbling as well as interlocking of grains. (From Shirai [10].)*

Table 20.2
Experimentally Observed Void Fractions at Incipient Fluidization, ε_i (from Leva [13])

Particles	ϕ_s	D-mm						
		0.02	*0.05*	*0.07*	*0.10*	*0.20*	*0.30*	*0.40*
Sharp sand	0.67		0.60	0.59	0.58	0.54	0.50	0.49
Round sand	0.86		0.56	0.52	0.48	0.44	0.42	
Mixed round sand					0.42	0.42	0.41	
Coal and glass powder		0.72	0.67	0.64	0.62	0.57	0.56	
Anthracite coal	0.63		0.62	0.61	0.60	0.56	0.53	0.51
Adsorbent carbon		0.74	0.72	0.71	0.69			
Fischer-Tropsch catalyst	0.58				0.58	0.56	0.55	
Carborundum			0.61	0.59	0.56	0.48		

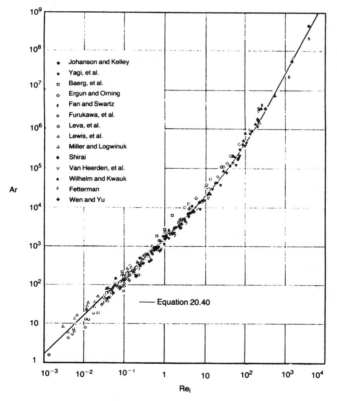

FIGURE 20–15 *Comparison of experimental data for the incipient velocity for fluidization with Equation 20.40. (From Wen and Yu [12].)*

The Mode of Fluidization

Wilhelm and Kwauk [4] observed two distinct modes of fluidization, as illustrated in Figure 20–16. With liquids, a uniform dispersion is maintained

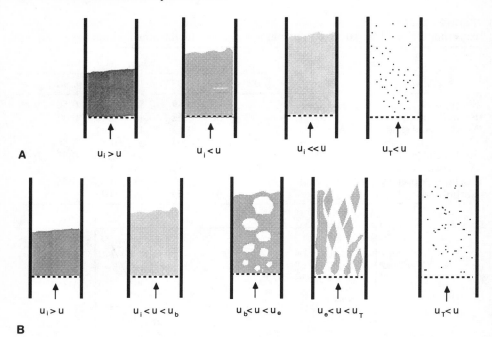

FIGURE 20-16   *Effect of increasing velocity on the behavior of a bed of particles: (A) particulate fluidization; (B) aggregative fluidization.*

as the velocity is increased, and the bed expands according to Equation 20.2, which can be rewritten as

$$\frac{u_\varepsilon}{u_i} = \left(\frac{\varepsilon}{\varepsilon_i}\right)^n \tag{20.42}$$

They called this type of behavior *particulate fluidization.*

Conversely, for the fluidization of fine powders with *gases*, the bed at first expands uniformly, then further additions of gas pass through this dispersion as discrete bubbles, much as does a gas through a liquid. Wilhelm and Kwauk observed that when bubbles formed, the solid particles tended to clump together; hence they called this behavior *aggregative fluidization.* They concluded that the criterion for particulate fluidization is

$$\mathrm{Fr}_b = \frac{u_b^2}{gD} < 0.13 \tag{20.43}$$

where Fr_b is the Froude number for bubbling based on the diameter and the superficial velocity. A value of 0.13 is not normally exceeded with liquids (except perhaps for large lead balls), but is usually exceeded with gases.

Oltrogge and Kadlec [14] used this criterion of Wilhelm and Kwauk to derive an expression for the corresponding critical void fraction, ε_b, for bubbling, choosing the slightly modified value of 0.1 for the critical Froude number. Combining Equation 20.22 with the laminar (viscous) term of the Ergun equation, Equation 19.41, in the notation of this chapter, gives

$$\frac{g\Delta\varrho\varepsilon^3 D}{\varrho u_\varepsilon^2} = \frac{150\mu(1-\varepsilon)}{Du_\varepsilon\varrho} \tag{20.44}$$

Then substituting $u_\varepsilon^2 = 0.1gD$ and rearranging give

$$\frac{\varepsilon_b^3}{1-\varepsilon_b} = \frac{47.4\mu}{g^{1/2}D^{3/2}\Delta\varrho} = \frac{47.4}{Ar^{1/2}}\left(\frac{\varrho}{\Delta\varrho}\right)^{1/2} \tag{20.45}$$

The experimental data in Figure 20–17 from Harriott and Simone [15] appear to confirm this criterion. ε_b also represents the maximum possible void fraction in the dense phase for gaseous fluidization. Each of these sets of data represents the effect of only one variable as indicated.

Foscolo and Gibilaro [16] recently derived an expression for ε_b without invoking the empirical observation represented by Equation 20.43. They started from the postulate of Wallis [17] that discontinuities (bubbles) are formed when the velocity of propagation u_d of a disturbance in the void fraction exceeds the velocity u_w of an elastic wave in the dispersion.

The velocity of propagation of a small disturbance due to a change in ε is

$$u_d = (1-\varepsilon)\frac{du_\varepsilon}{d\varepsilon} \tag{20.46}$$

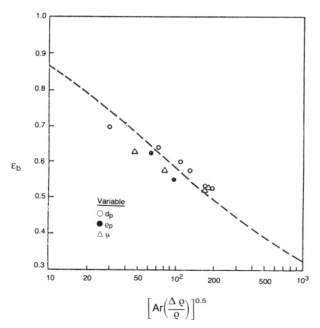

FIGURE 20–17   *Comparison of experimental data for the onset of bubbling in fluidized beds with Equation 20.45. (From Harriott and Simone [15].)*

(A detailed derivation of Equation 20.46 is given by Wallis.) Then, if u_ε is postulated to have the dependence on ε predicted by Equation 20.2,

$$u_d = n u_T (1 - \varepsilon)^{n-1} \tag{20.47}$$

On the other hand, the velocity of a weak wave in an elastic media is in general (see, for example, Churchill [18], p. 24)

$$u_w = \left(\frac{\partial p}{\partial \varrho} \right)^{1/2} \tag{20.48}$$

In this application the compressibility is that of the dispersion rather than of the liquid or the solid. That is, the density to be used in Equation 20.48 is

$$\varrho_d = (1 - \varepsilon)\varrho_s \tag{20.49}$$

Now consider a small perturbation in pressure δp on a horizontal cross section A_x of the dispersion. This produces a net force δF on each of the N_s particles in that cross section. That is,

$$\delta p \cdot A_x = N_s \delta F \tag{20.50}$$

A material balance for the solid phase indicates that, on-the-mean, if the particles are *spherical*,

$$N_s = \frac{4(1 - \varepsilon)A_x}{\pi D^2} \tag{20.51}$$

Hence,

$$\delta p = \frac{4(1 - \varepsilon)\delta F}{\pi D^2} \tag{20.52}$$

Then, from Equations 20.48 and 20.52,

$$u_w = \frac{2}{D} \left(\frac{1 - \varepsilon}{\pi} \right)^{1/2} \left(\frac{\delta F}{\delta \varrho_d} \right)^{1/2} \tag{20.53}$$

It follows from Equation 20.49 that

$$\frac{\delta F}{\delta \varrho_d} = \frac{\partial F}{\partial \varepsilon} \frac{\partial \varepsilon}{\partial \varrho_d} = -\frac{1}{\varrho_s} \frac{\partial F}{\partial \varepsilon} \tag{20.54}$$

The net force on a particle within the dispersion is equal to the difference between the drag force F_p and the buoyant weight W'_p; that is,

$$F = F_p - W'_p \tag{20.55}$$

The buoyant weight of a sphere in a dispersion is given by the right side of Equation 20.23, which is written for a particle at dynamic equilibrium (i.e., with $F \equiv 0$).

Foscolo and Gibilaro postulate that for a given fluid and a particle of given diameter the drag on a particle in a dispersion in all regimes of flow can be approximated by

$$F_p = \phi\{u_\varepsilon\}\varepsilon^{-3.8} \qquad (20.56)$$

This dependence of F_p on ε does not differ greatly from that of Equation 20.26 for creeping flow, or for the analogous expression for purely inertial flow (see problem 19 and note Equations 20.8 and 20.9). Substituting for F_p in Equation 20.55 from 20.56 and for W_p' from the right side of Equation 20.23 gives

$$F = \phi\{u_\varepsilon\}\varepsilon^{-3.8} - \frac{\pi D^2 \varepsilon g \Delta \varrho}{6} \qquad (20.57)$$

Then

$$\frac{\partial F}{\partial \varepsilon} = -3.8\phi\{u_\varepsilon\}\varepsilon^{-4.8} - \frac{\pi D^2 g \Delta \varrho}{6} \qquad (20.58)$$

Substituting for $\phi\{u_\varepsilon\}$ from Equation 20.56 and approximating F_p by the right side of Equation 20.23, in view of the small perturbation in F, give

$$\frac{\partial F}{\partial \varepsilon} = -0.8\pi D^3 g \Delta \varrho \qquad (20.59)$$

Then from Equations 20.53, 20.54, and 20.59,

$$u_w = \left(\frac{3.2gD(1-\varepsilon)\Delta\varrho}{\varrho_s}\right)^{1/2} \qquad (20.60)$$

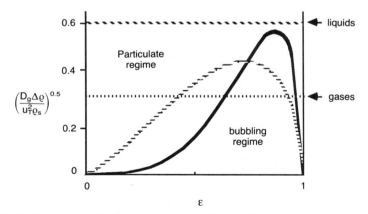

FIGURE 20–18 *Locus for the onset of bubbling in a fluidized bed according to Foscolo and Gibilaro [16]. ———— creeping flow (Equation 20.63); ·········· purely inertial flow (Equation 20.65).*

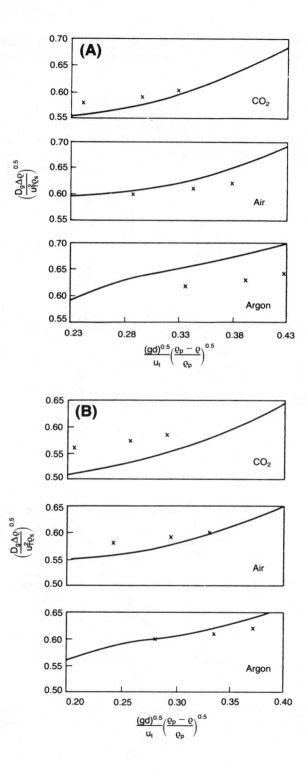

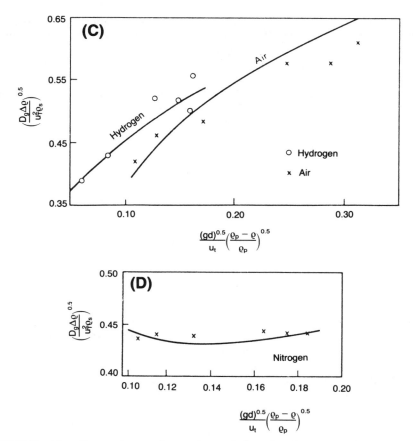

FIGURE 20–19 *Comparisons of experimental data for the point of transition from particulate to aggregative fluidization (the onset of bubbling) with the prediction of Equation 20.61. (From Foscolo and Gibilaro [16]): (A) data of Rowe et al. [19] for a cracking catalyst with $D = 70 \mu m$ and $\varrho_s = 819 \ kg/m^3$ at elevated pressures; (B) data of Rowe et al. [19] for a cracking catalyst with $D = 84 \mu m$ and $\varrho_s = 819 \ kg/m^3$ at elevated pressure; (C) data of Rietema [20] for particles of various sizes and densities at ambient pressure; (D) data of King and Harrison [21] for ballotini with $D = 61 \mu m$ and $\varrho_s = 2,400 \ kg/m^3$ at elevated pressure.*

Finally, equating u_d from Equation 20.46 with u_w from Equation 20.60 gives, on rearrangement,

$$0.559n(1 - \varepsilon_b)^{1/2} \varepsilon_b^{n-1} = \left(\frac{Dg\Delta\varrho}{u_T^2\varrho_s}\right)^{1/2} \qquad (20.61)$$

or

$$0.559n(1 - \varepsilon_b)^{1/2} \varepsilon_b^{n-1} = \left(\frac{3}{2}\frac{\varrho}{\varrho_s}\bar{C}_t\right)^{1/2} \qquad (20.62)$$

Equation 20.62 provides a relationship for the void fraction at the onset of bubbling as a function of ϱ/ϱ_s and Re_T only, since Table 20.1 gives n as a function of Re_T, and expressions such as Equation 16.62 approximate the dependence of $\bar{C}_T$ on Re_T. The principal empiricism in the expression arises from the dependence of F_p on ε as given by Equation 20.56 and the dependence of u_ε on ε as given by Equation 20.2.

For small Re_T, the terminal velocity is given by Equation 20.28, and from Table 20.1, $n = 4.65$. Equation 20.61 then reduces to

$$(1 - \varepsilon_b)^{1/2}\varepsilon_b^{3.65} = 6.92\left(\frac{\varrho}{\varrho_s\,\mathrm{Ar}}\right)^{1/2} \tag{20.63}$$

In the other limit of large Re_T, Equations 20.23 and 20.31 give

$$u_T = 2.72\left(\frac{Dg\Delta\varrho}{\varrho}\right)^{1/2} \tag{20.64}$$

and Table 20.1 gives $n = 2.4$. Equation 20.61 then reduces to

$$(1 - \varepsilon_b)^{1/2}\varepsilon_b^{1.4} = 0.454\left(\frac{\varrho}{\varrho_s}\right)^{1/2} \tag{20.65}$$

Equation 20.62 suggests that in the regime of creeping flow the onset of bubbling depends on the viscosity and hence on the temperature but not on the pressure, whereas Equation 20.65 suggests that in the regime of purely inertial flow the onset of bubbling is sensitive to the density and, for a gas, to both the temperature and pressure. Also, the onset of bubbling is implied to depend critically upon the particle size in the laminar regime but to be independent of this variable in the inertial regime. Actually most practical conditions fall in the intermediate regime of flow and can be expected to demonstrate behavior intermediate to these two extremes.

The right side of Equation 20.61 is plotted in Figure 20–18 versus ε_b for the two extremes of creeping flow ($n = 4.65$) and purely inertial flow ($n = 2.4$). Curves for intermediate flows fall between these two extremes. With a gas, as illustrated by the vertically dashed horizontal line, particulate fluidization occurs as the bed expands until the curve for the operative value of Re_T is reached, whereupon bubbling begins. With most liquids, as illustrated by the obliquely dashed horizontal line, the fluidization remains particulate for all Re_T and all degrees of expansion.

Experimental observations of the void fraction for the onset of bubbling under a variety of conditions are compared with the predictions of Equation 20.61 in Figure 20–19. The agreement is reasonably good in all of these cases. Equations 20.61 and 20.63 are compared with Equation 20.45 in problem 30.

PARTICULATE FLUIDIZATION

The gross behavior for particulate fluidization is completely circumscribed by the relationships of Chapter 16 and the first section of this chapter for isolated particles in free-fall, those of Chapter 19 for packed beds, and those of the previous sections for suspensions. The relevant relationships can be used in combination as follows.

Dimensional analysis of flow through packed beds, as well as all of the particular relationships of Chapter 19, indicate that the pressure drop can be represented functionally as

$$\frac{(-\Delta\mathcal{P})\varrho D^3}{L\mu^2} = \left\{\frac{Du_\varepsilon\varrho}{\mu}, \varepsilon_p\right\} \tag{20.66}$$

At the other extreme, the velocity required to suspend an isolated particle can be expressed functionally as

$$\frac{\varrho D^3 g \Delta\varrho}{\mu^2} = \phi\left\{\frac{Du_T\varrho}{\mu}\right\} \tag{20.67}$$

Examination of Equations 20.66 and 20.67 suggests expressing Equation 20.22 for the intermediate regime of fluidization in the form

$$\frac{(-\Delta\mathcal{P})\varrho D^3}{\mu^2 L(1-\varepsilon)} = \frac{\varrho D^3 g \Delta\varrho}{\mu^2} \tag{20.68}$$

Equation 20.66 can be rearranged to have the same left side as Equation 20.68; that is,

$$\frac{(-\Delta\mathcal{P})\varrho D^3}{\mu^2 L(1-\varepsilon_p)} = \phi\left\{\frac{Du_\varepsilon\varrho}{\mu}, \varepsilon_p\right\} \tag{20.69}$$

The various particular expressions of Chapter 19 for the functional relationship of Equation 20.69 indicate that a plot of $(-\Delta\mathcal{P})\varrho D^3/\mu^2 L(1-\varepsilon)$ versus $Du_\varepsilon\varrho/\mu$ will result in a monotonically increasing relationship for the packed-bed regime (for a fixed value of ε). The various particular expressions of Chapter 16 for the drag of a sphere indicate that a plot of $\varrho D^3 g\Delta\varrho/\mu^2$ versus $Du_T\varrho/\mu$ on the same coordinates will have a similar form for the regime of terminal setting of an isolated particle. Finally, Equation 20.68 indicates that in the fluidized-bed regime between these two limits a plot of $(-\Delta\mathcal{P})\varrho D^3/\mu^2 L(1-\varepsilon)$ versus $Du_\varepsilon\varrho/\mu$ should yield a fixed ordinate of $\varrho D^3 g\Delta\varrho/\mu^2$. This latter regime must extend from the curve corresponding to Equation 20.69 for the fixed-bed regime to that corresponding to Equation 20.67 for an isolated particle. As discussed earlier, some deviations from these three relationships are to be expected at the transition from the packed-bed to the fluidized-bed regime.

The data of Wilhelm and Kwauk [4] for large Socony beads in water are plotted in Figure 20–20 in the suggested form. The corresponding void fractions are plotted above on the same abscissa. Equation 19.41 with $\varepsilon_p = 0.368$, Equation 20.22, Equation 16.62, and Equation 20.2 with $n = 2.4$ corresponding to $\mathrm{Re}_T = 641$, are included in Figure 20–20 for comparison. The agreement appears to be reasonably good, except for ε, although the logarithmic coordinates and the compressed scales disguise the deviations somewhat.

A generalized correlation in the same form as the lower portion of Equation 20.20 is shown in Figure 20–21 with ε_p and $\mathrm{Ar} = \varrho D^3 g\Delta\varrho/\mu^2$ as parameters. The horizontal lines represent chosen values of Ar. Curves representing the fixed values of ε in the regime of fluidization could be added to Figure 20–21 (see problem 32).

Figure 20–22 is a plot analogous to Figure 20–21 prepared by Wilhelm and

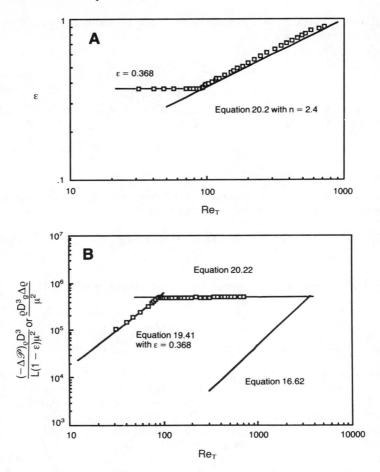

FIGURE 20–20   *Comparison in canonical form of the experimental data of Wilhelm and Kwauk [4] for large Socony beads (D = 0.174 inch, ϱ_s = 100 lb/ft³, ϱ = 62.4 lb/ft³, μ = 0.000672 lb/ft·s, ε_p = 0.368) with theoretical predictions: (A) void fraction; (B) pressure drop.*

Kwauk [4] to summarize their data for a number of solids in both water (particulate fluidization) and air (the particulate regime of aggregative fluidization).

The range of *operability* for particulate fluidization is necessarily constrained by the velocity for incipient fluidization, u_i, and the terminal velocity of a single particle, u_T. The ratio of these two velocities is

$$R = \frac{u_T}{u_i} = \frac{\text{Re}_T}{\text{Re}_i} = \left[\frac{\bar{C}_t\{\text{Re}_i\}}{\bar{C}_t\{\text{Re}_T\}}\right]^{1/2} \tag{20.70}$$

where $\bar{C}_t\{\text{Re}_i\}$ implies the drag coefficient of a single particle at the point of incipient fluidization based upon u_i. R can be evaluated as a function of either Re_T or Ar as follows. Choose Re_T and calculate the corresponding value of $\bar{C}_t$

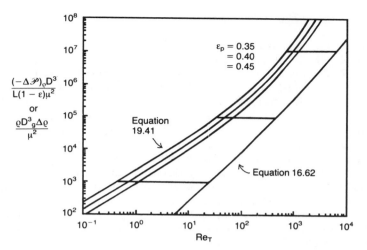

FIGURE 20–21   *Generalized correlation for pressure drop and expansion in particulate fluidization.*

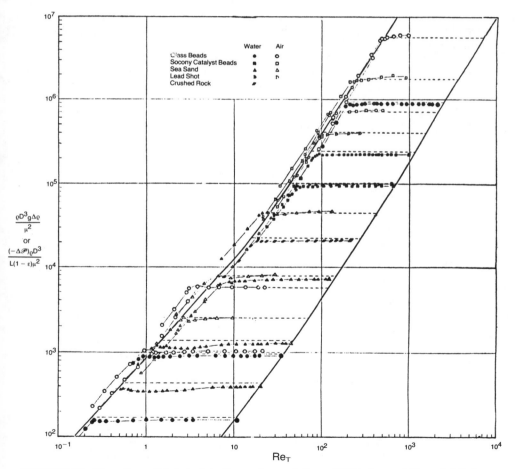

FIGURE 20–22   *Consolidated plot of experimental data of Wilhelm and Kwauk [4] for the pressure drop in the particulate regime of fluidization.*

from an expression such as Equation 16.62. Then calculate Ar from $\bar{C}_t$ and Re_T by Equation 16.71. Use this value of Ar to calculate Re_i from Equation 20.39 and, finally, R. The resulting dependence of R on Ar for spheres with $\varepsilon_i = 0.406$ is shown graphically in Figure 20–23. The limiting behavior for creeping flow can be determined by letting $\text{Ar} \to 0$ in Equation 20.39 and then substituting $(3/2)\,\bar{C}_t\text{Re}_T^2 = (3/2)(12/\text{Re}_T)\,\text{Re}_T^2 = 18\text{Re}_T$ for Ar, thereby obtaining the upper limiting value of $R = 74$. Similarly, for purely inertial flow, letting $\text{Ar} \to \infty$ in Equation 20.39 and noting from Equation 16.62 that $\bar{C}_t = 0.245$ give the lower limiting value of $R = 8.4$. These limiting values are indicated by the horizontal lines in Figure 20–23. A comparison by Bourgeois and Grenier [22] of experimental data of their own and others spheres for spheres of various materials in water with the relationship of Figure 20–23 is shown in Figure 20–24. The agreement is surprisingly good, particularly since, as they note, this prediction is very sensitive to the rather arbitrary choice of a mean value of 0.406 for ε_i. The prediction is also dependent on the use of Equation 16.62 for the drag of an isolated sphere and of Equation 19.41 for the pressure drop over a packed bed of spheres, both of which are empirical and subject to some uncertainty. (The segmental curve and segmental straight line in Figure 20.24 are an attempt by Bourgeois and Grenier to represent the behavior by a series of analytical correlating equations; with the advent of modern computers this procedure is unnecessary.)

A plot of data for other fluids by Richardson [23] in Figure 20–25 does appear to indicate a dependence on ε_i, as suggested by the theoretical curves for $\varepsilon_i = 0.38$, 0.40, and 0.42. (These curves are based on a relationship for $C_t = \phi\{\text{Re}_T\}$ differing slightly from Equation 16.62.)

Richardson [23] also noted that R can be used to calculate n in Equation 20.2 as follows. Taking the logarithm and rearranging give

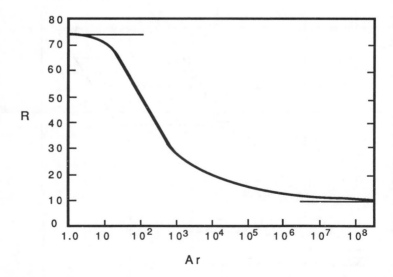

FIGURE 20–23　*Range of operability $R = u_T/u_i$ for particulate fluidization as predicted by Equations 16.62 and 20.39 for $\varepsilon_i = 0.406$. Horizontal lines indicate limiting behavior.*

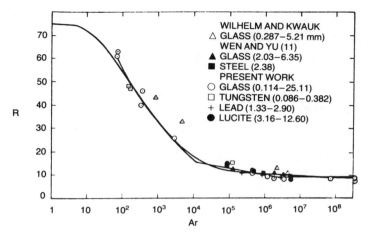

FIGURE 20-24 *Comparison of the predicted range of operability for particulate fluidization with experimental data for fluidization with water by Bourgeois and Grenier [22]. The continuous curve is that of Figure 20-23 and the segmental ones represent correlating equations proposed by these authors.*

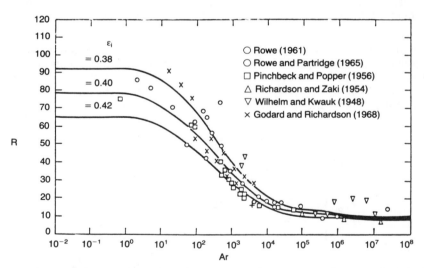

FIGURE 20-25 *Comparison of predicted range of operability for particulate fluidization at several values of ε_i with experimental data for a number of fluids and spherical particles. (From Richardson [23].)*

$$n = \frac{\ln\{u_\varepsilon/u_T\}}{\ln\{\varepsilon\}} \tag{20.71}$$

Equation 20.71 should be applicable at the point of incipient fluidization. Hence

$$n = \frac{\ln\{u_i/u_T\}}{\ln\{\varepsilon_i\}} = \frac{-\ln\{R\}}{\ln\{\varepsilon_i\}} \tag{20.72}$$

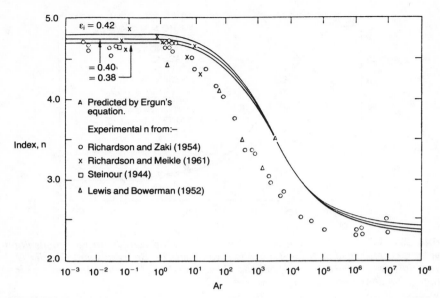

FIGURE 20–26   *Comparison of predictions of exponent n of Equation 20.2 from Equations 20.72 and 20.37 with experimental data. (From Richardson [23].)*

In Figure 20–26, values of n determined from the experimental data of various investigators are compared with curves representing values computed from Equation 20.72 for $\varepsilon_i = 0.38$, 0.40, and 0.42, using Equation 20.37 to compute R as a function of Ar. The data follow the trend of the prediction but fall consistently below the prediction for intermediate values of Ar and Re_T.

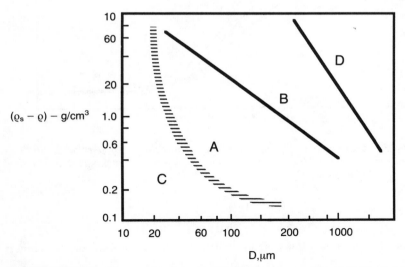

FIGURE 20–27   *Classes of aggregative fluidization according to Geldart. (After Harriott and Simone [15].)*

Richardson suggests that this discrepancy is due to the freedom of the particles to move somewhat at the point of incipient fluidization, thereby reducing the pressure drop below that predicted by Equation 19.41, which was used in the prediction of R and hence of n.

AGGREGATIVE FLUIDIZATION

As noted earlier, aggregative fluidization is much more complex than particulate fluidization owing to the formation, presence, and relative motion of the bubbles. Geldart [24] classified the fluidization of solids by gases in four groups depending on D and $\Delta\varrho$, as illustrated in Figure 20–27. For the very small particles of Group C, surface forces are important and fluidization is difficult at best. For Group A, $u_i < u_b$, and significant expansion occurs before the onset of bubbling. For Group B, bubbling occurs at the onset of fluidization—that is, at $u_b = u_i$. For the very large particles of Group D, $u_b < u_i$, and spouting begins at the onset of bed movement.

Bed Expansions

The expansion of a Group A and a Group B bed is illustrated in Figure 20–28. The catalyst powder expands uniformly with velocity, beginning at u_i, to a maximum height at u_b, falls as bubbling begins, and then again rises uniformly. In the bed of sand, bubbling begins with the onset of fluidization. In both cases the *emulsion phase* consists of a fairly uniform dispersion with a gas velocity only slightly greater than u_i, the balance of the gas passing through the bed as bubbles.

Experimental data for the particulate regime of fluidization of Group A powders with gases are plotted in Figure 20–29 in the form suggested by

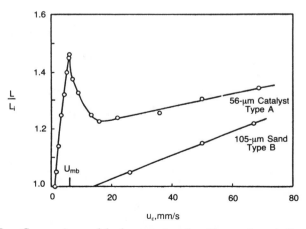

FIGURE 20–28 *Comparison of bed expansion for Group A and Group B powders. (From Harriott and Simone [15].)*

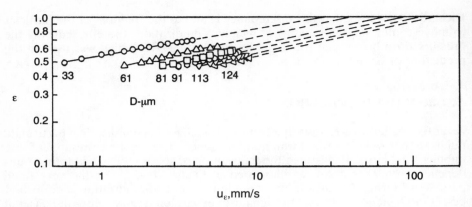

FIGURE 20–29   *Experimental data for expansion of beds of cracking catalyst in the particulate regime of aggregative fluidization plotted in the form suggested by Equation 20.2. (From Harriott and Simone [15].)*

Equation 20.2. The slopes are somewhat less than the values of n predicted by Table 20.1. Also, the straight lines do not extrapolate to u_T.

Harriott and Simone [15] suggest that the expansion in the narrow regime of flow between incipient fluidization and bubbling might be approximated by the relationship between void fraction and velocity for a packed bed. For the laminar regime, Equation 20.44 is then applicable. Rearrangement as

$$\frac{\varepsilon^3}{1 - \varepsilon} = \left(\frac{150\mu}{D^2 g \Delta\varrho}\right) u_\varepsilon \tag{20.44A}$$

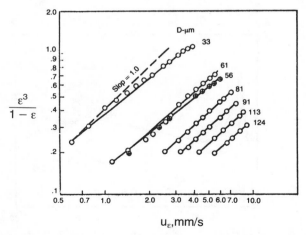

FIGURE 20–30   *Experimental data for expansion of beds of cracking catalyst in the particulate regime of aggregative fluidization plotted in the form suggested by Equation 20.44A. (From Harriott and Simone [15].)*

suggests plotting $\varepsilon^3/(1 - \varepsilon)$ versus u_ε on logarithmic coordinates. The experimental test of Equation 20.44A in Figure 20–30 indicates a somewhat lesser expansion.

Pressure Drop

As indicated above in Figure 20–14, the loss in dynamic pressure for fluidization of a Group A powder is approximated by Equation 20.22 up to the point at which entrainment begins owing to the gross disturbance of the dispersion by the bubbles. Thereafter the dynamic pressure drop falls rapidly to zero as the velocity increases to its terminal velocity.

The data of Wilhelm and Kwauk [4] for fluidization with air are seen in Figure 20–22 to follow the predicted behavior for particulate fluidization up to some velocity short of the terminal value, implying the onset of entrainment at that point.

Bubble Rise

The rate of rise of bubbles through the emulsion phase can be approximated by the expressions of Chapter 17 for the rise of bubbles through real liquids. This approach would appear to introduce the difficulty of predicting the effective viscosity of the emulsion. Fortunately, the bubbles encountered in fluidized beds are generally so large that their behavior can be represented by Equation 17.89, which, in this application, can be rewritten as

$$U_{bf} = 0.714\left(\frac{Dg(\varrho_e - \varrho)}{\varrho_e}\right)^{1/2} \tag{20.73}$$

$$\cong 0.714(Dg)^{1/2} \tag{20.74}$$

where U_{bf} = rate of rise of a bubble in the emulsion phase, m/s
ϱ_e = density of emulsion phase, kg/m^3

Equation 20.74 is applicable only for a single, isolated bubble. The velocity of a swarm of bubbles due to their displacement alone is $U_0 - U_{0i}$, and to this must be added the indicated buoyant motion. Thus the absolute rate of rise is predicted to be

$$U_{bf} = U_0 - U_{0i} + U_{bf} \tag{20.75}$$

The predictions of Equations 20.74 and 20.75 have been confirmed experimentally (see, for example, Kunii and Levenspiel [25], Chapter 4).

Velocity Fields

The actual velocity fields of the gas within the bubble and emulsion phases and of the individual solid particles are very complex. For example, a stream of gas

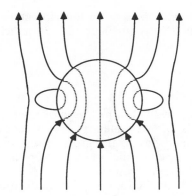

FIGURE 20–31   *Flow of emulsion around and up through a slowly rising bubble. (After Catepovic [26].)*

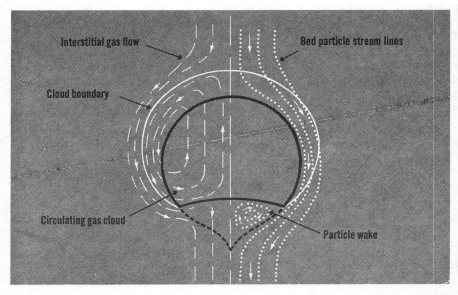

FIGURE 20–32   *Flow of gas and particles around and through a rapidly rising bubble [27].*

and particles actually flows continuously up into and out of a slowly rising bubble, as illustrated in Figure 20–31, whereas a fast-moving bubble retains its integrity but entrains particles near its surface and, particularly, in its dimpled wake, as illustrated in Figures 20–32 and 20–33. The details of this behavior (see, for example, Rowe [29] and Kunii and Levenspiel [25]) are very important in that the majority of applications involve transport of species between the particles and the gas stream.

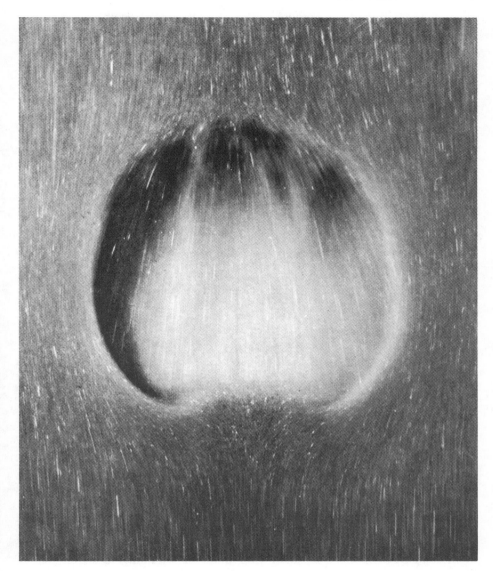

FIGURE 20–33   *Flow of particles around a rapidly rising bubble. (From Murray et al. [28].)*

Range of Operability

Figure 20–34 is the analog of Figure 20–24 for fluidization by air. The agreement is good. However, this plot is not a conservative guide to the range of operability, but an upper bound that cannot be approached closely, since the velocity for entrainment is far less than the terminal velocity.

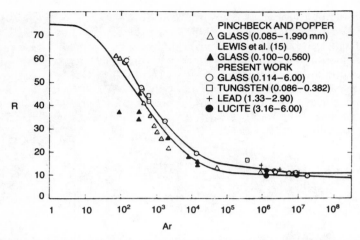

FIGURE 20–34   *Comparison of the predicted range of operability for aggregative fluidization with experimental data for spherical particles of various materials in air by Bourgeois and Grenier [22]. The lower curve is that of Figure 20.23 and the upper one represents a series of correlating equations proposed by the authors.*

SUMMARY

The relative motion of fluids and dispersed solids has been examined, although limited to a few of the many situations in which this behavior arises. Particular attention has been given to fluidized beds because of their importance and illustrative value.

Three principal complications impede the theoretical treatment of such flows: (1) the superposition of the effects of particle scale and container scale; (2) nonuniformity in the distribution of the solids; and (3) the irregularity and size distribution of the solid particles in most practical applications. Even so, the combination of simple mass, volume, and momentum balances together with the structure previously developed for single, geometrically regular particles and packed beds has been shown to provide considerable insight.

The gross structure of fluidized beds is actually quite well predicted by such elementary concepts. Thus, the onset of fluidization corresponds to the velocity that produces a pressure drop through the nonfluidized bed equal to the weight of the bed per unit cross-sectional area, and, for a liquid, the upper limit of fluidization corresponds to terminal settling of the individual particles. The expansion of the bed can be approximated by logarithmic interpolation between these limits, while the dynamic pressure drop remains approximately constant. The expansion and termination of gas-fluidized beds are complicated by the passage of most of the gas, above the minimum required for fluidization, through the bed as bubbles. However, the rate of rise of these bubbles through the emulsion phase equals that for asymptotically large bubbles through a normal liquid.

The important effects of nonsphericity, nonuniformity, compressibility, and so on, have not been examined here. Also, the velocity field of the gas and of the individual solid particles, which are of primary importance in most applications, have only been described superficially.

The foregoing global structure for fluidized beds was, remarkably, almost all elucidated quantitatively by Wilhelm and Kwauk [4] in virtually the first published article on fluidization. Furthermore, they confirmed their predictions with experimental data for many spherical particles and fluids. Most subsequent work, as summarized, for example, by Davidson and Harrison [30] and Kunii and Levenspiel [25], has been primarily concerned with secondary effects and the fine structure. Exceptions include the recent analyses of Foscolo et al. [6] and Foscolo and Gibilaro [16], which provide an improved theoretical structure for bed expansion and the onset of bubbling in particular.

PROBLEMS

1. Explain how the diameter of a particle can be determined graphically without trial and error using Figure 20–1 if u_T, ϱ, ϱ_s, and μ are known.
2. Explain how the terminal velocity of a particle can be determined graphically without trial and error using Figure 20–1 if D, ϱ, ϱ_s, and μ are known.
3. Repeat problem 1 for Figure 20–2.
4. Determine the minimum set of dimensionless variables for

$$u_\varepsilon = \phi\{D, \varrho, \mu, \Delta\varrho, g, D_t, \varepsilon\} \qquad (20.76)$$

Compare the result with Equation 20.11 and explain.
5. Explain on physical grounds why u_T appears to depend on D_t according to Equation 20.19 for fluidization but not for sedimentation.
6. Repeat the derivation of Equation 20.29 using Equation 19.23 in place of 19.25. Compare the result with Equation 20.29.
7. Repeat the derivation of 20.33 using 19.31 in place of 19.35. Compare the result with 20.33.
8. Repeat the derivation of 20.34 using 19.41 in place of 19.43. Compare the result with 20.34.
9. Foscolo et al. [6] used the expression

$$\frac{-\Delta\mathscr{P}}{L} = 17.3\frac{\mu u_\varepsilon}{D^2}(1 - \varepsilon)\varepsilon^{-4.8} + 0.336\frac{\varrho u_\varepsilon^2}{D}(1 - \varepsilon)\varepsilon^{-4.8} \qquad (20.77)$$

in place of Equation 19.43 to derive

$$\frac{u_\varepsilon}{u_T} = \frac{[0.0777\,\text{Re}_T(1 + 0.0194\,\text{Re}_T)\varepsilon^{4.8} + 1]^{1/2} - 1}{0.0388\,\text{Re}_T} \qquad (20.78)$$

Plot curves corresponding to the coordinates of Figure 20–7 for $\text{Re}_T = 1$, 10, and 100.
10. Compare Equation 20.78 with Equations 20.34 and 20.2 plus 20.15–20.17 for $\text{Re}_T = 1$, 10, and 100.
11. Explain why one expression for the drag on a single particle was combined with one for a packed bed in deriving Equations 20.29, 20.33, and 20.34, whereas only an expression for the packed bed was used in deriving Equation 20.78.
12. Compare the prediction of Equation 20.34 with the data of Figure 20.4.
13. Explain on physical grounds why the drag on a single sphere in a

dispersion, as given by Equation 20.23, equals the weight of the sphere minus the weight of the *dispersion* it displaces.

14. Combine Equations 20.24 and 20.25 according to Equation 19.45 with an arbitrary exponent p. Derive the resulting generalized form of Equation 20.29. Compare the resulting expressions for $p = 1/2$ and 2 with Equation 20.29.

15. Repeat the derivation of Equations 20.37 and 20.32 using Equation 19.43 instead of Equation 19.41.

16. Determine the limiting behavior of the solution of problem 15 for creeping and purely inertial flow.

17. Derive Equation 20.46.

18. Compare the dependence on ε of Equations 20.26 and 20.32 with that of Equation 20.56.

19. Repeat the derivation of Equation 20.61 using u_ε from Equation 20.29 rather than from Equation 20.2, and F_p from Equation 20.26 rather from Equation 20.56. Compare the result with Equation 20.63.

20. Repeat the derivation of Equation 20.61 using u_ε from Equation 20.33 rather than from Equation 20.2, and F_p from Equation 20.32 rather than from Equation 20.56. Compare the result with Equation 20.65.

21. Derive an interpolating equation from the results of problems 19 and 20 as an alternative to Equation 20.62 and compare.

22. Determine the minimum size of spheres of glass, titanium, steel, and lead for which bubbling will occur in water at 30°C. At what velocity and void fraction will this occur?

23. Prepare a plot of the void fraction for the onset of bubbling in a dispersion of glass spheres ($\varrho_s = 2.32$ Mg/m^3, $D = 5$ mm) in air at 30°C as a function of pressure.

24. Prepare a plot of the void fraction for the onset of bubbling in a dispersion of glass spheres ($\varrho_s = 2.32$ Mg/m^3, $D = 5$ mm) in air at 1 atm as a function of temperature.

25. Why does particulate fluidization generally occur for liquids but not for gases?

26. Derive Equation 20.63 directly rather than by reducing Equation 20.61.

27. Derive an expression analogous to Equation 20.63 by substituting $\bar{C}_t$ from Equation 16.17 in Equation 20.62. Explain the results.

28. Derive Equation 20.65 directly rather than by reducing Equation 20.61.

29. Derive Equation 20.65 by substituting for $\bar{C}_t$ in Equation 20.62.

30. Compare Equations 20.63 with 20.45 qualitatively and quantitatively.

31. Compare Equations 20.61 and 20.45 quantitatively. Interpret the results.

32. Prepare an extended version of Figure 20–21 with ε in the regime of fluidization as a parameter.

33. Show how R can be evaluated from the drag coefficients of Equation 20.70 and carry out such calculations for $0.1 < \mathrm{Re}_T < 3000$.

34. Derive explicit formulas for $R = u_T/u_i$ in terms of Ar, using Equation 16.63 instead of 16.62. Compare this expression with that based on Equation 16.62. Reexpress this solution in terms of Re_T.

35. Derive expressions for n from Equation 20.72, using $R = 74$ for creeping flow and $R = 8.4$ for purely inertial flow.

36. Calculate the effective value of n defined by Equations 20.34 and 20.72 as a function of Ar and Re_T.

37. Calculate the effective values of n defined by Equation 20.78 and 20.72 as a function of Re_T.

38. Determine the effective value of n defined by Equation 20.33 as a function of ε.

39. Compare the effective values of n determined in problems 35–38 with those given in Table 20.1.

40. Can a curve defining U_b be added to Figure 20–21? Explain.

41. A cylinder 100 mm in diameter and 0.3 m high is filled with a very dilute suspension of glass beads in water at 21°C. The suspension consists of 4.0 g/m^3 of spheres 7.5 μm in diameter and 10 g/m^3 of spheres 20 μm in diameter. The density of the glass is 2.23 Mg/m^3. The cylinder is thoroughly shaken until the beads are uniformly distributed throughout the suspension. The cylinder is then placed in a vertical position, and the beads are allowed to settle.

    a. How much time will elapse before there are no 20-μm spheres in the top three-fourths of the cylinder?

    b. What percentage of the original 7.5-μm spheres will be in the top three-fourths at that time?

    c. Outline a method for complete separation.

42. Crushed silica ranging in size from 100 to 1000 μm is to be analyzed using an elutriator whose analyzing zone is a cylinder 90 mm in diameter. Water at 13°C is to be used as the analyzing fluid. What rates of flow of the water, measured in cubic meters per second will be necessary to give cuts ranging from 100 to 300 μm and 300 to 700 μm?

43. Gold is to be recovered from a mixture of gold particles and sand. The use of a free-settling classifier has been suggested, with water to be used as the classifying fluid. The density of gold is 1205 lb/ft^3, of sand 165 lb/ft^3. The mixture analyzes 20% wt gold. Laboratory tests on a sample of the mixture produced the following data. What fraction of the gold in the original mixture can be recovered in the pure state by a single-stage classification? Spherical particles may be assumed. Suggest a procedure for separating the sand from the remainder of the gold.

Effective Particle Diameter (in.)	Wt. fraction with smaller diameter	
	Gold	Sand
0.13	1.00	1.00
0.10	0.855	0.800
0.08	0.750	0.660
0.065	0.660	0.530
0.050	0.550	0.390
0.040	0.460	0.280
0.030	0.360	0.165
0.023	0.280	0.080
0.017	0.195	0.000
0.012	0.120	
0.007	0.050	
0.004	0.000	

44. Calculate the velocities required to fluidize and to entrain the solids of the following problems. For b and c estimate the velocity for the onset of bubbling.

    a. Problem 14 with water
    b. Problem 15 with air
    c. Problem 17 with water
    d. Problem 18 with air

45. What is the maximum possible rate of flow that can be used to backwash the filter bed of problem 21 in Chapter 19?

46. Determine the mean value of n corresponding to the several parts of problem 44. Compare these values with the values proposed by Richardson and Zaki (Table 20.1).

47. It is proposed to reduce the diameter of the particles in a fluidized bed of catalyst in order to increase the rate of reaction.

    a. Will the pressure drop required to fluidize the bed increase or decrease significantly? Explain.
    b. Will the mass rate of flow required to fluidize the bed increase or decrease significantly? Explain.

48. Bartholomew [31] obtained the following data on a fluidized bed 4 in. in diameter and 30 in. high. Heat was transferred from the wall to the air–solid mixtures. The void fraction was maintained at 0.74 in all runs.

Air Velocity (lb/hr-ft²)	Bed Temperature (°F)	Particle Diam. (in.)	Particle Density (lb/ft³)	Measured $-\Delta\mathcal{P}$ (in. Hg)
420	376	0.009 56	166	1.91
322	388	0.006 70	166	1.76
232	390	0.004 72	166	1.69
177	391	0.003 38	166	1.55
410	378	0.009 84	166	1.49
292	388	0.006 81	166	1.90
239	391	0.004 73	166	1.92
438	375	0.009 55	167	0.66
271	390	0.006 22	166	1.72
230	387	0.004 78	166	1.72
290	382	0.006 82	167	1.64
279	386	0.006 20	160	1.60
219	587	0.006 26	166	1.62
188	596	0.004 72	166	1.57
219	595	0.006 07	167	1.60
219	598	0.006 32	160	1.66

Compare the measured pressure drops with those predicted from the measured velocities.

49. A catalyst bed made up of 6.35 mm by 6.35 mm cylinders is 0.6 m in diameter, 0.9 m deep, with a void fraction of 0.42. The bed is supported from below by a screen. Gas with the following properties passes up through the bed.

Inlet temperature	555 K
Inlet pressure	120 MPa
Viscosity	30 μPa · s
Molar mass	0.054

    a. Estimate the maximum allowable gas velocity if movement of the bed and consequent mechanical destruction of the pellets is to be avoided.
    b. Estimate the velocity at which bubbling will begin.
    c. Estimate the velocity at which pellets will be carried completely out of the reactor.

50. The following data were obtained for the pressure drop for air flowing up through a bed of particles 9 ft high and 2 ft in diameter confined between screens.

velocity *(ft/s)*	$-\Delta \mathscr{P}$ *(lb$_f$/in.^{2})*
0.1	0.02
0.5	0.15
5.0	7.2

Estimate the rate of flow required to fluidize the bed if the upper confining screen is removed and the bed has a bulk density of 48 lb/ft^3. Also estimate the velocities for the onset of bubbling and enhancement.

51. A fluidized-bed reactor is to be designed for 2.5 kg/s of feed having a viscosity of 20 μPa · s and a density of 11.2 kg/m^3 at the operating conditions of 533 K and 0.2 MPa. A space velocity of 50 kg feed/hr-kg catalyst is required for the desired conversion. The catalyst has a density of 1.4 Mg/m^3 and a mean particle diameter of 1.27 mm. Size the reactor (actual bed section only). Will bubbles occur? What is the minimum size of a catalyst particle that will be retained in the bed?

REFERENCES

1. E. B. Christiansen, *Effect of Particle Shape of Free Settling Rates*, Ph.D. Thesis, University of Michigan, Ann Arbor (1943).
2. R. M. Richards, "Velocity of Galena and Quartz Falling in Water," *Trans. Am. Inst. Met. Engr.*, *38* (1907) 210.
3. R. T. Hancock, "The Law of Motion of Particles in a Fluid," *Trans. Inst. Mining Engr.*, 94 (1937) 114.
4. R. H. Wilhelm and M. Kwauk, "Fluidization of Solid Particles," *Chem. Eng. Progr.*, *44*, 201 (1948).
5. W. K. Lewis, E. R. Gilliland, and W. C. Bauer, "Characteristics of Fluidized Particles," *Ind. Eng. Chem.*, *41*, (1949) 1104.

6.   P. U. Foscolo, L. G. Gibilaro, and S. P. Waldram, "A Unified Model for Parti-culate Expansion of Fluidized Beds and Flow in Fixed Porous Media," *Chem. Eng. Sci.*, *38* (1983) 1251.
7.   J. F. Richardson and W. N. Zaki, "Sedimentation and Fluidization," *Trans. Inst. Chem. Engr.* (London), *32* (1954) 35.
8.   J. F. Richardson and R. Meikle, "Sedimentation and Fluidization. Part III. The Sedimentation of Uniform Fine Particles and of Two-Component Mixtures of Solids," *Trans. Inst. Chem. Engr.* (London), *39* (1961) 348.
9.   J. L.-P. Chen and D. L. Keairns, "Particle Segregation in a Fluidized Bed," *Can. J. Chem. Eng.*, *53* (1975) 395.
10. T. Shirai, *Fluidized Beds*, Kagaku-gijutsu-sha, Kanazawa (1958).
11. C. van Heerden, A. P. P. Nobel, and D. W. van Krevelen, "Studies on Fluidization. I. The Critical Mass Velocity," *Chem. Eng. Sci.*, *1* (1951) 37.
12. C. Y. Wen and Y. H. Yu, "A Generalized Method for Predicting the Minimum Fluidization Velocity," *AIChE J.*, *12* (1966) 610.
13. Max Leva, *Fluidization*, McGraw-Hill, New York (1959).
14. R. D. Oltrogge and R. H. Kadlec, "Gas Fluidized Beds of Fine Particles," Paper 10a, Presented at 75th National Meeting of AIChE, Detroit, MI (1973).
15. P. Harriott and S. Simone, "Fluidizing Fine Powders," Chap. 25 in *Handbook of Fluids in Motion*, N. P. Cheremisinoff and R. Gupta, Eds., Ann Arbor Science, Ann Arbor, MI (1983).
16. P. U. Foscolo and L. G. Gibilaro, "A Fully Predictive Criterion for the Transition between Particulate and Aggregative Fluidization," *Chem. Eng. Sci.*, *39* (1984) 1667.
17. G. B. Wallis, *One-Dimensional Two-Phase Flow*, McGraw-Hill, New York (1969).
18. S. W. Churchill, *The Practical Use of Theory in Fluid Flow. Book I. Inertial Flows*, Etaner Press, Thornton, PA (1980).
19. P. N. Rowe, P. U. Foscolo, A. C. Hoffmann, and J. Yates, "Fine Powders Fluidized at Low Velocities at Pressures up to 20 bar with Gases of Different Viscosity," *Chem. Eng. Sci.*, *37* (1982) 1115.
20. K. Rietema, "The Effect of Interparticle Forces on the Expansion of a Homo-geneous Gas-Fluidized Bed," *Chem. Eng. Sci.*, *28* (1973) 1493.
21. D. F. King and D. Harrison, "The Dense Phase of a Fluidized Bed at Elevated Pressures," *Trans. Inst. Chem. Engr.* (London), *60* (1982) 26.
22. P. Bourgeois and P. Grenier, "The Ratio of Terminal Velocity to Minimum Fluidizing Velocity for Spherical Particles," *Can J. Chem. Eng.*, *46* (1968) 325.
23. J. F. Richardson, "Incipient Fluidization and Particulate Systems," Chap. 2 in *Fluidization*, J. F. Davidson and D. Harrison, Eds., Academic Press, New York (1971).
24. D. Geldart, "Types of Fluidization," *Powder Tech.*, *7* (1973) 285.
25. D. Kunii and O. Levenspiel, *Fluidization Engineering*, John Wiley, New York (1969).
26. N. M. Catipovic, G. N. Govanovic, and T. J. Fitzgerald, "Regimes of Fluidization for Large Particles," *AIChE J.*, *24* (1978) 543.
27. "Fluidized Beds Take on New Life," *Chem. Eng. News*, *48*, No. 52 (Dec. 14, 1970) 46.
28. J. D. Murray, "On the Mathematics of Fluidization. Part 2. Steady Motion of Fully Developed Bubbles," *J. Fluid Mech.*, *22* (1965) 57.
29. P. N. Rowe, "Experimental Properties of Bubbles," Chap. 4 in *Fluidization*, J. F. Davidson and D. Harrison, Eds., Academic Press, New York (1971).
30. J. F. Davidson and D. Harrison, Eds., *Fluidization*, Academic Press, New York (1971).
31. R. Bartholomew, *Heat Transfer from a Metal Surface to Fixed Fluidized Beds of Fine Particles*, Ph.D. Thesis, University of Michigan, Ann Arbor (1951).

Appendix

Table A.1
Viscosities and Densities of Common Gases at Atmospheric Pressure

Gas	Temperature (K)	$\mu \times 10^6$ (Pa·s)	ϱ (kg/m³)
air	300	18.53	1.183
N_2	300	17.84	1.1421
O_2	300	20.63	1.3007
CO_2	300	14.96	1.7973
CO	300	17.84	1.139
H_2	300	8.963	0.08185
CH_4	293	10.87	0.6679
Cl_2	300	13.50	2.88
He	300	20.08	0.163
H_2O	373	12.28	0.597

Table A.2
Viscosities and Densities of Common Liquids at 20°C

Liquid	$\mu \times 10^4$ (Pa·s)	ϱ (kg/m³)
acetone	3.31	791
ammonia	2.20	612
ethylacetate	4.49	900
ethylalcohol	12.0	790
glycerol	14,800	1260
mercury	15.48	13,579
toluene	5.86	866
water	9.93	998

Credits

The following publishers have generously given permission to reprint material from copyrighted sources. See references at the end of each chapter for complete citations.

Figures 2–13 and 2–15   Reprinted with permission from the Society of Rheology

Figures 2–14, 4–9, 4–10, 4–11, 17–9, 17–19, 17–28, 17–29, 19–10, 19–14, 19–15, 19–16, 19–17, 20–32, and Table 17.5   Reprinted with permission from American Chemical Society

Figures 3–3, 3–4, 4–4, 4–12, 5–4, 5–6, 5–7, 5–8, 17–26, 17–27, 17–30, 17–37, 17–38, 19–3, 19–5, 19–6, 19–11, 19–12, 19–18, 19–19, 20–15, 20–22, and Tables 5.1, 17.10, 17.11, and 17.15   Reprinted with permission from the American Institute of Chemical Engineers

Figures 3–7, 3–8, 4–7, 6–5, 16–21, 16–42, 17–6, 17–31, 17–32, 17–33, 17–34, 17–35, 17–36, 20–12, 20–13, 20–24, 20–34, and Tables 6.1 and 17.12   Reprinted with permission from the *Canadian Journal of Chemical Engineering*

Figures 6–2, 6–3, 17–17, 20–25, and 20–26   Reprinted with permission from Academic Press

Figures 7–6, 7–8, 10–3, 10–12   Reprinted with permission from H. Schlichting, *Boundary Layer Theory*, 1960, McGraw-Hill Book Company

Figures 13–5, 13–6, 13–7, 13–8, 16–8, 16–11, 16–11, 16–12, 16–25, 17–5, 17–7, 17–8, 17–15, 17–23, 17–24, 17–43, 17–44, 17–45, 17–46, 17–47, 17–48, 17–49, 19–7, 19–8, and Tables 16.5 and 17.7   Reprinted with permission from VCH Verlagsgesellschaft

Figure 10–21   Reprinted with permission from Taylor & Francis Ltd.

Figures 10–11, 10–27, 13–3, 13–4, 15–11, 15–16, 15–17, 15–19, 15–27, 15–28, 16–22, 16–23, 16–30, 16–31, 16–32, 16–33, 16–34, 16–35, 16–36, 17–10, 17–25, 20–33, and Tables 10.1, 13.1, 13.2, 15.5, 15.7, 16.2, 16.4, 17.1, 17.4, and 17.6   Reprinted with permission from the Cambridge University Press

Tables 14.1 and 14.2   Reprinted with permission from the Institute of High Speed Mechanics

Figures 15–1, 15–3, 15–8, 15–10, 15–20, 15–25, 17–14, 19–4, 19–21, 19–22, and Tables 15.2 and 15.9   Reprinted with permission from Springer-Verlag

585

Figures 17–2, 17–3, 17–4, 19–13, and Table 17.13   Reprinted with permission from the American Society of Civil Engineers

Table 15.3   Reprinted with permission from Addison-Wesley

Figures 15–9, 16–13, 16–14, 16–18, 16–20, 16–24, 18–2, and Table 15.6 Reprinted with permission from *Physical Fluids* of the American Institute of Physics

Figure 15–22   Reprinted with permission from Dover Publications

Figures 10–22 and 10–23   Reprinted with permission from Elsevier Science Publishers, B.V.

Figures 13–14, 13–15, 15–24, 16–29, and Tables 13.3 and 16.1 appear courtesy of NASA

Figure 13–10   Reprinted with permission from The American Society of Mechanical Engineers

Figures 15–14, 15–26, 17–18, 19–20, 20–19 and Tables 15.5, 15.8, 16.8 Reprinted with permission from Pergamon Journals, Ltd.

Figures 10–9, 18–3, 18–4, 18–5   Reprinted with permission from AIAA

Figures 15–30, 16–37, 16–38, 16–39, 16–40, 16–41, 17–16, 17–20, 17–21, 17–22, and Tables 17.3 and 17.8   Reprinted with permission from the Society of Chemical Engineers, Japan

Figures 15–13, 16–10, and Tables 11.11, 17.3, and 17.8   Reprinted with permission from The Royal Society

Figure 16–15   Reprinted with permission from Her Majesty's Stationery Office

Table 18.1   Reprinted with permission from H. Rouse and J. W. Howe, *Basic Mechanics of Fluids*, 1953, from Wiley Eastern Limited Publishers

Figure 10–2   Reprinted with permission from the *Journal de mathematiques pures et appliqués*

Figures 14–2 and 14–3   Reprinted with permission from E. R. G. Eckert and R. M. Drake, Jr., *The Analysis of Heat and Mass Transfer*, 1972, McGraw-Hill Book Company

Figures 15–18 and 16–19   Reprinted from *Perturbation Methods in Fluid Mechanics* with permission from The Parabolic Press

Figures 10–27, 15–2A, 15–2D, 15–4, 15–7, 15–18, 16–6, 16–7A, and 16–7B   Reprinted with permission from M. Van Dyke, *An Album of Fluid Motion*, 1982, from The Parabolic Press

Figures 10–14A and 10–29   Reprinted with permission from D. H. Peregrine from the University of Bristol

Figure 15–5 reproduced, with permission, from the *Annual Review of Fluid Dynamics*, Vol. 5 © 1973 by Annual Reviews, Inc.

Figure 16–9   Reprinted with permission from the Office National d'Études et de Recherches Aérospatiales

Figure 20–14   Reprinted with permission from T. Shirai, *Fluidized Beds*, 1958, from Kagaku-gijitsu-sha

Figures 7–2 and 7–3   Reprinted with permission from Akademic-Verlay GmbH

Figures 4–13, 15–6, 15–29, and Table 14.1   Reprinted with permission from McGraw-Hill Book Company

Figures 19–1, 19–2, and Table 19.1   Reprinted with permission from John Wiley & Sons, Inc.

Author Index

Subject Index

V	volumetric rate of flow, m³/s; volume, m³
v	component of velocity, normal to primary direction of flow, m/s
$\mathbf{v}$	velocity vector
v_o	superficial velocity through a porous plate, m/s
$\mathbf{v}_o$	superficial velocity vector normal to surface, m/s
We	Weber number $= \dfrac{\varrho u^2 D}{\sigma}$
w	mass rate of flow, kg/s
$w\{z\}$	complex velocity potential $= \phi + i\psi$, m²/s
x	Cartesian coordinate, usually in primary direction of flow, m
x_i	Cartesian coordinate in i-th direction, m
y	Cartesian coordinate, usually normal to surface, m
z	Cartesian coordinate; distance from central plane, m; complex variable $= x + iy$

Greek letters

α	power-law index (see Equation 2.10); angle, rad
α'	modified power-law index for pipe flow (see Equation 5.36)
β	fractional angle of wedge, rad; aspect ratio $= \dfrac{H}{D}$
Γ	absolute value of rate of shear $\left(= \left|\dfrac{du_x}{dy}\right| \text{ in one-dimensional flow}\right)$, s⁻¹; strength of a vortex, m²/s
$\Delta\varrho$	absolute value of density difference $= |\varrho_d - \varrho_c|$, kg/m³
δ	thickness of boundary layer, m
ε	eccentricity of an ellipse $= \left(1 - \dfrac{b^2}{a^2}\right)^{1/2}$; void fraction
ζ	fraction of more dense or inner fluid
η	effective viscosity, Pa·s
θ	azimuthal angle in cylindrical coordinates measured from the x-direction (see Figure 8.2); polar angle in spherical coordinates measured from the z-direction (see Figure 8.3), rad
θ'	azimuthal angle in cylindrical coordinates measured from the negative x-direction; polar angle in spherical coordinates measured from the negative z-direction; $= \pi - \theta$, rad
μ	dynamic viscosity, Pa·s
μ_o	dynamic viscosity of a Bingham plastic for $|\tau| > \tau_0$, Pa·s
ν	kinematic viscosity $= \dfrac{\mu}{\varrho}$, m²/s
ξ	ratio of dynamic viscosities $= \dfrac{\mu_1}{\mu_2}$ or $\dfrac{\mu_d}{\mu_c}$
ϱ	specific density of continuous fluid, kg/m³
ϱ_i	specific density of i-th phase, kg/m³
σ	surface tension, N/m
τ	shear stress, Pa
τ_w	absolute value of shear stress on the wall, Pa
τ_{ij}	shear stress in i-th direction on the j-th surface, Pa
τ_o	yield stress of a Bingham plastic, Pa
ϕ	velocity potential (see Equation 9.51), m²/s; Stokes potential function (see Equations 9.63 and 9.64), m²/s
$\tilde{\phi}$	azimuthal angle in spherical coordinates measured from x-direction (see Figure 8.3), rad
ϕ_o	integral defined by Equation 12.11
$\phi\{z\}$	function of z
ψ	Lagrange stream function (see Equations 9.1 and 9.2), m²/s